Wind Effects on Structures

SECOND EDITION

View of Chicago with Standard Oil Company (Indiana) building near center (Architects: Perkins and Will, and Edward Durell Stone and Associates)

WIND EFFECTS ON STRUCTURES

AN INTRODUCTION TO WIND ENGINEERING

SECOND EDITION

Emil Simiu
Center for Building Technology
National Bureau of Standards
Gaithersburg, Maryland

Robert H. Scanlan
Department of Civil Engineering
The Johns Hopkins University
Baltimore, Maryland
(Emeritus Professor, Princeton University)

A Wiley-Interscience Publication
JOHN WILEY & SONS
New York / Chichester / Brisbane / Toronto / Singapore

Library of Congress Cataloging in Publication Data :

Simiu, Emil.
    Wind effects on structures.

    "A Wiley-Interscience publication."
    Includes bibliographical references and index.
    1. Wind-pressure.   2.   Buildings—Aerodynamics.
3.   Structural dynamics.   I.   Scanlan, Robert H.
II.   Title.
TA654.5.S55   1985   624.1'76   85-10598
ISBN 0-471-86613-X

Printed in the United States of America

10 9 8 7 6 5 4 3 2 1

למשפחתי ולזכר פאול

E. S.

Preface

In the almost ten years that have elapsed since the writing of *Wind Effects on Structures* a number of significant advances have occurred in the wind engineering field. These include the development of the following: improved micrometeorological models, particularly for atmospheric flows over the ocean, which are of interest in the design of offshore structures; procedures for the estimation of extreme winds from short records; new information on the modeling of extreme winds in hurricane- and tornado-prone regions; improved procedures for estimating the along-wind response of structures; new procedures for estimating the across-wind and torsional response of tall buildings and the across-wind response of towers and stacks; simple and probabilistically rigorous methods for taking wind directionality effects into account in design; practical procedures for the risk-consistent design of cladding for wind loads, which make it possible to achieve more economical designs for any given safety level, or safer designs for any given cost; methods for estimating the response of offshore structures to wind in the presence of current and waves; and new information on wind effects on various types of structure, including trussed frameworks, hyperbolic cooling towers, and semisubmersible platforms.

The text has been expanded to reflect these and other advances. It now includes five new chapters, as well as a new appendix which is intended to provide the reader with a brief introduction to modern structural reliability concepts. The original chapter on wind tunnels was substantially revised, and much new material was added to the other chapters, particularly those on the atmospheric boundary layer, extreme wind climatology, bluff body aerodynamics, aeroelastic phenomena, tall buildings, and tornado effects. Most of the new material consists of practical design information and methods. As in the first edition, a consistent effort has been made to point out and discuss the uncertainties, limitations, and errors inherent in various data, methods, and techniques.

The authors would like to express their warm appreciation to Dr. R. D. Marshall, who initiated and developed the wind engineering program at the

National Bureau of Standards; Dr. N. Isyumov of the University of Western Ontario, for contributions to Sect. 9.3; and Professor D. A. Reed of the University of Washington, for contributions to Chapter 11. Special thanks are also due, for valuable comments and criticism, to Professor E. A. Arens of the University of California at Berkeley; Dr. R. I. Basu of H. G. Engineering, Inc.; Professor O. Ditlevsen of the Engineering Academy of Denmark; Dr. B. R. Ellingwood of the National Bureau of Standards; Professor Y. Fujino of the University of Tokyo; Dr. M. P. Gaus of the George Washington University; Dr. P. S. Jackson of the University of Auckland; Dr. P. Mahmoodi of the 3M Company; and Professor B. J. Vickery of the University of Western Ontario. However, the responsibility for any errors or omissions lies solely with the authors. We also wish to thank our Editor, E. W. Smethurst, Editorial Supervisor, Balwan R. Singh, Designer, Lee Davidson, and Production Supervisor, Linda Shapiro, all of John Wiley & Sons, and Technical Editor of the Russian translation (1984), Dr. B. E. Maslov.

The references to the authors' affiliations are for purposes of identification only. The book is not a U.S. Government publication, and the views expressed do not necessarily represent those of the U.S. Government or any of its agencies.

EMIL SIMIU
ROBERT H. SCANLAN

Rockville, Maryland
Lawrenceville, New Jersey
August 1985

Preface To First Edition

The wind loading of civil engineering structures involves, in certain cases, considerable complexities that must be taken into account in order to achieve safe and serviceable designs. Examples of wind engineering problems that require special attention include: the dynamic response of tall structures; the performance of exterior glass and curtain walls, particularly in high-rise buildings; the serviceability of pedestrian areas in certain types of built environments; the oscillations and flutter of suspension bridges; the action of tornadoes on nuclear power plants; the estimation of the probability of occurrence of extreme winds at a given site.

Motivated by the need to provide rational descriptions of the phenomena involved and to develop appropriate analytical and design tools, a vast specialized literature—not always easily accessible—has emerged in the last two decades. *An Introduction to Wind Engineering* is an attempt to present a synthesis of the main trends of this literature in the form of a text designed for use by advanced students of engineering and by practicing structural engineers and architects. The text develops its chosen topics independently and, as often as possible, from fundamental principles. In addition, extensive references are provided to a wide range of primary sources.

The level of preparation assumed of the reader corresponds approximately to that of the bachelor's degree in science or engineering. A consistent effort has been made to avoid unnecessarily elaborate mathematical formulations. Simple notions of probability theory, statistics and the theory of random processes employed in modern wind engineering analysis have been presented in appendices, in which intuitive approaches have been strongly emphasized.

The first part of the text discusses meteorological, micrometeorological, and climatological aspects of the wind environment that are of interest in wind engineering. The second part presents basic elements of aerodynamics, structural dynamics, and aeroelasticity, followed by applications to the design of various types of structures and structural members. Separate chapters are

ix

devoted to a discussion of wind-induced discomfort in and around buildings, and to assessments of the wind tunnel as a design tool.

Wind engineering is a new and rapidly developing field. Current procedures for estimating wind effects, and the information on which they are based, should therefore not be regarded as definitive. It is the authors' strong feeling that areas of uncertainty must be carefully defined, and that the limitations inherent in current procedures must be stated clearly. This has been done throughout the text.

The division of responsibility for the work has been as follows: E. Simiu has written Chapters 1–3, 5, 7, 9–11, and the Appendices, and R. H. Scanlan has written Chapters 4, 6 and 8.* The authors have, however, shared editorship and extensive critical exchange on all parts of the text.

The authors' sincere thanks are extended to the following persons who read portions of the manuscript and offered valuable criticisms: Professor H. A. Panofsky, Pennsylvania State University; Dr. N. J. Cook, Building Research Establishment, U.K.; Dr. J. F. Costello, U.S. Nuclear Regulatory Commission; Dr. H. L. Crutcher, National Climatic Center, National Oceanic and Atmospheric Administration; Dr. J. J. Filliben, Statistical Engineering Laboratory, National Bureau of Standards; Dr. J. C. R. Hunt, Cambridge University, U.K.; Dr. G. E. Mattingly, Institute for Basic Standards, National Bureau of Standards; Dr. J. M. Mitchell, Environmental Data Service, National Oceanic and Atmospheric Administration; Dr. R. N. Wright, Center for Building Technology, National Bureau of Standards; and Professor J. T. P. Yao, Purdue University. All of them should share the recognition for the many improvements their comments brought about. The responsibility for all errors or imperfections rests, however, wholly with the authors. Many thanks are also due to Devra Simiu and Robert N. Scanlan for careful reading and editing of the text, and to Mrs. Sue Murray, Mrs. Rebecca Hocker, and Mrs. Nora Scanlan for their capable typing effort. The authors also wish to express their indebtedness to the late R. S. Woolson, Editor, J. Frances Tindall, Editorial Supervisor, Joel L. Bromberg, Editorial Assistant, and Debbie Oppenheimer and Sandra Winkler, Production Supervisors, all of John Wiley & Sons.

EMIL SIMIU
ROBERT H. SCANLAN

Washington, D.C.
Princeton, New Jersey
June, 1977

*Chapters 4, 6, and 8 of the first edition correspond in the second edition to Chapters 4, 6, and 13. For the second edition, R. H. Scanlan has revised the chapter on wind tunnels, and E. Simiu has been responsible for the other revisions and additions to the text.

Contents

Wind Effects on Structures

SECOND EDITION

Introduction

The development of modern materials and construction techniques has resulted in the emergence of a new generation of structures that are often, to a degree unknown in the past, remarkably flexible, low in damping, and light in weight. Such structures generally exhibit an increased susceptibility to the action of wind. Accordingly, it has become necessary to develop tools enabling the designer to estimate wind effects with a higher degree of refinement than was previously required. Wind engineering is the discipline that has evolved, primarily during the last decade, from efforts aimed at developing such tools.

It is the task of the engineer to ensure that the performance of structures subjected to the action of wind will be adequate during their anticipated life from the standpoint of both structural safety and serviceability. To achieve this end, the designer needs information regarding (1) the wind environment, (2) the relation between that environment and the forces it induces on the structure, and (3) the behavior of the structure under the action of these forces.

THE WIND ENVIRONMENT

Information on the wind environment needed in design includes elements derived from meteorology, micrometeorology, and climatology.

Meteorology provides a description and explanation of the basic features of atmospheric flows. Such features may be of considerable significance from a structural design viewpoint. For example, in the case of the tornado, the presence of a region of low atmospheric pressure at the center of the storm is a factor of major importance in the design of nuclear power plants.

Micrometeorology attempts to describe the detailed structure of atmospheric flows near the ground. Topics of direct concern to the structural designer include the variation of mean speeds with height above ground, the description of atmospheric turbulence, and the dependence of the mean speeds and of turbulence upon roughness of terrain.

1

Climatology, as applied to the wind environment, is concerned with the prediction of wind conditions at given geographical locations. Probability statements on future wind speeds may be conveniently summarized in wind maps, such as are currently included in various building codes.

WIND-INDUCED FORCES ON STRUCTURES

A structure immersed in a given flow field is subjected to aerodynamic forces that, in general, may be estimated using available results of aerodynamic theory and experiments. However, if the environmental conditions or the properties of the structure are unusual, it may be necessary to conduct special wind tunnel tests.

Aerodynamic forces include drag (along-wind) forces, which act in the direction of the mean flow, and lift (across-wind) forces, which act perpendicularly to that direction. If the distance between the elastic center of the structure and the aerodynamic center (i.e., the point of application of the aerodynamic force) is large, the structure is also subjected to torsional moments that may significantly affect the structural design.

STRUCTURAL RESPONSE TO WIND LOADS

Because the aerodynamic forces are dependent on time, the methods of structural dynamics may have to be employed to determine the response. Furthermore, the random character of this dependence requires that elements of the theory of random vibrations be applied to the analysis. In certain cases, it may be necessary to perform an aeroelastic analysis, that is, a study of the interaction between the aerodynamic and the inertial, damping, and elastic forces, with the purpose of investigating the aerodynamic stability of the structure.

From the foregoing it is seen that the design of modern structures subjected to wind loads requires the use of information and methods derived from a broad spectrum of disciplines. It will not be suggested here that complete answers to the questions involved exist at the present time. However, considerable progress has been made toward an understanding of some of these questions. As a result, procedures and techniques have been developed that have significantly improved the designer's ability to estimate the effects of wind from the standpoint of both strength and serviceability. It is the aim of this text to present these procedures and techniques, to provide the background material required for understanding their rationale, and to examine critically their capabilities as well as their limitations as design tools.

PART A

THE ATMOSPHERE

1

Atmospheric Circulations

Wind, or the motion of air with respect to the surface of the earth, is fundamentally caused by variable solar heating of the earth's atmosphere. It is initiated, in a more immediate sense, by differences of pressure between points of equal elevation. Such differences may be brought about by thermodynamic and mechanical phenomena that occur in the atmosphere nonuniformly both in time and space.

The energy required for the occurrence of these phenomena is provided by the sun in the form of radiated heat. While the sun is the original source, the source of energy most directly influential upon the atmosphere is the surface of the earth. Indeed, the atmosphere is to a large extent transparent to the solar radiation incident over the earth, much in the same way as the glass roof of a greenhouse. That portion of the solar radiation that is not reflected or scattered back into space may therefore be assumed to be absorbed almost entirely by the earth. The earth, upon being heated, will emit energy in the form of terrestrial radiation, the characteristic wave lengths of which are long (of the order of $10 \, \mu$) compared to those of heat radiated by the sun. The atmosphere, which is largely transparent to solar, but not to terrestrial, radiation, absorbs the heat radiated by the earth and re-emits some of it toward the ground.

1.1 ATMOSPHERIC THERMODYNAMICS

1.1.1 Temperature of the Atmosphere

In order to illustrate the role of the temperature distribution in the atmosphere in the production of winds, a simplified model of atmospheric circulation will be presented. In this model, the effects of the vertical variation of air temperature, of the humidity of the air, of the rotation of the earth, and of friction will be ignored, and the surface of the earth will be assumed to be uniform and smooth.

It will be recalled that the axis of rotation of the earth is inclined at approximately 66°30′ to the plane of its orbit around the sun (plane of the ecliptic). Therefore, the average annual intensity of solar radiation and, consequently, the intensity of terrestrial radiation and the temperature of the atmosphere will be higher in the equatorial than in the polar regions. To explain the circulation pattern that arises as a result of this temperature difference, Humphreys [1-1] proposed the following ideal experiment (Fig. 1.1.1).

Assume that the tanks A and B are filled with fluid of uniform temperature up to level *a* and that tubes 1 and 2 are closed. If the temperature of the fluid in A is raised while the temperature in B is maintained constant, the fluid in A will expand and reach the level *b*. The expansion entails no change in the total weight of the fluid contained in A. The pressure at *c* remains therefore unchanged, and if tube 2 were opened, there would be no flow between A and B. If tube 1 is opened, however, fluid will flow from A to B, on account of the difference of head (*b* − *a*). Consequently, at level *c* the pressure in A will decrease while the pressure in B will increase. Upon opening tube 2, fluid will now flow through it from B to A. The circulation thus developed will continue as long as the temperature difference between A and B is maintained.

If tanks A and B are replaced conceptually by the column of air above the equator and above the pole, it can be seen that, in the absence of other effects, an atmospheric circulation would be developed that could be represented as in Fig. 1.1.2. In reality, the circulation of the atmosphere is vastly complicated by the factors neglected in the above model. The effect of these factors will be discussed later in this chapter.

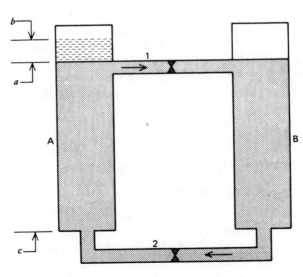

FIGURE 1.1.1. Circulation pattern due to temperature difference between two columns of fluid. From *Physics of the Air* by W. J. Humphreys. Copyright 1929, 1940 by W. J. Humphreys. Used with permission of McGraw-Hill Book Company.

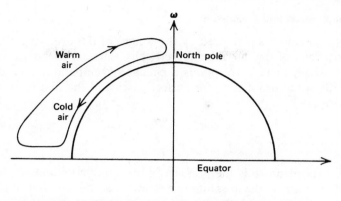

FIGURE 1.1.2. Simplified model of atmospheric circulation.

The temperature of the atmosphere is determined by the following processes [1-2, 1-3, 1-4, 1-5, 1-6]:

· solar and terrestrial radiation, as discussed previously in this chapter
· radiation in the atmosphere
· compression or expansion of the air
· molecular and eddy conduction
· evaporation and condensation of water vapor

1.1.2 Radiation in the Atmosphere

As a conceptual aid, consider the action of the following model. The heat radiated by the surface of the earth is absorbed by the layer of air immediately above the ground (or the surface of the ocean) and reradiated by this layer in two parts, one going downward and one going upward. The latter is absorbed by the next higher layer of air and again reradiated downward and upward. The transport of heat through radiation in the atmosphere, according to this conceptual model, is represented in Fig. 1.1.3.

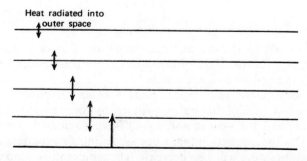

FIGURE 1.1.3. Transport of heat through radiation in the atmosphere.

1.1.3 Compression and Expansion

Atmospheric pressure is produced by the weight of the overlying air. A small mass (or particle) of dry air moving vertically thus experiences a change of pressure to which there corresponds a change of temperature. To determine the latter, the equation of state for perfect gases and the first law of thermodynamics are used:

$$pv = RT \tag{1.1.1}$$

$$dq = c_v \, dT + p \, dv \tag{1.1.2}$$

In these expressions p is the pressure, v the specific volume, R the specific constant for dry air, T the absolute temperature, dq the amount of heat transferred to the particle, and c_v the specific heat at constant volume.

Differentiating the first relation and substituting the quantity $p \, dv$ thus obtained in the second relation, there results

$$dq = (c_v + R) \, dT - v \, dp \tag{1.1.3}$$

Comparing this relation with

$$dq = c_p \, dT \tag{1.1.4}$$

which expresses the first law of thermodynamics in the particular case of an isobaric (constant pressure) process (c_p is the specific heat at constant pressure), it is easy to see that $c_v + R = c_p$. It is therefore possible to write, if the equation of state is used once more,

$$dq = c_p \, dT - RT \, \frac{dp}{p} \tag{1.1.5}$$

Processes for which $dq = 0$ are referred to as adiabatic. For such processes, the previous relation becomes

$$\frac{dT}{T} - \frac{R}{c_p} \frac{dp}{p} = 0 \tag{1.1.6}$$

which, after integration, yields the equation

$$\frac{T}{T_0} = \left(\frac{p}{p_0} \right)^{R/c_p} \tag{1.1.7}$$

known as Poisson's, or the dry adiabatic equation. For dry air, $R/c_p = 0.288$. A familiar example of the effect of pressure change on the temperature is the heating of compressed air in a tire pump.

If, in the atmosphere, the vertical motion of an air parcel is sufficiently rapid, the heat exchange of that parcel with its environment may be considered to be negligible and the assumption $dq = 0$ is approximately correct. It then follows from Poisson's equation that since ascending air experiences a pressure decrease, its temperature will also decrease. The temperature drop of adiabatically ascending dry air is known as the dry *adiabatic lapse rate* and is approximately 1°C/100 meters in the earth's atmosphere.

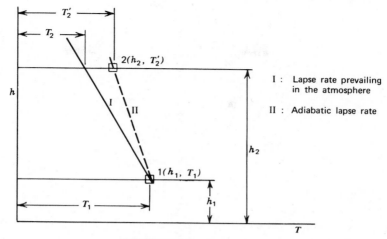

FIGURE 1.1.4.   Lapse rates.

Consider a small mass of dry air at position 1 (Fig. 1.1.4). Its elevation and temperature are h_1 and T_1, respectively. If the particle moves vertically upward at some reasonable speed, its temperature change will effectively be adiabatic, regardless of the lapse rate (temperature variation with height above ground) prevailing in the atmosphere. At position 2, while the temperature of the ambient air is T_2, the temperature of the element of air mass is $T_2' = T_1 - (h_2 - h_1)\gamma_a$, where γ_a is the adiabatic lapse rate. Since the pressure of the element and of the ambient air will be the same, it follows from the equation of state that to the temperature difference $T_2' - T_2$ there corresponds a difference of density between the element of air and the ambient air. This generates a buoyancy force that, if $T_2 < T_2'$, acts upwards and thus moves the element farther away from its initial position (superadiabatic lapse rate, as in Fig. 1.1.4), or, if $T_2 > T_2'$, acts downwards, thus tending to return the particle to its initial position. The stratification of the atmosphere is said to be *unstable* in the first case and *stable* in the second. If $T_2 = T_2'$, that is, if the lapse rate prevailing in the atmosphere is adiabatic, the stratification is said to be *neutral*.

A simple example of the stable stratification of fluids is provided by a layer of water underlying a layer of oil, while the opposite (unstable) case would have the water above the oil.

1.1.4   Molecular and Eddy Conduction

Molecular conduction is a diffusion process that effects a transfer of heat. It is achieved through the motion of individual molecules and is negligible insofar as atmospheric processes are concerned. Eddy heat conduction involves the transfer of heat by actual movement of air in which heat is stored.

1.1.5   Condensation and Evaporation of Water Vapor

The pressure of moist air is, according to Dalton's law, equal to the sum p of the partial pressure e of the water vapor and that of the dry air, $p - e$. It has

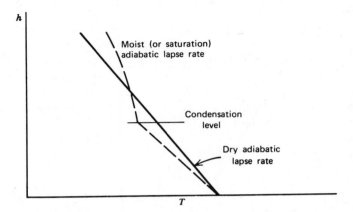

FIGURE 1.1.5.   Effects of condensation upon lapse rate.

been established experimentally that if the pressure e exceeds some value E, known as the saturation vapor pressure, condensation of the excess moisture will occur, and that the saturation pressure E increases exponentially as the temperature of the moist air increases.

An elementary mass of ascending unsaturated moist air (i.e., for which $e/E < 1$) will experience a temperature drop that can be shown to be essentially equal to the dry adiabatic lapse rate. As the element ascends and its temperature decreases, its saturation pressure will also decrease. If the element reaches a level at which the ratio e/E becomes unity, condensation will normally occur. Above this level, water vapor contained in the air element will continue to condense. In the process, heat of condensation is released. This is equal to the heat that was originally required to change the phase of water from liquid to vapor, that is, the latent heat of vaporization stored in the vapor.

The heat of condensation contributes to the mechanical work involved in the expansion of an ascending particle, which before saturation was performed only at the expense of the internal energy. The temperature drop of the saturated adiabatically ascending element of air is therefore slower than for dry or moist unsaturated air (Fig. 1.1.5). By furnishing energy that increases the temperature of a particle with respect to what it would have been under dry adiabatic conditions, the heat of condensation helps support convection of the air to higher levels of the atmosphere. This factor is important in the genesis of certain types of winds.

1.2   ATMOSPHERIC HYDRODYNAMICS

The motion of an elementary mass of air is determined by Newton's second law

$$\sum \mathbf{F} = m\mathbf{a} \tag{1.2.1}$$

where m is the mass, $\mathbf{a}$ is the acceleration, and $\sum \mathbf{F}$ is the sum of forces acting on the elementary mass of air. It is the purpose of this section to briefly describe the forces **F** and some of their effects upon the motion of air.

1.2.1   The Horizontal Pressure Gradient Force

Consider an infinitesimal volume of air $dx\,dy\,dz$ (Fig. 1.2.1) and let the mean pressures acting on the lower and upper face be p and $p+(\partial p/\partial z)\,dz$, respectively. In the absence of forces other than pressures, the net vertical force acting on the volume $dx\,dy\,dz$ will be $-(\partial p/\partial z)\,dx\,dy\,dz$, or $-\partial p/\partial z$ per unit volume. Similarly, the net forces per unit volume acting in the x and y direction will be denoted $-\partial p/\partial x$ and $-\partial p/\partial y$, respectively. The resultant of these forces is called horizontal pressure gradient and is denoted $-\partial p/\partial n$, where n is the normal to some contour of constant horizontal pressure. The horizontal pressure gradient is the driving force which initiates the horizontal motion of air. The net force per unit mass exerted by the horizontal pressure gradient, $(1/\rho)\,\partial p/\partial n$, is often referred to as the pressure gradient force (ρ is the air density.)

Air subjected solely to the action of pressure gradient forces will move from regions of high pressure to regions of low pressure. The direction of the pressure gradient force is indicated in Fig. 1.2.2, in which the isobars (lines contained

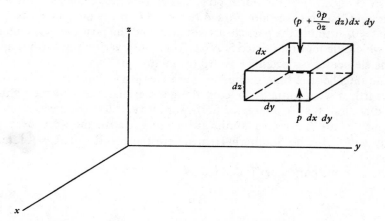

FIGURE 1.2.1.   Vertical pressures on an elementary mass of air.

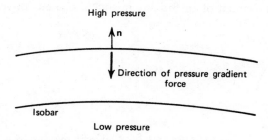

FIGURE 1.2.2.   Direction of pressure gradient force.

in the same horizontal plane and connecting points of equal pressure) are also shown.

1.2.2   The Deviating Force Due to the Earth's Rotation

If defined with respect to an absolute frame of reference, the motion of a particle not subjected to the action of an external force will follow a straight line. To an observer on the rotating earth, however, the path described by the particle will appear curved. The deviation of the particle motion from a straight line fixed with respect to the rotating earth may be attributed to an apparent force, the Coriolis force, the vector expression of which is [1-7]

$$\mathbf{F}_c = 2m(\mathbf{v} \times \boldsymbol{\omega}) \tag{1.2.2}$$

where m is the mass of the particle, ω is the angular velocity vector of the earth, and $\mathbf{v}$ is the velocity of the particle relative to a coordinate system rotating with the earth. $\mathbf{F}_c$ is perpendicular to ω and to $\mathbf{v}$, is directed according to the vector multiplication (right-hand) rule, and has the magnitude $2m|\omega||\mathbf{v}| \sin \alpha$, where α is the angle between ω and $\mathbf{v}$.

Let N (Fig. 1.2.3) be the north pole, and consider an element of air moving in a straight line in space along the direction NP. If the motion starts from N at time $t = 0$, at time t the particle arrives at P, and the position of the meridian along which the motion started is NP'. To an observer on the earth it appears that the element is deflected westward by an amount $P'P$.

It can thus be seen that, in the Northern Hemisphere, owing to the rotation of the earth, a wind initially directed along a meridian veers to the right of its initial direction; that is, if directed northward it veers toward the east (becomes a westerly wind). If directed southward it veers toward the west (becomes an easterly wind). In the Southern Hemisphere the reverse of these statements is true.

If the Coriolis parameter is defined as

$$f = 2\omega \sin \phi \tag{1.2.3}$$

where ϕ is the latitude of the point considered, it follows that the Coriolis force acting per unit of mass in a plane (P) parallel to the surface of the earth (Fig. 1.2.4) on an element of air moving in such a plane with velocity $\mathbf{v}$ relative

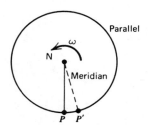

FIGURE 1.2.3. Apparent motion of an air particle due to the earth's rotation.

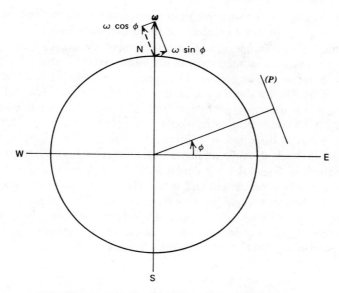

FIGURE 1.2.4.   Components of the rotation vector ω.

TABLE 1.2.1.   Coriolis Parameter.

Latitude (deg)	$f = 2\omega \sin \phi$ sec^{-1}	Latitude (deg)	$f = 2\omega \sin \phi$ sec^{-1}
0	0	50	1.1172×10^{-4}
5	0.1271×10^{-4}	55	1.1947
10	0.2533	60	1.2630
15	0.3775	65	1.3218
20	0.4988	70	1.3705
25	0.6164	75	1.4087
30	0.7292	80	1.4363
35	0.8365	85	1.4529
40	0.9375	90	1.4584
45	1.0313		

to the earth will have the magnitude

$$F_c = mfv \qquad (1.2.4)$$

The values of f are given in Table 1.2.1 as functions of latitude.

1.2.3   The Frictionless Wind Balance

At sufficiently great heights, the effects on the wind due to friction along the ground become negligible and the horizontal motion of air relative to the surface of the earth is determined, in unaccelerated flow, by the balance among the pressure gradient, the Coriolis, and the centrifugal force.

The effect of the forces acting on an elementary mass of air is shown in Fig. 1.2.5 (the mass is assumed to be in the Northern Hemisphere). If the particle started to move in the direction of the pressure gradient force (denoted **P**), it would be deflected by the Coriolis force $\mathbf{F}_{ca}$ (Fig. 1.2.5*a*). The particle would then move in the direction of the resultant of **P** and $\mathbf{F}_{ca}$, shown as direction II in Fig. 1.2.5*b*. The deflecting force would now become $\mathbf{F}_{cb}$, to which there would correspond a new direction of motion (direction III in Fig. 1.2.5*b*). When a steady state is reached, the wind flows along the isobars as shown in Fig. 1.2.5*c*.

The isobars in Fig. 1.2.5 are depicted as straight, which means that in the case shown there is no centrifugal force. However, in the more general case of a curved isobar centrifugal forces will be involved. This case is taken up below.

The steady velocity for which a balance between the pressure gradient force and the Coriolis force alone obtains is called the *geostrophic wind velocity G* and is related to the pressure gradient by the equation

$$2\omega G \sin \phi = P = \frac{dp/dn}{\rho} \qquad (1.2.5)$$

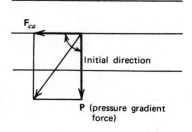

(*a*)

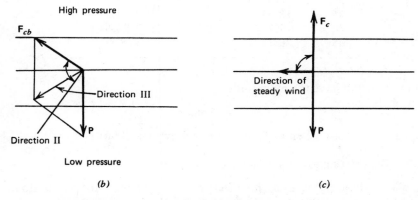

(*b*)   (*c*)

FIGURE 1.2.5.   Frictionless wind balance in geostrophic flow.

or

$$G = \frac{dp/dn}{\rho f} \qquad (1.2.6)$$

where P is the magnitude of the vector $\mathbf{P}$, f is the Coriolis parameter, and ρ is the air density.

If the isobars are curved (Fig. 1.2.6), the force $\mathbf{P}$ as well as the centrifugal force $\mathbf{C}$ will act on the elementary mass of air in the direction normal to the isobars, and the resulting steady wind will again flow along the isobars. Its velocity results from the relations

$$V_{gr} f \pm \frac{V_{gr}^2}{r} = P = \frac{dp/dn}{\rho} \qquad (1.2.7)$$

where, if the mass of air is in the Northern Hemisphere, the positive or the negative sign is used according as the circulation is cyclonic (around a center of low pressure) or anticyclonic (around a center of high pressure), and where r is the radius of curvature of the air trajectory.* The velocity V_{gr} is called the *gradient wind velocity*; it is equal to the geostrophic wind velocity in the particular case in which the curvature of the isobars is zero. If the radius of curvature is finite, in the Northern Hemisphere

$$V_{gr} = -\frac{rf}{2} + \left[\frac{r}{\rho}\frac{dp}{dn} + \left(\frac{rf}{2}\right)^2\right]^{1/2} \qquad (1.2.8)$$

for cyclonic winds, and

$$V_{gr} = +\frac{rf}{2} - \left[\left(\frac{rf}{2}\right)^2 - \frac{r}{\rho}\frac{dp}{dn}\right]^{1/2} \qquad (1.2.9)$$

for anticyclonic winds. The sign of the radicals is given by the condition that $V_{gr} = 0$ when $dp/dn = 0$. It follows from the expressions for V_{gr} that, for the same

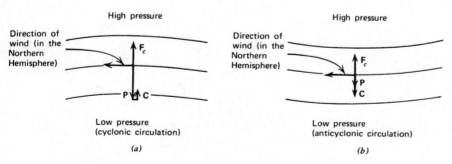

FIGURE 1.2.6.   Frictionless wind balance in cyclonic and anticyclonic flow.

*Strictly speaking, the radius of curvature of the trajectory may differ from the radius of curvature of the isobar. The difference may be neglected, however, if it can be assumed that the wind flow is approximately steady.

values of r, f and dp/dn, anticyclonic winds are weaker than cyclonic winds [1-1, p. 121].

The foregoing discussion explains *Buys-Ballot's law*, which states: If, in the Northern Hemisphere, a person stands with his back to the wind, the high pressure will be on his right and the low pressure will be on his left. In the Southern Hemisphere the reverse is true.

1.2.4   Effects of Friction

The surface of the earth exerts upon the moving air a horizontal drag force, the effect of which is to retard the flow. The effect of this force upon the flow decreases as the height above ground increases and, as indicated previously, becomes negligible above a height δ known as the height of the *boundary layer of the atmosphere*. Above this height the frictionless wind balance is established and the wind flows with the gradient wind velocity along the isobars. The atmosphere above the boundary layer is called the *free atmosphere* (Fig. 1.2.7).

It is the wind regime within the boundary layer of the atmosphere that is of direct interest to the designer of civil engineering structures. The questions of the boundary-layer height, of the variation of wind speed and direction with height above ground, and of the turbulence structure within the boundary layer are therefore discussed in more detail in Chapter 2.

It will be noted here that unlike the gradient wind velocity, the steady-state wind velocity within the boundary layer crosses the isobars. Consider a geostrophic flow (i.e., a flow in which the isobars may be assumed to be straight) and the balance of the forces acting on particles A and B, which move horizontally within its boundary layer (Fig. 1.2.8). If A (Fig. 1.2.8a) is at a higher level than B (Fig. 1.2.8b), its speed v and (by virtue of the relation $F_c = mfv$) its Coriolis force will be larger than those of B. The deviation angle α between the wind direction and the isobars will therefore be smaller for the higher (faster)

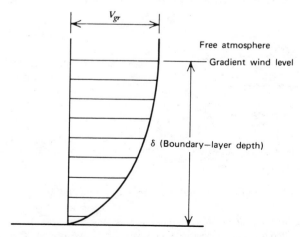

FIGURE 1.2.7.   The atmospheric boundary layer.

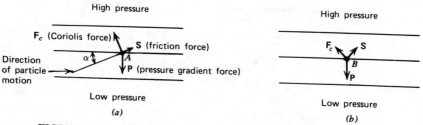

FIGURE 1.2.8.   Balance of forces in the atmospheric boundary layer.

particle. The angle α will be zero at the gradient level and will reach its maximum value α_0 near the ground. In the Northern Hemisphere the wind velocity in the boundary layer may thus be represented by a spiral, as in Fig. 1.2.9.

In the case of a cyclonic storm (or flow around a center of low pressure), near the ground, the wind will cross the isobars toward the center. The air will thus slowly converge and ascend. If the low-level convergence exceeds high-level divergence, the mass and weight of the air column at the center of the storm gradually increase and therefore the inward-directed pressure gradient force decreases. As a result of such a decrease the center of low pressure is dissipated and *filling* is said to occur.

In the case of an anticyclone, the wind near the ground will cross the isobars away from the center of high pressure. In the lower portions of a high, if low-level divergence exceeds high-level convergence, the atmosphere will then tend to spread out and sink, and dissipation of the center will thus occur.

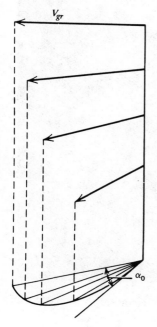

FIGURE 1.2.9.   Wind velocity spiral in the atmospheric boundary layer.

1.3   ATMOSPHERIC MOTIONS

Most atmospheric processes can be described in terms of the quantities briefly discussed in the preceding sections: wind velocity (i.e., horizontal and vertical wind), pressure, temperature, density, and moisture. The behavior of these six quantities is governed by six equations: the equation of state, the first law of thermodynamics, the equations of continuity of mass and moisture, and the horizontal and vertical equations of motion. Provided that an adequate data base exists, these equations can be integrated to yield a quantitative description of atmospheric conditions at some short time after the collection of the data. The calculated values of the six variables obtained by integration can then be used as initial conditions for a further integration step. This successive approximation process is the basis of numerical weather prediction techniques that came into being following the increased availability of observations—including, more recently, observations obtained by satellites (Fig. 1.3.1)—and the development of modern electronic computers.

Atmospheric motions may be described as superpositions of interdependent flows characterized by scales ranging from approximately one millimeter to thousands of kilometers. To analyze such motions, it is convenient to classify

FIGURE 1.3.1.   Satellite view of hurricane Fifi, Sept. 18, 1974 (courtesy National Oceanic and Atmospheric Administration).

them according to their horizontal scale. In meteorology three main groups of atmospheric scales are commonly defined: microscale, mesoscale, and synoptic scale. According to the classification of [1-8], the synoptic scale includes motions with characteristic dimensions exceeding 500 km or so and time scales of two days or more. The microscale includes motions with characteristic dimensions of less than 20 km or so and time scales of less than one hour. The mesoscale is defined by dimensions and periods between those characteristic of microscale and synoptic scale.

1.3.1 The General Circulation

The combined effects of the earth's rotation and of friction break the thermal circulation cell of Fig. 1.1.2 into a pattern that consists basically of three circulation cells as represented in Fig. 1.3.2 [1-2]. The theoretical pattern is compatible with the existence (at sea level) of a high pressure belt at the horse latitudes and of a low pressure belt at the polar front.

In reality, the tricellular meridional circulation model is complicated by seasonal and by geographical effects. Seasonal effects consist of variations of position and intensity of the pressure belts and are caused by the annual march of the sun north and south of the equator. Geographical effects are caused by the difference in physical properties, and by the uneven distribution over the globe, of water and land.

In summer, because the ocean surface warms up more slowly than the land, the air is colder over the ocean than over land. Just as in Fig. 1.1.1 fluid flows in tube 2 from the colder to the warmer tank, air near the surface will be driven in summer by a pressure gradient force directed from the ocean toward the land. On the other hand, in winter the air is colder over land and the oceans become heat sources.

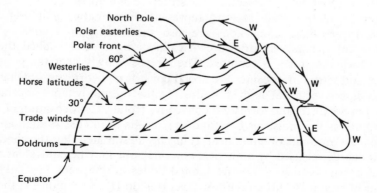

FIGURE 1.3.2. Tricellular meridional circulation model. After *General Meterology* by H. R. Byers. Copyright 1937, 1944 by the McGraw-Hill Book Company, Inc. Used with permission of McGraw-Hill Book Company.

1.3.2   Thermally Direct Secondary Circulations: Monsoons and Hurricanes

Secondary circulations are said to be of the thermally direct type if the centers of high or low pressure (i.e., the highs or the lows) around which they develop are formed by heating or cooling of the lower atmosphere.

Monsoons.   Monsoons are seasonal winds that form cells of the general circulation and develop around thermally produced continental highs in winter and lows in summer. Owing to the vast land mass of the Asian continent, monsoon effects are developed most strongly in Asia, where they have a considerable influence on the seasonal changes of weather patterns.

Hurricanes.   Tropical cyclones are storms that derive all their energy from the latent heat released by the condensation of water vapor and originate, generally, between the 5 and 20 latitude circles. Their diameters are usually of the order of several hundred kilometers. The depth of the atmosphere involved is of the order of ten kilometers. Hurricanes are defined as tropical cyclones with surface wind velocities exceeding about 120 km/hr. Spacecraft views of hurricanes are shown in Figs. 1.3.1 and 1.3.3.

Hurricanes (known as typhoons in the Far East and cyclones in the region of Australia and the Indian Ocean) occur most frequently during the late summer and early autumn months (August–September on the Northern Hemisphere, February–March on the Southern Hemisphere), except in the Northern Indian Ocean. Hurricanes normally travel as whole entities at speeds of 5 to 50 km/hr. The mean directions of hurricane motions are shown in Fig. 1.3.4. It is noted, however, that individual hurricanes may follow unusual, indeed erratic, paths. World tropical cyclone statistics are presented in Fig. 1.3.5 [1-9]. Data on tropical cyclones reaching the United States coastline are presented in some detail in Sect. 3.3.

As seen in a vertical plane section, the structure of a hurricane in the mature stage consists of five main regions, represented schematically in Fig. 1.3.6, in which approximate dimensions are also shown. Region I consists of a roughly circular, relatively dry core of calm or light winds, called the eye, around which the storm is centered. The air rises slowly near the perimeter of the eye and settles in its center. Region II consists of a vortex in which warm, moist air is convected at high altitudes (by the thermodynamic mechanism discussed in Sect. 1.1) and forms tall convective clouds. Condensation of water vapor occurs as the moist air rises and this results in intense rainfall and the release of vast amounts of latent heat. It has been estimated that the condensation heat energy released by a hurricane in one hour may be equivalent to the electrical energy used in the entire United States in one week [1-9]. The air flows out of region II into an outflow layer (region III). In region IV, the flow is vortex-like and settles very slowly into the boundary layer region V. Below region II, where strong updrafts are present, separation of the boundary layer may occur.

FIGURE 1.3.3. Hurricane Gladys as seen by the Apollo crew (courtesy National Oceanic and Atmospheric Administration).

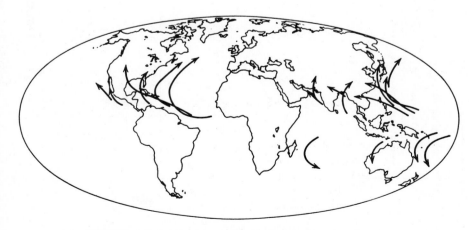

FIGURE 1.3.4. Mean directions of hurricane motions [1–9].

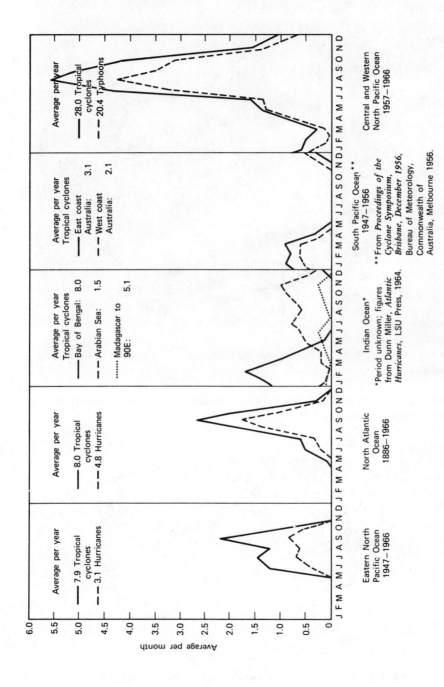

FIGURE 1.3.5. World tropical cyclone statistics [1–9].

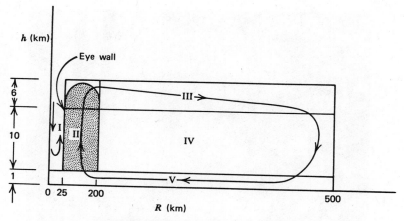

FIGURE 1.3.6.   Structure of a hurricane.

According to Graham and Hudson [1-10] an expression of the form

$$\frac{1}{\rho}\frac{dp}{dr}=\left(\frac{p_N-p_0}{\rho}\right)\frac{R_m}{r^2}\,e^{-R_m/r} \tag{1.3.1}$$

in which p_N is the pressure that is approached as the radius $r\to\infty$, p_0 is the pressure at the center of the hurricane eye, $(p_N-p_0)/\rho$ is assumed independent of height, and R_m is twice the radius of maximum dp/dr, is fairly representative of typical hurricane pressure fields. If this description of the pressure gradient field is used, the gradient wind velocity field results from the expression of the gradient wind (valid for cyclonic winds) derived in Sect. 1.2. There results from this expression that the gradient velocity reaches a maximum at a radius of the order of R_m. From this radius, the velocity decreases rapidly to zero at the center of the eye, and more slowly to the relatively small values which obtain at large distances from the center.

While the gradient wind velocity is directed along the isobars (see Fig. 1.2.5c), in the boundary layer the wind velocity has a radial component directed toward the low pressures, as was shown in Sect. 1.2. It is this component that effects the inflow of the warm moist air at the ocean surface into region II, thereby maintaining the supply of energy of the storm. Over land the dissipative effect of friction increases, while the supply of energy in the form of warm moist air tends to be cut off. As a result, tropical storms over land usually fill up within a few days at most.

The destructive effects of hurricanes are considerable and are due to the direct action of the wind—which may reach surface velocities of 250 km/hr or more—and, usually to an even larger extent, to the massive piling up of water by the wind known as storm surge, together with flooding by heavy rainfall (Fig. 1.3.7). In the United States, the average losses caused by hurricanes have been estimated to be of the order of $2 billion a year [1-11].

FIGURE 1.3.7.   Hurricane damage, Mississippi (courtesy National Oceanic and Atmospheric Administration).

1.3.3   The Extratropical Cyclone

Such circulations are produced either by the mechanical action of mountain barriers on large-scale atmospheric currents, or by the interaction of air masses along fronts. An example of damage caused by an extratropical storm is shown in Fig. 1.3.8.

Air masses are characterized by relatively uniform physical properties over horizontal distances comparable to the dimensions of oceans or continents. Their physical properties are acquired in the source region and may be modified during subsequent travel of the air mass. Air masses may be classified, according to the source region, into three main groups: arctic, polar, and tropical; each of these may in turn be divided into continental and maritime. Continental polar air, for example, is dry and cold, whereas maritime tropical air is moist and warm.

FIGURE 1.3.8. Damage caused by winter storm, Fire Island, N.Y., March 7, 1962 (courtesy National Oceanic and Atmospheric Administration).

Transition zones between air masses are called frontal zones. The variation of the physical properties of the atmosphere across frontal zones being fairly rapid, the latter may be idealized as surfaces of discontinuity known as frontal surfaces. The intersection of a frontal surface with a surface of equal elevation with respect to the sea level is called a front.

The equilibrium slope of the front between two air masses can be calculated approximately on the basis of simple hydrostatic considerations and varies normally between 1/50 and 1/400.

A front is referred to as a *cold front* or as a *warm front* (Fig. 1.3.9) according

FIGURE 1.3.9. Warm and cold front slopes.

as it moves in the direction of the warmer or colder air. Generally, a warm front moves slowly and is not associated with violent weather conditions. On the other hand, a cold front can move rapidly and cause severe weather. Frequently ahead of cold fronts squall lines develop that may be associated with huge thunderstorms and with tornadoes. The disturbance of the temperature, velocity, or pressure gradient field may cause wavelike perturbations on the front that propagate as waves in a continuous medium. Major disturbances may cause waves whose amplitudes increase with time and develop into intense vortices. The formation and development of the most intense large-scale circulations in middle latitudes, the extratropical cyclones, is connected with such unstable waves occurring predominantly along a front. On the average, the extratropical cyclones move eastward with velocities of the order of 20 km/hr in summer and 50 km/hr in winter.

1.3.4   Local Winds

The influence of small-scale local winds on the general circulation is negligible. However, their intensity may sometimes be considerable and in certain cases govern the design of buildings or structures.

Foehn Winds.   Air flowing across a mountain ridge is forced by the mountain slope to rise. If the air ascends to sufficiently great heights, condensation and precipitation due to adiabatic cooling will occur on the windward side. After having thus lost most of its initial water-vapor content, the air passes over the crest and is forced to descend. Consequent adiabatic compression results in high temperatures of the dry descending air. An example of a foehn wind is suggested in Fig. 1.3.10.

In the United States intense and highly turbulent winds of the foehn type, called chinook winds, develop on the slopes of the Rocky Mountains. In winter chinook winds are notable for bringing sudden high temperature rises, with unusually rapid dissipation of local snow.

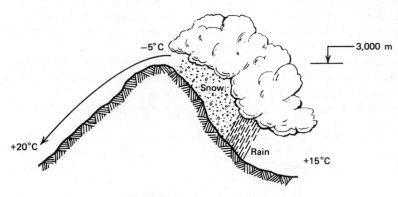

FIGURE 1.3.10.   Foehn wind.

The Bora.   The adiabatic heating during the descent of a very cold mass of air that has passed over a mountain barrier or a plateau may not be sufficient to change it into a warm wind of the foehn type. As the still cold air falls gravitationally into the warmer region on the lee side, its potential energy is converted into kinetic energy. Winds of extreme intensity may thus be produced, characterized by gusts of 150–200 km/hr separated by periods of calm.

Winds of the bora type occur in areas where a steep slope separates a cold plateau from a warm plain. Among the best known bora winds are those that occur at Trieste and Fiume on the northeast coast of the Adriatic.

Jet-Effect Winds.   The jet effect consists of an increase in wind intensity due to topographical configurations that produce a convergence of streamlines. The mistral wind of the lower Rhône Valley in southern France is a well-known example of a bora wind intensified by jet effects.

Thunderstorms.   A necessary condition for the occurrence of thunderstorms is the formation of tall convective clouds produced by the upward motion of warm, moist air. The motion may be started by thermal instability or by the presence of mountain slopes or of a front. Thunderstorms are classified accordingly as thermally convective, orographic, and frontal.

If condensation of the water vapor contained in the ascending air produces heavy precipitation, viscous drag forces exerted by the rain on the air through which it falls contribute to the initiation of a strong downdraft. Part of the falling water is evaporated in the underlying atmosphere that is thus cooled and therefore sinks. The cold downdraft spreads over the ground in the manner of a wall jet (i.e., a flow caused by a jet impinging on a wall), and produces squally winds. This stage in the life cycle of a thunderstorm associated with strong downdrafts usually lasts from 5 to 30 min and is called the mature stage [1-12]. As the energy supplied by the updraft is depleted, dissipation of the thunderstorm occurs. A schematic vertical cross section through a thunderstorm cell in the mature stage is shown in Fig. 1.3.11. Characteristic of thunderstorms is the sharp wind speed increase, known as first gust, which is associated with the passage of the discontinuity zone between the cold downdraft and the surrounding air.

Tornadoes.   Tornadoes contain the most powerful of all winds, causing damage estimated at $100 million a year in the United States alone [1-13]. A tornado consists of a vortex of air, typically of the order of 300 m in diameter, which develops within a severe thunderstorm and moves with respect to the ground with speeds of the order of 30–100 km/hr in a path, approximately 15 km long, directed predominantly toward the northeast. The maximum tangential speeds of tornadoes have been estimated to be of the order of 350 km/hr but the possibility that some may actually be considerably higher has not been ruled out.

Tornadoes are observed as funnel-shaped clouds (Fig. 1.3.12). The tangential

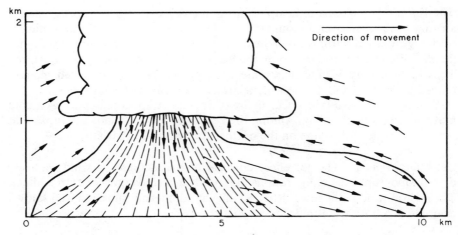

FIGURE 1.3.11.   Section through a thunderstorm in the mature stage [1–12].

FIGURE 1.3.12.   Tornado funnel (courtesy National Oceanic and Atmospheric Administration).

speeds are probably highest at the funnel edge and drop off toward the center and with increasing distance outside the funnel.

Since the centrifugal forces in the tornado vortex far exceed the Coriolis forces, the latter may be neglected and the gradient wind equation (see Sect. 1.2) may be written as

$$\frac{V^2}{r} = \frac{dp/dr}{\rho} \tag{1.3.2}$$

where V is known as the cyclostrophic wind velocity, r is the radial distance from the center of the vortex, ρ is the air density and dp/dr is the pressure gradient along the radius.

From Fig. 1.3.13, which represents the forces acting on a particle in a tornado

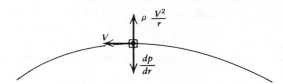

FIGURE 1.3.13.   Balance of forces in tornado vortex.

FIGURE 1.3.14.   Tornado damage in Rochester, Indiana (courtesy Professor U. F. Koehler, Ball State University).

vortex, it can be seen that the pressure in a tornado decreases toward its center. The difference between the pressure at the center and at a few hundred feet from the center of the vortex may be as high as 0.1 of one atmosphere, or about 200 psf.

Tornadoes have also been reported, although much less frequently than in the United States, in Australia, western Europe, India, and Japan. Tornadoes that occur in Japan are known as "tatsumakis." Typical diameters for tatsumakis are of the order of 50 m. Their forward speeds are of the order of 40–50 km/hr; the average length of their paths, which are directed generally toward the northeast, is about 3 km and their maximum tangential speeds are probably about 200 km/hr [1-14].

The destructive effects of tornadoes on buildings are illustrated by Fig. 1.3.14.

REFERENCES

1-1   W. J. Humphreys, *Physics of the Air*, McGraw-Hill Book Company, Inc., New York, 1940 (reprint, Dover, New York, 1964).

1-2   H. R. Byers, *General Meteorology*, McGraw-Hill, New York, 1944.

1-3   G. J. Haltiner and F. L. Martin, *Dynamical and Physical Meteorology*, McGraw-Hill, New York, 1957.

1-4   L. T. Matveev, *Physics of the Atmosphere*, TT67-51380, U.S. Department of Commerce, National Technical Information Service, Springfield, Va.

1-5   A. Miller, *Meteorology*, Charles E. Merril, Columbus, Ohio, 1971.

1-6   M. Neiburger, J. G. Edinger, and W. D. Bonner, *Understanding the Atmospheric Environment*, W. H. Freeman, San Francisco, 1973.

1-7   H. Goldstein, *Classical Mechanics*, Addison-Wesley, New York, 1950.

1-8   F. Fiedler and H. A. Panofsky, "Atmospheric Scales and Spectral Gaps," *Bull. Am. Meteorol. Soc.*, **51** (Dec. 1970) 1114–1119.

1-9   *Hurricane*, U.S. Department of Commerce, ESSA/PI 670009, 1969.

1-10 H. E. Graham and G. N. Hudson, *Surface Winds Near the Center of Hurricanes (and Other Cyclones)*, National Hurricane Research Project. Report No. 39, U.S. Department of Commerce, Washington, D.C., 1960.

1-11 G. T. Sav, *Natural Disasters: Some Empirical and Economic Considerations*, NBSIR 74-473, Center for Building Technology, National Bureau of Standards, Washington, D.C., 1974.

1-12 *Thunderstorm*, Report of the Thunderstorm Project, U.S. Department of Commerce, Washington, D.C., 1949.

1-13 E. Kessler, "Tornadoes," *Bull. Am. Meteoral. Soc.*, **51** (Oct. 1970) 926–936.

1-14 H. Ishizaki et al. " Disasters Caused by Severe Local Storms in Japan," *Bull. Disaster Prev. Res. Inst.*, Kyoto University, **20** (March 1971) 227–243.

2

The Atmospheric
Boundary Layer

As was indicated in Chapter 1, the Earth's surface exerts on the moving air a horizontal drag force, the effect of which is to retard the flow. This effect is diffused by turbulent mixing throughout a region referred to as the *atmospheric boundary layer*. The depth of the boundary layer normally ranges in the case of neutrally stratified flows from a few hundred meters to several kilometers, depending upon wind intensity, roughness of terrain, and angle of latitude. Within the boundary layer, the wind speed increases with elevation; its magnitude at the top of the boundary layer is often referred to as the *gradient speed*. Outside the boundary layer, that is, in the free atmosphere, the wind flows approximately with the gradient speed along the isobars.

This chapter is devoted to the study of aspects of atmospheric boundary-layer flow that are of interest in structural design. The theoretical and experimental results presented include descriptions of mean wind profiles, the relation between wind speeds in different roughness regimes, and the structure of atmospheric turbulence. Since the structural engineer is concerned primarily with the effect of strong winds, unless otherwise noted it will be assumed in the following that the flow is neutrally stratified. The justification of this assumption is that, in strong winds, mechanical turbulence* dominates the heat convection by far, so that thorough turbulent mixing tends to produce neutral stratification, just as in a shallow layer of incompressible fluid mixing tends to produce an isothermal state. Also, because wind speeds are considerably lower than the speed of sound, incompressibility may be assumed in the study of the dynamics of the flow.

---

*A qualitative description of the mechanical turbulence phenomenon is presented in Sect. 4.3.

2.1   GOVERNING EQUATIONS

2.1.1   Equations of Mean Motion

The motion of the atmosphere is governed by the fundamental equations of continuum mechanics that include the equation of continuity—a consequence of the principle of mass conservation—and the equations of balance of momenta, that is, Newton's second law. These equations must be supplemented by phenomenological relations, that is, empirical relations that describe the specific response to external effects of the continuous medium considered. (In the case of a linearly elastic body, for example, the phenomenological relations consist of the so-called Hooke's law.)

If the equation of continuity and the equation of balance of momenta are averaged with respect to time, and if terms that can be shown to be negligible are dropped [2-1, 2-2], the following equations describing the mean motion in the boundary layer of the atmosphere are obtained:

$$U\frac{\partial U}{\partial x}+V\frac{\partial U}{\partial y}+W\frac{\partial U}{\partial z}+\frac{1}{\rho}\frac{\partial p}{\partial x}-fV-\frac{1}{\rho}\frac{\partial \tau_u}{\partial z}=0 \qquad (2.1.1)$$

$$U\frac{\partial V}{\partial x}+V\frac{\partial V}{\partial y}+W\frac{\partial V}{\partial z}+\frac{1}{\rho}\frac{\partial p}{\partial y}+fU-\frac{1}{\rho}\frac{\partial \tau_v}{\partial z}=0 \qquad (2.1.2)$$

$$\frac{1}{\rho}\frac{\partial p}{\partial z}+g=0 \qquad (2.1.3)$$

$$\frac{\partial U}{\partial x}+\frac{\partial V}{\partial y}+\frac{\partial W}{\partial z}=0 \qquad (2.1.4)$$

where U, V, and W are the mean velocity components along the axes x, y, and z of a Cartesian system of coordinates, the z axis of which is vertical; p, ρ, f, and g are the mean pressure, the air density, the Coriolis parameter, and the acceleration of gravity, respectively; and τ_u and τ_v are shear stresses in the x and y directions, respectively. The x axis is selected, for convenience, to coincide with the direction of the shear stress at the surface, denoted τ_0 (Fig. 2.1.1).

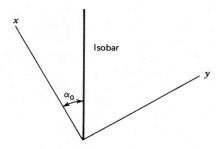

FIGURE 2.1.1.   Coordinate axes.

It can be seen, by differentiating Eq. 2.1.3 with respect to x or y, that the vertical variation of the horizontal pressure gradient depends upon the horizontal density gradient. For the purpose of this text it will be sufficient to consider only flows in which the horizontal density gradient is negligible (e.g., barotropic flows; see, for example, [2-2]). In this case, the horizontal pressure gradient does not vary with height and thus has, throughout the boundary layer, the same magnitude as at the top of the boundary layer, that is,

$$\frac{\partial p}{\partial n} = \rho \left[fV_{gr} \pm \frac{V_{gr}^2}{r} \right] \tag{2.1.5}$$

where V_{gr} is the gradient velocity, r is the radius of curvature of the isobars, and n is the direction of the gradient wind (see Eq. 1.2.7). If the geostrophic approximation may be applied, it follows from Eq. 1.2.6 that

$$\frac{1}{\rho} \frac{\partial p}{\partial x} = fV_g \tag{2.1.6a}$$

$$\frac{1}{\rho} \frac{\partial p}{\partial y} = -fU_g \tag{2.1.6b}$$

where U_g and V_g are the components of the geostrophic velocity G along the x and y axes.

The boundary conditions may be stated as follows: at the ground surface level the velocity vanishes, while at an elevation from the ground equal to the boundary-layer thickness, the shear stress vanish and the wind flows with the gradient velocity.

2.1.2 Mean Velocity Field Closure

To solve the equations of mean motion, it is necessary that phenomenological relations (also referred to as closure relations) be assumed defining the stresses τ_u and τ_v. A well-known assumption [2-1] is that an eddy viscosity K and a mixing length L may be defined such that

$$\tau_u = \rho K(x, y, z) \frac{\partial U}{\partial z} \tag{2.1.7a}$$

$$\tau_v = \rho K(x, y, z) \frac{\partial V}{\partial z} \tag{2.1.7b}$$

$$K(x, y, z) = L^2(x, y, z) \left[\left(\frac{\partial U}{\partial z} \right)^2 + \left(\frac{\partial V}{\partial z} \right)^2 \right]^{1/2} \tag{2.1.8}$$

The use of Eqs. 2.1.7–2.1.8 in conjunction with Eqs. 2.1.1–2.1.4 is referred to as the mean velocity field closure. In Eqs. 2.1.7, either the eddy viscosity or the mixing length field must be specified.

2.1.3   Mean Turbulent Field Closure

From the equations of balance of momenta for the mean motion, the following equation may be derived (see, for example, [2-3]):

$$\left[U \frac{\partial}{\partial x}\left(\overline{\frac{q^2}{2}}\right) + V \frac{\partial}{\partial y}\left(\overline{\frac{q^2}{2}}\right) + W \frac{\partial}{\partial z}\left(\overline{\frac{q^2}{2}}\right) \right]$$

$$- \left[\frac{\tau_u}{\rho}\frac{\partial U}{\partial z} + \frac{\tau_v}{\rho}\frac{\partial V}{\partial z} \right] + \frac{\partial}{\partial z}\left[\overline{w\left(\frac{p'}{\rho} + \frac{q^2}{2}\right)} \right] + \varepsilon = 0 \qquad (2.1.9)$$

where the bars indicate averaging with respect to time, u, v, w are turbulent velocity fluctuations in the x, y, z directions, respectively,

$$q = (u^2 + v^2 + w^2)^{1/2}$$

is the resultant fluctuating velocity, p' is the fluctuating pressure, and ε is the rate of energy dissipation per unit mass. Equation 2.1.9 is referred to as the *turbulent kinetic energy equation* and expresses the balance of turbulent energy advection (the terms in the first bracket), production (the terms in the second bracket), diffusion, and dissipation. The use of Eq. 2.1.9 and attendant phenomenological relations—in conjunction with Eqs. 2.1.1–2.1.4 is referred to as the mean turbulent field closure. Phenomenological descriptions of the quantities involved in Eq. 2.1.9 have been attempted by various authors [2-4, 2-5, 2-6]. Successful predictions of boundary-layer characteristics based on Eq. 2.1.9 and various phenomenological descriptions have been reported in the literature [2-7], although differences of opinion with regard to the relative merits of these descriptions still exist.

   In particular, the mean turbulent field closure appears to be advantageous in the study of three-dimensional boundary-layer flows. Following [2-8] and [2-9], the following relations were proposed in [2-10]:

$$\{\tau_u^2 + \tau_v^2\}^{1/2} = \rho a_1 \overline{q^2} \qquad (2.1.10)$$

$$\overline{w\left(\frac{p'}{\rho} + \frac{q^2}{2}\right)} = \frac{1}{Q_c}(\overline{q^2})_{max}(\overline{q^2})a_2\left(\frac{y}{\delta}\right) \qquad (2.1.11)$$

$$\varepsilon = \frac{(\overline{q^2})^{3/2}}{L_d(y/\delta)} \qquad (2.1.12)$$

$$\frac{\tau_u}{\partial U/\partial z} = \frac{\tau_v}{\partial V/\partial z} \qquad (2.1.13)$$

in which $a_1 \simeq 0.16$, δ is the boundary-layer thickness, and Q_e is the resultant velocity at the edge of the boundary layer (or the gradient velocity in atmospheric flow).

   In the case of the mean turbulent field closure in which Eqs. 2.1.9–2.1.13 are used, the empirical functions that have to be specified are the diffusion functions $a_2(y/\delta)$, and the dissipation length $L_d(y/\delta)$. Reference 2-10 proposes for these functions the form represented in Fig. 2.1.2.

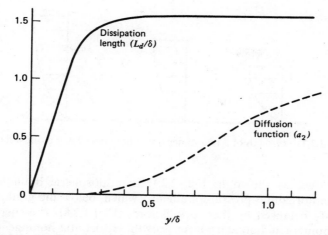

FIGURE 2.1.2. Empirical functions. From J. F. Nash, "The Calculation of Three-Dimensional Turbulent Boundary Layers in Incompressible Flow," *Journal of Fluid Mechanics*, **37** (1969), Cambridge University Press, New York, p. 629.

2.1.4 Second-Order Closure

The second-order closure consists in supplementing the equations of balance of momenta and of continuity by the Reynolds equations, which govern the behavior of the stress tensor components and are derived from first principles [2-11]. Reynolds' equations contain unknown terms, including triple velocity correlations, for which suitable phenomenological relations must be sought. To obtain such relations, the method of invariant modeling has been proposed, which is based upon the following requirements. The modeled terms must: (a) exhibit the tensor and symmetry properties of the original terms in Reynolds' equations; (b) be dimensionally correct; (c) be invariant under a Galilean transformation, that is, a translation of the coordinate axes; (d) satisfy all the general conservation laws [2-11, 2-12]. The second-order closure has been applied, for example, to the study of the flow structure in the boundary layer near a sudden change of surface roughness [2-13].

2.2 MEAN VELOCITY PROFILES IN HORIZONTALLY HOMOGENEOUS FLOW

It may be assumed that, in large-scale storms, within a horizontal site of uniform roughness over a sufficiently large fetch a region exists over which the flow is horizontally homogeneous. The existence of horizontally homogeneous atmospheric flows is supported by observations and distinguishes atmospheric boundary layers from two-dimensional boundary layers such as occur along flat plates. Indeed, it is known that in the latter case the flow in the boundary layer is decelerated by the horizontal stresses, so that the boundary-layer

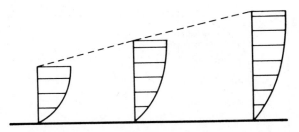

FIGURE 2.2.1.   Growth of a two-dimensional boundary layer along a flat plate.

thickness grows, as shown in Fig. 2.2.1. In atmospheric boundary layers, however, the horizontal pressure gradient—which, below the gradient height, is only partly balanced by the Coriolis force (Fig. 1.2.8)—"re-energizes" the fluid and counteracts boundary-layer growth. Horizontal homogeneity of the flow is thus maintained [2-14].

Under equilibrium conditions, in horizontally homogeneous flow Eqs. 2.1.1 and 2.1.2, in which Eqs. 2.1.6 are used, become

$$V_g - V = \frac{1}{\rho f} \frac{\partial \tau_u}{\partial z} \tag{2.2.1a}$$

$$U_g - U = -\frac{1}{\rho f} \frac{\partial \tau_v}{\partial z} \tag{2.2.1b}$$

2.2.1   The Ekman Spiral

If in the above model the shear stresses are represented by Eqs. 2.1.7 and if, in addition, it is assumed that the eddy viscosity is constant, the model obtained is called the Ekman spiral. Equations 2.2.1 then become a system with constant coefficients. With the boundary conditions $U = V = 0$ for $z = 0$ and $U = U_g$, $V = V_g$ for $z = \infty$, the solution of the system is

$$U = \frac{1}{\sqrt{2}} G[1 - e^{-az}(\cos az - \sin az)] \tag{2.2.2a}$$

$$V = \frac{1}{\sqrt{2}} G[1 - e^{-az}(\cos az + \sin az)] \tag{2.2.2b}$$

where $a = (f/2K)^{1/2}$.

Equations 2.2.2, which describe the Ekman spiral, are represented schematically in Fig. 1.2.9. The agreement of these equations with observations has been found to be unsatisfactory, however. For example, while according to Eqs. 2.2.2 the angle α_0 between the surface stress τ_0 and the geostrophic wind direction (Figs. 2.1.1 and 1.2.9) is 45°, observations indicate that, in barotropic flows, depending chiefly upon roughness of terrain, this angle may range between approximately 6° and 30°. The cause of the discrepancies is the assumption, mathematically convenient but physically incorrect, that the eddy viscosity does not depend on height.

2.2.2 The Turbulent Ekman Layer

Meteorologists have attempted to solve Eqs. 2.2.1 using assumptions on the variation of eddy viscosity with height that are more plausible than the assumption of constancy. A survey of corresponding solutions can be found in [2-2] and [2-15].

A different type of approach was recently developed in [2-14] in which, rather than resorting to a mean velocity field, closure is based on similarity considerations analogous to those used in the theory of two-dimensional boundary layer flows. In this approach, the boundary layer is divided into two regions, a surface layer and an outer layer. It is logical to assert that the surface shear τ_0 must depend upon the flow velocity at some small distance z from the ground, the roughness of the terrain (i.e., a roughness length z_0), and the density ρ of the air. Thus τ_0 may be expressed as a function F of these quantities, that is,

$$\tau_0 = F(U\mathbf{i} + V\mathbf{j}, z, z_0, \rho) \qquad (2.2.3)$$

where $\mathbf{i}$ and $\mathbf{j}$ are unit vectors in the x and y directions, respectively. It is convenient to write Eq. 2.2.3 in nondimensional form as follows:

$$\frac{U\mathbf{i} + V\mathbf{j}}{u_*} = f_1\left(\frac{z}{z_0}\right) \qquad (2.2.4)$$

where the quantity

$$u_* = \left(\frac{\tau_0}{\rho}\right)^{1/2} \qquad (2.2.5)$$

is known as the shear velocity of the flow and f_1 is some function of the ratio z/z_0. Equation 2.2.4 is a form of the well-known "law of the wall" and describes the flow in the surface layer.

In the outer layer, it can be similarly asserted that the reduction of velocity $[(U_g\mathbf{i} + V_g\mathbf{j}) - (U\mathbf{i} + V\mathbf{j})]$ at height z must depend upon the surface shear τ_0, the height to which the effect of the wall stress has diffused in the flow, that is, the boundary-layer thickness δ, and the density ρ of the air. The expression of this dependence in nondimensional form is known as the "velocity defect law":

$$\frac{U\mathbf{i} + V\mathbf{j}}{u_*} = \frac{U_g\mathbf{i} + V_g\mathbf{j}}{u_*} + f_2\left(\frac{z}{\delta}\right) \qquad (2.2.6)$$

where f_2 is some function to be defined.

If it is postulated that a gradual change occurs from conditions near the ground to conditions in the outer layer, it may be assumed that a region of overlap exists in which both laws are valid. Let Eq. 2.2.4 be written in the form

$$\frac{U\mathbf{i} + V\mathbf{j}}{u_*} = f_1\left[\left(\frac{z}{\delta}\right)\left(\frac{\delta}{z_0}\right)\right] \qquad (2.2.7)$$

From the form of Eqs. 2.2.6 and 2.2.7, and the condition that their right-hand sides be equal in the overlap region, it follows that a multiplying factor inside the function f_1 must be equivalent to an additive quantity outside the function

f_2. In the case of the analogous two-dimensional problem, it is well known that the two functions must be logarithms [2-16, 2-17]. The requirements of the problem at hand will be satisfied if f_1 and f_2 are defined as follows [2-14]:

$$f_1(\xi) = (\ln \xi^{1/k})\mathbf{i} \tag{2.2.8}$$

$$f_2(\xi) = (\ln \xi^{1/k})\mathbf{i} + \frac{B}{k}\mathbf{j} \tag{2.2.9}$$

where B and k are constants. Substituting Eqs. 2.2.8 and 2.2.9 in Eqs. 2.2.7 and 2.2.6, respectively,

$$\frac{U\mathbf{i} + V\mathbf{j}}{u_*} = \frac{1}{k}\left(\ln \frac{z}{\delta} + \ln \frac{\delta}{z_0}\right)\mathbf{i} \tag{2.2.10}$$

$$\frac{U\mathbf{i} + V\mathbf{j}}{u_*} = \frac{U_g\mathbf{i} + V_g\mathbf{j}}{u_*} + \frac{1}{k}\left(\ln \frac{z}{\delta}\right)\mathbf{i} + \frac{B}{k}\mathbf{j} \tag{2.2.11}$$

If Eqs. 2.2.10 and 2.2.11 are now equated in the overlap region, there results

$$\frac{U_g}{u_*} = \frac{1}{k}\ln \frac{\delta}{z_0} \tag{2.2.12}$$

$$\frac{V_g}{u_*} = -\frac{B}{k} \tag{2.2.13}$$

from which there follows

$$G = \left(B^2 + \ln^2 \frac{\delta}{z_0}\right)^{1/2} \frac{u_*}{k} \tag{2.2.14}$$

It can further be shown that the boundary-layer thickness δ may be expressed as

$$\delta = c\frac{u_*}{f} \tag{2.2.15}$$

where c is a constant. To prove this relation, let Eqs. 2.2.1a and 2.2.1b be multiplied by the unit vectors $\mathbf{j}$ and $\mathbf{i}$, respectively. From the expressions thus obtained, and remembering that $\tau_u = \tau_0$, $\tau_v = 0$ at the surface, and $\tau_u = \tau_v = 0$ at $z = \delta$, it follows that

$$\int [U\mathbf{i} + V\mathbf{j} - (U_g\mathbf{i} + V_g\mathbf{j})]\,dz = \frac{\tau_0}{\rho f}\mathbf{i} \tag{2.2.16}$$

where the integration is carried out over the boundary-layer depth. Since the bulk of the mass transport takes place in those parts of the boundary layer where Eq. 2.2.6 holds—which include the overlap part of the surface layer down presumably to a very small height—the velocity profile in Eq. 2.2.16 may be approximately described by Eq. 2.2.6. If Eq. 2.2.15 is now substituted into Eq. 2.2.6 and Eq. 2.2.5 is used, the left-hand side of Eq. 2.2.16 becomes

$$\int u_* f_2\left(\frac{zf}{cu_*}\right) dz = \frac{cu_*^2}{f} \int f_2(\zeta)\, d\zeta$$

$$= \text{constant} \; \frac{\tau_0}{\rho f} \qquad (2.2.16a)$$

that is, Eq. 2.2.16 is verified and the validity of Eq. 2.2.15 is established [2-14]. Equation 2.2.14 may then be written as

$$G = \left[B^2 + \left(\ln \frac{u_*}{f z_0} - A \right)^2 \right]^{1/2} \frac{u_*}{k} \qquad (2.2.17)$$

Equation 2.2.17 was obtained independently in [2-18] and [2-5]. The derivation of [2-5] is based on the turbulent energy equation and the assumption of a mixing length proportional to z. The quantities A and B are universal constants. From the analysis of observations it was found that $4.3 < B < 5.3$ and $0 < A < 2.8$ [2-14, 2-15, 2-18, 2-19, 2-20, 2-21, 2-22, 2-23, 2-24]. On the basis of experiments in the wind tunnel and in the atmosphere, the well-known *von Kármán's constant* is generally assumed to be $k \simeq 0.4$.* The coefficient c in Eq. 2.2.15 is of the order of 0.25–0.3 [2-20, 2-26].

2.2.3 The Logarithmic Law

Equation 2.2.10 may be written as

$$U(z) = \frac{1}{k} u_* \ln \frac{z}{z_0} \qquad (2.2.18)$$

($k \simeq 0.4$), where z is the height above the surface, z_0 is the roughness length, and $U(z)$ is the mean wind speed. Equation 2.2.18 is known as the logarithmic law.

Recent micrometeorological research has established that the height above ground z_l up to which Eq. 2.2.18 may be assumed to be approximately valid, is defined by the relation

$$z_l = b \frac{u_*}{f} \qquad (2.2.19)$$

where b is a constant, the order of magnitude of which is 0.015–0.03 [2-26, 2.27]. As noted in [2-26], Eq. 2.2.19 expresses the fact, well known from laboratory experiments—including experiments conducted in rotating wind tunnels [2-28, p. 148, 2-29]—that the logarithmic layer extends to some fraction (of the order of 10 percent) of the boundary layer depth δ (see Fig. 2.2.2). Figure 2.2.3 [2-30] represents averages of 14 mean wind profiles (average mean speed at 9.1 m above ground $U(9.1) = 5.3$ m/sec) measured in nearly neutral flow near Dallas, Texas. It is seen that for the profiles of Fig. 2.2.3 the logarithmic

---

*The actual value of k has in recent years become the object of some debate [2-25]. However, calculations of interest in engineering applications described in this text are not affected significantly by the actual value of k.

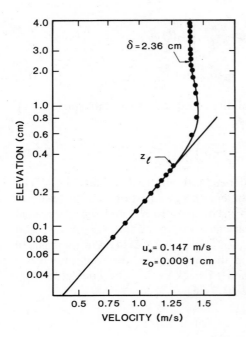

$\delta = 2.36$ cm

z_ℓ

$u_* = 0.147$ m/s
$z_0 = 0.0091$ cm

FIGURE 2.2.2. Mean wind profile as measured in a rotating wind tunnel [2–29]. Copyright © 1975 by D. Reidel Publishing Company.

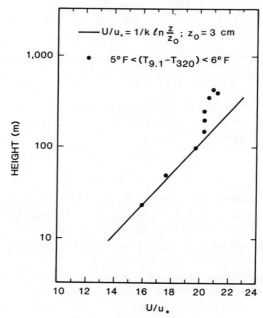

$U/u_* = 1/k \, \ell n \frac{z}{z_0}$; $z_0 = 3$ cm

$5°F < (T_{9.1} - T_{320}) < 6°F$

FIGURE 2.2.3. Average of 14 mean wind profiles recorded near Dallas, Texas. After R. H. Thuiller and U. O. Lappe, "Wind and Temperature Profile Characteristics from Observations on a 1400 ft Tower," *J. Appl. Met.* **3**(1964), 299–306, American Meteorological Society.

law provides a good description of the data up to at least 100 m elevation. This is in agreement with Eq. 2.2.19. Indeed, for $U(9.1) = 5.3$ m/sec, $z_0 = 0.03$ m, $f \simeq 0.77 \times 10^{-4}$ (Table 1.2.1), and $b \simeq 0.02$, Eqs. 2.2.18 and 2.2.19 yield $z_l \simeq$ 100 m. Note in Figs. 2.2.2 and 2.2.3 that the use of the logarithmic law for heights exceeding z_l is conservative from a structural design viewpoint.

Equation 2.2.19 may also be shown to follow from the assumption that, in the region $0 < z < z_l$, the shear stress τ_u differs little from the surface stress τ_0 (see, for example, [2-1, p. 489]), and the component V of the velocity is small. Integration of Eq. 2.2.1a over the height z_l yields

$$\tau_u = \tau_0 + \rho f \int_0^{z_l} (V_g - V)\, dz \simeq \tau_0 + \rho f V_g z_l \qquad (2.2.20a)$$

or

$$|\rho f V_g z_l| = \eta \tau_0 \qquad (2.2.20b)$$

where η is a small number. Using Eqs. 2.2.5 and 2.2.13,

$$z_l = \frac{\eta u_*^2}{f V_g} = \frac{\eta k}{f B}\, u_* = b\, \frac{u_*}{f} \qquad (2.2.21)$$

It is shown in [2-26] and [2-31] that the logarithmic law holds, for practical purposes, even beyond heights at which η is of the order of 30%.

If, for example, $f = 10^{-4}$ sec^{-1}, $U = 30$ m/sec at 10 m above ground, $z_0 = 0.05$ m (open terrain), and $b = 0.02$, it follows then from Eqs. 2.2.18 and 2.2.19 that $z_l \simeq 400$ m. In the case of strong winds, the validity of the logarithmic law up to elevations of the order of 200 m has been confirmed by measurements reported in [2-32] and [2-33], as well as by observations at Sale [2-34] and Cranfield [2-35] analyzed in [2-22].

On account of the finite height of the roughness elements, the following empirical modification of Eq. 2.2.18 is required [2-36]. The quantity z, rather than denoting height above ground, is defined as

$$z = z_g - z_d \qquad (2.2.22)$$

where z_g is the height above ground and z_d is a length known as the zero plane displacement [2-37]. The quantity z will be referred to as the effective height. The flow parameters z_0 and z_d are determined empirically and are functions of the nature, height, and distribution of the roughness elements [2-38]. The roughness length z_0 is a measure of the eddy size at the ground. It is suggested in [2-33] that reasonable values of the zero plane displacement in cities may be obtained using the formula

$$z_d = \bar{H} - \frac{z_0}{k} \qquad (2.2.22a)$$

where $\bar{H}$ is the general roof-top level.

Typical values of z_0 for various types of terrain, and the corresponding values of the surface drag coefficient (defined as

$$\kappa = \left[\frac{k}{\ln\left(\dfrac{10}{z_0}\right)} \right]^2 \tag{2.2.23}$$

in which z_0 is expressed in meters) are given in Table 2.2.1 [2-39, 2-40, 2-41, 2-42]. Table 2.2.1 also includes suggested values of z_0 for built-up terrain. The determination of representative wind profiles in built-up terrain is generally difficult on account of local flow inhomogeneities (e.g., those associated with wake effects). For this reason values of z_0 in built-up terrain may differ considerably from experiment to experiment. The values listed in Table 2.2.1 are intended for use in structural engineering calculations in conjunction with the assumption $z_d = 0$. They are based on a careful analysis of full-scale data, as shown in [2-42].

The surface drag coefficient κ (Eq. 2.2.23) for *wind flow over water surfaces* depends upon wind speed. On the basis of a large number of measurements, the following empirical relations were proposed for the range $4 < U(10) < 20$ m/sec [2-43]:

$$\kappa = 5.1 \times 10^{-4} [U(10)]^{0.46} \tag{2.2.24a}$$

or

$$\kappa = 10^{-4} [7.5 + 0.67 U(10)] \tag{2.2.24b}$$

TABLE 2.2.1. Values of Surface Roughness Length (z_0) and of Surface Drag Coefficients for Various Types of Terrains

Type of Surface	z_0 (cm)		$10^3 \kappa$	
Sand[a]	0.01–	0.1	1.2–	1.9
Snow surface	0.1 –	0.6	1.9–	2.9
Mown grass (~ 0.01 m)	0.1 –	1	1.9–	3.4
Low grass, steppe	1 –	4	3.4–	5.2
Fallow field	2 –	3	4.1–	4.7
High grass	4 –	10	5.2–	7.6
Palmetto	10 –	30	7.6–	13.0
Pine forest (mean height of trees: 15 m; one tree per 10 m^2; $z_d \simeq 12$ m [2-40])	90 –	100	28.0–	30.0
Sparsely built-up suburbs[b]	20 –	40	10.5–	15.4
Densely built-up suburbs, towns[b]	80 –	120	25.1–	35.6
Centers of large cities[b]	200 –	300	61.8–	110.4

[a]Ref. 2-38.
[b]Values of z_0 to be used in conjunction with the assumption $z_d = 0$ [2-42].

where $U(10)$ is the mean wind speed in m/sec at 10 m above the mean water level. According to [2-44], for $U(10) > 20$ m/sec or so κ is constant.

A more recent evaluation of existing measurements led to the expression proposed in [2-45] for wind speeds $U(10)$ up to 40 m/sec:

$$\kappa = 0.0015 \left[1 + \exp\left(-\frac{U(10) - 12.5}{1.56} \right) \right]^{-1} + 0.00104 \qquad (2.2.25)$$

For example, if $U(10) = 20$ m/sec, it follows from Eqs. 2.2.25 and 2.2.23 that $\kappa = 2.5 \times 10^{-3}$ and $z_0 = 0.35$ cm. It can be verified that errors in the estimation of wind speeds due to uncertainties associated with differences among Eqs. 2.2.24a, 2.2.24b, and 2.2.25 are insignificant. Additional information on the surface drag for wind flow over the ocean is presented in [2-46] and [2-132].

According to [2-133], the influence on the waves on the wind profile appears to be restricted to elevations below three wave heights; in this zone wind speeds are lower than indicated by the logarithmic profile.

2.2.4 The Power Law

Historically, the first representation of the mean wind profile in horizontally homogeneous terrain has been the power law, proposed in [2-47] in 1916:

$$U(z_{g1}) = U(z_{g2}) \left(\frac{z_{g1}}{z_{g2}} \right)^{\alpha} \qquad (2.2.26)$$

where α is an exponent dependent upon roughness of terrain and z_{g1} and z_{g2} denote heights above ground.

In [2-48] it is assumed (1) that the power law holds with constant exponent α up to the gradient height δ, and (2) that δ itself is a function of α alone. The first of these assumptions implies

$$\frac{U(z_g)}{G} = \left(\frac{z_g}{\delta} \right)^{\alpha} \qquad (2.2.27)$$

The second assumption represents in effect an engineering simplification of the boundary-layer depth description given by Eq. 2.2.15. Values of δ and α recommended for design purposes in [2-48] and [2-49] are shown in Table 2.2.2. It is noted that values of δ (in open terrain and centers of large cities) similar to those given in Table 2.2.2 were proposed in 1935 by Pagon [2-50, p. 744].

TABLE 2.2.2. Values of δ and α Recommended in [2-48] and [2-49]

	Coastal Areas		Open Terrain		Suburban Terrain		Centers of Large Cities	
	α	δ (meters)	α	δ (meters)	α	δ (meters)	α	δ (meters)
Ref. 2-48	—	—	0.16	275	0.28	400	0.40	520
Ref. 2-49	1/10	215	1/7	275	1/4.5	400	1/3	460

Currently, the logarithmic law is regarded by meteorologists as a superior representation of strong wind profiles in the lower atmosphere [2-26, 2-51, 2-52, 2-53, 2-54, 2-55]; consequently, the power law is no longer employed in micrometeorological practice.

2.2.5   Relation between Wind Speeds in Different Roughness Regimes

Consider two adjacent terrains, each of uniform roughness and of sufficiently large fetch. Let the roughness lengths for the two terrains be denoted by z_{01} and z_0, and assume $z_{01} < z_0$. The retardation of the flow by surface friction will be more effective over the rougher terrain; therefore, if the geostrophic speed is the same over both sites, at equal elevations the mean wind speeds will be lower over the rougher site. A schematic representation of the respective wind profiles is shown in Fig. 2.2.4.

The profiles of Fig. 2.2.4 suggest the following procedure for relating wind speeds in different roughness regimes. To calculate the wind speed $U(z_g, z_0)$ over the rougher terrain if the speed $U(z_{g1}, z_{01})$ is known, Eq. 2.2.27 is applied to each profile; then the quantity G is eliminated from the two relations thus obtained, and

$$U(z_g, z_0) = \left(\frac{z_g}{\delta(z_0)}\right)^{\alpha(z_0)} \left(\frac{\delta(z_{01})}{z_{g1}}\right)^{\alpha(z_{01})} U(z_{g1}, z_{01}) \qquad (2.2.28)$$

where $\alpha(z_0)$, $\delta(z_0)$ and $\alpha(z_{01})$, $\delta(z_{01})$ correspond to the roughness lengths z_0 and z_{01}, respectively. Equation 2.2.28 was proposed in [2-48] and will be referred to as the *power law model*.

Recently, an alternative procedure has been proposed, which is based on results of both theoretical and experimental studies [2-22]. If the speed

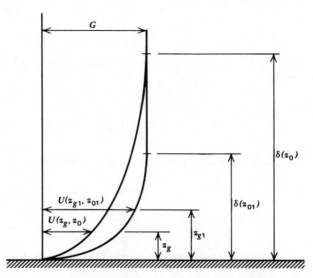

FIGURE 2.2.4.   Wind velocity profiles.

$U(z_{g1}, z_{01})$ is known, it follows from Eq. 2.2.18

$$u_{*1} = \frac{U(z_{g1}, z_{01})}{2.5 \ln(z_1/z_{01})} \qquad (2.2.29)$$

where the notation of Eq. 2.2.22 is used. Applying now Eq. 2.2.29 to the two profiles represented in Fig. 2.2.4 and eliminating G,

$$\left[B^2 + \left(\ln \frac{u_*}{fz_0} - A \right)^2 \right] u_* = \left[B^2 + \left(\ln \frac{u_{*1}}{fz_{01}} - A \right)^2 \right]^{1/2} u_{*1} \qquad (2.2.30)$$

Equation 2.2.30 determines the value of the friction velocity u_*. Then

$$U(z_g, z_0) = 2.5 u_* \ln \frac{z_g}{z_0} \qquad (2.2.31)$$

Equations 2.2.29, 2.2.30, and 2.2.31 will be referred to as the *similarity model*.

As has been shown in [2-22], the uncertainty with regard to the exact values of the constants A and B in Eq. 2.2.30 turns out to be of little consequence insofar as estimates of wind speeds in the lower atmosphere are concerned. With possible errors of the order of 3% or less, it may be assumed $A = 1.4$ and $B = 4.7$. Also, the dependence of the results on u_* and f is insignificant and may be neglected. For practical purposes, therefore, the ratios u_*/u_{*1} may be calculated simply as functions of the roughness lengths z_0 and z_{01}. The dependence of u_*/u_{*1} upon z_0 and z_{01} can be represented by the relation [2-56]

$$\frac{u_*}{u_{*1}} = \left(\frac{z_0}{z_{01}} \right)^{0.0706} \qquad (2.2.32)$$

However, subsequent research has shown that the similarity model must be subjected to empirical adjustments in the case of terrain for which $z_0 > 0.30$ m or so. Table 2.2.3 lists ratios u_*/u_{*1} based on full-scale measurements, corresponding to $z_{01} = 0.07$ m and various values z_0 of practical interest [2-42].

The application of the similarity model will now be illustrated by a numerical example. The data used in the example were obtained by measurements in and near London and were reported in [2-33]. At Heathrow, $z_{01} = 0.08$ m, $z_{d1} \simeq 0$, and the measured mean wind at a height above ground $z_{g1} = 10$ m is $U(z_{g1}, z_{01}) = 11.7$ m/sec. The mean wind $U(z_g, z_0)$ at a height above ground $z_g = 195$ m is sought at the Post Office Tower in London, where $z_0 = 2.5$ m ($z_d = 0$).

From Eq. 2.2.29, $u_{*1} = 0.968$ m/sec. From Table 2.2.3, $u_*/u_{*1} = 1.46$, that is, $u_* = 1.41$ m/sec. Using Eqs. 2.2.31, $U(z_g, z_0) = 15.3$ m/sec. It is noted that this result coincides with the actual measured speed [2-33].

TABLE 2.2.3. Ratios u_*/u_{*1} for $z_{0_1} = 0.07$ m and Various Values z_0 [2-42]

z_0(m)	0.005	0.07	0.30	1.00	2.50
u_*/u_{*1}	0.83	1.00	1.15	1.33	1.46

If the mean speed near the Post Office Tower at $z_g = 195$ m is calculated using the power law model (Eq. 2.2.28) with the parameters α and δ suggested in [2-48], there results $U(z_g, z_0) = 13.4$ m/s versus the measured 15.3 m/sec speed.

2.2.6   Effect of Thermal Convection on Mean Speed Profiles in Strong Winds

It is of interest to estimate the extent to which the effect of thermal convection is significant in structural engineering and extreme wind climatological calculations. To do this we use the following expression, based on the work of Monin and Obukhov [2-2, p. 282; 2-5]:

$$U(z) = \frac{u_*}{k} \left[\ln \frac{z}{z_0} - \psi \left(\frac{z}{L} \right) \right] \tag{2.2.33}$$

where $u_* =$ friction velocity, $k =$ von Kármán's constant, $z_0 =$ roughness length, $\psi =$ Monin-Obukhov function, and $L =$ Monin-Obukhov length. If the stratification is neutral, $L = \infty$, $\psi = 0$, and Eq. 2.2.33 becomes the well-known logarithmic law (Eq. 2.2.18).

The length L is defined by the following expression [1-4, p. 281]:

$$L = - \frac{u_*^3}{k \frac{g}{T} \frac{Q_0}{c_p \rho}} \tag{2.2.34}$$

where $g =$ acceleration of gravity ($g = 9.81$ m/sec^2), $T =$ absolute temperature, $c_p =$ specific heat at constant pressure ($c_p = 240$ cal/kg degree [1-4, p. 132]), $\rho =$ air density ($\rho \simeq 1.2$ kg/m^3), and $Q_0 =$ eddy heat flux (usual orders of magnitude for Q_0 are 10 to 60 cal/m^2/sec [1-4, p. 276]).

Unstable Stratification.   In unstable air the following expression will be used for $\psi(z/L)$:

$$\psi \left(\frac{z}{L} \right) = \int_{z_0/L}^{z/L} [1 - (1 - 16\zeta)^{-1/4}] \frac{d\zeta}{\zeta} \tag{2.2.35}$$

Equation 2.2.35 was proposed in Ref. 2-25. According to Ref. 2-57, it provides a very good fit to experimental data over uniform terrain and for $0 > z/L > -2$. (Note that L is by definition negative if the stratification is unstable.) Equation 2.2.35 is represented in Fig. 2.2.5.

Stable Stratification.   In the case of stable stratification it may be assumed

$$\psi \left(\frac{z}{L} \right) \simeq -5 \frac{z}{L} \tag{2.2.36}$$

[2-25, 2-57]. The length L is defined by Eq. 2.2.34; however, empirical studies reported in Ref. 2-58 suggest that under stable stratification conditions it may be assumed that

$$L \simeq 1.1 \times 10^3 u_*^3 \tag{2.2.37}$$

where u_* is expressed in m/sec.

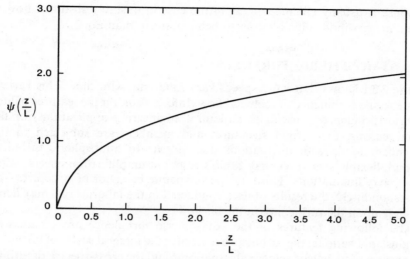

$$\psi\left(\frac{z}{L}\right)$$

$$-\frac{z}{L}$$

FIGURE 2.2.5.   Function $\psi(z/L)$ for unstably stratified flow. From E. Simiu, "Thermal Convection and Design Wind Speeds," *Journal of the Structural Division*, ASCE, **108** (July 1982), 1671–1675.

Table 2.2.4 lists estimated deviations from the logarithmic profile (Eq. 2.2.18) for three representative cases of interest in structural engineering applications†. The results of Table 2.2.4 show that such deviations may indeed be neglected when estimating wind pressures on structures (see Case 1) or when reducing to a common elevation largest monthly or yearly wind speeds recorded at a weather station (see Case 2). However, for wind speeds $U(10)$ of the order of 5 m/sec the deviations from a logarithmic profile are significant (see Case 3). The latter conclusion is of interest for the design of structures, such as smoke stacks, that exhibit a significant across-wind response at low wind speeds. This

TABLE 2.2.4.   Deviation of Mean Wind Speeds from Logarithmic Profile [2-59]

	Case 1[a]		Case 2[b]	Case 3[c]
Elevation	50 m	200 m	15 m	15 m
Unstable stratification	-1%	-4%	-4%	$\sim -22\%$
Stable stratification	1%	4%	2%	$\sim 12\%$

[a]Hourly wind speed at 10 m elevation over open terrain: $\simeq 25$ m/sec.
[b]Hourly wind speed at 10 m elevation over open terrain: $\simeq 12$ m/sec.
[c]Hourly wind speed at 10 m elevation over open terrain: $\simeq 5$ m/sec.

†The estimates were based on the assumption that in unstably stratified flows $Q = 50$ kcal/m² sec and $T = 290°$.

response is usually enhanced if, as in the case in unstably stratified flow, the actual mean wind profile is closer to being uniform than Eq. 2.2.18.

2.3   ATMOSPHERIC TURBULENCE

Figure 2.3.1 shows that wind speeds vary randomly with time. This variation is due to the turbulence of the wind flow. Information on the features of atmospheric turbulence is useful in structural engineering applications for three main reasons. First, rigid structures and members are subjected to time-dependent loads with fluctuations due in part to atmospheric turbulence. Second, flexible structures may exhibit resonant amplification effects induced by velocity fluctuations. Third, the aerodynamic behavior of structures—and, correspondingly, the results of tests conducted in the laboratory—may depend strongly upon the turbulence in the air flow.

The following features of the atmospheric turbulence are of interest in various applications: the turbulence intensity; the integral scales of turbulence; the spectra of turbulent velocity fluctuations; and the cross-spectra of turbulent velocity fluctuations. Also of interest to structural designers is the dependence of the largest wind speeds in a record upon averaging time.

2.3.1   Turbulence Intensity

The simplest descriptor of atmospheric turbulence is the turbulence intensity.

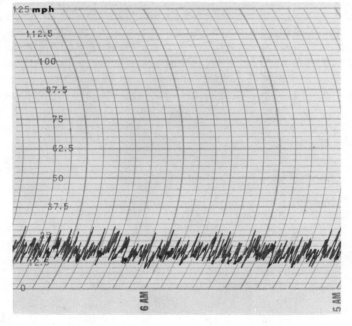

FIGURE 2.3.1.   Wind speed record.

Let $u(z)$ denote the velocity fluctuations parallel to the direction of the mean speed in a turbulent flow passing a point with elevation z (Fig. 2.3.1). The longitudinal turbulence intensity is defined as

$$I(z) = \frac{\overline{u^2(z)}^{1/2}}{U(z)} \qquad (2.3.1)$$

where $U(z) =$ mean wind speed at elevation z and $\overline{u^2}^{1/2} =$ root mean square value of u.[†] Vertical and lateral turbulence intensity may be similarly defined.

The longitudinal turbulence fluctuations can be written as

$$\overline{u^2} = \beta u_*^2 \qquad (2.3.2)$$

where $u_* =$ friction velocity (see Eq. 2.2.18). It is commonly assumed that β does not vary with height.[‡] Values of β suggested for structural design purposes on the basis of a large number of measurements are listed in Table 2.3.1 [2-42].

The averaging time in Eqs. 2.3.1 and 2.3.2 should be equal to the duration of the strong winds in a storm. It is commonly assumed that this duration is between 10 minutes and 1 hour.

For example, if $z = 30$ m, $z_0 = 0.07$ m, and $U(30) = 20$ m/sec, it follows from Eqs. 2.1.18, 2.3.1, 2.3.2, and Table 2.3.1 that the turbulence intensity is $I(30) = 0.162$.

2.3.2 Integral Scales of Turbulence

The velocity fluctuations in a flow passing a point (Fig. 2.3.1) may be considered to be caused by a superposition of conceptual eddies transported by the mean wind. Each eddy is viewed as causing at that point a periodic fluctuation with circular frequency $\omega = 2\pi n$, where n is the frequency. By analogy with the case of the traveling wave, we define the eddy wave length as $\lambda = U/n$, where $U =$ wind speed, and the eddy wave number, $K = 2\pi/\lambda$. The wave length is a measure of eddy size.

Integral scales of turbulence are measures of the average size of the turbulent eddies of the flow. There are altogether nine integral scales of turbulence, corresponding to the three dimensions of the eddies associated with the longitudinal, transverse, and vertical components of the fluctuating velocity, u, v,

TABLE 2.3.1.   Values of β Corresponding to Various Roughness Lengths

z_0(m)	0.005	0.07	0.30	1.00	2.50
β	6.5[a]	6.0	5.25	4.85	4.00

[a]Based on measurements reported in [2-79] and used in conjunction with Eqs. 2.2.23 and 2.2.25. See also [2-132].

[†]The alternative notation $\sigma_u = \overline{u^2}^{1/2}$ is also commonly used.
[‡]This use of the notation β should not be confused with its use as the safety index (Appendix A3).

and w. For example, L_u^x, L_u^y, and L_u^z are, respectively, measures of the average longitudinal, transverse, and vertical size of the eddies associated with the longitudinal velocity fluctuations (x is the direction of the mean wind U and of the longitudinal fluctuations u).

Mathematically, L_u^x is defined as

$$L_u^x = \frac{1}{\overline{u^2}} \int_0^\infty R_{u_1 u_2}(x)\, dx \qquad (2.3.3)$$

where $R_{u_1 u_2}(x)$ is the cross-covariance function of the longitudinal velocity components $u_1 \equiv u(x_1, y_1, z_1, t)$ and $u_2 \equiv u(x_1 + x, y_1, z_1, t)$, defined in a manner analogous to Eq. A2.29, $t =$ time, and $\overline{u^2}^{1/2}$ is the root mean square value of u_1 (and u_2). Note that in horizontally homogeneous flow, L_u^x is independent of x_1 and y_1. Similar definitions apply to the other integral turbulence scales.

From their mathematical definition it follows that integral scales are small if the cross-covariance functions are rapidly decaying functions of distance, and conversely. Velocity fluctuations separated by a distance considerably larger than the integral scales are uncorrelated, and will therefore act on a structural element at cross-purposes. For example, values of L_u^y and L_u^z that are small compared to the dimensions of a panel normal to the mean wind indicate that the effect of the longitudinal velocity fluctuations upon the overall wind loading is small. However, if L_u^y and L_u^z are large, the eddy will envelop the entire panel, and that effect will be significant.

Equation 2.3.3 can be transformed if it is assumed that the flow disturbance travels with the velocity $U(z)$ and, therefore, that the fluctuation $u(x_1, \tau + t)$ may be identified with $u(x_1 - x/U, \tau)$, where $t =$ time (Taylor's hypothesis). Then

$$L_u^x = \frac{U}{\overline{u^2}} \int_0^\infty R_u(\tau)\, d\tau \qquad (2.3.4)$$

where $R_u(\tau)$ is the autocovariance function of the fluctuation $u(x_1, t)$. The length of the record from which $R_u(\tau)$ is estimated should be the same as that used to estimate U and $\overline{u^2}$ (that is, about one hour—see Sect. 2.3.1).

Estimates of turbulence scales depend significantly upon the length and the degree of stationarity of the record being analyzed, and usually vary widely from experiment to experiment. For example, for open exposure, measured values of L_u^x reported in [2-60] (Part 2, pp. 31 and 32) vary between 120 m and 630 m at 150.8 m elevation (the average value being 400 m); between 110 m and 690 m at 110.8 elevation (average value: 350 m); between 60 m and 650 m at 80.8 m elevation (average value: 300 m); between 130 m and 450 m at 50.8 m elevation (average value: 200 m); and between 60 m and 460 m at 30.8 m elevation (average value: 200 m). Data reviewed in [2-61] suggest that L_u^x is a decreasing function of terrain roughness. For example, the following data are listed in [2-61]:

Site	$z(m)$	$z_0(m)$	$L_u^x(m)$
Cardington	15	0.01	82
Round Hill	17	0.04–0.10	55
Brookhaven	16	1.00	36

The following empirical expression was proposed in [2-61] for the height range $z = 10$–240 m:

$$L_u^x = Cz^m \tag{2.3.5}$$

where C and m are given in Fig. 2.3.2 and z is the elevation (L_u^x and z in meters). The application of Eq. 2.3.5 to the data just listed yields, approximately, the values $L_u^x = 150$ m (Cardington), 140–120 m (Round Hill), 70 m (Brookhaven), which are about twice as high as the measured values.

According to [2-61] the integral scales L_u^y and L_u^z are, respectively, about one-third and one-half the integral scale L_u^x as given by Eq. 2.3.5. However, according to [2-62], a better estimate of L_u^y is obtained from the expression:

$$L_u^y \simeq 0.2 L_u^x \tag{2.3.6}$$

It has been suggested that

$$L_u^z \simeq 6z^{0.5} \tag{2.3.7}$$

(z in meters) [2-136]. The expression

$$L_w^x \simeq 0.4z \tag{2.3.8}$$

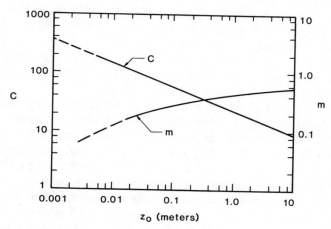

FIGURE 2.3.2.   Values of C and m as functions of z_0 [2–61]. Reprinted with permission from J. Counihan, "Adiabatic Atmospheric Boundary Layers: A Review and Analysis of Data from the Period 1880–1972," *Atmospheric Environment*, **9** (1975), 871–905, Pergamon Press.

was proposed in [2-64] and confirmed by subsequent measurements, as indicated in [2-61].

2.3.3   Spectra of Longitudinal Velocity Fluctuations

The Energy Cascade.   It was mentioned in Sect. 2.3.2 that the turbulent velocity fluctuations may be considered to be caused by a superposition of eddies, each characterized by a periodic motion of circular frequency $\omega = 2\pi n$ (or by a wave number $K = 2\pi/\lambda$, where λ is the wave length). The total kinetic energy of the turbulent motion may, correspondingly, be regarded as a sum of contributions by each of the eddies of the flow. The function $E(K)$ representing the dependence upon wave number of these energy contributions is defined as the energy spectrum of the turbulent motion.

If the equations of motion of the turbulent flow are suitably transformed, it can be shown that the inertial terms in these equations are associated with transfer of energy from larger eddies to smaller ones, while the viscous terms account for energy dissipation [2-63]. The latter is effected mostly by the smallest eddies in which the shear deformations, and therefore the viscous stresses, are large. In the absence of sources of energy, the kinetic energy of the turbulent motion will decrease—that is, the turbulence will decay—faster if the viscosity effects are large, more slowly if these effects are small.

More precisely, in the latter case, the decay time is long if compared to the periods of the eddies in the high wave number range. The energy of these eddies may therefore be considered to be approximately steady. This can only be the case if the energy fed into them through inertial transfer from the larger eddies is balanced by the energy dissipated through viscous effects. The small eddy motion is then determined solely by the rate of energy transfer (or, equivalently, by the rate of energy dissipation, denoted ε (see Eq. 2.1.9) and by the viscosity. The assumption that this is the case is known as *Kolmogorov's first hypothesis*. It follows from this assumption that, since small eddy motion is dependent solely upon internal parameters of the flow, it is independent of external conditions such as boundaries and that, therefore, *local isotropy*—the absence of preferred directions of small eddy motion—obtains.

It may further be assumed that the energy dissipation is produced almost in its entirety by the very smallest eddies of the flow. Thus, at the lower end of the higher wave number range to which Kolmogorov's first hypothesis applies the influence of the viscosity is small. In this subrange, known as the *inertial subrange*, the eddy motion may be assumed to be independent of viscosity and, thus determined solely by the rate of energy transfer (which, in turn, is equal to the rate of energy dissipation). From this assumption, known as *Kolmogorov's second hypothesis*, it follows that a relation involving $E(K)$ and ε holds for sufficiently high K, that is,

$$F[E(K), K, \varepsilon] = 0 \tag{2.3.9}$$

where $E(K)$ is the energy per unit wave number.

---

†A detailed discussion of spectra is presented in Appendix A2.

The dimensions of the quantities within brackets in Eq. 2.3.9 are $[L^3 T^{-2}]$, $[L^{-1}]$, and $[L^2 T^{-3}]$, respectively. From dimensional considerations (see Sect. 7.1) it follows immediately

$$E(K) = a_1 \varepsilon^{2/3} K^{-5/3} \tag{2.3.10}$$

in which a_1 is a universal constant. On account of the isotropy, the expression of the longitudinal velocity fluctuation spectrum [which will be denoted $S(K)$] is, to within a constant, similar to Eq. 2.3.10. Thus,

$$S(K) = a \varepsilon^{2/3} K^{-5/3} \tag{2.3.11}$$

in which it has been established by measurements that $a \simeq 0.5$ [2-21].

Spectra in the Inertial Subrange. Measurements carried out in the surface layer of the atmosphere confirm the assumption that in horizontally homogeneous, neutrally stratified flow the energy production (see Eq. 2.1.9) is approximately balanced by the energy dissipation [2-3]. The expression of this balance may be written as

$$\varepsilon = \frac{\tau_0}{\rho} \frac{dU(z)}{dz} \tag{2.3.12}$$

where

$$U(z) = \frac{1}{k} u_* \ln \frac{z}{z_0} \tag{2.3.13}$$

If Eqs. 2.2.5, 2.3.12, and 2.3.13 are used

$$\varepsilon = \frac{u_*^3}{kz} \tag{2.3.14}$$

Substituting Eq. 2.3.14 into Eq. 2.3.11, if it is assumed that

$$K = \frac{2\pi n}{U(z)} \tag{2.3.15}$$

there results

$$\frac{nS(z, n)}{u_*^2} = 0.26 f^{-2/3} \tag{2.3.16}$$

where the nondimensional quantity†

$$f = \frac{nz}{U(z)} \tag{2.3.17}$$

is known as the Monin (or similarity) coordinate, and

$$S(z, n) \, dn = S(z, K) \, dK \tag{2.3.18}$$

---

†This use of the standard notation f should not be confused with its previous use as the Coriolis parameter.

Equation 2.3.15 implies the validity of Taylor's hypothesis (see Sect. 2.3.2).

The left member of Eq. 2.3.16 is called the reduced spectrum of the longitudinal velocity fluctuations and is seen to be a function of height. Although individual samples may deviate considerably from the predicted values, Eq. 2.3.16 is, on the average, a very good representation of spectra in the high frequency range [2-51, 2-52, 2-53, 2-64, 2-65, 2-67] and may, for engineering purposes, be conservatively assumed to be valid for $f > 0.2$ [2-64, p. 27, 2-67, 2-69]. As in the case of the logarithmic law, for high wind speeds such as are assumed in structural design (of the order of 20 m/sec, say or more), it is reasonable to apply Eq. 2.3.16 throughout the height range of interest to the structural engineer.

Spectra in the Lower Frequency Range.   The lower frequency range is defined between $n = 0$ and the lower end of the inertial subrange. As noted in [2-51], [2-52], and [2-65], in the lower frequency range similarity breaks down and the spectra cannot be described by a universal relation. However, descriptions that are useful for engineering purposes may be obtained by noting that:

1.   The value of the spectra for $n = 0$ is

$$S(0) = \frac{4\overline{u^2}L_u^x}{U} \tag{2.3.19}$$

where $\overline{u^2}$ is the mean square value of the longitudinal fluctuations, U is the mean velocity, and L_u^x is the longitudinal integral scale.† Eq. 2.3.19 follows from Eqs. 2.3.4 and A2.25.

2.   The derivative of $S(n)$ with respect to n vanishes at $n = 0$. (This follows from Eq. A2.25.)

3.   The spectrum $S(n)$ is monotonically decreasing.

4.   The spectrum $S(n)$ is continuous at the lower end of the inertial subrange with the curve $S(n)$ given by Eq. 2.3.16.

5.   The area under the spectral curve in the lower frequency range is equal to the mean square value of the longitudinal velocity fluctuations (Eq. 2.3.2) less the area under the spectral curve $S(n)$ represented by Eq. 2.3.16. (This follows from Eq. A2.15.)

Two comments on lower frequency spectra are in order. First, as in the case of the mean speed U, the mean square value $\overline{u^2}$, and the integral scale L_u^x, estimates of spectra in the lower frequency range depend upon the length of record being used. For consistency, the length of the record from which $S(n)$ is estimated must be the same as that for U, $\overline{u^2}$, and L_u^x. As indicated in Sects. 2.3.1 and 2.3.2, for structural engineering purposes this length should be equal to the duration of the strong winds in a typical storm. Commonly this is assumed

---

†By virtue of the definition of the spectral density, Eq. 2.3.19 implies a vanishingly small, rather than a finite, contribution of fluctuating components with zero frequency to the mean square value of the fluctuations.

to be 1 hour, although record lengths as low as 10 minutes are used by some workers. The 1 hour period beyond which winds in a typical storm may be assumed to become relatively weak is sometimes referred to as the "spectral gap" (or quiescent period) in a conventional representation of wind activity corresponding to a continuous range of periods, including daily, monthly, seasonal, yearly, and secular periodicities [2-68].

A second comment pertains to the relation between the frequency n_{peak} at which the curve $nS(n)$ reaches a maximum and the integral scale L_u^x. As shown in [2-61], the assumption has been used in the literature that

$$L_u^x = \frac{1}{2\pi} \frac{U}{n_{peak}} \tag{2.3.20}$$

However, it was pointed out in [2-73] that the estimation of L_u^x based on measured values of U and n_{peak} can be in error several fold, owing to the sensitivity of L_u^x to the assumptions concerning the spectral shape between $n=0$ and $n=n_{peak}$. This shape is in general unknown and, therefore, so is the relationship between n_{peak} and L_u^x.

Expressions for the Spectrum Used for Structural Design Purposes. The curve

$$\frac{nS(z, n)}{u_*^2} = \frac{200f}{(1 + 50f)^{5/3}} \tag{2.3.21}$$

whose form was proposed in [2-66], approximates very closely Eq. 2.3.16 in the inertial subrange (z is the height above ground, n is the frequency in Hertz, u_* and f are given by Eqs. 2.2.18 and 2.3.17, respectively). It can be verified that Eq. 2.3.21 implies

$$\overline{u^2} = 6u_*^2 \tag{2.3.22}$$

which, for built-up terrain ($z_0 > 0.30$ m, see Table 2.3.1) may result in an over-estimation of structural response of the order of 5%. Requirements previously listed pertaining to the value of $S(n)$ and $dS(n)/dn$ at $n=0$ are not satisfied. However, this is inconsequential as far as the design of most land-based structures is concerned, since their fundamental frequencies of vibration are usually higher than the frequency corresponding to the lower end of the inertial subrange. Therefore, provided that Eq. 2.3.22 is satisfied, the response of such structures does not depend significantly upon the shape of the spectrum in the lower frequency range (see Sect. 9.1.4).

The development of Eq. 2.3.21 [2-70] was motivated by criticism of the following expression, proposed in [2-71] and used in the National Building Code of Canada [2-72]:

$$\frac{nS(z, n)}{u_*^2} = 4.0 \frac{x^2}{(1 + x^2)^{4/3}} \tag{2.3.23}$$

in which $x = 1,200n/U(10)$; n is expressed in Hertz and $U(10)$ is the mean wind speed, in meters per second, at $z = 10$ m. Equation 2.3.23 was obtained by

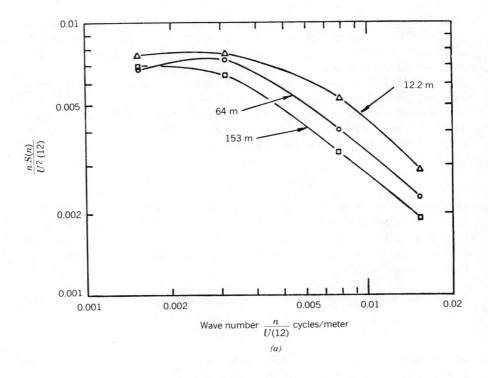

(a)

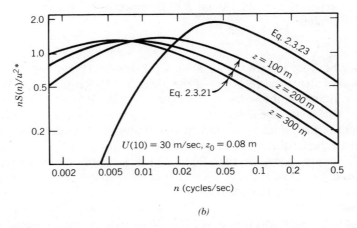

(b)

FIGURE 2.3.3. (*a*) Longitudinal turbulence spectra measured at Sale, Australia (based on 20 records) [2–71]. From A. G. Davenport, "The Spectrum of Horizontal Gustiness Near the Ground in High Winds," *Quarterly Journal of the Royal Meterological Society*, **87** (1961), 202. (*b*) Comparison of spectra given by Eqs. 2.3.21 and 2.3.23. From E. Simiu, "Wind Spectra and Dynamic Alongwind Response," *J. Struc. Div.*, ASCE, **100** (1974), 1897–1910.

averaging results of measurements obtained at various heights above ground and does not, therefore, reflect the dependence of spectra on height. In the absence of models capable of describing this dependence—such models were only developed subsequently in the 1960s—Eq. 2.3.23 and similar expressions proposed in the literature have provided useful first approximations of the longitudinal turbulence spectra in the atmospheric boundary layer. It is noted that the dependence of spectra on height is clearly suggested by data published in [2-71] (Fig. 2.3.3a).

As mentioned earlier, the spectral distribution in the lower frequency range has little influence on building response; however, the magnitude of the turbulent fluctuation components at frequencies equal, or close, to the natural frequencies of a tall structure may affect its response very significantly. It is therefore of interest to compare the higher frequency components in Eq. 2.3.23 to those of Eq. 2.3.16 (or, equivalently, Eq. 2.3.21). Such a comparison shows that Eq. 2.3.23 may overestimate the longitudinal spectra of turbulence in the higher frequency range by as much as 100–400%, as can be seen in Table 2.3.2 and Fig. 2.3.3b.

It is also noted that Eq. 2.3.23 yields $\overline{u^2} = 6u_*^2$, and that it implies $S(0) = 0$, or $L_u^x = 0$ (see Eq. 2.3.19), which is physically not possible [2-3].

The von Kármán spectrum [2-134]

$$\frac{nS(n)}{u_*^2} = \frac{4\beta \dfrac{nL_u^x}{U}}{\left[1 + 70.8\left(\dfrac{nL_u^x}{U}\right)^2\right]^{5/6}} \tag{2.3.24}$$

was proposed before the development of Eq. 2.3.16. Equation 2.3.24 satisfies the conditions $S(n) = 0$ and $dS(n)/dn = 0$ for $n = 0$. However, for Eq. 2.3.24 to be consistent with Eq. 2.3.16, it can easily be shown that it would be necessary to have $L_u^x \simeq 0.3\beta^{3/2}z$, which does not appear to be the case in the atmosphere. That Eq. 2.3.24 is, in general, not consistent with Eq. 2.3.16 can be explained

TABLE 2.3.2. Values of $nS(n)/u_*^2$ for $z_0 = 0.08$ m and $U(10) = 30$ m/sec [2-70]

n cycles per second (1)	$z = 100$ m		$z = 300$ m		All Values of z Eq. 2.3.23 (6)
	f (2)	Eq. 2.3.16 or 2.3.21 (3)	f (4)	Eq. 2.3.16 or 2.3.21 (5)	
0.1	0.255	0.70	0.586	0.37	1.47
0.2	0.450	0.43	1.172	0.23	0.98
0.5	1.125	0.24	2.930	0.13	0.54
1.0	2.250	0.15	5.860	0.08	0.34

physically by the fact, discussed earlier in connection with Kolmogorov's hypotheses, that the higher frequency spectrum is independent of the large-scale features of the turbulence which determine L_u^x. Equation 2.3.24 is not used in applications where the magnitude of the higher frequency components of the longitudinal velocity fluctuations is of interest. However, it can be used in applications in which the effect of the low frequency component could be important, such as the analysis of structures with very long natural periods of vibration (e.g., compliant offshore platforms, which have motions with periods or about 50 to 120 sec).

For the purpose of studying the sensitivity of tall building response to changes in the value of various parameters determining spectral shape, an alternative expression for the spectrum, consistent with Eq. 2.3.16, was proposed in [2-70]. This expression depends upon the parameter β and an additional parameter allowing the modification of the shape of the lower frequency part of the spectrum, and is subject to the constraint imposed by Eq. 2.3.2. A similar expression was developed in [2-74] to study the sensitivity of compliant structures to changes in the values of the parameters β and L_u^x, and to changes in the shape of the lower frequency portion of the spectrum consistent with Eq. 2.3.2. The expression of [2-74] is:

$$\frac{nS(z, n)}{u_*^2} = \begin{cases} a_1 f + b_1 f^2 + d_1 f^3 & f \leqslant f_m & (2.3.25a) \\ c_2 + a_2 f + b_2 f^2 & f_m < f < f_s & (2.3.25b) \\ 0.26 f^{-2/3} & f \geqslant f_s & (2.3.25c) \end{cases}$$

where u_* and f are given by Eqs. 2.2.18 and 2.3.17, n is expressed in Hertz, z is the height above the surface (in the case of flow over the ocean, the height above the mean water level), f_s is the lower limit of the inertial subrange ($f_s \simeq 0.2$), f_m is a parameter allowing changes in the shape of the spectral curve for $f < f_s$, and

$$a_1 = \frac{4 L_u^x(z) \beta}{z} \tag{2.3.25d}$$

$$\beta_1 = 0.26 f_s^{-2/3} \tag{2.3.25e}$$

$$b_2 = \frac{\frac{1}{3} a_1 f_m + \left(\frac{7}{3} + \ln \frac{f_s}{f_m}\right) \beta_1 - \beta}{\frac{5}{6} (f_m - f_s)^2 + \frac{1}{2} (f_m^2 - f_s^2) + 2 f_m (f_s - f_m) + f_s (f_s - 2 f_m) \ln \frac{f_s}{f_m}} \tag{2.3.25f}$$

$$a_2 = -2 b_2 f_m \tag{2.3.25g}$$

$$d_1 = \frac{2}{f_m^3} \left[\frac{a_1 f_m}{2} - \beta_1 + b_2 (f_m - f_s)^2\right] \tag{2.3.25h}$$

$$b_1 = \frac{-a_1}{2 f_m} - 1.5 f_m d_1 \tag{2.3.25i}$$

$$c_2 = \beta_1 - a_2 f_s - b_2 f_s^2 \tag{2.3.25j}$$

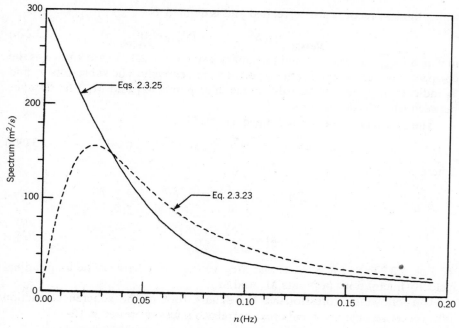

FIGURE 2.3.4. Spectra of longitudinal velocity fluctuations (Eqs. 2.3.25).

Equations 2.3.25 are plotted in Fig. 2.3.4 for $k=0.4$, $z_0=0.001266$ m, $z=35$ m, $U(35)=45$ m/sec ($u_*=1.76$ m/sec), $\beta=6.0$, $f_s=0.2$, $L_u^x=180$ m, and $f_m=0.07$. Also plotted in Fig. 2.3.4 is Eq. 2.3.23 (interrupted lines). Unlike Eq. 2.3.24, Eqs. 2.3.25 are consistent with Eq. 2.3.16. However, they do not satisfy the requirement $dS(n)/dn=0$ for $n=0$. This requirement could be satisfied by modifying Eq. 2.3.25a in the immediate vicinity of $n=0$. However, such a modification is not necessary in practice since its effect on results of engineering calculations would be negligible. Finally, it is seen in Fig. 2.3.4 that Eq. 2.3.23 significantly underestimates the spectral ordinates at very low frequencies. This is due to the fact, noted earlier, that Eq. 2.3.23 implies $L_u^x=0$.

Finally, we mention the spectrum proposed by Harris in 1968 [2-137]:

$$\frac{nS(n)}{u_*^2}=4.0\frac{x}{(2+x^2)^{5/6}} \tag{2.3.26}$$

where $x=1{,}800n/U(10)$. Like Eq. 2.3.23, Eq. 2.3.26 does not reflect the variation of the spectrum with height above ground. However, it has over Eq. 2.3.23 the advantage that it implies a nonzero integral scale of turbulence $L_u^x=1{,}000[U(z)/U(10)]/\beta$ (in meters).

2.3.4. Cross-Spectra of Longitudinal Velocity Fluctuations

The cross-spectrum† of two continuous records is a measure of the degree to

---

†A detailed discussion of cross-spectra is presented in Appendix A2.

which the two records are correlated and is defined as

$$S_{u_1u_2}^{cr}(r, n) = S_{u_1u_2}^{C}(r, n) + iS_{u_1u_2}^{Q}(r, n) \tag{2.3.27a}$$

in which $i = \sqrt{-1}$. The real and imaginary parts in Eq. 2.3.27a are known as the co-spectrum and the quadrature spectrum, respectively. The subscripts u_1 and u_2 indicate that the two records are taken at points M_1 and M_2, the distance between which is denoted by r.

The coherence function is defined as [2-75]

$$\mathscr{C}(r, n) \equiv [\text{Coh}(r, n)]^2 = c_{u_1u_2}^2(r, n) + q_{u_1u_2}^2(r, n) \tag{2.3.27b}$$

where

$$c_{u_1u_2}^2(r, n) = \frac{[S_{u_1u_2}^{C}(r, n)]^2}{S(z_1, n)S(z_2, n)} \tag{2.3.27c}$$

$$q_{u_1u_2}^2(r, n) = \frac{[S_{u_1u_2}^{Q}(r, n)]^2}{S(z_1, n)S(z_2, n)} \tag{2.3.27d}$$

In Eqs. 2.3.27c and d, $S(z_1, n)$ and $S(z_2, n)$ are the spectra of the longitudinal velocity fluctuations at points M_1 and M_2.

The following expression for the square root of the coherence function (also known as narrow-band cross-correlation) was proposed in [2-76]:

$$\text{Coh}(r, n) = e^{-f} \tag{2.3.28}$$

where

$$\hat{f} = \frac{n[C_{1z}^2(z_1 - z_2)^2 + C_{1y}^2(y_1 - y_2)^2]^{1/2}}{U(10)} \tag{2.3.29}$$

or, alternatively [2-76]

$$\hat{f} = \frac{n[C_z^2(z_1 - z_2)^2 + C_y^2(y_1 - y_2)^2]^{1/2}}{\frac{1}{2}[U(z_1) + U(z_2)]} \tag{2.3.30}$$

In Eqs. 2.3.29 and 2.3.30, y_1, y_2 and z_1, z_2 are the coordinates of points M_1, M_2, the line M_1, M_2 is assumed to be perpendicular to the direction of the mean wind, $U(10)$ is the wind velocity at $z = 10$ m, and the exponential decay coefficients C_y, C_z (or C_{1y}, C_{1z}) are determined experimentally.

In homogeneous turbulence the quadrature spectrum vanishes [2-64]. In the atmosphere it appears that the ratio of quadrature spectrum to co-spectrum is small and that the square root of the coherence function may therefore be assumed, for engineering purposes, to be approximately equal to the reduced co-spectrum $c_{u_1u_2}$. On the basis of wind tunnel measurements, it has been suggested in [2-77] that it is reasonable to assume in engineering calculations

$$S_{u_1u_2}^{C}(r, n) = S^{1/2}(z_1, n)S^{1/2}(z_2, n)e^{-\hat{f}} \tag{2.3.31}$$

where $\hat{f}$ is defined by Eq. 2.3.30 and $C_z = 10$, $C_y = 16$. It appears, however, that the exponential decay coefficients C_z, C_y (or C_{1z}, C_{1y}), rather than being independent of roughness, are generally larger for rough surface conditions such as urban areas than for smooth surfaces [2-64]. Moreover, full-scale measurements indicate that the exponential decay coefficients depend on height above ground and, quite strongly, on wind speed, as shown in Figs. 2.3.5 and 2.3.6 [2-60, 2-78]. The dependence of the exponential decay coefficients upon wind speed is illustrated in Figs. 2.3.7a and 2.3.7b, which represent Eq. 2.3.28 and measured values of the square root of the coherence function for records (taken at points of equal elevation) with $U(10) = 20.8$ m/sec ($C_{1y} = 3.5$) and $U(10) = 35.2$ m/sec ($C_{1y} = 8.8$) [2-60]. The dependence of the exponential

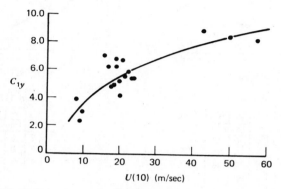

FIGURE 2.3.5. Variation of exponential decay coefficient C_{1y} with wind speed (open terrain) [2–78].

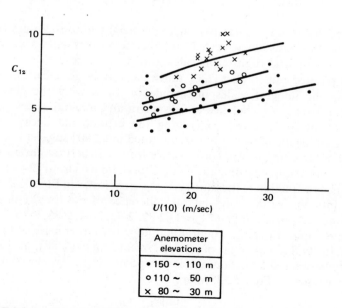

FIGURE 2.3.6. Variation of C_{1z} with wind speed and height (open terrain) [2-78].

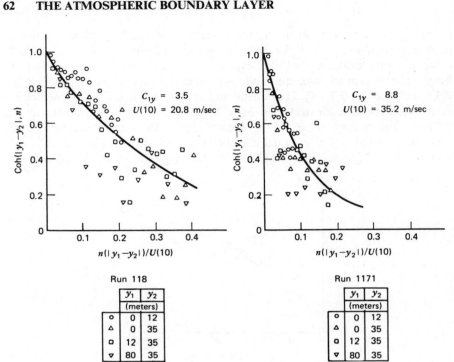

Run 118

	y_1	y_2
	(meters)	
o	0	12
△	0	35
□	12	35
▽	80	35

(a)

Run 1171

	y_1	y_2
	(meters)	
o	0	12
△	0	35
□	12	35
▽	80	35

(b)

FIGURE 2.3.7.   Measured values of $\text{Coh}(|y_1 - y_2|, n)$ [2-60].

decay coefficients upon terrain roughness, height above ground, and wind speed is insufficiently documented and therefore represents a source of uncertainty in structural engineering calculations.

It was pointed out in Sect. 2.3.2 that relatively large uncertainties remain concerning the integral scales of turbulence. In view of the close physical relationship between turbulence cross-spectra and integral scales, similar uncertainties can be expected concerning the exponential decay coefficients. Nevertheless, results of recent research quoted in [2-80] suggest that the value $C_z \simeq 10$ is acceptable or even conservative from a structural design viewpoint. A similar conclusion regarding the value $C_y \simeq 16$ follows from [2-81], according to which C_y is a function of the ratio $|y_1 - y_2|/z$, as shown in Fig. 2.3.8. Additional research into the vertical and lateral coherence of the longitudinal velocity fluctuations is reported in [2-62] and [2-82, 2-83, 2-84, 2-85, 2-86, 2-87].

In some applications the longitudinal (along-wind) coherence of the longitudinal velocity fluctuations is of interest. According to [2-88], the longitudinal coherence between the fluctuations at two points $M_1(x_1, y, z)$ and $M_2(x_2, y, z)$ can be expressed by Eq. 2.3.28, where

$$\hat{f} = \frac{nC_x|x_1 - x_2|}{U(z)} \tag{2.3.32}$$

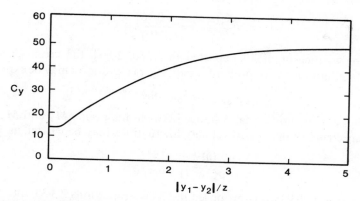

FIGURE 2.3.8. Dependence of C_y upon $|y_1 - y_2|/z$ according to [2-80]. Copyright © 1981 by D. Reidel Publishing Company.

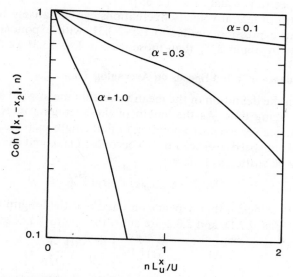

FIGURE 2.3.9. Longitudinal coherence as a function of nL_u^x/U for three values of $\alpha = I(z)|x_1 - x_2|/L_u^x(z)$ [2-89]. Copyright © 1979 by D. Reidel Publishing Company.

with $C_x \simeq 3.0$ over water and $C_x \simeq 6.0$ over land. A theoretical approach reported in [2-89] suggests that the longitudinal coherence depends upon the turbulence intensity $I(z)$, the distance $|x_1 - x_2|$, and integral scale $L_u^x(z)$, as shown in Fig. 2.3.9.

2.3.5 Spectra and Cross-Spectra of Vertical and Lateral Velocity Fluctuations

It is shown in [2-3] that the spectra of vertical fluctuations up to about 50 m may be estimated by the formula

$$\frac{nS_w(z, n)}{u_*^2} = \frac{3.36f}{1 + 10f^{5/3}} \tag{2.3.33}$$

According to measurements reported in [2-60, 2-80], the cross-spectrum of vertical fluctuations at two points M_1 and M_2, of elevation z may be expressed as

$$S_{w_1 w_2}(\Delta y, n) = S_w(z, n)e^{-8n\Delta y/U(z)} \tag{2.3.34}$$

in which Δy is the horizontal distance between the points M_1 and M_2.

The spectrum of the lateral velocity fluctuations may be written as

$$\frac{nS_v(n)}{u_*^2} = \frac{15f}{(1 + 9.5f)^{5/3}} \tag{2.3.35}$$

The form of Eq. 2.3.35 was proposed in [2-66]. Equations 2.3.33 and 2.3.35, in which the parameter f is given by Eq. 2.3.17, are consistent with the requirement that, in the higher frequency range, the ratio of the vertical and lateral to the longitudinal spectra is equal to 4/3 [2-65].

Cross-spectra of lateral velocity fluctuations can tentatively be assumed to be given by an expression similar to Eq. 2.3.31, with exponential decay coefficients lower by about 33% than those used in Eq. 2.3.31 [2-80, 2-90].

2.3.6   Dependence of Wind Speeds on Averaging Time

It follows from the definition of the mean value that mean wind speeds depend upon the averaging time. As the length of the averaging interval decreases, the maximum mean speed corresponding to that length increases. The relation between the wind speed averaged over t seconds, $U_t(z)$, and the hourly speed, $U_{3600}(z)$, may be written as follows:

$$U_t(z) = U_{3600}(z) + c(t)\overline{u'^2}^{1/2} \tag{2.3.36}$$

where $c(t)$ is a coefficient that depends on t and u' is the longitudinal turbulent fluctuation. If Eqs. 2.2.18 and 2.3.2 are substituted into Eq. 2.3.36.

$$U_t(z) = U_{3600}(z)\left(1 + \frac{\beta^{1/2}c(t)}{2.5 \ln(z/z_0)}\right) \tag{2.3.37}$$

The coefficient $c(t)$ is determined on the basis of statistical studies of wind speed records. Results of such studies were reported by Durst [2-91] and are plotted in Fig. 2.3.10, which corresponds to open terrain conditions ($z_0 \simeq 0.05$ m) and an elevation $z = 10$ m. Values of $c(t)$ consistent with Fig. 2.3.10 are listed in Table 2.3.3.

Experimental results presented in [2-93] suggest that Eq. 2.3.36 is applicable, with the values of the coefficient $c(t)$ of Table 2.3.3, to wind speeds over terrains with roughness lengths of up to $z_0 = 2.50$ m.

Mean speeds used in the design of tall buildings are hourly averages, while information on wind intensities is currently provided in terms of fastest mile wind speeds at about 10 m above ground in open terrain. Fastest mile wind speeds are averaged over the time required for the passage over the anemometer

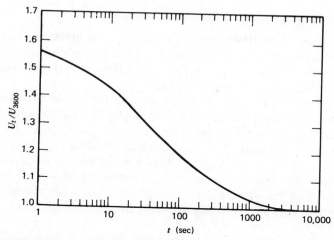

FIGURE 2.3.10. Ratio of probable maximum speed averaged over period t to that averaged over one hour [2-92].

TABLE 2.3.3. Coefficient $c(t)$

	1	10	20	30	50	100	200	300	600	1000	3600
$c(t)$	3.00	2.32	2.00	1.73	1.35	1.02	0.70	0.54	0.36	0.16	0.00

of a volume of air with a horizontal length of one mile. From this definition it follows that for the fastest mile U_f the averaging time in seconds is $t = 3600/U_f$, where U_f is given in miles per hour. For example, if $U_f = 90$ mph, then $t = 40$ sec and the corresponding hourly mean is, from Fig. 2.3.10, $90/1.28 \simeq 70$ mph (31 m/sec). The relationship between fastest mile wind speeds and winds of longer duration was studied for coastal locations in the United States in [2-135]. According to the results of [2-135], estimates of the hourly mean speeds based on fastest mile speeds and Fig. 2.3.10 are slightly conservative from a design viewpoint.

2.4 HORIZONTALLY NONHOMOGENEOUS FLOWS

Horizontal nonhomogeneities of atmospheric flows may be ascribed either to conditions at the Earth's surface (e.g., changes in surface roughness, topographic features of the terrain) or to the meteorological nature of the flow (as in the case of tropical cyclones or of thunderstorms). While the structure of horizontally homogeneous flows is basically well understood, results obtained in the study of horizontally nonhomogeneous flows are to a large extent still incomplete or tentative. Some of these results are, nevertheless, of interest to the designer and will therefore be discussed herein.

2.4.1 Flow Near a Change in Surface Roughness

In the case dealt with in the preceding sections, of a horizontally homogeneous flow, it is assumed that the surface roughness is uniform over an infinite plane. In reality, a site is limited in size; the flow near its boundaries is therefore affected by the surface roughness of adjoining sites.

Useful information on the flow structure in the transition zones may be obtained by considering the simple case of an abrupt roughness change along a line perpendicular to the direction of the mean flow [2-13, 2-94, 2-95, 2-96, 2-97, 2-98] (Fig. 2.4.1). Upwind of the discontinuity, the flow is horizontally homogeneous and, near the ground, governed by the parameters z_{01} and u_{*1}. Downwind of the discontinuity, the flow will be disturbed over a height $h(x)$. This height, known as the depth of the *internal boundary layer*, increases with the distance x until the entire flow adjusts to the roughness length z_{02} of the terrain downwind of the discontinuity.

If the investigation is limited to the lower portion of the boundary layer, it may be assumed that the flow is two-dimensional. For steady flow, and neglecting the pressure gradient force—the effect of which was shown to be insignificant [2-98]—the equations of continuity and of balance of momenta may be written as

$$U \frac{\partial U}{\partial x} + W \frac{\partial U}{\partial z} = \frac{1}{\rho} \frac{\partial \tau}{\partial z} \tag{2.4.1}$$

$$\frac{\partial U}{\partial x} + \frac{\partial W}{\partial z} = 0 \tag{2.4.2}$$

Since Eqs. 2.4.1 and 2.4.2 contain three unknowns, a third equation is required to close the system. In the solution of [2-96] the mean turbulent field closure was used (Eq. 2.1.9), which, for two-dimensional flow and with phenomenological relations similar to those proposed in [2-9] and [2-10] (see Eqs. 2.1.10–2.1.13) takes the form

$$\frac{U}{0.16} \frac{\partial(\tau/\rho)}{\partial x} + \frac{W}{0.16} \frac{\partial(\tau/\rho)}{\partial z} = \frac{\tau}{\rho} \frac{\partial U}{\partial z} + \frac{\partial}{\partial z} \left(\frac{\tau/\rho}{0.16} \frac{\partial(\tau/\rho)}{\partial z} \bigg/ \frac{\partial U}{\partial z} \right) - \frac{(\tau/\rho)^{3/2}}{L} = 0 \tag{2.4.3}$$

in which L is the mixing length.

In horizontally homogeneous flow, the validity in the surface layer of the logarithmic law implies the following expression for the mixing length [2-1]:

$$L = kz \tag{2.4.4}$$

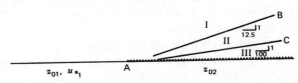

FIGURE 2.4.1. Flow zones downwind of a change in roughness of terrain.

($k \simeq 0.4$). Following Monin [2-99], it is assumed in [2-96] that in Eq. 2.4.3 the same expression for L holds near the ground throughout the flow, including the disturbed flow downwind of the discontinuity.

The boundary conditions for Eqs. 2.4.1–2.4.3 are:

$$U = 2.5u_{*1} \ln \frac{z}{z_{01}} \qquad (2.4.5)$$

$$W = 0 \qquad \left.\right\} (\dot{x} < 0) \qquad (2.4.6)$$

$$\tau = \rho u_{*1}^2 \qquad (2.4.7)$$

$$U = 0 \qquad (2.4.8)$$

$$W = 0 \qquad (2.4.9)$$

$$\tau = \rho \left[0.4 z_{02} \left(\frac{\partial U}{\partial x} \right)_{z_{02}} \right]^2 \qquad \left.\right\} (x > 0; \, z = z_{02})$$

(see Eqs. 2.1.7–2.1.8).

Equations 2.4.1–2.4.3 with the boundary conditions, Eqs. 2.4.5–2.4.9, were solved numerically in [2-96] for various values of the parameter $m = \ln(z_{01}/z_{02})$. In the case of the smooth-to-rough transition, the calculations indicate that three regions may be distinguished downwind of the discontinuity (Fig. 2.4.1). In region I (above line AB, approximately defined by a slope of 1 : 12.5), the velocity is essentially equal to the velocity upwind of the discontinuity. This result is consistent with conclusions reached independently by other authors [2-94] and [2-95]. In region III (below line AC, defined by a slope of about 1 : 100) it may be assumed, at least very roughly, that the flow is adjusted to the new roughness conditions, that is, is determined by the same parameters z_{02}, u_{*2} that would control the flow if the roughness length were everywhere z_{02}. In region II, as the distance downwind from the discontinuity increases, the velocity profiles deviate increasingly from the profile given by Eq. 2.4.5 and the turbulence energy varies gradually from line AB, where it is presumably nearly the same as upwind of the discontinuity, to line AC, where it may be described in terms of the parameters z_{02}, u_{*2}. For practical purposes it may be assumed that: (1) the profile corresponding to these parameters is completely established at distances of more than 5 km downward from the roughness change; (2) for a distance downwind of the roughness change of less than 500 m the profile is the same as upwind of the discontinuity; and (3) in the interval 500 m $< x <$ 5 km the profile is logarithmic below line AB, with zero speed at the ground surface, and a speed at elevation $x/12.5$ equal to the speed at that elevation upwind of the roughness change [2-42].

A more "exact" model of the internal boundary layer growth is

$$h(x) = 0.28 z_{0_r} \left(\frac{x}{z_{0_r}} \right)^{0.8} \qquad (2.4.10)$$

where z_{0_r} is the larger of z_{0_1} and z_{0_2} [2-100]. Equation 2.4.10 was based on the analysis of a considerable number of data and holds for both smooth-to-rough and rough-to-smooth transition. It is approximately valid for values $h(x) < 0.2\delta$ where δ is the boundary layer depth. For additional references on flows near a change in surface roughness, see [2-100] and [2-138].

2.4.2   Wind Flow Over Hills

Wind tunnel investigations of simulated flows over ramps and escarpments are reported in [2-101, 2-102, 2-103]. For open terrain conditions, ratios $(U_2/U_1)^2$ at various stations given in [2-101] are represented in Figs. 2.4.2 and 2.4.3. (U_2 and U_1 denote wind speeds at height z above ground downwind and upwind of the ramp, respectively.) Measurements of [2-102] tend to corroborate these results. The results of [2-101] and [2-102] also suggest that for ramps with slopes of about 20% to 35%, the ratios U_2/U_1 are, for practical purposes, independent of slope. However, for a ramp with a 10% slope, the ratios $(U_2 - U_1)/U_1$ are only about one half as large as in the case of a 20% slope [2-101]. More detailed wind tunnel measurements of ratios U_2/U_1 for escarpments with 25%, 50%, and 100% slopes and for a cliff, as well as measurements of the root mean square of the longitudinal turbulence fluctuations, are reported

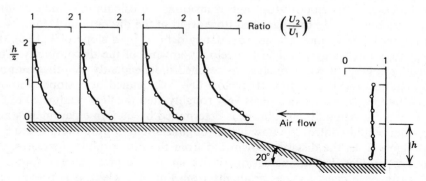

FIGURE 2.4.2.   Wind profiles over an escarpment [2-101].

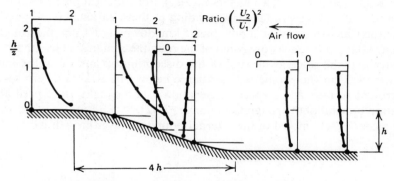

FIGURE 2.4.3.   Wind profiles over an escarpment [2-101].

in [2-103]. The ratios U_2/U_1 of [2-103] are similar to those of Fig. 2.4.2, except at low elevations (about 5 m above ground) where they are larger by about 20%.

Recently, results of theoretical and numerical studies of wind flows over hills have been reported in [2-104, 2-105, 2-106, 2-107, 2-108]. For a hill with maximum height h, a longitudinal scale $L(L \gg h)$ and a profile $hf(x/L)$, where $f(x/L) \leqslant 1$ (Fig. 2.4.4), the following result was obtained in [2-104]:

$$\frac{U_2}{U_1} = 1 + \frac{h\sigma \ln^2(L/z_0)\hat{u}^{(0)}(x, z)}{L \ln(l/z_0)[\ln(z/l) + \ln(l/z_0)]} \tag{2.4.11}$$

in which U_2 is the wind speed at (x, z), U_1 is the wind speed at $(x = -\infty, z)$, z_0 is the roughness length, x is the horizontal distance (see Fig. 2.4.4), z is the height above surface of the hill at the point considered, $\hat{u}^{(0)}$ is the approximate value of a dimensionless quantity representing the perturbation to the upwind velocity due to the presence of the hill,

$$\sigma = -\frac{1}{\pi} \int_{-\alpha}^{\alpha} \frac{f'(x/L)d(x/L)}{(x/L)} \tag{2.4.12}$$

and

$$l/z_0 = \frac{1}{8}\left(\frac{L}{z_0}\right)^{0.9} \tag{2.4.13}$$

(The quantity l is the thickness of the internal boundary layer created by the change in surface shear stress as the air flows over the hill. This internal boundary layer is similar to that caused by changes in terrain roughness.) For any hill symmetric about $x = 0$, $\hat{u}^{(0)}$ can be expressed in terms of Kelvin functions as shown in [2-104]. In the particular case

$$f\left(\frac{x}{L}\right) = \frac{1}{1 + (x/L)^2} \tag{2.4.14}$$

in which L is the horizontal distance from the top of the hill to the point at which the height is half the maximum height h, the quantity $\sigma = 1$. Values of $\hat{u}^{(0)}$ corresponding to the profile 2.4.14 are represented in Fig. 2.4.5 at $x/L = 0$ (top of the hill), $x/L = -0.5$ and $x/L = 0.5$, for $l/z_0 = 10^3$, $L/z_0 = 2.1 \times 10^4$ (curves A), $l/z_0 = 10^4$, $L/z_0 = 3.2 \times 10^5$ (curves B), and $l/z_0 = 10^5$, $L/z_0 = 3.6 \times 10^6$ (curves C). Values of $[(U_2 - U_1)/U_1](L/h)$ calculated in [2-104] are listed in Table 2.4.1.

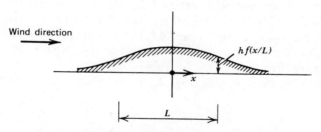

FIGURE 2.4.4. Profile of a low hill.

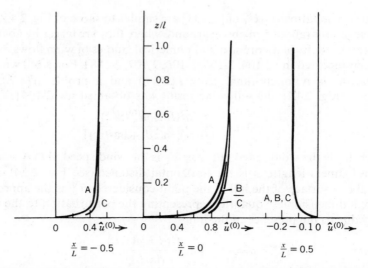

FIGURE 2.4.5. Values of $\hat{u}^{(0)}$. From P. S. Jackson and J. C. R. Hunt, "Turbulent Flow Over a Low Hill," *Quarterly Journal of the Royal Meterological Society*, **101** (1975), 929–955.

TABLE 2.4.1. Values of $[(U_2 - U_1)/U_1](L/h)$ at Top of Hill

z/l	$z_0/l = 10^{-3}$	$z_0/l = 10^{-4}$	$z_0/l = 10^{-5}$
0.0	2.09	1.87	1.72
0.1	2.46	2.13	1.92
0.3	2.33	2.07	1.85
0.6	2.20	1.97	1.78
1.0	2.08	1.87	1.72
1.5	1.97	1.79	1.66
2.1	1.88	1.73	1.62

The analysis and results of [2-104] are valid for hills in rural terrain ($z_0 \simeq 0.03$ m) with $0.1 \gtrsim L \gtrsim 10$ km and with ratios $h/L \gtrsim \frac{1}{8}(z_0/L)^{0.1}$. For example, if $z_0 = 0.025$ m, $L = 500$ m, and $h = 25$ m, then, from Eq. 2.4.13, $z_0/l = 1.0 \times 10^{-3}$, to which there corresponds, from Table 2.4.1, $U_2/U_1 \simeq 1.12$ at $z/l = 0.1$ (or $z \simeq 2.5$ m). The theory becomes less accurate in rough terrain ($z_0 \simeq 0.5$ m), the actual speeds U_2 being lower than those given by Table 2.4.1.

For flow over escarpments (Fig. 2.4.6), the following relation is derived in [2-109]:

$$\frac{U_2}{U_1} \simeq 1 + \frac{h}{L}\frac{1}{4\pi}\frac{\ln(L/z_0)}{\ln(z/z_0)} \ln \frac{(z/L)^2 + [1+(x/L)]^2}{(z/L)^2 + [1-(x/L)]^2} \qquad (2.4.15)$$

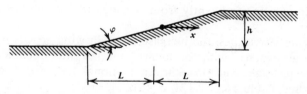

FIGURE 2.4.6.   Flow over escarpments—notations [2-109].

in which notations similar to those of Eq. 2.4.11 are used. It is suggested in [2-109] that Eq. 2.4.15 may be applied to flow over escarpments with $L \ll 5$ km and with slopes as large as 20° or so. For example, if $L = 250$ m, $h = 50$ m, and $z_0 = 0.025$ m, for $x = L$ and $z = 10$ m the ratio $U_2/U_1 = 1.19$. According to [2-103], Eq. 2.4.15 provides useful indications of the trends of the variation of U_2/U_1 with x and z, rather than dependable quantitative results.

Full-scale and wind tunnel measurements of flows over two- and three-dimensional hills and over embankments are reported in [2-108] (which extends the analytical approach of [2-104] to three-dimensional hills), and [2-110, 2-111, 2-112, 2-113, 2-114, 2-115]. As noted in [2-110], estimates obtained independently in [2-104], [2-105], and [2-106] agree well with each other and with the full-scale measurements of [2-110].

2.4.3 The Hurricane Boundary Layer

The horizontal inhomogeneity of a hurricane wind flow over a uniform, horizontal surface is associated with the variation of the pressure gradient with distance from the center of the storm (see Eq. 1.3.1). In deriving the logarithmic description of the mean velocity profiles near the ground (Eq. 2.2.18) it was assumed that the flow in the free atmosphere is geostrophic (Sect. 2.2). This assumption does not hold in the region of highest winds of the mature hurricane; the question therefore arises as to whether or not Eq. 2.2.18 is applicable in this region.

Several analytical solutions of the hurricane boundary-layer problem have been attempted so far [2-116, 2-117, 2-118, 2-119, 2-120], all of which apply to steady, axisymmetric mean flows. The solutions of [2-116] through [2-119] are based on the assumption that the eddy viscosity is constant, and cannot therefore provide a reliable detailed description of the flow near the ground. A considerably more realistic modeling of the turbulence effects is used in [2-120], in which the equations of motion and continuity are supplemented by the turbulence closure relations discussed in Sect. 2.1 (Eqs. 2.1.9–2.1.13). The system of equations thus obtained—in which the expression for the pressure gradient field given by Eq. 1.3.1 was used—was solved numerically assuming values of the surface roughness of 0.002 m to 0.90 m, differences between the high pressure in the far field and the low pressure at the storm center of 60 mb to 140 mb, and radii at which the gradient wind has a maximum value of 30 km to 50 km. According to [2-120], in the lowest 400 m of the boundary layer the mean wind profiles differ only insignificantly from the logarithmic profiles described by Eq. 2.2.18.

For decaying hurricanes, the increase of mean wind speeds with height in approximate accordance with the logarithmic law was documented in 1954 following the passage over Brookhaven National Laboratory of hurricanes Carol and Edna (Table 2.4.2 [2-121, p. 46]).

More recently, [2-122] reported extensive observations of mean wind speeds recorded at elevations from 9.1 m to 390 m during the passage of four decaying tropical cyclones over northwestern Australia. The mean wind profiles were in most cases irregular and, as noted in [2-122], a wind speed maximum was often observed at 60–200 m. Nevertheless, the profiles corresponding to the largest 10-min wind speed observed during each storm at 9.1 m were by and large consistent with the logarithmic law and a roughness length of 1 to 4 cm, as can be seen in Table 2.4.3, in which the only significant anomaly is the speed observed during cyclone Karen at 59.7 m elevation.

Whether the logarithmic profile holds in the case of mature hurricanes remains an open question. Implicit in the provisions of the 1975 Southern

TABLE 2.4.2. Variation of Wind Speeds With Height in Hurricanes Carol and Edna

Height Above Ground (meters)	Carol		Edna	
	Mean	Max. 1-min	Mean	Max. 1-min
11.3	14.5	22.8	11.8	17.0
22.9	18.7	29.1		
45.7	24.1	35.8	20.3	25.9
108.2	29.1	42.9		
125.0			25.9	30.8

TABLE 2.4.3. 10-min Speeds at Various Elevations Corresponding to Maximum 10-min Speed at 9.1 m During Four Tropical Cyclones[a]

Height Above Ground (meters)	Wind Speed (meters/second)			
	Beryl (12/73; 12:00)	Trixie (2/75; 18:30)	Beverly (3/75; 21:00)	Karen (3/77; 19:00)
9.1	21	22	32.5	30.5
59.7		28	39.5	51
191.4	32	31	47	43.5
279.2				48.5
390.1	36	34	57.5	48.5

[a]Numbers in parentheses indicate the month, year, and hour (GMT).

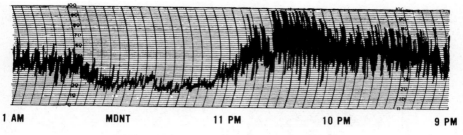

1 AM   MDNT   11 PM   10 PM   9 PM

FIGURE 2.4.7. Hurricane wind speed record.

Building Code [2-123] is the assumption that hurricane wind profiles are considerably flatter than would be indicated by the logarithmic law. To date there is no conclusive evidence that this is the case. Since design wind speeds specified in building codes correspond to an elevation of 10 m or so, the use of this assumption in the design of tall structures might be imprudent. For this reason, the American National Standard A58.1 [2-49] does not differentiate between profiles of hurricanes and extratropical storms.

Two more notes on hurricane winds are in order. First, in the immediate proximity of the eye, flow separation occurs and the boundary-layer assumptions break down (see Sect. 1.3). The implications of this phenomenon to the designer are not yet well understood. Second, as the hurricane moves inland, filling occurs (see Sect. 1.3) and the maximum winds tend to decrease. An empirical description of the wind intensity reduction as a function of distance from the coastline was proposed in [2-124] and [2-125]. According to [2-125], the ratios of peak gusts at 50 km, 100 km, and 150 km inland to peak gusts at the coastline are, approximately, 0.90, 0.80, and 0.70, respectively.

A hurricane wind speed record, which clearly indicates the passage of the eye, is shown in Fig. 2.4.7.

2.4.4 Thunderstorm Winds

The cold air flow which, in a thunderstorm, spreads horizontally over the ground was compared in Sect. 1.3 to a wall jet. Just as in the case of the wall jet, the surface friction retards the spreading flow, which may thus be expected to be similar, near the ground, to an ordinary boundary layer [2-126, 2-127, 2-128].

Of particular interest to the designer is the so-called *first gust* (or gust *front*), that is, the wind occurring in a thunderstorm that exhibits a considerable and relatively rapid change of speed and direction (Fig. 2.4.8). Following [2-129] and [2-130], the wind speed increase and the time interval during which this increase takes place will be referred to as the gust size ΔV and the gust length Δt, respectively. Depending upon thunderstorm intensity, the gust size may vary approximately from 3 m/sec to 30 m/sec, while the gust length may range from a few minutes to 20 minutes or so.

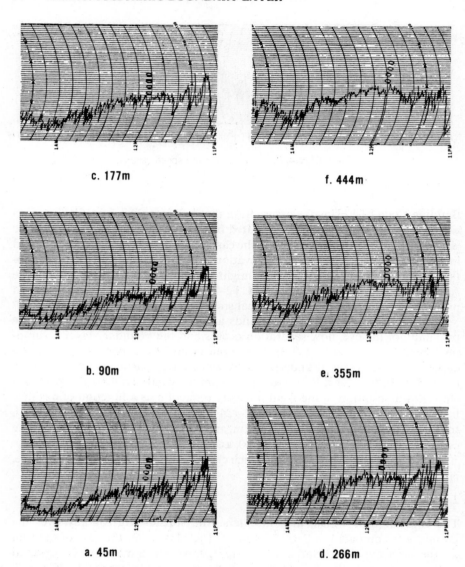

c. 177m

f. 444m

b. 90m

e. 355m

a. 45m

d. 266m

FIGURE 2.4.8. Thunderstorm wind speeds recorded simultaneously at six elevations from 45 m to 444 m above ground near Oklahoma City (courtesy of National Severe Storms Laboratory, National Oceanic and Atmospheric Adminsistration).

The thunderstorm wind records reported in [2-130] suggest that during the interval Δt: (1) up to 100 m above ground wind speeds vary with height in accordance with the logarithmic law, and (2) above 100 m the variation of wind speeds with height is negligible. (This is reasonably compatible with the records of Fig. 2.4.8.) No relation between wind speeds in different roughness regimes, based on a rational model of the thunderstorm wind flow, has been derived so far. To convert thunderstorm wind speeds recorded over open terrain into wind speeds over built-up terrain, the American National Standard A58.1 uses the same procedure that it applies to extratropical cyclone winds (Eq. 2.2.26), even though the notions of gradient height and gradient speed have no meaning in the case of thunderstorms. Whether or not this practice is acceptable for structural engineering purposes is a question that merits investigation, particularly if it is recalled that, according to [2-131], about one-third of the extreme wind speeds recorded in the United States are associated with thunderstorms.

REFERENCES

2-1   H. Schlichting, *Boundary Layer Theory*, McGraw-Hill, New York, 1960.

2-2   L. T. Matveev, *Physics of the Atmosphere*, TT 67-51380, National Technical Information Service, Springfield, Va., 1967.

2-3   J. L. Lumley and H. A. Panofsky, *The Structure of Atmospheric Turbulence*, Wiley, New York, 1964.

2-4   H. A. Panofsky and J. A. Dutton, *Atmospheric Turbulence: Models and Methods for Engineering Applications*, Wiley, New York, 1984.

2-5   A. S. Monin and A. M. Yaglom, *Statistical Fluid Mechanics: Mechanics of Turbulence*, Vol. 1, MIT Press, Cambridge, Mass., 1971.

2-6   J. F. Nash and V. C. Patel, *Three-Dimensional Turbulent Boundary Layers*, S.B.C. Technical Books, Atlanta, 1972.

2-7   *Computation of Turbulent Boundary Layers*, Proceedings of the AFOSR-IFP Stanford Conference, Vols. 1 and 2, Stanford Univ., 1968.

2-8   A. A. Townsend, *The Structure of Turbulent Shear Flow*, Cambridge Univ. Press. Cambridge, U.K., 1955.

2-9   P. Bradshaw, D. H. Ferris and N. P. Atwell, "Calculation of Boundary Layer Development Using the Turbulent Energy Equation," *J. Fluid Mech.*, **28** (1967) 593–616.

2-10   J. F. Nash, "The Calculation of Three-Dimensional Turbulent Boundary Layers in Incompressible Flow," *J. Fluid Mech.*, **37** (1969) 625–642.

2-11   C. duP. Donaldson, "Calculation of Turbulent Shear Flows for Atmospheric and Vortex Motions." *AIAA J.*, **10**, 1 (Jan. 1972) 4–12.

2-12   J. L. Lumley and B. Khajeh-Nouri, "Computational Modeling of Turbulent Transport," *Adv. Geophys.*, **18A**, Academic, New York (1974) 169–192.

2-13   K. S. Rao, J. C. Wyngaard and O. R. Coté, "The Structure of the Two-Dimensional Internal Boundary Layer over a Sudden Change of Surface Roughness," *J. Atmos. Sci.*, **31**, 3 (April 1974) 738–746.

2-14   G. T. Csanady, "On the Resistance Law of a Turbulent Ekman Layer," *J. Atmos. Sci.*, **24** (Sept. 1967) 467–471.

2-15   S. R. Hanna, *Characteristics of Winds and Turbulence in the Planetary Boundary Layer*, Technical Memorandum No. ERLTM-ARL 8, ESSA Research Laboratories, Oak Ridge, Tenn., 1969.

2-16   G. B. Schubauer and C. M. Tchen, *Turbulent Flow*, Princeton Univ. Press, Princeton, N.J., 1961.

2-17   C. B. Millikan, "A Critical Discussion of the Turbulent Flows in Channels and Circular Tubes," in Proceedings of the Fifth International Congress of Applied Mechanics, Cambridge, Mass., 1938.

2-18   A. E. Gill, "Similarity Theory and Geostrophic Adjustment," *J. Royal Meteorol. Soc.*, **94** (1968) 586–588.

2-19   R. H. Clarke, "Observational Studies in the Atmospheric Boundary Layer," *J. Royal Meteorol. Soc.*, **96** (1970) 91–114.

2-20   E. J. Plate, *Aerodynamic Characteristics of Atmospheric Boundary Layers*, U.S. Atomic Energy Commission Critical Review Series, 1972.

2-21   H. Tennekes and J. L. Lumley, *A First Course in Turbulence*, MIT Press, Cambridge, Mass., 1972.

2-22   E. Simiu, "Logarithmic Profiles and Design Wind Speeds," *J. Eng. Mech. Div.*, ASCE, **99**, No. EM5, Proc, Paper 10100 (Oct. 1973) 1073–1083.

2-23   E. L. Deacon, "Geostrophic Drag Coefficients," *Bound Layer Meteorol.*, **5**, 3 (July 1973) 321–340.

2-24   A. K. Blackadar (personal communication, Feb. 1976).

2-25   J. A. Businger et. al. "Flux Profile Relationships in the Atmospheric Surface Layer," *J. Atmos. Sci.*, **28** (1971), 788–794.

2-26   H. Tennekes, "The Logarithmic Wind Profile," *J. Atmos. Sci.*, **30** (1973) 234–238.

2-27   J. Wyngaard, "Notes on Surface Layer Turbulence," in Proceedings of the American Meteorological Society Workshop on Micrometeorology, Boston, 1972.

2-28   D. R. Caldwell, C. W. Van Atta, and K. N. Helland, "A Laboratory Study of the Turbulent Ekman Layer," *Geophys. Fluid Dyn.*, **3** (1972), 125–160.

2-29   G. C. Howroyd and P. R. Slawson, "The Characteristics of a Laboratory Produced Turbulent Ekman Layer," *Bound. Layer Meteorol.*, **8** (1975), 201–219.

2-30   R. H. Thuillier and V. O. Lappe, "Wind and Temperature Profile Characteristics from Observations on a 1400 ft Tower," *J. Appl. Meteorol.*, **3** (June 1964), 299–306.

2-31   A. K. Blackadar and H. Tennekes, "Asymptotic Similarity in Neutral Barotropic Boundary Layers," *J. Atmos. Sci.*, **25** (Nov. 1968) 1015–1020.

2-32   D. M. Carl, T. C. Tarbell and H. A. Panofsky, "Profiles of Wind and Temperature from Towers over Homogeneous Terrain," *J. Atmos. Sci.*, **30** (July 1972) 788–794.

2-33   N. C. Helliwell, "Wind Over London," in *Proceedings of the Third International Conference on Wind Effects on Buildings and Structures*, Tokyo, 1971, Saikon, 1972, pp. 23–32.

2-34   E. L. Deacon, "Gust Variation with Height up to 150 m," *J. Royal Meteorol. Soc.*, **91** (1955) 562–573.

2-35   R. I. Harris, "Measurements of Wind Structure at Heights up to 598 ft Above Ground Level," in Proceedings of the Symposium on Wind Effects on Buildings and Structures, Loughborough University of Technology, Leicestershire, U.K., 1968.

2-36   O. G. Sutton, *Atmospheric Turbulence*, Methuen, London and Wiley, New York, 1960.

2-37   P. S. Jackson, "On the displacement height in the logarithmic velocity profile," *J. Fluid Mech.*, **111** (1981), 15–25.

2-38   A. C. Chamberlain, "Roughness Length of Sea, Sand, and Snow," *Bound. Layer Meteorol.*, **25** (1983), 405–409.

2-39   G. E. Daniels, (Ed.), *Terrestrial Environment (Climatic) Criteria Guidelines for Use in Space Vehicle Development*, 1971 Revision, NASA Technical Memorandum TM X-64589, George C. Marshall Space Flight Center, Ala.

2-40   H. R. Oliver, "Wind Profiles in and Above a Forest Canopy," *J. Royal Meteorol Soc.*, **97** (1971) 548–553.

2-41   P. Duchêne-Marullaz, "Full-Scale Measurements of Atmospheric Turbulence in a Suburban Area," in *Proceedings of the Fourth International Conference on Wind Effects on Buildings and Other Structures*, London, 1975, Cambridge Univ. Press. Cambridge, U.K., 1976, pp. 23–31.

2-42   J. Biétry, C. Sacré, and E. Simiu, "Mean Wind Profiles and Changes of Terrain Roughness," *J. Struct. Div.*, ASCE, **104** (Oct. 1978), 1585–1593.

2-43   S. D. Smith and E. G. Banke, "Variation of the Sea Surface Drag Coefficient with Wind Speed," *J. Royal Meteorol. Soc.*, **101** (1975), 665–673.

2-44   J. Wu, "Wind Stress and Surface Roughness at Air-Water Interface," *J. Geophys. Res.*, **74** (1969) 444–445.

2-45   J. Amorocho and J. J. deVries, "A New Evaluation of the Wind Stress Coefficient over Water Surfaces," *J. Geophys. Res.*, **85** (1980), 433–442.

2-46   J. R. Garratt, "Review of Drag Coefficients over Oceans and Continents," *Monthly Weather Rev.*, **105** (1977), 915–929.

2-47   G. Hellman. "Über die Bewegung der Luft in den untersten Schichten der Atmosphäre." *Meteorol. Z.*, **34** (1916) 273.

2-48   A. G. Davenport. "The Relationship of Wind Structure to Wind Loading." in *Proceedings of the Symposium on Wind Effects on Buildings and Structures*. Vol. 1. National Physical Laboratory. Teddington. U.K., Her Majesty's Stationery Office. London. 1965, pp. 53–102.

2-49   *American National Standard A58.1–1982, Minimum Design Loads for Buildings and Other Structures*, American National Standards Institute, Inc., New York, 1982.

2-50   W. W. Pagon, "Wind Velocity in Relation to Height Above Ground," *Eng. News Rec.*, **114** (May 1935), 742–745.

2-51   F. Pasquill. "Wind Structure in the Atmospheric Boundary Layer." in *A Discussion on Architectural Acrodynamics, Phil. Trans. Roy. Soc. London*, **A269** (1971) 439–456.

2-52   F. Pasquill, "Some Aspects of Boundary Layer Description," *J. Royal Meteorol. Soc.*, **98** (1972) 469–494.

2-53   P. R. Owen. "Buildings in the Wind." *J. Royal Meteorol. Soc.*, **97** (1974) 396–413.

2-54   H. C. Shellard, "The Estimation of Design Wind Speeds, in *Proceedings of the Symposium on Wind Effects on Buildings and Structures*, Vol. 1, National Physical Laboratory, Teddington, U.K., Her Majesty's Stationery Office, London, 1965, pp. 30–51.

2-55   N. Sisserwine, P. Tattleman, D. D. Grantham and I. I. Gringorten, *Extreme Wind Speeds, Gustiness and Variations with Height for MIL-STD210B*. Technical Report No. AFCRL-TR-0560, Air Force Cambridge Research Laboratories, 1973.

2-56   J. Biétry, personal communication, July 1976.

2-57   S. Schotz and H. A. Panofsky, "Wind Characteristics at the Boulder Atmospheric Observatory," *Bound. Layer Meteorol.*, **19**, (1980), 155–164.

2-58   A. Venkatram, "Estimating the Monin-Obukhov Length in the Stable Boundary Layer for Dispersion Calculations," *Bound. Layer Meteorol.*, **19** (1980), 481–485.

2-59   E. Simiu, "Thermal Convection and Design Wind Speeds," *J. Struct. Div.*, ASCE, **108** (July 1982), 1671–1675.

2-60   M. Shiotani, *Structure of Gusts in High Winds*, Parts 1-4, The Physical Sciences Laboratory, Nikon University, Furabashi, Chiba, Japan, 1967–1971.

2-61   J. Counihan, "Adiabatic Atmospheric Boundary Layers: A Review and Analysis of Data from the Period 1880–1972," *Atmos. Environ.*, **9** (1975), 871-905.

2-62   P. Duchêne-Marullaz, "Effect of High Roughness on the Characteristics of Turbulence in the Case of Strong Winds, *Proceedings Fifth International Conference on Wind Engineering*, Vol. 1, Pergamon Press, 1980.

2-63   J. O. Hinze, *Turbulence*, McGraw-Hill, New York, 1959.

2-64   H. W. Teunissen, *Characteristics of the Mean Wind and Turbulence in the Planetary Boundary Layer*, Review No. 32, Institute for Aerospace Studies, University of Toronto, 1970.

2-65   N. E. Busch and H. A. Panofsky, "Recent Spectra of Atmospheric Turbulence." *J. Royal Meteorol. Soc.*, **94** (1968) 132–148.

2-66   J. C. Kaimal et al., "Spectral Characteristics of Surface-Layer Turbulence." *J. Royal Meteorol. Soc.*, **98** (1972) 563–589.

2-67   I. A. Singer, N. E. Busch and J. A. Frizzola, "The Micrometeorology of the Turbulent Flow Field in the Atmospheric Boundary Surface Layer," in *Proceedings of the International Research Seminar on Wind Effects on Buildings and Structures*, Ottawa, Vol. 1, University of Toronto Press, Toronto, 1968, pp. 557–594.

2-68   F. Fiedler and H. A. Panofsky, "Atmospheric Scales and Spectral Gaps," *Bull. Am. Meteorol. Soc.*, **51**, 12 (Dec. 1970) 1114–1119.

2-69   G. H. Fichtl and G. E. McVehil, "Longitudinal and Lateral Spectra of Turbulence in the Atmospheric Boundary Layer at the Kennedy Space Center," *J. Appl. Meteor.* (Sept. 1970) 51–63.

2-70   E. Simiu, "Wind Spectra and Dynamic Alongwind Response," *J. Struct. Dir.*, ASCE, **100**, No. ST9, Proc. Paper 10815 (Sept. 1974) 1897–1910.

2-71   A. G. Davenport, "The Spectrum of Horizontal Gustiness Near the Ground in High

Winds," *J. Royal Meteorol. Soc.*, **87** (1961) 194–211.

2-72 *Canadian Structural Design Manual*, Supplement No. 4 to the National Building Code of Canada, Associate Committee on the National Building Code and National Research Council of Canada, Ottawa, 1971.

2-73 F. Pasquill and H. E. Butler, "A Note on Determining the Scale of Turbulence," *J. Royal Meteorol. Soc.*, **90** (1964), 79–84.

2-74 E. Simiu and S. D. Leigh, *Turbulent Wind Effects on Tension Leg Platform Surge*, Building Science Series BSS 151, National Bureau of Standards, Washington, D.C., March 1983.

2-75 H. A. Panofsky and I. A. Singer, "Vertical Structure of Turbulence," *J. Royal Meteorol. Soc.*, **91** (1965) 339–344.

2-76 A. G. Davenport, "The Dependence of Wind Load upon Meteorological Parameters," in *Proceedings of the International Research Seminar on Wind Effects on Buildings and Structures*, University of Toronto Press, Toronto, 1968, pp. 19–82.

2-77 B. J. Vickery, "On the Reliability of Gust Loading Factors," in *Proceedings of the Technical Meeting Concerning Wind Loads on Buildings and Structures*, National Bureau of Standards, Building Science Series 30, Washington, D.C., 1970, pp. 93–104.

2-78 M. Shiotani and Y. Iwatani, "Correlations of Wind Velocities in Relation to the Gust Loadings," in *Proceedings of the Third International Conference on Wind Effects on Buildings and Structures*, Tokyo, 1971, Saikon, Tokyo, 1972, pp. 57–67.

2-79 S. SethuRaman, "Structure of Turbulence Over Water During High Winds," *J. Appl. Meteorol.*, **18**, (1979), 324–328.

2-80 L. Kristensen and N. O. Jensen, "Lateral Coherence in Isotropic Turbulence and in the Natural Wind," *Bound. Layer Meterol.*, **17** (1979), 353–373.

2-81 L. Kristensen, H. A. Panofsky, and S. D. Smith, "Lateral Coherence of Longitudinal Wind Components in Strong Winds," *Bound. Layer Meteorol.*, **21** (1981) 199–205.

2-82 C. F. Ropelewski, H. Tennekes, and H. A. Panofsky, "Horizontal Coherence of Wind Fluctuations," *Bound. Layer Meterol.*, **5** (1975), 353–363.

2-83 S. Berman and C. R. Stearns, "Near-Earth Turbulence and Coherence Measurements at Aberdeen Proving Ground, Md.," *Bound. Layer Meteorol.*, **11** (1977), 485–506.

2-84 J. Kanda and R. Royles, "Further Consideration of the Height-Dependence of the Root Coherence in the Natural Wind," *Build. Environ.*, **13** (1978), 175–184.

2-85 R. R. Brook, "A Note on Vertical Coherence of Wind Measured in an Urban Boundary Layer," *Bound. Layer Meteorol.*, **9** (1975), 247.

2-86 G. R. Stegen and R. L. Thorpe, "Vertical Coherence in the Atmospheric Boundary Layer", *Proceedings Fifth International Conference on Wind Engineering*, Vol. 1, Pergamon Press, 1980.

2-87 H. A. Panofsky and T. Mizuno, "Horizontal Coherence and Pasquill's Beta," *Bound. Layer Meteorol.*, **9** (1975), 247–256.

2-88 H. A. Panofsky et al., "Two-Point Velocity Statistics over Lake Ontario," *Bound. Layer Meteorol.*, **7** (1974), 309–321.

2-89 L. Kristensen, "On Longitudinal Spectral Coherence," *Bound. Layer Meteorol.*, **16**, (1979), 145–153.

2-90 A. K. Blackadar, H. A. Panofsky and F. Fiedler, *Investigation of the Turbulent Wind Field Below 500 Feet Altitude at the Eastern Test Range, Florida*, NASA CR-2438, National Aeronautics and Space Administration, Washington, D.C., 1974.

2-91 C. S. Durst, "Wind Speeds Over Short Periods of Time," *Meteorol. Mag.*, **89** (1960) 181–186.

2-92 J. Vellozzi and E. Cohen, "Dynamic Response of Tall Flexible Structures to Wind Loading," *Proceedings*, Technical Meeting Concerning Wind Loads on Buildings and Structures, Building Science Series BSS 30, National Bureau of Standards, Washington, DC, 1970, pp. 115–128.

2-93 P. Sachs, *Wind Forces in Engineering*, Pergamon, New York, 1972.

2-94 H. A. Panofsky and A. A. Townsend, "Change of Terrain Roughness and the Wind Profile," *J. Royal Meteorol. Soc.*, **90** (1964) 147–155.

2-95   P. A. Taylor, "On Wind and Shear Stress Profiles above a Change in Surface Roughness," *J. Royal Meteorol. Soc.*, **95** (1969) 77–91.

2-96   E. W. Peterson, "Modification of Mean Flow and Turbulent Energy by a Change in Surface Roughness under Conditions of Neutral Stability," *J. Royal Meteorol. Soc.*, **95** (1969) 561–575.

2-97   R. A. Antonia and R. E. Luxton, "The response of a turbulent boundary layer to a step change in surface roughness, Part I, Smooth to rough," *J. Fluid Mech.*, **48**, (1971) 721–761.

2-98   C. C. Shir, "A Numerical Computation of Air Flow Over a Sudden Change of Surface Roughness," *J. Atmos. Sci.*, **29** (March 1972) 304–310.

2-99   A. S. Monin, "Smoke Propagation in the Surface Layer of the Atmosphere," in *Adv. Geophys.*, **6**, Academic, New York (1959), 331–343.

2-100   D. H. Wood, "Internal Boundary-Layer Growth Following a Step Change in Surface Roughness," *Bound. Layer Meteorol.*, **22** (1982), 241–244.

2-101   B. G. De Bray, "Atmospheric Shear Flows over Ramps and Escarpments," *Ind. Aerodyn. Abstr.*, **4**, 5 (Sept.–Oct. 1973) 1–4.

2-102   C. Sacré, *Influence d'une colline sur la vitesse du vent dans la couche limite de surface*, Centre Scientifique et Technique du Bâtiment, Nantes, France, 1973.

2-103   A. J. Bowen and D. Lindley, "A Wind-Tunnel Investigation of the Wind Speed and Turbulence Characteristics Close to the Ground Over Various Escarpment Shapes," *Bound. Layer Meteorol.*, **12** (1977), 259–271.

2-104   P. S. Jackson and J. C. R. Hunt, "Turbulent Flow Over a Low Hill." *J. Royal Meteorol. Soc.*, **101** (1975) 929–955.

2-105   W. Frost, J. R. Maus, and G. H. Ficht, "A Boundary-Layer Analysis of Atmospheric Motion Over a Semi-Elliptical Surface Obstruction," *Bound. Layer Meteorol.*, **7** (1974), 165–184.

2-106   P. A. Taylor, "Numerical Studies of Neutrally Stratified Planetary Boundary Layer Flow Above Gentle Topography," *Bound. Layer Meteorol.*, **12** (1977), 37–60.

2-107   H. Nørstrud, "Wind Flow Over Low Arbitrary Hills," *Bound. Layer Meteorol.*, **23** (1982) 115–124.

2-108   P. J. Mason and R. I. Sykes, "Flow Over an Isolated Hill of Moderate Slope," *J. Royal Meteorol. Soc.*, **105** (1979), 383–395.

2-109   P. S. Jackson, "A Theory for Flow Over Escarpments," in *Proceedings of the Fourth International Conference on Wind Effects on Buildings and Structures*, London, 1975, Cambridge Univ. Press, Cambridge, U.K., 1976, pp. 33–40.

2-110   N. O. Jensen and E. V. Peterson, "On the Escarpment Wind Profile," *J. Royal Meteorol. Soc.*, **104** (1978), 719–728.

2-111   F. Bradley, "An Experimental Study of the Profiles of Wind Speed, Shearing Stress and Turbulence at the Crest of a Large Hill," *J. Royal Meteorol. Soc.*, **106** (1980), 101–123.

2-112   R. E. Britter, J. C. R. Hunt, and K. J. Richard, "Air Flow Over a Two-Dimensional Hill. Studies of Velocity Speed-Up, Roughness Effects and Turbulence," *J. Royal Meteorol. Soc.*, **107**, (1981), 91–110.

2-113   G. J. Jenkins et al., "Measurements of the Flow Structure Around Ailsa Craig, a Steep, Three-Dimensional Isolated Hill," *J. Royal Meteorol. Soc.*, **107** (1981), 833–851.

2-114   C. Sacré, "An Experimental Study of the Air Flow Over a Hill in the Atmospheric Boundary Layer," *Bound. Layer Meteorol.*, **17** (1979), 381–401.

2-115   T. Hauf and G. Neumann-Hauf, "Turbulent Wind Flow Over an Embankment," *Bound. Layer Meteorol.*, **24** (1982), 357–369.

2-116   S. L. Rosenthal, *A Theoretical Analysis of the Field of Motion in the Hurricane Boundary Layer*, National Hurricane Research Project, Report No. 56, U.S. Department of Commerce, Washington, D.C., 1962.

2-117   R. K. Smith, "The Surface Boundary Layer of a Hurricane," *Tellus*, **20** (1968) 473–484.

2-118   G. F. Carrier, A. L. Hammond and O. D. George, "A Model of the Mature Hurricane," *J. Fluid Mech.*, **47** (1971) 145–170.

2-119   E. Simiu, "Variation of Mean Winds with Hurricanes," *J. Eng. Mech. Div.*, ASCE, **100**,

No. EM4, Proc. Paper 10692 (Aug. 1974) 833–837.

2-120   E. Simiu, V. C. Patel and J. F. Nash, "Mean Wind Profiles in Hurricanes," *J. Eng. Mech. Div.*, ASCE, **102**, No. EM2, Proc. Paper 12044 (April 1976) 265–273.

2-121   *Survey of Meteorological Factors Pertinent to Reduction of Loss of Life and Property in Hurricane Situations*, National Hurricane Research Project Report No. 5, U.S. Department of Commerce, Weather Bureau, Washington, D.C., March 1957.

2-122   K. J. Wilson, "Characteristics of the Subcloud Layer Wind Structure in Tropical Cyclones," Bureau of Meteorology, Dept. of Science and Technology, Melbourne, Prepared for International Conference on Tropical Cyclones, Perth, Australia, Nov. 1979.

2-123   *Southern Standard Building Code*, Birmingham, Ala., 1965, p. 12–5.

2-124   W. Malkin, *Filling and Intensity Changes in Hurricanes over Land*, National Hurricane Research Project, Report No. 34, U.S. Department of Commerce, Washington, D.C., 1959.

2-125   J. L. Goldman and T. Ushiyima, "Decrease in Maximum Hurricane Winds after Landfall," *J. Struct. Div.*, ASCE, **100**, No. STI, Proc. Paper 10295 (Jan. 1974) 129–141.

2-126   M. B. Glauert, "The Wall Jet," *J. Fluid Mech.*, **1** (1956) 625.

2-127   P. Bakke, "An Experimental Investigation of a Wall Jet," *J. Fluid Mech.*, **2** (1957) 467.

2-128   J. Burnham and M. J. Colmer, *On Large Rapid Wind Fluctuations Which Occur When the Wind Had Previously Been Light*, Technical Report No. 69261, Royal Aircraft Establishment, Farnborough, U.K., 1969.

2-129   M. J. Colmer, "On the Character of Thunderstorm Gust-Fronts," Technical Report Aero 1316, Royal Aircraft Establishment, Farnborough, U.K., 1971.

2-130   R. W. Sinclair, R. A. Anthes and H. A. Panofsky, *Variation of the Low Level Winds During the Passing of a Thunderstorm Gust Front*, NASA Contractor Report No. CR-2289, 1973.

2-131   H. C. S. Thom, "New Distributions of Extreme Wind Speeds in the United States, *J. Struct. Div.*, ASCE, No. ST7, Proc Paper 6038 (July 1968) 1787–1801.

2-132   S. D. Smith, "Wind Stress and Heat Flux Over the Ocean in Gale Force Winds," *J. Phys. Oceanography*, **10** (May 1980), 709–726.

2-133   L. Krügermeyer, M. Grünewald, and M. Dunckel, "The Influence of Sea Waves on the Wind Profile," *Bound. Layer Meteorol.*, **14** (1978), 403–414.

2-134   T. von Kármán, "Progress in the Statistical Theory of Turbulence," *Proc. Nat. Acad. Sci.*, Washington D.C. (1948), 530–539.

2-135   J. Hawxhurst, *Relationship Between Fastest-Mile and Coincident Extreme Winds Over Larger Durations For Coastal Locations*, NUREG Report, Nuclear Regulatory Commission, Washington, D.C., 1985 (in print).

2-136   J. Biétry, Personal communication, 1981.

2-137   R. I. Harris, "The nature of wind", in *The Modern Design of Wind-Sensitive Structures*, Construction Industry Research and Information Association, London, U.K., 1971.

2-138   D. M. Deaves, "Computations of Wind Flow Over Changes in Surface Roughness," *J. Wind Eng. Ind. Aerdyn.*, **7** (1981), 65–94.

3

Extreme Wind Climatology

Climatology may be defined as a set of probabilistic statements on long-term weather conditions. The branch of climatology that specializes in the study of winds is referred to as wind climatology. Wind climatology provides the designer and the code writer with information on the extreme winds that might affect a structure during its lifetime.* Such information is required for making rational decisions on the magnitude of the wind loads to be used in design.

This chapter is devoted to a review of problems involved in the description of the wind climate for structural design purposes and in the development of criteria for the definition of design wind speeds. Procedures for estimating extreme winds are presented, and the uncertainties inherent in these procedures are discussed. Some of the material included herein is heavily dependent upon probabilistic and statistical notions and tools. These are presented in some detail in Appendix A1.

The reliability of climatological statements based on the analysis of extreme wind speed data is clearly dependent upon the quality of the data. This topic is discussed in Sect. 3.1. The question of the prediction of extreme wind speeds in well-behaved wind climates and in hurricane-prone regions is dealt with in Sects. 3.2 and 3.3, respectively. The dependence of extreme wind speeds upon direction is discussed in Sect. 3.4. Information on the frequency of occurrence of tornado winds of various intensities in the United States is presented in Sect. 3.5.

In the United States surface wind speeds reported by the Weather Service have traditionally been expressed in miles per hour (1 mph = 0.447 m/sec). In hurricane-related work, lengths are frequently expressed in nautical miles (1 nmi = 1.15 mile). For convenience, where appropriate, these units will also be used herein.

---

*Winds other than those of interest from a structural safety viewpoint will be dealt with in Chapter 15.

3.1 WIND SPEED DATA

To provide useful information on the wind climate at a given location, wind speed data recorded at that location must be *reliable* and must constitute a *micrometeorologically homogeneous set.*

3.1.1 Reliability of Wind Speed Data

Wind speed data may be considered to be reliable if:

1. The instrumentation used for obtaining the data (i.e., the sensor and the recording system) may be assumed to have performed adequately and was properly calibrated. If it can be determined that the calibration was not adequate, the data must be adjusted—whenever the information needed for that purpose is available.

Example

The following information is excerpted from [3-1] regarding the 5-min winds given in the original U.S. Weather Bureau records taken before 1932: "Up to 31 December 1927, all recorded wind speeds were the uncorrected readings of 4-cup anemometers. From 1928 through 1931, all speeds from the older 4-cup anemometers were corrected to agree with the readings of the 3-cup instruments, then being introduced, readings from which were not corrected to true speeds. From 1 January 1932 onward all readings, whether from 3- or 4-cup anemometers, were already corrected to true speed in the original records." Official U.S. Weather Bureau instructions for the correction of 3- and 4-cup anemometer readings are given in Table 3.1.1, which is excerpted from [3-2], and the use of which will now be illustrated. At Williston, N.D., the original readings of the maximum 5-min. wind in 1922 and 1930 on record at the National Oceanic and Atmospheric Administration are 56 mph and 37 mph, respectively. Using the corrections of Table 3.1.1, the true speeds (according to U.S. Weather Bureau calibrations) are $56 - 12 = 44$ mph and $37 - 2 = 35$ mph, respectively.

2. The sensor was exposed in such a way that it was not influenced by local flow effects due to the proximity of an obstruction (e.g., building top, or instrument support). For most U.S. weather stations, the existence of such an obstruction during the period of record is noted, in principle, in Local Climatological Data Summary sheets (LCD Summaries) issued by the Environmental Data Service of the National Oceanic and Atmospheric Administration [3-3].

3. The atmospheric stratification may be assumed to have been neutral. This assumption is acceptable for wind speeds at 10 m above ground in open terrain in excess of 25 mph or so (see Sect. 2.2.5).

3.1.2 Micrometeorological Homogeneity of Wind Speed Data

A set of wind speed data is referred to herein as micrometeorologically homogeneous if all the data belonging to the set may be considered to have been obtained under identical or equivalent micrometeorological conditions.

TABLE 3.1.1. Corrections to Indicated Wind Speeds [3-2]

	Speeds Indicated	
By 3-cup "S" Type Anemometer, mph. [1928–1931[a]]	By 4-cup Anemometer, mph. [Up to 31 Dec., 1927[a]]	Corrections in Whole Miles per Hour
0[b]–16	0[b]–8	+1
17–26	9–12	0
27–35	13–16	−1
36–44	17–20	−2
45–52	21–24	−3
53–61	25–28	−4
62–70	29–32	−5
71–79	33–36	−6
80–87	37–39	−7
88–96	40–43	−8
97–105	44–47	−9
106–114	48–51	−10
115–122	52–54	−11
123–132	55–58	−12
133–139	59–62	−13
140–149	63–65	−14
150–157	66–69	−15
158–166	70–73	−16
167–174	74–77	−17
175–184	78–80	−18
185–192	81–84	−19
193–200	85–88	−20
	89–91	−21
	92–95	−22
	96–99	−23
	100–103	−24
	104–106	−25
	107–110	−26
	111–114	−27
	115–117	−28
	118–121	−29
	122–125	−30
	126–128	−31
	129–132	−32
	133–136	−33
	137–140	−34
	141–143	−35

[a]Ref. 3-1.
[b]Movement of anemometer cups observed.

These conditions are determined by the following factors, which will be briefly discussed below:

- averaging time (i.e., whether highest gust, fastest mile, one-minute average, five-minute average, etc. was recorded)
- height above ground
- roughness of surrounding terrain (exposure)

1. *Averaging Time.* If various averaging times have been used during the period of record, the data must be adjusted to a common averaging time. This can be done by using Eq. 2.3.37 and Tables 2.2.1 and 2.3.3.

Data averaged over short time intervals, such as highest gusts or fastest miles, may in certain cases be affected by stronger than usual local turbulence effects, and thus provide a somewhat distorted picture of the intensity of the mean winds. In principle, it is desirable, therefore, that the data used for the description of the wind climate be averages over relatively long periods, say five minutes or so.

2. *Height Above Ground.* If during the period of record the elevation of the anemometer has been changed, the data must be adjusted to a common elevation as follows. Let the roughness length and the zero plane displacement be denoted by z_0 and z_d, respectively (z_0 and z_d are parameters that define the roughness of terrain; see Sect. 2.2). For strong winds (i.e., with speeds exceeding 10 m/sec or so), the relation between the mean speeds $U(z_1)$ and $U(z_2)$ over horizontal terrain of uniform roughness at elevation z_1 and z_2 above ground, respectively, can be written as

$$\frac{U(z_1)}{U(z_2)} = \frac{\ln[(z_1 - z_d)/z_0]}{\ln[(z_2 - z_d)/z_0]} \tag{3.1.1}$$

Equation 3.1.1 follows directly from Eqs. 2.2.18 and 2.2.22. For open terrain $z_d \simeq 0$, and the values of the roughness length z_0 can be taken from Table 2.2.1. As noted in Sect. 2.2.3, considerable uncertainties subsist with regard to the values of the roughness parameters in built-up terrain. Good judgment and experience are required to keep the errors inherent in the subjective estimation of the roughness parameters within reasonable bounds. It is clearly advisable to investigate in individual cases the effect of such possible errors upon the predictions of extreme wind speeds.

3. *Roughness of Surrounding Terrain.* In many cases anemometer locations have been changed during the period of record, for example from a town to a neighboring airport station. The corresponding records can, in principle, be adjusted to a common terrain roughness by using the similarity model (Eqs. 2.2.29 and 2.2.31 and Table 2.2.3) described in Sect. 2.2. As indicated in Sect. 2.4.1, this model may be assumed to be applicable in horizontal terrain if at each station the terrain roughness is reasonably uniform over a distance from the anemometer of about 100 times the anemometer elevation. In terrain

in which sheltering effects by small-scale obstacles are present, the data may be adjusted by using a procedure presented in [3-4].

A situation commonly encountered in practice is one in which, while the anemometer may not have been moved, the roughness of the terrain surrounding the anemometer has changed significantly over the years as a result of extensive land development. In such situations, the adjustment of the data to a common roughness may pose insurmountable problems, unless detailed information on the phases of the land development is available.

Anemometer elevation and location changes are listed for most U.S. weather stations in Local Climatological Data Summaries [3-3].

3.2 ESTIMATION OF EXTREME WIND SPEEDS IN WELL-BEHAVED CLIMATES

Infrequent winds (e.g., hurricanes) that are meteorologically distinct from and considerably stronger than the usual annual extremes are referred to herein as extraordinary winds. Climates in which extraordinary winds may not be expected to occur are referred to as well behaved. In such climates it is reasonable to assume that each of the data in a series of the largest annual wind speeds contributes to the description of the probabilistic behavior of the extreme winds. A statistical analysis of such a series can therefore be expected to yield useful predictions of long-term wind extremes.

Thus, in a well-behaved climate, at any given station a random variable may be defined, which consists of the largest yearly wind speed. If the station is one for which wind records over a number of consecutive years are available, then the cumulative distribution function (CDF) of this random variable may be estimated to characterize the probabilistic behavior of the largest annual wind speeds. The basic design wind speed is then defined as the speed corresponding to a specified value p of the CDF or, equivalently, to a specified mean recurrence interval $\bar{N}$.* A wind corresponding to an $\bar{N}$-year mean recurrence interval is commonly referred to as the $\bar{N}$-year wind.

This section is devoted to the question of estimating (a) the CDF of the largest annual speeds and (b) errors inherent in the wind speed predictions. Such errors include, in addition to those associated with the quality of the data (see Sect. 3.1), *modeling* errors and *sampling* errors. Modeling errors are due to an inadequate choice of the probabilistic model itself. Sampling errors are a consequence of the limited size of the samples from which the distribution parameters are estimated and become, in theory, vanishingly small as the sample size increases indefinitely.

3.2.1 Probabilistic Modeling of Largest Yearly Wind Speeds

Several probability distributions have been proposed to model extreme wind behavior. These include: the Type I distribution of the largest values (Eq. A1.39),

*Recall that $\bar{N} = 1/(1 - p)$ (see Appendix A1, Eq. A1.45).

the Type II distribution of the largest values (Eq. A1.42), and the Weibull distribution (Eq. A1.65). Extreme wind speeds inferred from any given sample of wind speed data depend on the type of distribution on which the inferences are based. For large mean recurrence intervals ($\bar{N} > 50$ years, say) estimates based on the assumption that a Type II distribution is valid are higher than corresponding estimates obtained by using a Type I distribution, while estimates based on a Weibull distribution with tail length parameter $\gamma \geqslant 2$ are lower.*

According to [3-5], extreme winds in well-behaved climates may be assumed to be best modeled by a Type II distribution with $\mu = 0$ and $\gamma = 9$. However, subsequent research has shown that this assumption is not borne out by analyses of extreme wind speed data [3-6, 3-7, 3-8]. In [3-6], 37 year-series of 5 minute largest yearly speeds measured at stations with well-behaved climates were subjected to the probability plot correlation coefficient test (see Sect. A1.6) to determine the tail length parameter of the best fitting distribution of the largest values. Of these series, 72% were best fit by Type I distributions or by Type II distributions with $\gamma = 13$ (which differ insignificantly from the Type I distribution); 11% by Type II distributions with $7 \leqslant \gamma < 13$; and 17% by Type II distribution with $2 \leqslant \gamma < 7$. Virtually the same percentages were obtained in [3-7] from the analysis of sets of 37 data generated by the Monte Carlo simulation from a population with a Type I distribution. On the other hand, the analysis of sets generated by Monte Carlo simulation from a Type II distribution with tail length parameter $\gamma = 9$ led to percentages differing significantly from those corresponding to the actual wind speed data. On the basis of these results it can be confidently stated that in well-behaved climates extreme wind speeds are modeled more realistically by the Type I than by the Type II distribution with $\gamma = 9$. This conclusion was reinforced by studies reported in [3-8], in which techniques similar to those of [3-7] were used in conjunction with wind speed data at one hundred United States weather stations obtained from [3-9].

As indicated earlier, the Type I distribution results in lower estimates of the extreme wind speeds than the Type II distribution with $\gamma = 9$. An interesting result obtained in [3-8] is that at most stations in the United States even the Type I distribution appears to be an unduly severe model of the wind speeds corresponding to large mean recurrence intervals; at these stations a better fit to the data is obtained by Weibull distributions with $\gamma \geqslant 2$. Thus, structural reliability calculations based on the assumption that the Type I distribution holds are in most cases likely to be conservative [3-10].

3.2.2 Estimation of and Confidence Intervals for the $\bar{N}$-year Wind: Numerical Example

It is shown in Sect. A1.6 that, given a set of data with a Type I extreme value underlying distribution, several techniques can be used to estimate the param-

---

*The differences between speeds estimated on the basis of Type II distributions and the Type I distribution increase as γ decreases. Differences between speeds based on the Type I distribution and Weibull distributions increase as γ increases.

eters of the distribution and, hence, the value of the variate corresponding to a given mean recurrence interval.* However, inherent in these estimates are sampling errors. A measure of the magnitude of the latter can be obtained by calculating *confidence intervals* for the quantity being estimated, that is, intervals of which it can be stated—with a specified confidence that the statement is correct—that they contain the true, unknown value of that quantity. Techniques that can be used to estimate the $\bar{N}$-year wind, and confidence intervals for the $\bar{N}$-year wind, are discussed in some detail in Sect. A1.6. One of these techniques is presented and illustrated below.

Using the approximation $-\ln[-\ln(1-1/\bar{N})] \simeq \ln \bar{N}$, it follows from Eq. A1.74 (which is based on the method of moments) that the estimated value $\hat{v}_{\bar{N}}$ of the $\bar{N}$-year wind $v_{\bar{N}}$ is

$$\hat{v}_{\bar{N}} \simeq \bar{X} + 0.78(\ln \bar{N} - 0.577)s \qquad (3.2.1)$$

where $\bar{X}$ and s are, respectively, the sample mean and the sample standard deviation of the largest yearly wind speeds for the period of record.

As previously noted, inherent in the estimates of $v_{\bar{N}}$ are sampling errors. It follows from Eqs. A1.76 and A1.70 (which are based on the method of moments) that the standard deviation of the sampling errors in the estimation of $v_{\bar{N}}$ can be written as

$$SD(\hat{v}_{\bar{N}}) \simeq 0.78[1.64 + 1.46(\ln \bar{N} - 0.577) + 1.1(\ln \bar{N} - 0.577)^2]^{1/2} \frac{s}{\sqrt{n}} \quad (3.2.2)$$

where n is the sample size.

Example

At Great Falls, Montana, the largest yearly fastest-mile wind speeds at 10 m above ground during the period 1944–1977 (sample size $n = 34$) were [3-9]:

$$57, 65, 62, 58, 64, 65, 59, 65, 59, 60, 64, 65, 73, 60, 67, 50, 74$$
$$60, 66, 55, 51, 60, 55, 60, 51, 51, 62, 51, 54, 52, 59, 56, 52, 49$$

(mph). The sample mean and the sample standard deviation for these data are $\bar{X} = 59$ mph and $s = 6.41$ mph. From Eqs. 3.2.1 and 3.2.2 it follows that for $\bar{N} = 50$ years and $\bar{N} = 1,000$ years,

$$\hat{v}_{50} \simeq 76 \text{ mph} \qquad SD(\hat{v}_{50}) \simeq 3.7 \text{ mph}$$
$$\hat{v}_{1000} \simeq 91 \text{ mph} \qquad SD(\hat{v}_{1000}) \simeq 6.4 \text{ mph}$$

If it is assumed that the largest yearly wind speeds are described by a Rayleigh distribution,† the $\bar{N}$-year wind, denoted by $v_{\bar{N}}^R$, can be obtained from Eq. A1.65

---

*In Appendix A1 this value is denoted by $G_X(p)$, where $p = 1 - 1/\bar{N}$ and $\bar{N}$ is the mean recurrence interval. In this chapter the notation $G_X(1 - 1/\bar{N}) = v_{\bar{N}}$ is used.

†It is recalled that the Weibull distribution with tail length parameter $\gamma = 2$ is commonly referred to as the Rayleigh distribution. Note that of all Weibull distributions with $\gamma \geq 2$, the Rayleigh distribution is the closest to the Type I distribution (i.e., it has the longest tail).

(with $\gamma = 2$) as follows:

$$v_N^R \simeq \bar{X} + \frac{s}{0.463} [(\ln \bar{N})^{1/2} - 0.886] \tag{3.2.3}$$

where $\bar{X}$ and s are defined as in Eq. 3.2.1. In the case of Great Falls, $\bar{X} = 59$ mph and $s = 6.41$ mph, so that $v_{50}^R = 74$ mph and $v_{1000}^R = 83$ mph, versus $v_{50} = 76$ mph and $v_{1000} = 91$ mph, as estimated in the preceding example by assuming the validity of the Type I distribution. As indicated previously, in engineering calculations it is prudent to assume the validity of the Type I distribution (Eq. 3.2.1), rather than using Eq. 3.2.3. This conservative approach was adopted in developing the map of basic design wind speeds (i.e., fastest-mile wind speeds at 10 m above ground in open terrain, with a 50-year mean recurrence interval) included in the American National Standard A58.1 [2-49] (Fig. 3.2.1).

As shown in Sect. A1.6, the probabilities that $v_{\bar{N}}$ is contained in the intervals $\hat{v}_{\bar{N}} \pm SD(\hat{v}_N)$, $\hat{v}_{\bar{N}} \pm 2SD(\hat{v}_N)$, and $\hat{v}_{\bar{N}} \pm 3SD(\hat{v}_{\bar{N}})$ are approximately 68%, 95%, and 99%, respectively. These intervals are referred to as the 68%, 95%, and 99% confidence intervals for $v_{\bar{N}}$, and are shown for the 34-year Great Falls sample in line (1) of Table 3.2.1.

It is also shown in Sect. A1.6 that the width of the confidence intervals can be reduced if a more efficient estimator is used; however, the intervals cannot be narrower than those obtained by using the Cramér–Rao (C.R.) lower bound (Eq. A1.77). For the Great Falls sample, the confidence intervals based on the latter are shown in line (2) of Table 3.2.1. It is seen that the differences between the results of lines (1) and (2) of Table 3.2.1 are small. This is consistent with the conclusion of Sect. A1.6 that the efficiency of the method of moments (Eq. 3.2.1) is generally adequate for structural design purposes.

It is noted that, in Table 3.2.1, the errors in the estimation of the 50-year wind are of the order of 10% at the 95% confidence level. Since the wind pressures are proportional to the wind speeds (see Chapter 4), the corresponding errors in the estimation of the pressures are of the order of 20%.

TABLE 3.2.1. Confidence Intervals for the $\bar{N}$-year Wind at Great Falls

Confidence level	68%		95%		99%	
Mean recurrence interval, $\bar{N}$ (years)	50	1000	50	1000	50	1000
(1) Estimated by method of moments	76 ± 3.7	91 ± 6.4	76 ± 7.4	91 ± 12.8	76 ± 11.1	91 ± 19.2
(2) Estimated using C.R. lower bound	76 ± 3.1	91 ± 5.0	76 ± 6.2	91 ± 10.0	76 ± 9.3	91 ± 15.0

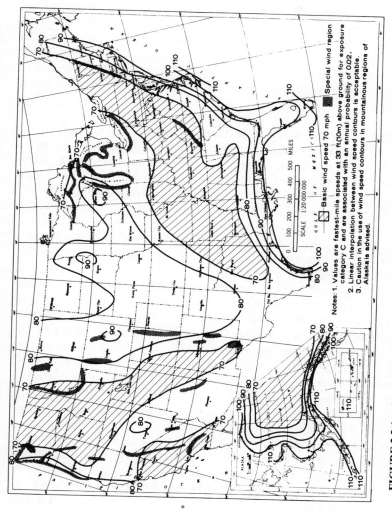

FIGURE 3.2.1. Map of basic design wind speeds. Reproduced with permission from American National Standard A58.1 *Building Code Requirements for Minimum Design Loads in Buildings and Other Structures*, copyright 1982 by the American National Standards Institute. Copies of this standard may be purchased from the American National Standards Institute at 1430 Broadway, New York, NY 10018.

Notes: 1. Values are fastest-mile speeds at 33 ft(10m) above ground for exposure category C and are associated with an annual probability of 002.
2. Linear interpolation between wind speed contours is acceptable.
3. Caution in the use of wind speed contours in mountainous regions of Alaska is advised.

Basic wind speed 70 mph   Special wind region

SCALE 1:20 000 000

0   100   200   300   400   500
MILES

An alternative approach to accounting for sampling errors, which applies the theorem of total probability, is suggested in [3-51].

3.2.3   Methods for Estimating the Extreme Speeds at Locations with Insufficient Largest Yearly Wind Speed Data

There are about one hundred U.S. weather stations for which reliable and relatively long wind speed records are available (i.e., records over periods of, say, 20 years or more). Some of these stations cover areas of tens of thousands of square miles, over which—for meteorological reasons or owing to topographic effects—the extreme wind climate is not necessarily uniform. There arises therefore in practice the problem of estimating extreme wind speeds at various locations where long-term records of the largest yearly wind speed data do not exist.

Estimates of Extreme Wind Speeds in a Marine Environment. Reference 3-11 lists three methods that are in principle available to carry out such estimates for marine environments where the extreme speeds are associated with extra-tropical storms. The first method makes use of climatological information on various parameters of the storm and of physical models relating those parameters to the surface wind speeds. It is shown in Sect. 3.3 that such a method can be applied to estimate extreme wind speeds in hurricane-prone regions. However, as noted in [3-11], owing to the complexity of the surface wind patterns in extratropical storms, the usefulness of this method appears to be uncertain in regions where such storms are dominant.

A second method listed in [3-11] is the use of objective analysis schemes. These consist of: (a) an initial guess at the surface wind on a regular grid, (b) an automated procedure for screening wind reports from ships to eliminate erroneous readings, and (c) a procedure for correcting the initial guess on the basis of the usable set of ship reports, which involves relations among the surface wind speeds, sea-level pressures, and air and sea temperatures. Details on objective analysis schemes and of errors currently inherent in such schemes (which may range from 10% to 30%) are given in [3-11].

The third method listed in [3-11] is referred to as direct kinematic analysis. The method, which involves subjective judgment by experienced analysts, consists of synthesizing discrete meteorological observations to obtain a continuous field represented in terms of streamlines and isotachs. Objective or kinematic analyses applied to a sufficient number of strong storms make it possible to provide estimates of extreme winds that may occur at any one location. As indicated in [3-11], one of the major difficulties in conducting such analyses is that much of the vast store of existing data is currently not accessible in readily usable form.

Estimation of Extreme Wind Speeds from Short-Term Records. A practical procedure for estimating extreme wind speeds at locations where long-term data are not available is described in [3-12]. The method, whose applicability was tested for a large number of U.S. weather stations, makes it possible to infer the probabilistic behavior of extreme winds from data consisting of the largest monthly wind speeds recorded over a period of three years or longer. Estimates based on the monthly speeds, denoted by $\hat{v}_{\bar{N},m}$, are obtained by re-writing Eq. A1.74 as follows:

$$\hat{v}_{\bar{N},m} \simeq \bar{X}_m + 0.78[\ln(12\bar{N}) - 0.577]s_m \tag{3.2.4}$$

where $\bar{X}_m$ and s_m are, respectively, the sample mean and the sample standard deviation of the largest monthly wind speed data, and $\bar{N} = $ mean recurrence interval in years.

The standard deviation of the sampling error in the estimation of $\hat{v}_{\bar{N},m}$ is obtained from Eqs. A1.76 and A1.70 as

$$SD(\hat{v}_{\bar{N},m}) = 0.78\{1.64 + 1.46[\ln(12\bar{N}) - 0.577] + 1.1[\ln(12\bar{N}) - 0.577]^2\}^{1/2} \frac{s_m}{\sqrt{n_m}}$$

$$\tag{3.2.5}$$

where $n_m = $ sample size.

Example

At Great Falls, the sample mean and the sample standard deviation of the largest monthly fastest-mile wind speeds at 10 m above ground for the period September 1968 through August 1971* (sample size $n_m = 36$) are $\bar{X}_m = 42$ mph, $s_m = 6.96$ mph. From Eqs. 3.2.4 and 3.2.5, the estimates for $\bar{N} = 50$ years and $\bar{N} = 1000$ years are:

$$\hat{v}_{50,m} \simeq 74 \text{ mph} \qquad SD(\hat{v}_{50,m}) \simeq 6.23 \text{ mph}$$

$$\hat{v}_{1000,m} = 90 \text{ mph} \qquad SD(\hat{v}_{1000,m}) = 8.85 \text{ mph}$$

It is seen that the estimated speeds based on the set of 36 largest monthly data are only slightly lower than those obtained from the set of 34 largest yearly speeds ($\hat{v}_{50} = 76$ mph and $\hat{v}_{1000} = 91$ mph; see Sect. 3.2.2); however, the sampling errors are larger.

Similar calculations carried out for 67 sets of records taken at 36 stations are reported in [3-12], where it was found that the differences $\hat{v}_{50,m} - \hat{v}_{50}$, where $\hat{v}_{50}$ is the 50 year wind speed estimated from long-term largest yearly data, were less than $SD(\hat{v}_{50,m})$ in 66% of the cases and less than twice the value of $SD(\hat{v}_{50,m})$ in 95% of the cases. This remarkable result, confirmed by additional calculations reported in [3-13], indicates that the estimates based on largest monthly wind speeds recorded over three years or more provide a useful description of the extreme wind speeds in regions with a well-behaved wind climate.

Inferences concerning the probabilistic model of the extreme wind climate

---

*For the actual data, see the Local Climatological Data summaries for the years 1968–1971.

have also been attempted from data consisting of largest daily wind speeds [3-12], or of wind speeds measured at 1-hour intervals [3-14]. One problem that arises in this respect is that data recorded on two successive days are generally strongly correlated. Nevertheless, as shown in [3-14], in practice such correlation has a negligible effect on the statistical estimates, and the assumption of statistical independence among the data can therefore be used. However, a second and more serious problem is that the daily (or hourly) data reflect a large number of events (e.g., morning breezes) that are altogether unrelated meteorologically to the storms associated with the extreme winds. These events can be viewed as noise that obscures the information relevant to the description of the extreme wind climate. Indeed, it was verified in [3-12] that estimates of extreme winds based on daily data differ significantly from estimates obtained for long-term records of largest yearly speeds. This conclusion is *a fortiori* true for inferences based on hourly data.

3.3   ESTIMATION OF EXTREME WIND SPEEDS IN HURRICANE-PRONE REGIONS

We now consider the prediction of extreme winds in climates characterized by the occurrence of hurricanes. It was suggested in Sect. 3.2 that in a well-behaved wind climate each of the data in a series of the largest yearly speeds contributes to the description of the probabilistic behavior of the extreme winds. However, in a hurricane-prone region most of the speeds in a series of the largest yearly winds are considerably lower than the extreme speeds associated with hurricanes; they may therefore be irrelevant from a structural safety point of view. This situation is illustrated by the plot of Fig. 3.3.1, which shows the 5-min largest speeds recorded at Corpus Christi, Texas between 1912 and 1948 [3-6]. It may then be argued that in hurricane-prone regions the series of the largest yearly speeds cannot provide useful statistical information on winds of interest to the structural designer, much in the same way as the population of a first-grade classroom—which might include a teacher—is of little use in a statistical study of the height of adults. That this is the case is suggested below.

The abscissa in Fig. 3.3.1 represents the reduced variate

$$y = -\ln\left[-\ln\left(1 - \frac{1}{\bar{N}}\right)\right]$$

where $\bar{N}$ is the mean recurrence interval. In virtue of Eqs. A1.43 and A1.45, a Type I extreme value cumulative distribution function would be represented in Fig. 3.3.1 by a straight line, the intercept and slope of which would be equal to the distribution parameters μ and σ, respectively. To the extent that the population of largest yearly speeds would be described by a Type I distribution, the actual data would then fit, approximately, a straight line. In Fig. 3.3.1 this is roughly the case as far as the winds of less than hurricane force are concerned. However, if—as in Fig. 3.3.1—the hurricane-force winds are included in the

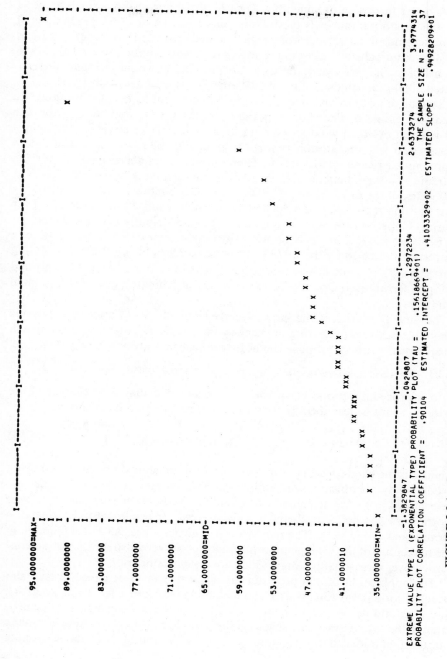

FIGURE 3.3.1. Probability plot of 1912–1948 annual largest speeds at Corpus Christi, Texas [3-6].

set being analyzed, the fit of a Type I distribution to the data is, clearly, extremely poor.

A better fit can be obtained if a Type II distribution with a small value of the tail length parameter is used. However, as shown in [3-6], predictions of extreme winds in hurricane-prone regions based on Type II distributions are in most cases unrealistic. For example, fitting such a distribution to the 1917–1936 record of the largest yearly speeds at Corpus Christi would yield, for the estimated 1000-year wind, a value of 1950 mph—a ridiculous result [3-6].

Serious difficulties also arise if mixed Fréchet probability distributions are used [3-5]. Indeed, since hurricanes are rare events, the number of hurricane (or tropical cyclone) wind speed data in a record of the largest yearly winds observed at any one station is small (e.g., in Fig. 3.3.1 only two of the data represent hurricane wind speeds). Therefore, the confidence intervals for the extreme wind predictions are, in general, unacceptably wide (e.g., for $\bar{N} = 100$ years, of the order of $\hat{v}_{100}(1 \pm 0.6)$ at the 68% confidence level; see [3-15]). It is for this reason that the 50-year fastest-mile wind estimated in [3-5] for Corpus Christi on the basis of a mixed Fréchet distribution is only 76 mph at 30 ft above ground in open terrain. This value appears to be severely low; indeed, in the period 1916–1970 Corpus Christi was hit by three devastating hurricanes [3-16] with fastest-mile winds of up to 120 mph at 23 ft above ground in open terrain (see Corpus Christi 1970 Local Climatological Data Annual Summary).

Because the series of the largest yearly winds does not appear to provide a suitable basis for predicting hurricane wind speeds, alternative bases for such predictions have been proposed in the literature, which are now briefly discussed.

3.3.1   Procedure Based on the Maximum Average Monthly Speed

In this procedure, proposed in [3-17], it is assumed that the behavior of the extreme winds is described by the cumulative distribution function

$$F(v) = p_T \exp\left[-\left(\frac{v}{\sigma}\right)^{-4.5} \right] + (1 - p_T) \exp\left[-\left(\frac{v}{\sigma}\right)^{-9} \right] \qquad (3.3.1)$$

where v is the wind speed, p_T is the probability of an annual extreme wind being produced by a tropical storm, and σ is a scale parameter. The parameter p_T, determined in [3-17] as an empirical function of the mean number of tropical storm passages per year through a five-degree longitude–latitude square, is represented in Fig. 3.3.2. The parameter σ is given in Fig. 3.3.3 as a function of the maximum of the average monthly wind speeds recorded at the station concerned over a reasonably long period (say, ten years or so).

The application of this procedure is illustrated in three cases: West Palm Beach (Florida), Boston (Mass.), and Columbia (Missouri), for which $p_T \simeq 0.43$, $p_T \simeq 0.12$, and $p_T = 0$, respectively (Fig. 3.3.2). At West Palm Beach, the maximum of the average monthly speeds in the period 1952–1974 (obtained from the Local Climatological Data Summaries) was 13.9 mph at 30 ft above ground.

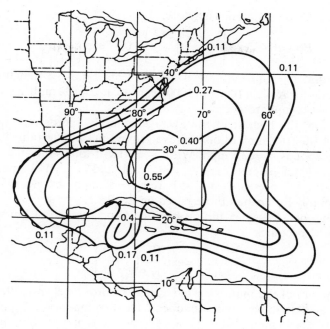

FIGURE 3.3.2. Probability p_T of an annual extreme wind being produced by a tropical storm. From H. C. S. Thom, "Toward a Universal Climatological Extreme Wind Distribution," in *Proceedings*, International Research Seminar on Wind Effects on Buildings and Structures, Vol. 1, p. 682. Copyright, Canada, 1968, University of Toronto Press.

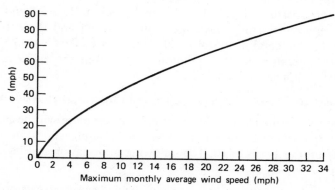

FIGURE 3.3.3. Scale parameter σ. From H. C. S. Thom, "Toward a Universal Climatological Extreme Wind Distribution," in *Proceedings*, International Research Seminar on Wind Effects on Buildings and Structures, Vol. 1, p. 682. Copyright, Canada, 1968, University of Toronto Press.

From Fig. 3.3.3, $\sigma = 51$ mph. Therefore,

$$F(v) = 0.43 \exp\left[-\left(\frac{v}{51}\right)^{-4.5}\right] + 0.57 \exp\left[-\left(\frac{v}{51}\right)^{-9}\right] \qquad (3.3.2)$$

Recalling that $\bar{N} = 1/[1 - F(v_{\bar{N}})]$, it follows from Eq. 3.3.2 that the estimated $\bar{N}$-year winds for $\bar{N} = 50$, 100, and 1000 years are $v_{50} = 102$ mph, $v_{100} = 120$ mph, and $v_{1000} = 198$ mph, respectively.

At Boston, the highest of the average monthly speeds recorded between 1950–1974 was 18.8 mph at 30 ft above ground in open terrain. To this value there corresponds $\sigma = 63.6$ mph. With $p_T = 0.12$, it follows from Eq. 3.3.1 that the extreme wind estimates are $v_{50} = 106$ mph and $v_{100} = 119$ mph. It is noted that the estimates presented in [3-5] are $v_{50} = 88$ mph and $v_{100} = 93$ mph, that is, considerably lower than those based on Eq. 3.3.1.

At Columbia, Missouri the probability of occurrence of hurricanes is nil and Eq. 3.3.1 becomes

$$F(v) = \exp\left[-\left(\frac{v}{\sigma}\right)^{-9}\right] \qquad (3.3.3)$$

The maximum of the monthly wind speeds recorded between 1951–1974 was 15.7 mph at 30 ft above ground in open terrain so that $\sigma = 57.0$ mph and $v_{50} = 88$ mph, $v_{100} = 95$ mph and $v_{1000} = 123$ mph. It is noted that the estimated extremes of [3-5] are lower, that is, $v_{50} = 70$ mph and $v_{100} = 85$ mph. The extreme speeds at Columbia were also estimated assuming the validity of the Type I distribution, with parameters inferred from the 1951–1974 series of the largest yearly speeds at 30 ft above ground in open terrain. The results thus obtained were $v_{50} = 66$ mph, $v_{100} = 69$ mph, and $v_{1000} = 81$ mph, versus $v_{50} = 88$ mph, $v_{100} = 95$ mph, and $v_{1000} = 123$ mph, as estimated on the basis of Eq. 3.3.1 with the attendant assumptions of [3-17].

Among these assumptions is the relation implicit in Eq. 3.3.1 and Figs. 3.3.2 and 3.3.3 between maximum average monthly speed and the extreme wind speeds. No fundamental meteorological grounds are offered in [3-17] or elsewhere in the literature for this relation which, from the evidence available so far, does not appear to be justified.

3.3.2   Procedure Based on Climatological and Physical Models of Hurricanes

To illustrate the principle of this procedure, an estimate will be made of the probability that hurricane winds in excess of 155 mph will occur at any one specific site on the Texas coast. The following information will be used in the estimation:

- average number per year of hurricanes with wind speeds in excess of 155 mph moving inland in the United States, v_A^{155}. According to the National Weather Service, there have been two such hurricanes in the past 75 years or so, the Labor Day Florida Keys hurricane in 1935 and hurricane Camille in 1969 [3-18]. A reasonable estimate is then $v_A^{155} \simeq 2$ hurricanes/(75 years) = 0.027 hurr/year.

- average number per year of hurricanes moving inland in the United States, v_A. This quantity can be estimated from Fig. 3.3.4: $v_A \simeq 116$ hurricanes/ (63 years) $\simeq 1.84$ hurr/year.
- average number per year of all hurricanes moving inland in Texas, v_T. From Fig. 3.3.4, $v_T \simeq 27$ hurricanes/(63 years) $\simeq 0.43$ hurr/year.
- average width of area swept by winds in excess of 155 mph in one hurricane, W. According to [3-20], the path of destruction of the Labor Day Florida Keys hurricane was 35–40 miles wide. It will be assumed conservatively that winds in excess of 155 mph affected a width $W = 30$ miles of that path. In the case of hurricane Camille it appears that it may be assumed conservatively $W = 20$ miles [3-21]. A reasonable value to be used in the calculations is then $W = (30 + 20)/2 = 25$ miles.

It will be assumed that the average number per year of hurricanes with speeds in excess of 155 mph moving inland in Texas is

$$v_T^{155} = \frac{v_T}{v_A} v_A^{155} \qquad (3.3.4)$$

(Implicit in Eq. 3.3.4 is the assumption that the probability distribution of the hurricane intensities, given that a hurricane has occurred, is the same throughout the U.S. Gulf and Atlantic coasts.) The length of the (smoothed) Texas coast being about 375 miles, the probability sought is

$$P(v > 155 \text{ mph}) = v_T^{155} \frac{W}{375} = 0.00042 \qquad (3.3.5)$$

that is, approximately 1/2500 per year.

The estimate just presented has several significant weaknesses. First, the errors in the estimate of v_A^{155} could conceivably be large, the estimate being based on a 75-year-long record containing just two relevant data. Second, the assumption that the rate of arrival of hurricanes is uniformly distributed over the length of the Texas coastline overestimates the probability of hurricane strikes over the coastline segment adjacent to the Mexican border (by about 25%), and underestimates that probability (by about 25%) nearer the Louisiana border (Figs. 3.3.5 and 3.3.6). Third, the reliability of the estimate of v_T is difficult to ascertain. Indeed, according to Fig. 3.3.6, $v_T \simeq 1.6/(100 \times 10)$ entries/year/nmi of coast $\times 0.53$ hurricanes/entry $\times 330$ nmi of coast $= 0.28$ hurr/year, versus 0.43 hurr/year, as obtained from the data of Fig. 3.3.4. (This discrepancy is possibly due to the counting of certain tropical cyclones as hurricanes in [3-19]*). Fourth, the estimates of W are largely subjective, since no measurements were taken in the field; the error may therefore be significant in the estimation of W as well.

In spite of these weaknesses, the approach just outlined may be of some use

---

*Reference 3-19 was revised in 1978 and is updated annually by the National Hurricane Center [3-52]. Additional information on North Atlantic hurricanes is available in [3-53] and, on tape, in [3-54]. For information on Western North Pacific tropical cyclones, see [3-55].

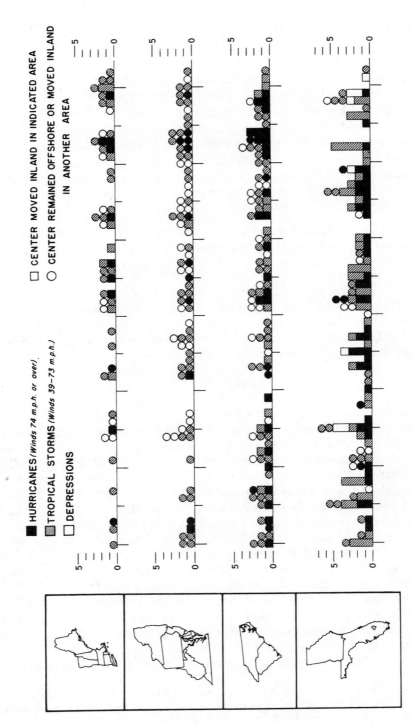

CENTER MOVED INLAND IN INDICATED AREA

■ HURRICANES (*Winds 74 m.p.h. or over*).

▨ TROPICAL STORMS (*Winds 39-73 m.p.h.*)

□ DEPRESSIONS

□ CENTER MOVED INLAND IN INDICATED AREA

○ CENTER REMAINED OFFSHORE OR MOVED INLAND
IN ANOTHER AREA

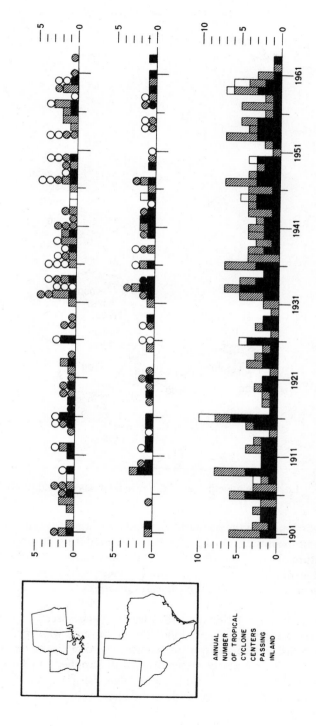

FIGURE 3.3.4. Annual frequency of tropical cyclones moving inland over the United States or passing close enough offshore to affect significantly various sections of the coast with tropical storm or hurricane winds, heavy rainfall, or high storm tides, 1901–1963 [3-19].

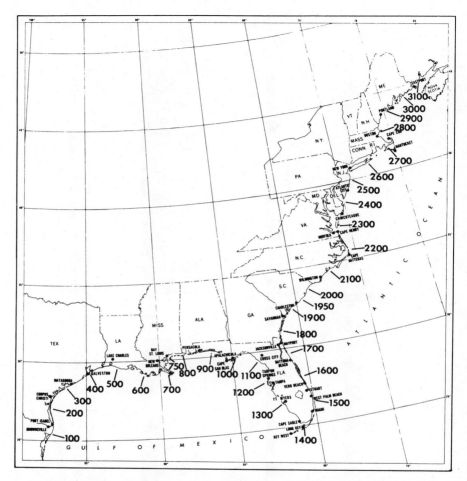

FIGURE 3.3.5.   Locator map with coastal distance intervals marked (nmi) [3-22].

for making gross estimates of the probability of failure of structures in hurricane-prone regions, for example, for the purpose of calculating insurance rates. Also, a useful feature of Eqs. 3.3.4 and 3.3.5 is that, unlike the model based on the maximum average monthly speed, they offer a clear, physically meaningful picture of the uncertainties involved in the estimation.

Monte Carlo Procedure for Estimating Hurricane Wind Speeds.   A comprehensive and effective approach to the modeling of extreme wind probabilities at a site on the basis of information on tropical cyclone characteristics was developed in [3-23]. This approach was subsequently applied in [3-24], where extreme wind speeds associated with hurricanes were estimated on the basis of the climatological and physical models described below.

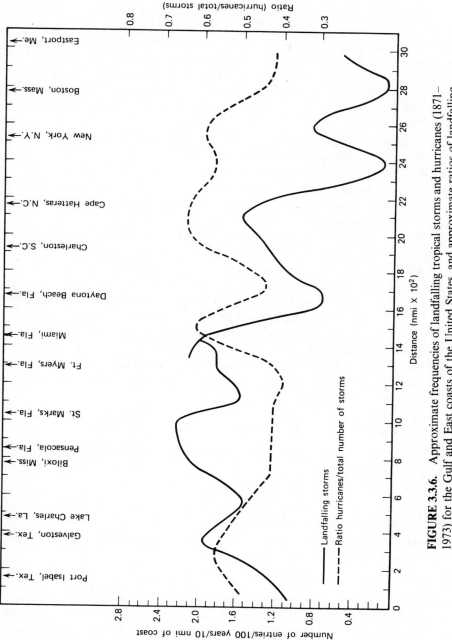

FIGURE 3.3.6. Approximate frequencies of landfalling tropical storms and hurricanes (1871–1973) for the Gulf and East coasts of the United States, and approximate ratios of landfalling hurricanes to total number of landfalling tropical storms and hurricanes (1886–1973) [3-22].

Climatological Models

1.   The hurricane frequency of occurrence is modeled by a Poisson process with a constant rate.

2.   The probability distribution of the pressure difference between center and periphery of the storm, Δp_{max}, is lognormal. To eliminate values of Δp_{max} judged, in the light of historical data, to be unrealistically high, the distribution is censored so that $\Delta p_{max} < 101.6$ mm (4.00 in) of mercury [3-23]. (Note that $\Delta p_{max} \simeq 101.6$ mm corresponds to the lowest atmosphere pressure ever recorded worldwide [3-25].)

3.   The probability distribution of the radius of maximum wind speeds, R, is lognormal. This distribution is censored so that 8 km $< R < 100$ km to avoid unrealistically "tight" or "broad" storms [3-23].

4.   The average correlation coefficient between R and Δp_{max} is about -0.3. (See [3-22], pp. 68 and 69.) All other climatological characteristics of hurricanes are statistically independent.

5.   The probability distribution of the speed of translation, s, is normal. This distribution is censored so that 2 km/hr $< s < 65$ km/hr [3-23].

6.   The probability distributions of the distance between any specified point on the coast and the hurricane crossing point along the coast (or on a line normal to the coast) are curves matching the historical data. Separate curves are defined for entering, exiting, upcoast heading, and downcoast heading storms.

7.   For entering storms the probability distribution of the direction of storm translation is a curve matching the historical data. For exiting upcoast heading and downcoast heading curves the distributions are uniform between $\pm 30°$ of the mean directions of storm translation. In all cases, the storm path is assumed to be a straight line.

Physical Models

1.   The maximum gradient speed is given by Eq. 1.2.8 in which $r = R$, and in which it is assumed that

$$\frac{dp}{dn} = \alpha \Delta p_{max} \tag{3.3.6}$$

where α is obtained for empirical data [3-26, 3-28].

2.   The maximum wind speed at 10 m above the ocean surface, averaged over 10 minutes, is given by the empirical relation

$$U(10, R) = 0.865 V_{gr}(R) + 0.5s \tag{3.3.7}$$

[3-26]. This relation corresponds to the average of data observed during the 1949 hurricane that crossed Lake Okeechobee, Florida [3-27, 3-28]. Whether Eq. 3.3.7 can be assumed to be generally valid is uncertain. However, it is likely that its use is conservative from a structural engineering viewpoint.

3.   Let the center of the storm be denoted by O, and consider a line OM

that makes an angle of 115° clockwise with the direction of motion of the storm. The 10-minute wind speed at 10 m above the ocean surface at a distance r from O along line OM is denoted by $U(10, r)$. The ratio $U(10, r)/U(10, R)$ is assumed to depend on r as shown in Fig. 3.3.7 [3-26]. Let the angle between a line ON and line OM be denoted by θ. The 10-minute wind speed $U(10, r, \theta)$ at a distance r from O along line ON is given by the expression [3-26]:

$$U(10, r, \theta) = U(10, r) - \frac{s}{2}(1 - \cos\theta) \qquad (3.3.8)$$

4.   The wind velocity vector has a component directed toward the center of the storm, O. The angle between that vector and the tangent to the circle centered at O varies linearly between 0° and 10° in the region $O < r < R$ and between 10° and 25° in the region $R \leqslant r < 1.2R$, and is equal to 25° in the region $r \geqslant 1.2R$ [3-26].

5.   The storm decay results from a decrease with time of the difference between pressure at the center and pressure at the periphery of the storm in accordance with the relation

$$\Delta p(t) = \Delta p_{max} - 0.02[1 + \sin\phi]t \qquad (3.3.9)$$

where t = travel time in hours, $\Delta p(t)$ and Δp_{max} are given in inches, and ϕ = angle between coast and storm track ($0 < \phi < 180°$). This model is consistent with measurements reported in [2-124].

6.   The reduction of wind speeds due to increased surface friction over land is given by the ratio $U^l(10)/U^w(10) = 0.85$, where $U^l(10)$ and $U^w(10)$ are the 10-

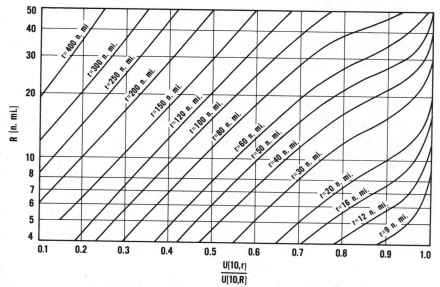

FIGURE 3.3.7.   Ratios $U(10, r)/U(10, R)$ [3-26].

minute speeds at 10 m elevation over land and over water, respectively. It can be verified that the model developed for extratropical storms (Eqs. 2.2.29 and 2.2.31, and Table 2.2.3) would yield a somewhat smaller ratio $U^l(10)/U^w(10)$.

7.   The dependence of wind speeds upon averaging time is modeled as in Sect. 2.3.6.

Note that physical models proposed in [3-28] are in some cases slightly modified with respect to the corresponding models of [3-26].

Estimates of the probabilities of occurrence of hurricane wind speeds were obtained in [3-24] by assuming each of the areas adjoining 56 mileposts (Fig. 3.3.5) to be hit by $m = 1000$ hurricanes. The climatological characteristics of the hurricanes were determined by Monte Carlo simulation from the respective probabilistic models as fitted to historical data. For each of the m hurricanes, the climatological characteristics used in conjunction with the physical models described earlier define a wind field which depends upon the position of the hurricane. To each position of the hurricane with respect to the site of interest there corresponds a wind speed at that site. Wind speeds caused by a hurricane at the site are calculated for a sufficiently large number of such positions. The largest among these speeds is the maximum wind speed caused by the hurricane at the site. A set of m speeds is thus obtained, which is used as the basic set of data for the estimation of the probability of occurrence of hurricane wind speeds. These speeds are ranked by magnitude. The i-th smallest speed in a set of m wind speeds is denoted by u_i.

Let the probability that the wind speed in any one storm is less than some value, u, be denoted by F_u. The probability that the highest wind U in n storms is less than u can be written as

$$F(U < u|n) = F_u^n \qquad (3.3.10)$$

The probability that $U < u$ in τ years is denoted by $F(U < u, \tau)$. The total probability theorem (Eq. A1.5) yields

$$F(U < u, \tau) = \sum_{n=0}^{\infty} F(U < u|n)p(n, \tau) \qquad (3.3.11)$$

where $p(n, \tau)$ denotes the probability that n storms will occur in τ years. Assuming that $p(n, \tau)$ is a Poisson process (Eq. A1.34), Eq. 3.3.11 becomes

$$F(U < u, \tau) = \sum_{n=0}^{\infty} F_u^n \frac{(\lambda\tau)^n e^{-\lambda\tau}}{n!} \qquad (3.3.12a)$$

$$= e^{-\lambda\tau} \sum_{n=0}^{\infty} \frac{(\lambda\tau F_u)^n}{n!} \qquad (3.3.12b)$$

$$= e^{-\lambda\tau(1 - F_u)} \qquad (3.3.12c)$$

where λ is the annual rate of occurrence of hurricanes in the area of interest for the site being considered. For $\tau = 1$, $F(U < u, \tau)$ is the probability of occurrence of wind speeds less than u in any one year.

Consider now the wind speed, u_i. The probability that $U < u_i$ in any one storm is

$$F_{u_i} = \frac{i}{m+1}$$

(3.3.13)

Thus,

$$F(U < u_i, 1) = e^{-\lambda \left(1 - \frac{i}{m+1}\right)}$$

(3.3.14)

For each of the mileposts in Fig. 3.3.5, estimates of hurricane wind speeds corresponding to various probabilities of occurrence (or mean recurrence intervals) were obtained in [3-24] both at the coastline and at various distances inland from the coastline. These estimates were used to develop the wind speed map included in [2-49] (Fig. 3.2.1). A summary of estimated wind speeds over land near the coastline is listed in Table 3.3.1, which also includes a set of estimates obtained independently in [3-29]. The estimates of [3-29] are in most cases close to those of [3-24], although there are a few exceptions: for example, near milestone 250, where the 50-year speed in [3-24] is higher by about 10% than in [3-29]; and near milestones 800, 1100, and 1400, where the 50-year speeds in [3-24] are lower by about 10% than in [3-29].

Estimated hurricane wind speeds differ from the "true" speeds owing to observation, probabilistic modeling, physical modeling, and sampling errors (i.e., errors due to the limited size of the data samples being used). Sensitivity studies reported in [3-24] suggest that the effect of probabilistic modeling errors is not significant. For wind speeds with mean recurrence intervals of the order of 50 years, it was shown in [3-30] that the standard deviation of the sampling errors is about 10% of the estimated speeds.

Parameters of the Weibull distributions that best fit coastline wind speeds generated by Monte Carlo simulation at each of 56 mileposts (Fig. 3.3.5) are given in [3-50] in knots. (Owing to a misprint, the estimated shape parameter for milepost 2200 is listed in [3-50] as $\gamma = 3.0$. The authors have verified that the correct value for the estimate is $\gamma = 5.0$.)

Procedures for estimating hurricane wind speeds that are similar conceptually to [3-23] were developed in [3-31] and [3-32] for the purpose of studying hurricane-generated waves.

Mixed Distributions.   Hurricane-prone regions are also subjected to winds not associated with hurricanes (or tropical cyclones), whose effects can be accounted for by developing mixed distributions of hurricane and non-hurricane wind speeds. Since the occurrence of hurricane winds and the occurrence of non-hurricane winds are independent events, it is possible to write

$$F(U < u) = F_H(U < u) F_{NH}(U < u)$$

(3.3.15)

where $F(U < u)$ is the probability that the wind speeds U associated with any storm are less than u in any one year, and $F_H(U < u)$ and $F_{NH}(U < u)$ are the probabilities that hurricane wind speeds and non-hurricane wind speeds are

TABLE 3.3.1. Estimated Hurricane Wind Speeds at 10 m above Ground in Open Terrain near Coastline for Various Mean Recurrence Intervals, $\bar{N}$

Milestone[a]	$\bar{N} = 25$ yr		$\bar{N} = 50$ yr		$\bar{N} = 100$ yr		$\bar{N} = 2000$ yr	
	Ref 3-24 mph[b](m/sec[c])	Ref 3-29[d] (m/sec[c])	Ref 3-24 mph[b](m/sec[c])	Ref 3-29[d] (m/sec[c])	Ref 3-24 mph[b](m/sec[c])	Ref 3-29[d] (m/sec[c])	Ref 3-24 mph[b](m/sec[c])	Ref 3-29[d] (m/sec[c])
150	85(30)	30	98(34)	33	107(37)	36	140(47)	51
200	84(29)	27	97(34)	31	106(36)	34	140(47)	48
250	85(30)	27	97(34)	30	106(36)	34	139(46)	45
300	84(29)	28	95(33)	31	105(36)	35	136(46)	47
400	81(28)	30	92(32)	32	101(35)	36	134(45)	48
500	78(27)	30	90(31)	33	100(34)	36	133(45)	49
600	80(28)	31	91(32)	33	101(35)	37	131(44)	49
700	82(29)	31	92(32)	33	101(35)	37	132(44)	50
800	82(29)	31	91(32)	35	101(35)	39	135(45)	51
850	81(28)	31	91(32)	35	101(35)	39	136(46)	52
900	78(27)	30	89(31)	34	97(34)	37	128(44)	50
1000	76(27)	28	83(29)	31	90(31)	34	117(40)	46
1050	77(27)	27	85(30)	30	91(32)	34	116(40)	45
1100	77(27)	29	87(30)	34	93(32)	36	127(43)	48
1200	85(30)	32	95(33)	35	105(36)	39	139(46)	51
1300	94(33)	32	105(36)	36	112(38)	40	137(46)	55

1400	97(34)	35	106(36)	39	114(39)	43	141(47)	57
1500	94(33)	34	103(35)	38	112(38)	43	140(47)	56
1600	89(31)	31	99(34)	35	107(37)	40	133(45)	54
1700	82(29)	27	91(32)	31	98(34)	33	118(40)	45
1750	77(27)	26	86(30)	29	92(32)	31	112(38)	44
1800	77(27)	27	87(30)	31	94(33)	36	121(41)	46
1900	82(29)	29	94(33)	33	105(36)	35	141(47)	49
2000	86(30)	29	97(34)	32	106(36)	35	136(46)	48
2100	87(30)	30	98(34)	33	107(37)	36	137(46)	49
2200	87(30)	30	98(34)	33	106(36)	36	134(45)	48
2300	76(27)	26	86(30)	30	95(33)	33	130(44)	45
2400	65(23)	25	78(27)	29	91(32)	31	126(43)	43
2500	63(22)	27	86(30)	30	97(34)	32	134(45)	47
2600	82(29)	29	96(33)	32	106(36)	35	141(47)	50
2700	81(28)	27	94(33)	31	106(36)	35	141(47)	48
2800	68(24)	24	84(29)	28	95(33)	31	133(45)	45
2900	59(21)	21	74(26)	25	86(30)	28	129(43)	41

[a] See Fig. 3.3.5.
[b] Fastest-mile wind speeds.
[c] Hourly wind speeds.
[d] P. N. Georgiou, A. G. Davenport, and B. J. Vickery, "Design Wind Speeds in Regions Dominated by Tropical Cyclones," *J. Wind Eng. Industr. Aerodyn.,* **13** (1983), 139–152.

less than u in any one year. The probability F_H is determined as shown previously in this section. The probability F_{NH} is determined as shown in Sect. 3.2.

Calculations reported in [3-24] show that the probability $F(U < u)$ is virtually the same as $F_H(U < u)$ for mean recurrence intervals $\bar{N} > 50$ years. For $\bar{N} \simeq 20$ years, estimated wind speeds that include the effect of non-hurricane winds exceed the estimated hurricane wind speeds by about 5%. Note that these conclusions are not necessarily applicable north of Cape Hatteras, where non-hurricane winds may control the design at certain locations [3-33].

3.4   WIND DIRECTIONALITY

Wind effects on various structures and components depend not only on the magnitude of the wind speeds, but on the associated wind directions as well. For this reason, knowledge of continuous joint probability distributions of extreme wind speeds and directions would be useful for design and code development purposes. However, so far no credible models for such distributions have been proposed in the literature.

In the absence of such models, wind effects and their probability distributions may be estimated in well-behaved wind climates on the basis of information consisting of largest yearly wind speed data recorded for each octant over periods of 20 years, say, or longer (see Sects. 8.1.2 and 8.1.3). Such data have been published for a number of U.S. weather stations in [3-34]. Summary statistics of largest yearly wind data recorded at Sheridan, Wyoming in the period 1958–1977 (see Table 8.1.2) are shown in Fig. 3.4.1. It is seen that in this case winds blowing from the northeast are considerably weaker than northwest or southwest winds.

As shown in Sect. 8.1.3, there are important practical applications in which information is needed on the univariate probability distributions of the largest yearly wind speeds associated with each of the principal compass directions, and on the correlation coefficients for the largest yearly winds blowing from any two directions. In well-behaved climates the largest yearly wind speeds for any given direction are in most cases—though not always—best fitted by Type I distributions of the largest values. As indicated in [8-14], the correlation between wind speeds occurring in any two of the eight principal compass directions is in most cases weak. For example, the estimated correlation coefficients between wind speeds from directions i and $j(i, j = 1, 2, \ldots, 8)$ are shown for Sheridan, Wyoming in Table 3.4.1. These values are fairly typical. However, there are stations where the values of the correlation coefficients are higher (e.g., Detroit, Michigan, where 8 of the 28 estimated values are larger than 0.45, although none exceeds about 0.6).

An important practical problem faced by the designer is obtaining the largest yearly wind speed data for each of the eight principal compass directions at locations not covered in [3-34]. There are two such sources of data. One source consists of the original unpublished records stored by the National Oceanic

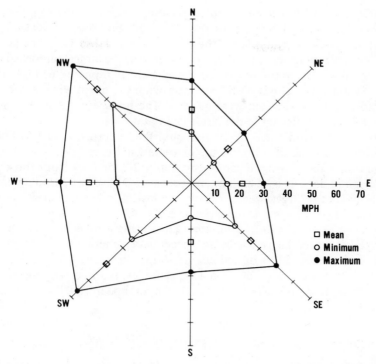

FIGURE 3.4.1. Summary statistics of largest yearly wind speeds by direction at Sheridan, Wyoming (1958–1977).

TABLE 3.4.1. Estimated Correlation Coefficients for Directional Wind Speeds in Sheridan, Wyoming

Direction	1	2	3	4	5	6	7	8
1	1	−0.05	−0.35	0.12	0.16	−0.031	−0.22	0.07
2		1	0.01	−0.15	−0.34	−0.03	0.07	−0.01
3			1	0.04	0.34	0.10	−0.12	−0.43
4				1	0.17	0.03	−0.16	0.40
5		symmetric			1	0.03	−0.16	−0.41
6						1	0.20	0.01
7							1	0.32
8								1

and Atmospheric Administration (NOAA). Obtaining and extracting the needed data from these records is both inconvenient and time consuming. A second source consists of published Local Climatological Data summaries issued monthly by NOAA. Directional largest yearly speeds in the published data differ in a few cases from the corresponding speeds in the original records. The reason for these differences is that the published data consist of (a) the largest daily speed for every day of the year and (b) the direction for that speed. Consider, for example, the case where in a given year the largest published speeds for winds blowing from the north and the east are 70 mph and 65 mph, respectively. It is conceivable that on the same day that the north wind occurred, the winds blowing from the east were 69 mph. The highest wind speed from the east would not be reflected in the published data.

An exhaustive study of original and published data listed in [3-34] for 24 stations showed conclusively that the extreme wind speed estimates based on published data differ insignificantly (by about 3% or less) from those based on the original data. It is, therefore, appropriate to base structural engineering calculations on the largest yearly directional fastest-mile wind speeds obtained from Local Climatological Data summaries [8-14].

In hurricane-prone regions estimates of hurricane wind effects can be carried out on the basis of hurricane wind speed data generated by Monte Carlo simulation for each of 16 directions, as shown in Sects. 3.3 and 8.1.3 (Eqs. 8.1.21–8.1.23). Such data—used in [3-24] for estimating extreme hurricane winds blowing from any direction—are listed on tape in [8-9] for 56 mileposts (Fig. 3.3.5).

3.5 PROBABILITIES OF OCCURRENCE OF TORNADO WINDS

Consider an area A_0, say, a one-degree longitude-latitude square, and let the tornado frequency in that area (i.e., the average number of tornado occurrences per year) be denoted by $\bar{n}$. The probability that a tornado will strike a particular location during one year is assumed to be

$$P(S) = \bar{n}\,\frac{\bar{a}}{A_0} \tag{3.5.1}$$

where $\bar{a}$ is the average individual tornado area. In certain applications, for example, the design of nuclear power plants, rather than the probability $P(S)$, it is of interest to estimate the probability $P(S, V_0)$ that a tornado with maximum wind speeds higher than some specified value V_0 will strike a location in any one year. This probability can be written as

$$P(S, V_0) = P(V_0)P(S) \tag{3.5.2}$$

where $P(V_0)$ is the probability that the maximum wind speed in a tornado will be higher than V_0.

Estimates of probabilities $P(S)$ in the United States are shown in Fig. 3.5.1, which is taken from [3-35]. Figure 3.5.1 is based on Eq. 3.5.1 in which $\bar{n}$ was estimated from 13-year frequency data, $\bar{a} = 2.82$ sq. miles (as estimated in [3-36] for the state of Iowa), and $A_0 = 4780\cos\phi$, where ϕ is the latitude at the center of the one-degree square considered. Estimated probabilities $P(V_0)$ are shown in Fig. 3.5.2, also taken from [3-35]. These estimates are based upon

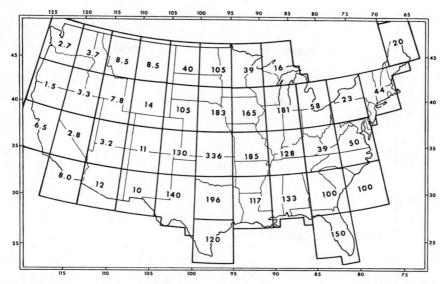

FIGURE 3.5.1.   Tornado strike probability within five-degree squares in the contiguous United States (units are 10^{-5} probability per year) [3-35].

observations of 1612 tornadoes during 1971 and 1972, and the rating (largely subjective) of these tornadoes according to an intensity scale proposed in [3-37].* It is noted that in estimating the probabilities of Fig. 3.5.2 it was assumed that tornado path areas are the same throughout the contiguous United States.

The maximum speed of the tornado corresponding to a specified probability of occurrence can be estimated using Figs. 3.5.1 and 3.5.2. According to [3-35], "in order to adequately protect public health and safety, the determination of the design basis tornado is based on the premise that the probability of occurrence of a tornado that exceeds the Design Basis Tornado (DBT) should be on the order of 10^{-7} per year per nuclear power plant." The required probability $P(V_0)$ is then determined from the relation

$$P(V_0)P(S) = 10^{-7} \tag{3.5.3}$$

where the value of $P(S)$ for the location considered is taken from Fig. 3.5.1. The wind speed corresponding to the probability $P(V_0)$ so determined is then obtained from Fig. 3.5.2. The average tornado intensity with a 10^{-7} probability per year for each 5-degree square in the contiguous United States, based on Eq. 3.5.3 and Figs. 3.5.1 and 3.5.2, is shown in Fig. 3.5.3 [3-35].

For nuclear power plant design purposes, the contiguous United States

---

*According to this scale tornadoes may be divided into the following classes: F0 (maximum wind speed <72 mph), F1 (73–112 mph), F2 (113–157 mph), F3 (158–206 mph), F4 (207–260 mph), F5 (261–318 mph), and F6 (319–380 mph).

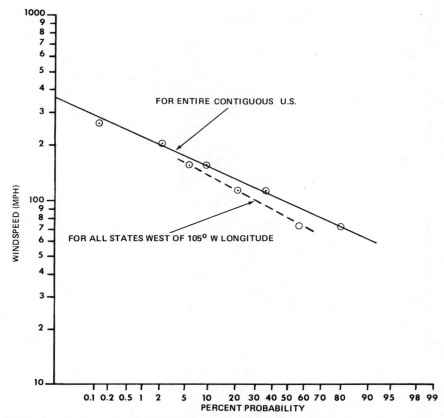

FIGURE 3.5.2. Percent probability of exceeding ordinate value of the wind speed [3-35].

are divided, in [3-35], into three tornado intensity regions shown in Fig. 3.5.4. The corresponding tornado winds are given in Table 3.5.1.

The pressure drop due to the passage of tornadoes can be estimated from the equation for the cyclostrophic wind. Using the relation $V_{tr} = dr/dt$, Eq. 1.3.2

TABLE 3.5.1. Regional Tornado Winds

Region	Maximum Speed V_{max} (mph)	Rotational Speed V_{rot} (mph)	Translational Speed V_{tr} (mph)	Radius of Maximum Rotational Wind Speed R_m (ft)
I	360	290	70	150
II	300	240	60	150
III	240	190	50	150

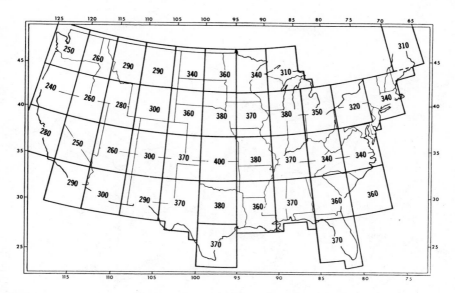

FIGURE 3.5.3. Calculated tornado wind speed by five-degree squares for 10^{-7} probability per year [3-35].

TABLE 3.5.2. Regional Pressure Drops and Pressure Drop Rate

Region	Total Pressure Drop (psi)	Rate of Pressure Drop (psi/sec)
I	3.0	2.0
II	2.25	1.2
III	1.5	0.6

can be written as

$$\frac{dp}{dt} = \frac{V_{tr}}{R_m} \rho V_t^2 \tag{3.5.4}$$

where p is the pressure, t is the time, V_{tr} is the translational speed, ρ is the air density, R_m is the radius of maximum rotational wind speed, and V_t is the maximum tangential wind speed* [3-35]. Assuming R_m is typically 150 ft for intense tornadoes and that $V_t \simeq V_{rot}$, Eq. 3.5.4, in which the parameters of Table 3.5.1 are used, yields approximately the values of Table 3.5.2 [3-35].

Following the development in [3-35] of the estimates summarized in Tables 3.5.1 and 3.5.2, various attempts to improve the probabilistic and physical description of tornado winds have been reported [3-38, 3-39, 3-40, 3-41, 3-42,

---

*The rotational speed V_{rot} is the resultant of the tangential and radial velocities.

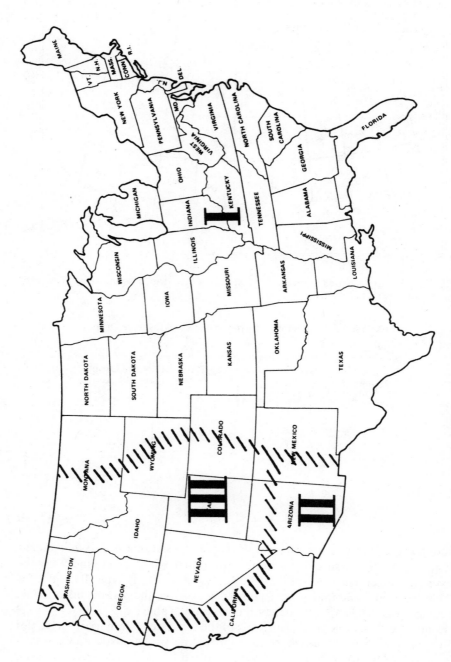

FIGURE 3.5.4. Tornado intensity regions [3-35].

3-43, 3-44, 3-45, 3-46, 3-47]. Using as a point of departure tornado risk maps presented in [3-46], a regionalization of tornado risks which divides the contiguous United States into four areas was proposed in [3-45] (see also [3-44, p. 480]. Regional tornado occurrence rate (per mi^2 per year) were estimated in [3-45] from a 29-year (1950-1978) data bank maintained by the National Severe Storms Forecast Center and comprising about 20,000 reported tornadoes. These regional occurrence rates are corrected in [3-43] and [3-45] to account for:

1. Failure to record tornado intensity, which affects about 10% of the total number of reported tornadoes. This correction is based on the assumption that unrated tornadoes may be apportioned among the various intensity categories according to the reported tornado frequencies for those categories.

2. Temporal variations in tornado reporting efficiency. The number of reported annual tornado occurrences in the United States has increased from about 250 in 1950 to 850 in 1979. The growing trend in the number of reported tornadoes during this period has been ascribed to a corresponding increase in population density. An explicit relation to this effect has been proposed in [3-47]. Corrections accounting for tornado reporting efficiencies were effected in [3-45] by averaging the 1971–1978, 1970–1978, 1969–1978, and 1950–1978 data and assuming that the true occurrence rates are equal to the largest of these estimates.

3. Possible errors in the rating of tornado intensities on the basis of observed damage. The reason for the occurrence of such errors is that maximum tornado winds are in practice not measured, but inferred, largely on the basis of professional judgment, from observations of damage to buildings, signs, and so forth [3-42].

4. Inhomogeneous distribution along the tornado path of buildings and various other objects susceptible of being damaged. In the possible absence of such objects over the portions of the tornado path where the winds are highest—or even over the entire tornado path—the rating of the tornado is bound to be in error. The effect of corrections for such errors is to increase the estimated probability of occurrence of tornadoes with higher intensities.

5. Variation of tornado intensity along the tornado path. Accounting to this factor results in smaller estimated risks of high tornado winds than would be the case if the maximum tornado winds (by which tornado intensities are rated) were uniform along the entire path. Corrections effected in [3-45], based upon the analysis of documented tornadoes, led to risk reductions by a factor of about five for F4 tornadoes and about ten for F6 tornadoes.

The corrections for the factors listed involve subjective judgments that may be formalized by Bayesian techniques (see Eq. A1.6). In [3-45] the corrected rates of occurrence differ insignificantly from the uncorrected (prior) rates, with the following exceptions. For the three areas of the regionalization map proposed in [3-45] in which the most intense tornadoes recorded in the period 1950–1978 were F5, it was estimated in [3-45] that rates of occurrence of F6 tornadoes, rather than being zero, are about 1/20 times the rates of occurrence of F5 tornadoes. For the fourth area of that map, in which the most intense tornadoes recorded in the same period were F4, it was estimated that the corrected rate of occurrence of F4 tornadoes is about six times the uncorrected rate, and that the rate of occurrence of F5 tornadoes, rather than being zero, is 1/20 times the corrected rate of occurrence of F4 tornadoes.

Reference 3-43 suggests that the velocity ranges associated in [3-37] with the tornado ratings F1 through F6 (see p. 111) are overconservative by amounts varying from about 5% for F1 tornadoes to about 20% or more for F6 tornadoes. The wind speed reductions proposed in [3-43] are used in [3-45] as a basis for suggesting a reduction of the 360 mph, 300 mph, and 240 mph wind speeds, specified in [3-35] for regions I, II, and III of Fig. 3.5.4, to 300 mph, 225 mph, and 200 mph, respectively. In the authors' opinion, the arguments adduced in [3-43] in favor of such reductions are largely tentative, in some instances at least. For example, to support the contention that the maximum wind speeds in a tornado classified as F3 are lower than the values proposed in [3-37], [3-43] interprets a 133–167 mph estimate of the velocity causing the collapse of a chimney during the Xenia, Ohio tornado of 3 April, 1974 [3-42, p. 1715] simply as a 133 mph estimate [3-43, p. 1625]. On the other hand, it should be noted that the estimates of [3-35] and [3-37] are also tentative.

A position that is to some extent a compromise between [3-35] and [3-45] was adopted in the American National Standard ANSI/ANS-2.3-1983 [3-48], which divides the contiguous United States into three zones, denoted as 1, 2, and 3. Zones 1 and 2 cover, approximately, region I of Fig. 3.5.4, while zone 3 covers approximately regions II and III. Table 3.5.3 lists the maximum tornado wind speeds V_{max}, the translational wind speeds V_{tr}, the radius of the maximum wind speed R_m, and the maximum atmospheric pressure drop p_a,

TABLE 3.5.3. Standard Tornado Characteristics (Extracted from American National Standard ANSI/ANS-2.3-1983 with permission of the publisher, the American Nuclear Society)

Probability of Exceedance, per Year	Zone	V_{max} (mph)	V_{tr} (mph)	R_{max} (ft)	p_a (psi)
10^{-7}	1	320	70	540	1.96
	2	250	55	435	1.35
	3	180	40	320	0.70
10^{-6}	1	260	57	453	1.46
	2	200	45	355	0.85
	3	140	32	253	0.41
10^{-5}	1	200	45	355	0.85
	2	150	33	270	0.47
	3	100	25	185	0.20

given in [3-48] for tornadoes corresponding to various probabilities of exceedance.

It was noted in [3-40] that probabilities of a target being hit by a tornado wind in excess of any specified threshold depend upon the size of that target. This topic is analyzed in detail in [3-44, 3-45], where the estimates are based upon statistics of tornado intensities, path lengths, and path widths on the one hand, and on the geometric characteristics of the target on the other. It is suggested in [3-49] that tornado wind loads dominate the design of most transmission lines over 10 miles in length over wide areas of the United States.

REFERENCES

3-1   A. Court, "Wind Extremes as Design Factors," *J. Franklin Inst.*, **256** (July 1953) 39–55.

3-2   *Manual of Surface Observations*, U.S. Weather Service, Washington, D.C., 1951, p. 92.

3-3   *Selective Guide to Climatic Data Sources*, Key to Meteorological Records Documentation No. 4.11, Environmental Data Service, U.S. Department of Commerce, Washington, D.C., 1969.

3-4   J. Wierenga, "An Objective Exposure Correction Method for Average Wind Speeds Measured at a Sheltered Location," *J. Royal Meteorol. Soc.*, **102** (1976) 241–253.

3-5   H. C. S. Thom, "New Distributions of Extreme Wind Speeds in the United States," *J. Struct. Div.*, ASCE, **94**, No. ST7, Proc. Paper 6038 (July 1968) 1787–1801.

3-6   E. Simiu and J. J. Filliben, *Statistical Analysis of Extreme Winds*, Technical Note No. 868, National Bureau of Standards, Washington, D.C., 1975.

3-7   E. Simiu, J. Biétry, and J. J. Filliben, "Sampling Errors in the Estimation of Extreme Winds," *J. Struct. Div.*, ASCE **104**, No. ST3 (March 1978) 491–501.

3-8   E. Simiu and J. J. Filliben, "Weibull Distributions and Extreme Wind Speeds," *J. Struct. Div.*, ASCE, No. ST12 (Dec. 1980), 2365–2374; (errata in *J. Struct. Div.*, ASCE, **107**, No. ST4, Proc. Paper 16152 (April 1981), 716–717.

3-9   E. Simiu, M. Changery, and J. J. Filliben, *Extreme Wind Speeds at 129 Stations in the Contiguous United States*, NBS Building Science Series 118, U.S. Department of Commerce, National Bureau of Standards, Washington, D.C., March 1979.

3-10 E. Simiu, J. Shaver, and J. J. Filliben, "Wind Speed Distributions and Reliability Estimates," *J. Struct. Div.*, ASCE, **107**, No. ST5, Proc. Paper (May 1981), 1003–1007; (errata in *J. Struct. Div.*, ASCE, **107**, No. ST10, Proc. Paper 16527 (Oct. 1981), 2052.

3-11 V. L. Cardone, A. J. Broccoli, C. V. Greenwood, and J. A. Greenwood, "Error Characteristics of Extratropical-Storm Wind Fields Specified from Historical Data," *J. Petroleum Technol.*, (May 1980) 872–880.

3-12 E. Simiu, J. J. Filliben, and J. R. Shaver, "Short-Term Records and Extreme Wind Speeds," *J. Struct. Div.*, ASCE, **108**, No. ST11 (Nov. 1982), 2571–2577.

3-13 M. Grigoriu, "Estimation of Extreme Winds from Short Records," *J. Struct. Eng.*, **110**, No. 7 (July 1984) 1467–1484.

3-14 M. Grigoriu, "Estimates of Design Winds from Short Records," *J. Struct. Div.*, ASCE, **108**, No. ST5 (May 1982), 1034–1048.

3-15 H. L. Crutcher, "Wind Extremes," in *Proceedings of the Second U.S. National Conference on Wind Engineering Research*, Colorado State University, Fort Collins, Colo., 1975.

3-16 A. L. Sugg, L. G. Pardue, and R. L. Carrodus, *Memorable Hurricanes of the United States*, National Weather Service, Southern Region, NOAA Technical Memorandum NWS SR-56, Fort Worth, Tex., 1971.

3-17 H. C. S. Thom, "Toward a Universal Climatological Extreme Wind Distribution," in *Proceedings of the International Research Seminar on Wind Effects on Buildings and Structures*, Vol. 1, University of Toronto Press, Toronto, Canada, 1968.

3-18   "The Hurricane Disaster Potential Scale," *Weatherwise* **27**, 4, (Aug., 1974) 169, 186.

3-19   C. W. Cry, *Tropical Cyclones of the North Atlantic Ocean—Tracks and Frequencies of Hurricanes and Tropical Storms, 1871–1963*, Technical Paper No. 55, U.S. Department of Commerce, Weather Bureau, Washington, D.C., 1965.

3-20   G. E. Dunn and B. J. Miller, *Atlantic Hurricanes*, Louisiana State Univ. Press, Baton Rouge, La., 1960.

3-21   H. C. S. Thom and R. D. Marshall, "Wind and Surge Damage Due to Hurricane Camille," *J. Waterways, Harbors, and Coastal Eng. Div.*, ASCE, **97**, WW5(May 1971) 355–363.

3-22   F. P. Ho, R. W. Schwerdt, and H. V. Goodyear, *Some Climatological Characteristics of Hurricanes and Tropical Storms, Gulf and East Coasts of the United States*, NOAA Technical Report No. NWS 15, National Oceanic and Atmospheric Administration, Washington, D.C., May 1975.

3-23   L. R. Russell, "Probability Distributions for Hurricane Effects," *J. Waterways, Harbours, and Coastal Eng. Div.*, ASCE, **97**, WW2(Feb. 1971) 139–154.

3-24   M. E. Batts, L. R. Russell, and E. Simiu, "Hurricane Wind Speeds in the United States," *J. Struct. Div.*, ASCE, **100**, No. ST10(Oct. 1980), 2001–2015.

3-25   *Worldwide Extremes of Temperature, Precipitation, and Pressures Recorded by Continental Area*, ESSA/PI 680032, Environmental Data Service, U.S. Department of Commerce, October 1968.

3-26   *Revised Standard Project Hurricane Criteria for the Atlantic and Gulf Coasts of the United States*, Memorandum HUR7-120, U.S. Department of Commerce, National Oceanic and Atmospheric Administration, Washington, D.C., June 1972.

3-27   V. A. Myers, *Characteristics of United States Hurricanes Pertinent to Levee Design for Lake Okeechobee, Florida*, Hydrometeorological Report No. 32, U. S. Weather Bureau, Department of Commerce and U.S. Army Corps of Engineers, Washington, D.C., 1952.

3-28   R. W. Schwerdt, F. P. Ho, and R. R. Watkins, *Meteorological Criteria for Standard Project Hurricane and Probable Maximum Hurricane Windfields, Gulf and East Coasts of the United States*, NOAA Technical Report NWS23, U.S. Department of Commerce, National Oceanic and Atmospheric Administration, Washington, D.C., Sept. 1979.

3-29   P. N. Georgiou, A. G. Davenport, and B. J. Vickery, "Design Wind Loads in Regions Dominated by Tropical Cyclones," *Proceedings Sixth International Conference on Wind Engineering*, Feb. 1983, Gold Coast, Australia, in *J. Wind Eng. Ind. Aerod.*, **13** (1983), 139–152.

3-30   M. E. Batts, M. R. Cordes, and E. Simiu, "Sampling Errors in Estimation of Extreme Hurricane Winds," *J. Struct. Div.*, ASCE, **106**, No. ST10(Oct. 1980), 211–2115.

3-31   V. J. Cardone, W. J. Pierson, and E. G. Ward, "Hindcasting the Directional Spectra of Hurricane-Generated Waves," *J. Petroleum Technol.*, (April 1976), 385–394.

3-32   E. G. Ward, L. E. Borgman, and V. J. Cardone, "Statistics of Hurricane Waves in the Gulf of Mexico," *J. Petroleum Technol.*, (May 1979), 632–646.

3-33   C. S. Gilman and V. A. Myers, "Hurricane Winds for Design Along the New England Coast," *J. Waterways and Harbors Div.*, ASCE 87, WW5(May 1961) 45–65.

3-34   M. J. Changery, E. J. Dumitriu-Valcea, and E. Simiu, *Directional Extreme Wind Speeds for the Design of Buildings and Other Structures*, Building Science Series BSS 160, National Bureau of Standards, March 1984.

3-35   E. H. Markee, J. G. Beckerley, and K. E. Sanders, *Technical Basis for Interim Regional Tornado Criteria*, WASH-1300 (UC-11), U.S. Atomic Energy Commission, Office of Regulation, Washington, D.C., 1974.

3-36   H. C. S. Thom, "Tornado Probabilities," *Mon. Weather Rev.*, **17** (Dec. 1973) 730–736.

3-37   T. T. Fujita, *Proposed Characterization of Tornadoes and Hurricanes by Area and Intensity*, Satellite and Mesometeorology Research Project (University of Chicago), Research Paper No. 89, 1970.

3-38   Y. K. Wen and S. L. Chu, "Tornado Risks and Design Wind Speed," *J. Struct. Div.*, ASCE (Dec. 1973), 2409–2421.

3-39 J. R. Eagleman, V. U. Muirhead, and W. Willems, *Thunderstorms, Tornadoes and Building Damage*, Lexington Books, Heath, Lexington, Mass., 1975.

3-40 R. G. Garson, J. M. Catalan, and C. A. Cornell, "Tornado Design Winds Based on Risk," *J. Struct. Div.*, ASCE (Sept. 1975), 1883–1897.

3-41 R. F. Abbey, Jr., "Risk Probabilities Associated with Tornado Wind Speeds," *Proceedings Symposium on Tornadoes*, R. E. Peterson (Ed.), Texas Tech. University, Lubbock, Texas, June 22–24, 1976.

3-42 K. C. Mehta, J. E. Minor, and J. R. McDonald, "Wind Speed Analysis of April 3–4, 1974 Tornadoes," *J. Struct. Div.*, ASCE (Sept. 1976), 1709–1724.

3-43 L. A. Twisdale, "Tornado Characterization and Wind Speed Risk," *J. Struct. Div.*, ASCE (Oct. 1978), 1611–1630.

3-44 L. A. Twisdale and W. L. Dunn, "Probabilistic Analysis of Tornado Wind Risks," *J. Struct. Div.*, ASCE (Feb. 1983), 468–488.

3-45 L. A. Twisdale et al., *Tornado Missile Simulation and Design Methodology*, EPRI NP-2005, Electrical Power Research Institute, Palo Alto, California, Aug. 1981.

3-46 R. F. Abbey, Jr. and T. T. Fujita, "Regionalization of the Tornado Hazard," *Tenth Conference on Severe Local Storms*, American Meteorological Society, Oct. 1977, Omaha, Nebraska.

3-47 R. F. Abbey, Jr. and T. T. Fujita, "The Dapple Method for Computing Tornado Hazard Probabilities: Refinements and Theoretical Considerations," *Eleventh Conference on Severe Local Storms*, American Meteorological Society, Oct. 1979, Kansas City.

3-48 *American National Standard for Estimating Tornado and Extreme Wind Characteristics at Nuclear Power Sites*, ANSI/ANS-2.3-1983, American Nuclear Society, La Grange Park, Ill., 1983.

3-49 L. A. Twisdale, "Wind Loading Underestimates in Transmission Line Design," *Transmission and Distribution* (Dec. 1982), 40–46.

3-50 M. E. Batts, "Probabilistic Description of Hurricane Wind Speeds," *J. Struct. Div.*, ASCE (July 1982), 1643–1647.

3-51 Y. K. Wen and K. B. Rojiani, Discussion to "Sampling Errors in Estimation of Extreme Winds" by E. Simiu et al., *J. Struct. Div.*, ASCE **104** (1978), 1815–1817.

3-52 C. J. Neumann, G. W. Cry, E. L. Caso, and B. R. Jarvinen, *Tropical Cyclones of the North Atlantic Ocean*, 1871–1977, National Oceanic and Atmospheric Administration, National Climatic Center, Asheville, N.C., June 1978.

3-53 C. J. Neumann and M. J. Prsylak, *Frequency and Motion of Atlantic Tropical Cyclones*, NOAA Technical Report NWS 26, National Oceanic and Atmospheric Administration, Washington, D.C., March 1981.

3-54 B. R. Jarvinen, and E. L. Caso, *A Tropical Cyclone Data Tape for the North Atlantic Basin, 1886–1977: Contents, Limitations, and Uses*, NOAA Technical Memorandum NWS NHC 6, National Hurricane Center, National Oceanic and Atmospheric Administration, Coral Gables, Florida, 1978.

3-55 Z. Xue and C. J. Neumann, *Frequency and Motion of Western North Pacific Tropical Cyclones*, National Hurricane Center, National Oceanic and Atmospheric Administration, Miami, Florida, May 1984.

WIND LOADS AND THEIR EFFECTS ON STRUCTURES

I FUNDAMENTALS

4

Bluff-Body Aerodynamics

The subject of aerodynamics covers a very wide range. Of necessity, therefore, only a few highlights can be emphasized in the present chapter. The field received its great initial impulse from the efforts in the early twentieth century to achieve heavier-than-air flight. Since that time it has continuously received strong contributions from a great variety of aerospace studies, and from the sustained, intensive development of machines with internal flows, such as jet engines, pumps, and turbines.

In addition to these, interesting new advances in applications of aerodynamics to civil engineering structures have also occurred in the last three decades. Dealing as they do with the natural wind, these applications of aerodynamics are limited mainly to relatively low-speed, incompressible flow phenomena. In this application, aerodynamics is also closely associated with meteorology and concerned in particular with turbulent flows in the boundary layer of the earth's atmosphere.

In addition to a primary concern with the mean velocity of the wind, two aspects of these turbulent flows are of interest to the structural engineer: the state of turbulence of the natural wind approaching a structure and the local or "signature" turbulence provoked in the wind by the structure itself. Since most structures in civil engineering present bluff forms to the wind, emphasis is placed, in wind engineering, upon bluff-body aerodynamics. This fact, characteristic of a new situation not emphasized as strongly in aeronautical and other previous studies, has occasioned new research on the details of flow effects around bluff forms typical of such structures as buildings, towers, and bridges. In this context, interest centers particularly on details of the development of body pressures by the given flows.

In this chapter a few basic theoretical principles and experimental facts are reviewed that lay a foundation for the study of wind engineering.

4.1 GOVERNING EQUATIONS

4.1.1 Equations of Motion and Continuity

Consider a fixed elemental volume dV in a fluid. The vector velocity* of the fluid is commonly expressed by

$$\mathbf{u} = u\mathbf{i} + v\mathbf{j} + w\mathbf{k} \tag{4.1.1}$$

where $\mathbf{i}, \mathbf{j}, \mathbf{k}$ are unit vector components along the usual three fixed rectangular coordinate axes x, y, z. For compactness of notation let x, y, z be replaced respectively by x_1, x_2, x_3; u, v, w by u_1, u_2, u_3, and the unit vectors $\mathbf{i}, \mathbf{j}, \mathbf{k}$ by $\mathbf{i}_1, \mathbf{i}_2, \mathbf{i}_3$ so that Eq. 4.1.1 may be rewritten

$$\mathbf{u} = \sum_{i=1}^{3} u_i \mathbf{i}_i \tag{4.1.2}$$

The force acting on the fluid contained in the volume dV consists of two parts. The first part, referred to as the body force and caused by some force field, such as gravity, will be denoted by $\mathbf{F}\rho\, dV$, where ρ is the fluid density. The second part is due to the net action on the fluid of the internal stresses $\sigma_{ij}(i, j = 1, 2, 3)$. For example, the contribution to this action of the normal stress σ_{11} (see Fig. 4.1.1) is

$$-\sigma_{11}\, dx_2\, dx_3 + \left(\sigma_{11} + \frac{\partial \sigma_{11}}{\partial x_1}\, dx_1\right) dx_2\, dx_3 = \frac{\partial \sigma_{11}}{\partial x_1}\, dx_1\, dx_2\, dx_3$$

$$= \frac{\partial \sigma_{11}}{\partial x_1}\, dV \tag{4.1.3}$$

It can similarly be shown that the net force component in the i direction due to the action of all the stresses σ_{ij} is

$$\sum_{j=1}^{3} \frac{\partial \sigma_{ij}}{\partial x_j}\, dV \tag{4.1.4}$$

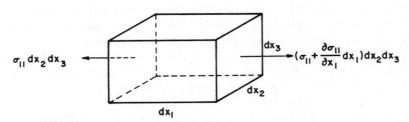

FIGURE 4.1.1.   Forces on an elementary volume of fluid.

---

*In applications where there exists a single important mean flow velocity accompanied by variable components, the mean flow is often taken as being in the x direction, with velocity designated as U, the respective x, y, z components then being designated as $U + u, v, w$.

Denoting, then, the components of **F** by $F_i (i = 1, 2, 3)$ the force balance equations, given by Newton's second law, are

$$\frac{Du_i}{Dt} \rho \, dV = F_i \rho \, dV + \sum_{j=1}^{3} \frac{\partial \sigma_{ij}}{\partial x_j} dV \qquad (i = 1, 2, 3) \qquad (4.1.5)$$

where the operator D/Dt, known as the *substantial* or the *material* derivative, is defined as follows:

$$\frac{D}{Dt} = \frac{\partial}{\partial t} + \sum_{j=1}^{3} u_j \frac{\partial}{\partial x_j} \qquad (4.1.6)$$

Since Eq. 4.1.5 is true for all volume elements, the factor dV may be divided out of Eq. 4.1.5 and the *equations of motion*, in component form, of a fluid particle can be written as

$$\rho \frac{Du_i}{Dt} = \rho F_i + \sum_{j=1}^{3} \frac{\partial \sigma_{ij}}{\partial x_j} \qquad (i = 1, 2, 3) \qquad (4.1.7)$$

Various forms of this basic equation can be derived depending upon the nature of the forces F_i and stresses σ_{ij} acting upon the fluid particle.

Before examining these particular cases it will be useful to recall the principle of mass conservation. This principle states that the rate of increase of the fluid mass contained within a fixed closed surface must be equal to the difference between the rates of influx to and efflux from the volume enclosed by that surface. The equations of continuity can then be shown to be [4-1, 4-2]:

$$\sum_{i=1}^{3} \frac{\partial (\rho u_i)}{\partial x_i} = -\frac{\partial \rho}{\partial t} \qquad (4.1.8)$$

For an incompressible fluid wherein no change in density ρ occurs, this reduces to

$$\sum_{i=1}^{3} \frac{\partial u_i}{\partial x_i} = 0 \qquad (4.1.9)$$

4.1.2 The Navier–Stokes Equations

Unlike a solid, a fluid under static conditions is incapable of supporting any steady-state stresses other than normal pressure. In dynamic situations, on the other hand, it may support shear in a time-dependent manner. Most often, in fluid-mechanical applications, it has been adequate to assume then that the stresses involved are either normal pressures or ascribable to *viscosity* only. Fluids with internal shear stress proportional to the rate of change of velocity with distance normal to that velocity are termed *viscous* or *Newtonian*. For example the shear stress σ_{12} in the simple two-dimensional flow pictured in Fig. 4.1.2 is expressed as

$$\sigma_{12} = \mu \frac{\partial u_1}{\partial x_2} \qquad (4.1.10)$$

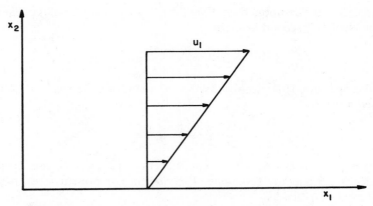

FIGURE 4.1.2.   Linear velocity increase with distance from a wall.

where the proportionality factor is defined as the fluid viscosity.*

Further, by dividing the whole stress *tensor* σ_{ij} at a fluid point into pressure stress (or simply *pressure*, that is, normal stress) p and *deviatoric* stress, defined as

$$d_{ij} = 2\mu \left(e_{ij} - \tfrac{1}{3}\delta_{ij} \sum_{k=1}^{3} e_{kk} \right) \quad (i, j = 1, 2, 3) \tag{4.1.11}$$

where

$$e_{ij} = \frac{1}{2}\left(\frac{\partial u_i}{\partial x_j} + \frac{\partial u_j}{\partial x_i} \right) \tag{4.1.12}$$

and

$$\delta_{ij} = \begin{cases} 1, & i = j \\ 0, & i \neq j \end{cases} \tag{4.1.13}$$

the following breakdown of stress σ_{ij} can be obtained:

$$\sigma_{ij} = -p\delta_{ij} + 2\mu \left(e_{ij} - \tfrac{1}{3}\delta_{ij} \sum_{k=1}^{3} e_{kk} \right) \tag{4.1.14}$$

Using this form of stress for a Newtonian fluid results in the equations of motion†

$$\rho \frac{Du_i}{Dt} = \rho F_i - \frac{\partial p}{\partial x_i} + \sum_{j=1}^{3} \frac{\partial}{\partial x_j} \left\{ 2\mu \left(e_{ij} - \tfrac{1}{3}\delta_{ij} \sum_{k=1}^{3} e_{kk} \right) \right\} \tag{4.1.15}$$

*The units of viscosity are:

$$\mu \equiv \frac{\text{force}}{\text{area}} \times \frac{\text{length}}{\text{velocity}} = \frac{\text{force time}}{\text{length}^2} = \frac{\text{mass}}{\text{length time}}$$

Typical values of μ for example for air and water at 20° are

$$\mu_{\text{air}} = 1.81 \text{ gm/cm sec}, \quad \mu_{\text{H}_2\text{O}} = 1.002 \text{ gm/cm sec}$$

Common units are *poises*, where 1 poise = 1 gm/cm sec = 0.002089 lbf sec/ft^2.

†For a more detailed account see, for example, [4-1], [4-2], [4-3], or [4-4].

Equations 4.1.15 ($i, j = 1, 2, 3$) are the well-known *Navier–Stokes equations*. If Eq. 4.1.12 is used, and if the viscosity μ may be considered to be constant throughout the fluid, then Eqs. 4.1.15 become

$$\rho \frac{Du_i}{Dt} = \rho F_i - \frac{\partial p}{\partial x_i} + \mu \left(\sum_{j=1}^{3} \frac{\partial^2 u_i}{\partial x_j^2} + \frac{1}{3} \frac{\partial \displaystyle\sum_{k=1}^{3} (\partial u_k / \partial x_k)}{\partial x_i} \right) \qquad (4.1.16)$$

Further simplification occurs in the case of an incompressible fluid, that is, one for which Eq. 4.1.9 holds. Equations 4.1.16 can then be written in the vector form

$$\rho \frac{D\mathbf{u}}{Dt} = \rho \mathbf{F} - \sum_{i=1}^{3} \frac{\partial p}{\partial x_i} \mathbf{i}_i + \mu \sum_{j=1}^{3} \frac{\partial^2 \mathbf{u}}{\partial x_j^2} \qquad (4.1.17)$$

4.1.3 Bernoulli's Equation

In the case of a fluid that, in addition to being incompressible, is also *inviscid* ($\mu = 0$) and acted upon by negligible body forces. Equation 4.1.17 reduces to

$$\rho \frac{D\mathbf{u}}{Dt} = -\sum_{i=1}^{3} \frac{\partial p}{\partial x_i} \mathbf{i}_i \qquad (4.1.18)$$

If the coordinate axes are so oriented that x_1 corresponds to the direction of motion, and if the flow is steady, it follows immediately from the integration of Eqs. 4.1.18 that

$$\tfrac{1}{2}|\mathbf{u}|^2 + \frac{p}{\rho} = \text{const} \qquad (4.1.19)$$

at every point of a streamline. Equation 4.1.19 is a special form of *Bernoulli's theorem* and is most commonly written as

$$\tfrac{1}{2}\rho u^2 + p = \text{const} \qquad (4.1.20)$$

where u is the flow velocity along a streamline. The quantity $\tfrac{1}{2}\rho u^2$ has the dimensions of pressure and is referred to as the *dynamic pressure*.

This important equation is widely used to interpret the relation between pressure and velocity in atmospheric and wind tunnel flows. Detailed comments Bernoulli's equation and its applicability in fluid flows—including flows in which viscosity is present—are provided in Sect. 3.5 of [4-3].

4.2 FLOW IN A CURVED PATH. VORTEX FLOW

Consider a two-dimensional flow between two locally concentric streamlines a distance dr apart and having radius of curvature r (Fig. 4.2.1). In order for the flow to maintain its curved path it must experience an acceleration toward the center of curvature of the streamlines of amount u^2/r, where u is here used to designate the local tangential velocity of the flow. Let the pressure acting on the fluid element under consideration be denoted by p. The pressure differential from one streamline to the next along r, which is responsible for this acceleration,

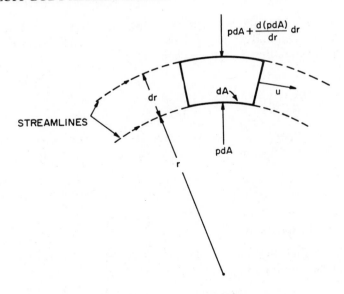

FIGURE 4.2.1.   Flow in a curved path.

is dp. The equation of motion for the fluid element is then

$$dp\,dA = \rho\,dr\,dA\,\frac{u^2}{r}$$

where ρ is the fluid density and dA is the area of the element in a plane normal to the plane of the figure. This relation indicates that the pressure change normal to the streamlines of a curved flow in the absence of any other forces are

$$dp = \rho u^2\,\frac{dr}{r} \tag{4.2.1}$$

Bernoulli's equation (4.1.20) then permits calculation of the pressure along a curved path of such a streamline flow.

In particular, one may consider the case wherein the flow is completely circular and the value of p_0 in Eq. 4.1.20 is the same on all streamlines. This is the case of *vortex flow*. Differentiation of Eq. 4.1.20 yields

$$\rho u\,\frac{du}{dr} + \frac{dp}{dr} = 0 \tag{4.2.2}$$

which, when combined with Eq. 4.2.1, yields

$$\frac{du}{u} = -\frac{dr}{r} \tag{4.2.3}$$

Equation 4.2.3 can then be integrated to yield

$$ur = C = \text{const} \tag{4.2.4}$$

This simple law states for an incompressible, inviscid fluid the theoretical (hyperbolic) relation between positional radius r and tangential velocity u in a *free vortex.*

In an actual free vortex, however, the effects of viscosity are present as well. They have not been included in the simple derivation above. These will have, in part, the effect of "locking" some portion of the fluid (near the center) together and causing it to rotate as a rigid body instead of as the perfect fluid described by Eq. 4.2.4. Thus locally, near the center of a free vortex, the velocity u *increases* with radius, whereas according to Eq. 4.2.4 it *decreases* with increasing r. This latter condition actually holds outward from a *transition region* in which u attains its maximum value. The value of u in such a region is dependent on the values of the fluid viscosity and of the total angular momentum of the vortex. Figure 4.2.2 illustrates qualitatively the pressure and velocity relations that hold in a free vortex occurring in a real fluid. It should be noted that the free vortex here described differs from the forced or *constrained vortex* that may develop in a fluid held in a rotating container.

The free vortex is of interest in many flows that occur in engineering applica-

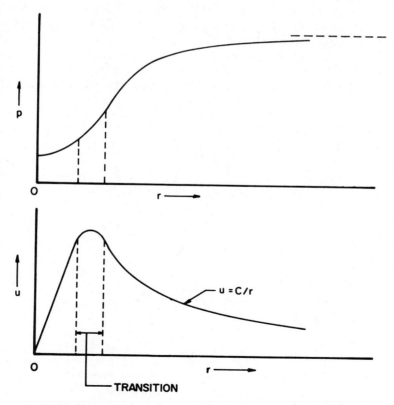

FIGURE 4.2.2.   Pressure and velocity distribution in a vortex flow.

tions. For example, atmospheric flows along the curved isobars of the weather map are described by generalizations of Eq. 4.2.1. These have been described in Sect. 1.2, where additional Coriolis forces have been included.

4.3   BOUNDARY LAYERS AND SEPARATION

The range of viscosity values to be found among various fluids is very great. The viscosity of air at normal meteorological pressures and temperatures however has a relatively small value. Nonetheless, in some circumstances this small viscosity plays an important role. An important manifestation of the viscous effects of air occurs in the formation of *boundary layers*.

Consider an air flow over and along a stationary smooth surface. It is an experimental fact that the air in contact with the surface adheres to it. This causes a retardation of the air motion in a layer near the surface referred to as the boundary layer. Within the boundary layer the velocity of the air increases from zero at the surface (no slip) to its full value, which corresponds to the external (as opposed to boundary layer) flow [2-1]. A boundary layer velocity profile is depicted in Fig. 4.3.1.

Air, since it has mass, evidences inertial effects according to Newton's second law (or, more specifically, the Navier–Stokes equations). The two most influential effects in an air flow are then viscous and inertial, and the relation of these to each other becomes an index of the type of flow characteristics or phenomena that may be expected to occur. This index can be expressed as a nondimensional parameter $\mathscr{R}e$, the Reynolds number, which is a measure of the

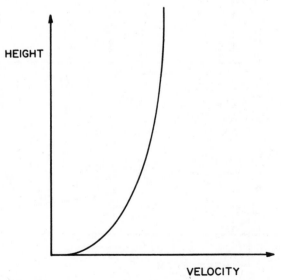

FIGURE 4.3.1.   Typical boundary-layer velocity profile.

ratio of inertial to viscous forces. For example, consider a volume of fluid with a typical surface dimension L. Then, by Bernoulli's theorem, the net pressure $p - p_0$ caused by fluid flow at velocity U, which is of the order of $\frac{1}{2}\rho U^2$, creates inertial forces on the fluid element enclosed by that volume which are of the order of $\rho U^2 L^2$. On the other hand, the viscous stresses on the element are of the order of $\mu U/L$, so that viscosity-related forces are of the order of $\mu U/L \cdot L^2$. The ratio of inertial to viscous forces is then of the order of

$$\mathcal{R}e = \frac{\rho U^2 L^2}{\mu U L^2/L} = \frac{\rho U L}{\mu} = \frac{UL}{v} \tag{4.3.1}$$

where $v = \mu/\rho$ is called the *kinematic viscosity*.* (See also Sect. 7.1.) Thus, when $\mathcal{R}e$ is large, inertial effects predominate; when it is small, the viscous effects are the stronger ones. It is noted that the concept of Reynolds number is, in relation to the boundaries affecting a flow, a very local thing; that is, the selection of the representative length L for the calculation of $\mathcal{R}e$ depends upon the interest of the investigator in local details. Thus a flow over a given object may develop a wide variety of Reynolds numbers, depending upon the particular region focused on for study. When discussing—as a whole—the flow that envelops a given body, it is usual to select for the length L some overall representative dimension of that body.

Boundary layer separation occurs if fluid particles in the boundary layer are sufficiently decelerated by inertial forces that the flow near the surface becomes reversed. These deceleration effects occur as a result of the presence in the flow of adverse pressure gradients. Such severe adverse pressure gradients as can be produced, for example, by the flow over the corner of a bluff body generally cause flow separation. Through processes that are not well understood, the separation layers generate discrete vortices, which are shed into the wake flow behind the bluff body (Fig. 4.3.2). Such vortices can cause extremely high suctions near separation points such as corners or eaves.

Flows of practical interest have Reynolds numbers ranging from nearly zero to as high as 10^8 or 10^9. Steadily increasing the Reynolds number of the flow over an obstacle generally produces a widely varying sequence of flow phenomena for which the Reynolds number provides a convenient index, as is seen, for example, in Sect. 4.4.

---

*Typical values of kinematic viscosity for air and water are, respectively,

$$v_{air} = 0.150 \text{ cm}^2/\text{sec at } 20°C$$

$$v_{H_2O} = 0.01 \text{ cm}^2/\text{sec at } 20°C$$

A common unit for kinematic viscosity is the *stoke*:

$$1 \text{ stoke} = 1 \text{ cm}^2/\text{sec} = 0.0010764 \text{ ft}^2/\text{sec}$$

A useful approximate formula for the Reynolds number in air at about 20°C and atmospheric pressure is $67\,000UL$, where U is in m/sec and L in meters. This becomes $6230UL$ for U in ft/sec and L in feet.

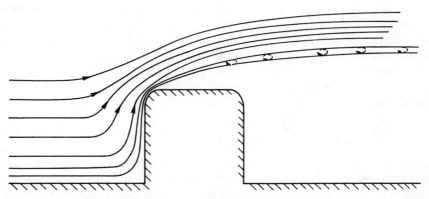

FIGURE 4.3.2.   Flow separation at corner of obstacle.

If, as is true in most cases, the flow over a body has separated at some point,[*] the wake will contain the effects of vortex formation. Depending upon the magnitude of the Reynolds number, the flow will be *turbulent* to a greater or lesser extent. Many turbulent flows may thus be typically viewed as wake flows in which upstream objects have already "stirred" the flow in some such manner as has been described. Turbulence can be caused by means other than the stirring mechanisms mentioned above (for example, by thermally-induced convection), but for the majority of flows of importance to wind engineering, turbulence can be considered to be initiated mechanically, as described. Thus, for example, trees, buildings, or terrain upstream of a given point play an important role in developing the turbulence of the wind observed in the atmospheric boundary layer at that point. Descriptions of turbulence in the natural wind are given in Sect. 2.3.

When turbulence is present, one turbulent layer of the fluid tends to produce turbulent motions in adjacent layers, as, for example, in a wake or boundary layer. This takes place through transfer of momentum from one layer to another. A similar phenomenon occurs in the absence of turbulence when a laminar, as opposed to turbulent, boundary layer is created. The difference between a laminar and a turbulent boundary layer is that, in the former, the transfer of momentum occurs at the molecular rather than the macroscopic scale. The fluid viscosity μ is in fact the result of such molecular transfers of momentum. As noted in Sect. 2.1 in the context of atmospheric flows, turbulent boundary layers may be viewed as being governed by an equivalent kinematic viscosity called *eddy viscosity*, the value of which reflects the large momentum transfers induced by turbulence.

---

[*]In the case of airfoils occurrence of this separation is usually desired as late as possible along the body, in accordance with the aim of controlling pressure distributions to increase lift and reduce drag by means of geometric form.

4.4 WAKE AND VORTEX FORMATIONS IN TWO-DIMENSIONAL FLOW

In the following discussion, the flow is assumed to be smooth (laminar) and two-dimensional, that is, independent of the coordinate normal to the plane of viewing. Consider a two-dimensional flow around the sharp-edged flat plate shown in Fig. 4.4.1. At a very low Reynolds number (for example, $UL/v \cong 0.3$, where L is the dimension of the plate across the flow), the flow turns the sharp corner and follows both front and rear contours of the plate (Fig. 4.4.1a). At a slightly higher Reynolds number ($\mathscr{R}e \cong 10$) obtained by merely increasing the flow velocity over the same plate, the flow separates at the corners and creates two large, symmetric vortices behind the plate that remain attached to the back of the plate (Fig. 4.4.1b). At increased Reynolds number ($\mathscr{R}e \cong 250$) the symmetrical vortices are broken and replaced by cyclically alternating vortices that form by turns at the top and bottom edges and are swept downstream (Fig. 4.4.1c). A full cycle of this phenomenon is defined as the activity between the occurrence of some instantaneous flow configuration about the body and the next identical configuration. At still higher Reynolds numbers, say $\mathscr{R}e \gtrsim 1000$ (Fig. 4.4.1d), the inertia forces predominate; large, distinct vortices have little possibility of forming and, instead, a generally turbulent wake is formed behind the plate, its two outer defining edges forming a "shear layer" consisting of a long series of smaller vortices that accommodate the wake region to the adjacent smooth flow region. Overall, these results dramatically illustrate the changes in the flow with Reynolds number, proceeding from predominantly viscous effects to predominantly inertial effects.

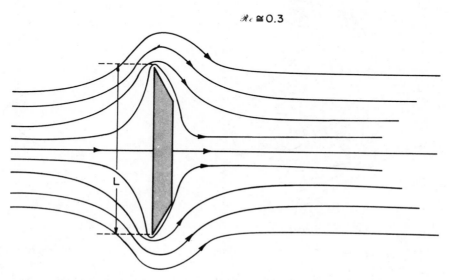

$\mathscr{R}_\epsilon \cong 0.3$

FIGURE 4.4.1a.   Flow past a sharp-edged plate $\mathscr{R}e \cong 0.3$.

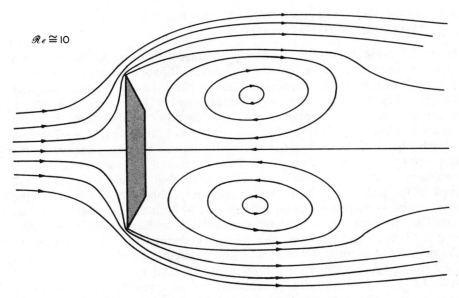

FIGURE 4.4.1b.   Flow past a sharp-edged plate $\mathscr{R}e \cong 10$.

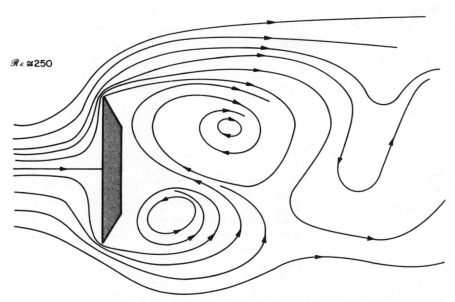

FIGURE 4.4.1c.   Flow past a sharp-edged plate $\mathscr{R}e \cong 250$.

Next the renowned case of two-dimensional flow about a circular cylinder (Fig. 4.4.2) is briefly examined. A number of flow situations can be created by increasing the flow velocity, each situation being identified by a specific Reynolds number range. At extremely low values of Reynolds number ($\mathscr{R}e \cong 1$) the flow (assumed laminar as it approaches) remains attached to the cylinder

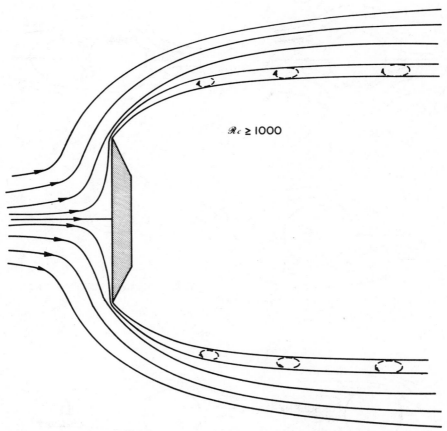

$\mathscr{R}e \geq 1000$

FIGURE 4.4.1d.   Flow past a sharp-edged plate $\mathscr{R}e \gtrsim 1000$.

throughout its complete periphery, as shown in Fig. 4.4.2a. At $\mathscr{R}e \cong 20$, the flow form remains symmetrical but flow separation occurs and large wake eddies are formed which reside near the downstream surface of the cylinder, as suggested in Fig. 4.4.2b. For $30 \leqslant \mathscr{R}e \leqslant 5000$, alternating vortices are shed from the cylinder and form a clear "vortex trail" downstream. This phenomenon was first reported by Bénard [4-5] and von Kármán [4-6] (Fig. 4.4.2c). The finer details of this striking occurrence are still not fully understood, and the process continues to be the focus of many studies, both experimental and theoretical [4-21]. Behind the cylinder there is established a staggered, stable arrangement of vortices that moves off downstream at a velocity somewhat less than that of the surrounding fluid. In this range of Reynolds number the wake flow is relatively smooth and regular apart from the vortices themselves. Figure 4.4.3 depicts the streamlines of the wake flow behind a circular cylinder in a water tunnel [4-7] within the above-mentioned $\mathscr{R}e$ range. The flow in this photograph was made visible by the emission of dye from the cylinder.

As Reynolds number further increases into the range $5000 \leqslant \mathscr{R}e \leqslant 200\,000$,

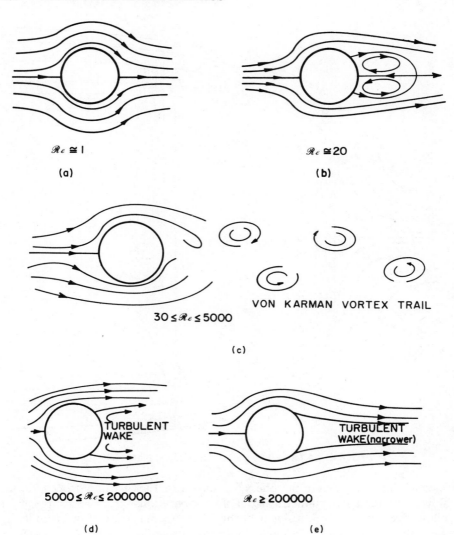

FIGURE 4.4.2. (*a*) Flow past circular cylinder $\mathscr{R}e \cong 1$. (*b*) Flow past circular cylinder $\mathscr{R}e \cong 20$. (*c*) Flow past circular cylinder $30 \leqslant \mathscr{R}e \leqslant 5000$. (*d*) Flow past circular cylinder $5000 \leqslant \mathscr{R}e \leqslant 200{,}000$. (*e*) Flow past circular cylinder $\mathscr{R}e \geqslant 200{,}000$.

the attached flow upstream of the separation point is laminar. In the separated flow three-dimensional patterns are observed, and transition to turbulent flow occurs in the wake—farther downstream from the cylinder for the lower Reynolds numbers and nearer the cylinder surface as the Reynolds numbers increase [4-19]. For the largest Reynolds numbers in this range, the cylinder wake undergoes transition to turbulence immediately after separation, and a turbulent wake is produced between the separated shear layers (Fig. 4.4.2*d*).

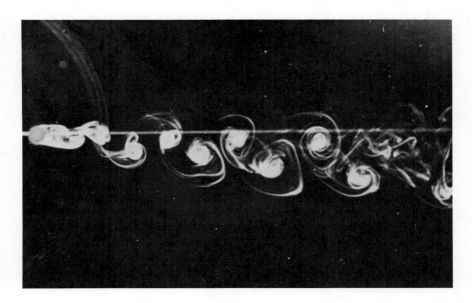

FIGURE 4.4.3. Vortex trail in water tunnel. Courtesy of the National Aeronautical Establishment, National Research Council of Canada.

Beyond $\mathscr{R}e \cong 200\,000$ (Fig. 4.4.2e) the wake narrows appreciably (giving rise to less drag; see p. 144) and vortex shedding appears much more random. However, with increasing velocity, at $\mathscr{R}e \cong 4 \times 10^6$ regular shedding again appears, though the wake now retains considerable turbulence. The highest Reynolds number for which the phenomenon has been experimentally explored is about 10^8 [9-31].

Other bluff bodies, notably triangles, squares, rectangles, and other regular and irregular prisms give rise to analogous vortex shedding phenomena.

The pronounced regularity of such wake effects was first reported by Strouhal [4-8] who pointed out that the vortex-shedding phenomenon is describable in terms of a nondimensional number (the Strouhal number):

$$\mathscr{S} = \frac{N_s D}{U}$$

where N_s is the frequency of full cycles of vortex shedding, D is a characteristic dimension of the body projected on a plane normal to the mean flow velocity, and U is the velocity of the oncoming flow, assumed laminar. The number $\mathscr{S}$ takes on different characteristic constant values depending upon the cross-sectional shape of the prism being enveloped by the flow. Figure 4.4.4 [4-9] summarizes the relation of $\mathscr{S}$ to $\mathscr{R}e$ for a circular cylinder in the range $10^5 \leqslant \mathscr{R}e \leqslant 10^7$. Table 4.4.1 [4-10] also lists a number of values of $\mathscr{S}$ for different cross-sectional shapes for Reynolds numbers in the clear vortex-shedding range, the approaching flow being laminar.

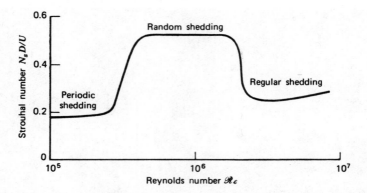

FIGURE 4.4.4. Relation of Strouhal number to Reynolds number for circular cylinder. From L. R. Wooton and C. Scruton, "Aerodynamic Stability," in *The Modern Design of Wind-Sensitive Structures*, Construction Industry Research and Information Association, London, U.K., 1971, pp. 65–81. By permission of the Director of the National Physical Laboratory, U.K., and the Director of the Construction Industry Research and Information Association, U.K.

A certain amount of debate continues on the question of whether or not periodic vortex shedding can still be exhibited at extremely large Reynolds numbers, say $\mathscr{R}e \gg 10^8$. If one substitutes an effective eddy viscosity (see Sect. 2.2) for the actual kinematic viscosity of the fluid, it is conceivable that a new Reynolds number range can be calculated in which alternating vortex shedding from extremely large bluff objects can once more be forecast. In this way the occasionally observed large vortex trails in ocean currents downstream of islands may possibly be reconciled with smaller-scale experimental observations. Figure 4.4.5, for instance, is a reproduction of a satellite photograph [4-11] of a vortex trail in the atmosphere made visible by cloud presence in the vortices shed from the mountain peak of Guadalupe Island over 1200 meters high off the Pacific coast of Mexico. The photograph spans some 250 km. Assuming, as in [2-117], an effective value of (kinematic) eddy viscosity $v_e \cong 50$ m²/sec, a full-scale Reynolds number of the order of 10^{10} for the phenomenon (based on $v \cong 1.5 \times 10^{-5}$ m²/sec) would be reduced to an effective value of $(\mathscr{R}e)_{\text{eff}} \cong 3000$, which falls well within the laminar vortex-shedding range. Assuming the island to be about 20 km long, the distance between successive periodic vortex centers is roughly 55 km. Further assuming a Strouhal number for the island peak as $\mathscr{S} \cong 0.12$, a mean wind velocity of $U = 30$ m/sec, and an effective island dimension of $D \cong 6000$ m yields the vortex-shedding frequency:

$$N_s = \frac{0.12(30)}{6000} = 6 \times 10^{-4} \text{ Hz}$$

which in turn gives a shedding period of $T = 1/N_s = 1667$ sec. Employing

TABLE 4.4.1. Strouhal number for a variety of shapes. From "Wind Forces on Structures," *Trans.* ASCE, **126** (1961), 1124–1198.

Wind	Profile dimensions, in mm	Value of $\mathscr{S}$	Wind	Profile dimensions, in mm	Value of $\mathscr{S}$
→ ↓	$t = 2.0$, 50, 50, t (I-section)	0.120 / 0.137	↓ ↓	$t = 1.0$, 12.5, 12.5, 25, 50 (channel)	0.147
→	$t = 0.5$, 25, 25 (H-section)	0.120	↓ ↓	$t = 1.0$, 12.5, 12.5, 12.5, 50 (channel)	0.150
↓ ↓	$t = 1.0$, 25, 50 (H-section)	0.144	← ↑ ↙	$t = 1.0$, 50, 50 (angle)	0.145 / 0.142 / 0.147
↓ ↓	$t = 1.5$, 12.5, 50 (I-section)	0.145	← ↑ ↙	$t = 1.0$, 25, 25 (angle)	0.131 / 0.134 / 0.137
↓ ↑	$t = 1.0$, 25, 50 (channel)	0.140 / 0.153	→ ↓	$t = 1.0$, 25, 25, 25, 25	0.121 / 0.143
↓ ↑	$t = 1.0$, 12.5, 50 (channel)	0.145 / 0.168	→	$t = 1.0$, 25, 25, 25, 12.5	0.135
→ ↓	$t = 1.5$, 50 (flat plate)	0.156 / 0.145	→	$t = 1.0$, 50, 100 (T-section)	0.160
	Cylinder $11\,800 < \mathscr{R}e < 19\,100$, 25	0.200	→ ↑	$t = 1.0$, 25, 50 (T-section)	0.114 / 0.145

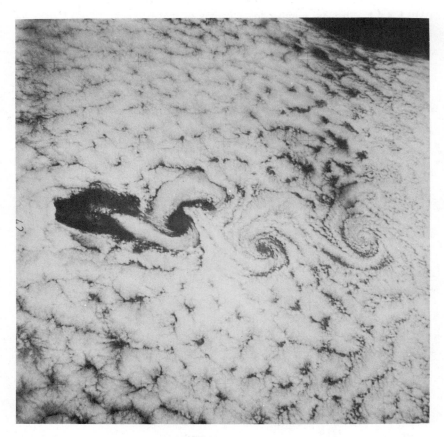

FIGURE 4.4.5. Satellite photo of cloud vortices downstream of Guadalupe Island (off Baja California) [4-11]. Courtesy of the National Aeronautics and Space Administration.

$S = UT$ yields a calculated vortex separation of

$$S = 30 \times 1667 = 50000 \text{ m} = 50 \text{ km}$$

a distance consistent with rough measurement of the photograph. Another interesting photograph of large-scale vortex shedding is presented in Fig. 4.4.6 [4-12].*

When conditions are such that a distinct vortex trail is present in the wake, a flow crossover aft of the body occurs that has a component normal to the approaching flow direction. Thus it becomes possible to inhibit the establishment of a vortex trail by placing a "splitter plate" in the near wake of the generating body, as first pointed out in [4-13]. (See Fig. 4.4.7.) The action of this

---

*It has been brought to the attention of the authors that a similar problem is treated in [4-70].

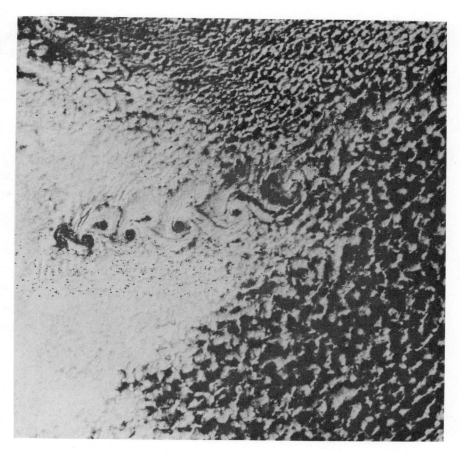

FIGURE 4.4.6. Satellite photo of Jan Mayen Island (Arctic Ocean). From *Weather*, **31**, 10 (Oct. 1976), 346.

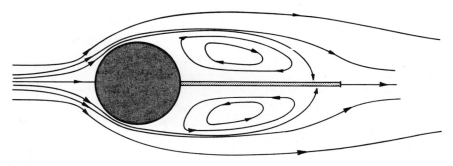

FIGURE 4.4.7. Effect of splitter plate on flow behind a circular cylinder [4-13, 4-61].

plate is to prevent the flow crossover and thus to quiet the entire wake flow. Qualitatively, the presence of the plate has the same type of effect as lengthening the body in the stream direction and causing it to approach, to some approximation, the form of a symmetrical airfoil. Following this type of approach it can be seen that elongated bodies, oriented with their long dimension parallel to the main flow, tend to elicit relatively narrow wakes, many without appreciable vortex production.

If flows about square and rectangular prisms are compared (Fig. 4.4.8), the square is seen (at reasonably high $\mathscr{R}e$) to produce flow separation followed by a wide, turbulent wake, whereas the more elongated rectangular form (depending on length-to-width ratio) may exhibit separation at leading corners that is followed downstream by flow reattachment and finally, once more, by flow separation at the trailing edge. Thus, it is seen that not only does the bluff face of the body presented to the fluid affect the resulting wake, but the streamwise length and general form of the body also play important roles in the wake form.

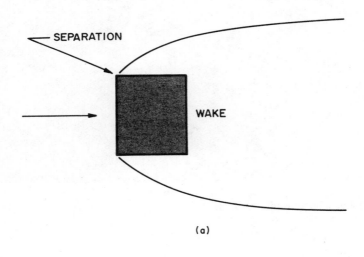

(a)

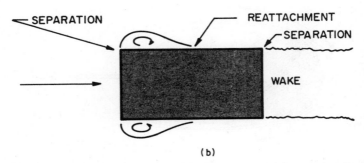

(b)

FIGURE 4.4.8. Flow separation and wake regions of square and rectangular cylinders.

In sharp distinction to the case of Fig. 4.4.8b, if the rectangle is placed with its long dimension normal to the flow, the wake exhibits a strong vortex-shedding characteristic, followed at higher $\mathcal{R}e$ by a turbulent wake not unlike that produced by the sharp-edged flat plate (see Figs. 4.4.1c and 4.4.1d).

4.5 PRESSURE, LIFT, DRAG, AND MOMENT EFFECTS ON TWO-DIMENSIONAL STRUCTURAL FORMS

Figure 4.5.1 suggests a section of a bluff body immersed in a flow of velocity U. The flow will develop local pressures p over the body in accordance with Bernoulli's equation:

$$\tfrac{1}{2}\rho U^2 + p = \text{const} \tag{4.5.1}$$

where the constant holds along a streamline and U represents the velocity on the streamline in the immediate vicinity of the body (i.e., immediately outside the boundary layer that forms on its surface). The integration of the pressures over the body surface results in a net force and a moment. The components of the force in the along-flow and across-flow directions are referred to as *drag* and *lift*, respectively. The drag, lift, and moment are quite obviously affected by both the shape of the body and the Reynolds number.

The body may, for example, be contoured with the express purpose of minimizing drag and maximizing lift, resulting in an airfoil-like shape. Again, as in many civil engineering applications, the shape of the body may not be amenable to such special adjustment; its form will most likely have been fixed by other design objectives than purely aerodynamic ones. Nevertheless, the lift, drag, and moment developed by the fluid flows about the structure will remain of strong interest because these are effects that must be designed against.

It is usual to refer all pressures measured at a structural surface to the mean dynamic pressure $\tfrac{1}{2}\rho U^2$ of the far upstream wind or the free-stream wind at

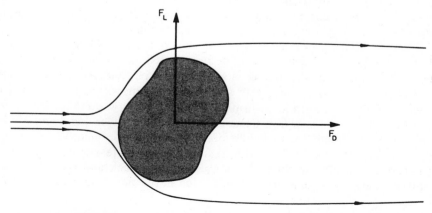

FIGURE 4.5.1.   Lift and drag on an arbitrary bluff body.

some distance from the structure (for example, at a point well above it out of the boundary layer). Thus nondimensional *pressure coefficients* C_p are defined by

$$C_p = \frac{p - p_0}{\frac{1}{2}\rho U^2} \tag{4.5.2}$$

where U is the mean value of the reference wind and $p - p_0$ represents the pressure difference between local and far upstream pressure p_0. Such non-dimensional forms enable the transfer of model experimental results to full scale, and the establishment of reference values for cataloguing the aerodynamic properties of given geometric forms.

Analogously, the net wind-pressure forces (per unit of span) F_L and F_D in the lift and drag direction, respectively, can be rendered dimensionless and expressed in terms of *lift* and *drag coefficients* C_L and C_D as follows:

$$C_L = \frac{F_L}{\frac{1}{2}\rho U^2 B} \tag{4.5.3}$$

$$C_D = \frac{F_D}{\frac{1}{2}\rho U^2 B} \tag{4.5.4}$$

where B is some typical reference dimension of the structure. For the net flow-induced moment M the corresponding coefficient is

$$C_M = \frac{M}{\frac{1}{2}\rho U^2 B^2} \tag{4.5.5}$$

When the flow is fluctuating as a consequence of oncoming turbulence, vortex-associated flow changes, or signature (body-induced) turbulence, the above quantities become time-dependent. In such cases, when time-varying forces and moments occur, mean values of force coefficients as well as spectral density distributions of these quantities are required for their fuller description.* (Note that in two-dimensional flow F_L, F_D, and M represent corresponding values per unit of dimension normal to the plane of observation. In three-dimensional cases, correct dimensionality is preserved by including an additional factor B in the denominator of each expression.)

Returning to the prism of circular cross section in smooth flow, the variation of its mean drag coefficient C_D may be represented as in Fig. 4.5.2, where the dependence on Reynolds number is shown. Note particularly how C_D drops sharply in the range of about $2 \times 10^5 \leqslant \mathcal{R}e \leqslant 5 \times 10^5$. This region of sharp drop is called the *critical* region and corresponds to a condition wherein the transition from laminar to turbulent flow occurs in the boundary layer that forms on the surface of the cylinder. The turbulent mixing that thus takes place in the boundary layer helps transport fluid with higher momentum toward the surface of the cylinder. Separation then occurs much farther back and the wake consequently narrows, finally producing a value of the time-averaged C_D that is

---

*See Appendix A2.

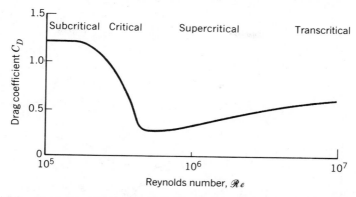

FIGURE 4.5.2. Evolution of mean drag coefficient with Reynolds number for a circular cylinder. After L. R. Wooton and C. Scruton, "Aerodynamic Stability," in *The Modern Design of Wind-Sensitive Structures*, Construction Industry Research and Information Association, London, U.K., 1971, pp. 65–81 and [4-22].

only about $\frac{1}{3}$ of its highest value. As $\mathcal{R}e$ increases into the supercritical and then the transcritical range ($\mathcal{R}e \gtrsim 4 \times 10^6$), C_D increases once more but remains much lower than its subcritical values.

Figure 4.5.3 depicts a typical distribution of the mean pressure coefficient about the circular cylinder in smooth flow as a function of angular position. The results are evidently sensitive to Reynolds number.*

The drag coefficient of an elongated rectangular-section body in smooth flow (Fig. 4.5.4) [4-14, 4-23] is also a function of the narrowness of its wake, but the lower limit of wake width is approximately the full width of the body. The wake width at somewhat lower $\mathcal{R}e$ is much greater than the body width, and this is accompanied by higher C_D; then, when flow reattachment to the body begins to occur, the drag coefficient drops. This is a function mainly of the elongation b/h of the body, as shown in the figure. Flow in the critical region is accompanied by turbulence, and therefore this region is shown as a shaded band of possible values in Fig. 4.5.4.

Figure 4.5.5 [4-15] illustrates the evolution with Reynolds number of the mean drag coefficient of a square in smooth flow during successive modifications of its corners. Note that only the sharp-cornered square exhibits practically unchanging drag with change of Reynolds number. This is simply accounted for by the early separation of the flow at the upstream corners and the shortness of the afterbody that practically precludes the possibility of flow reattachment, whereas squares with rounded corners tend to possess the same kind of critical region for the drag coefficient as seen earlier for the circular cylinder. Note also, in the case of the circular section, the dependence of the drag upon the roughness

---

*The pressures corresponding to $\theta = 0°$ and $\theta = 180°$ are referred to as the pressure at the stagnation point and the base pressure, respectively.

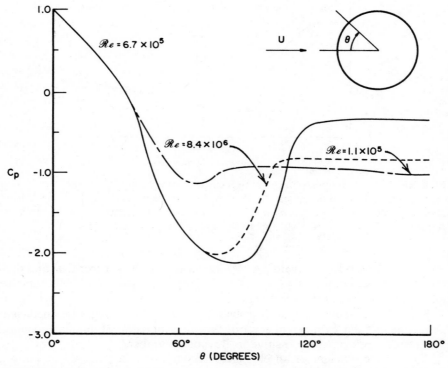

FIGURE 4.5.3.   Influence of Reynolds number on pressure distribution over a circular cylinder (after [4-22]).

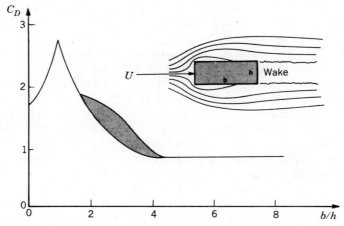

FIGURE 4.5.4.   Effect of afterbody upon drag of a rectangular cylinder [4-14], [4-23].

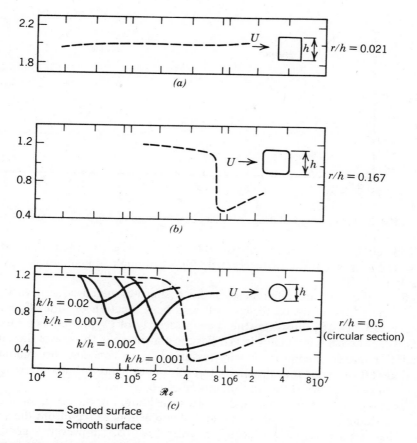

FIGURE 4.5.5. Influence of Reynolds number, corner radius, and surface roughness on drag coefficient, square to circular cylinders (r is the corner radius; k is the grain size of sand). After [4-15].

of the cylinder surface. This dependence was studied in detail in [4-24]. (See also Sect. 11.1.1.)

Because of such effects, certain features of the flow in tests over wind tunnel models can be expected to be Reynolds-number-independent while others may be quite sensitive thereto. Thus it can be argued that certain Reynolds-number-insensitive flow phenomena may be encountered in tests in which the flow will always break cleanly away at the same identifiable points. Certain types of bodies such as the circular cylinder offer extended regions of possible flow separation in which the location of the actual separation points depends upon Reynolds number. With such bodies the entire structure of the flow will be highly Reynolds-number-sensitive (see Sect. 7.3.2).

For extremely low Reynolds numbers the drag coefficient increases greatly as a result of viscous effects. This is illustrated in Fig. 4.5.6 [4-14], which depicts

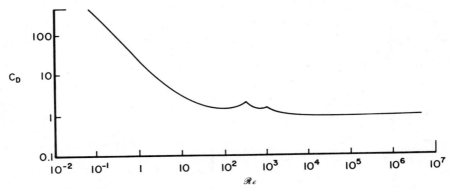

FIGURE 4.5.6.   Typical drag coefficient as a function of Reynolds number [4-14].

C_D for circular and square flat plates for $10^{-2} \leqslant \mathscr{R}e \leqslant 10^7$. (Analogous effects on lift and moment do not necessarily follow, though some distortion is very likely.)

Since the pressure differences across a sharp-cornered square vary with time, the sectional lift coefficient will also be a function of time: $C_L = C_L(t)$. Figure 4.5.7 [4-16] illustrates the spectral density of C_L plotted as a function of nB/U, where n is frequency in Hz, B is the dimension of the side of the square, and U is mean oncoming velocity (assumed to be constant throughout the region of flow under consideration). In both smooth and turbulent flow, a high spectral peak occurs at the Strouhal number $nB/U = 0.12$.

This is clear evidence of periodic vortex shedding. For any given bluff body, this shedding is not a purely sinusoidal phenomenon, as seen from the spread to other frequencies of the spectral peak in Fig. 4.5.7; however, a good first approximation to the lift force per unit span occurring at the peak Strouhal number is given by

$$F_L = \tfrac{1}{2}\rho U^2 B \bar{C}_L \sin \omega t \qquad (4.5.6)$$

where $\bar{C}_L$ is a mean lift coefficient that depends on the particular cross section shape and $\omega = 2\pi n$, n satisfying the Strouhal relation.

The root mean square (rms) value of the fluctuating normal force coefficient $C_{N_{\text{rms}}}$ on the square section is shown in Fig. 4.5.8 [4-16] as a function of angle of attack α with respect to the mean wind direction. Here the turbulence* is seen to lower the highest normal force below, and to raise the lowest normal force slightly above, the respective laminar values.

Figure 4.5.9 [4-7] presents two photographs of flow over proposed bridge deck sectional forms as visualized in a water tunnel flow containing fine aluminum particles. Figure 4.5.9a shows a section that produces severe flow separa-

---

*The turbulence characteristics in the experiment of Fig. 4.5.8 were the following: longitudinal scale 1.4B; lateral scale 0.4B; turbulence intensity 10%.

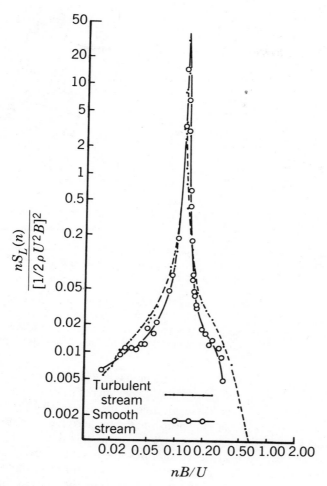

FIGURE 4.5.7. Spectrum of lift fluctuations on a square-section cylinder for flow normal to a face ($\mathscr{R}e = 10^5$). From B. J. Vickery, "Fluctuating Lift and Drag on a Long Cylinder of Square Cross-Section in a Smooth and in a Turbulent Flow," *Journal of Fluid Mechanics*, **25** (1966), Cambridge University Press, New York, pp. 481–494.

tion; Fig. 4.5.9b portrays the flow-smoothing effect of a modified section providing lower lift and drag.

Reference 4-10 presents mean values of C_D and C_L obtained under laminar flow conditions for a large number of sectional shapes common in construction, as taken from [12-2] and [4-18]; see Table 4.5.1, [4-17], and [4-62].

The results of Table 4.5.1 are applicable to members with large aspect ratio (ratio of length to width) λ, or to members with end plates (abutments). For members with small aspect ratio (say, $\lambda < 10$) and no end plates (abutments), end flow effects are significant, and the drag coefficients are smaller than in

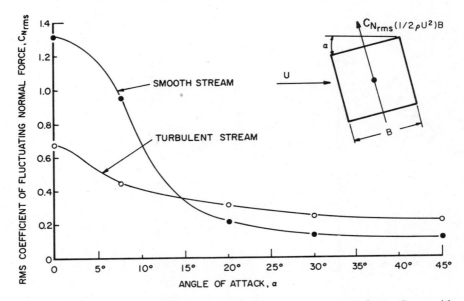

FIGURE 4.5.8. Variation of the coefficient of fluctuating normal force $C_{N_{rms}}$ with angle of attack for a rectangular prism. From B. J. Vickery, "Fluctuating Lift and Drag on a Long Cylinder of Square Cross-Section in a Smooth and in a Turbulent Flow," *Journal of Fluid Mechanics*, **25** (1966), Cambridge University Press, New York, pp. 481–494.

Table 4.5.1 (see Sect. 4.6.2). The drag coefficients are also modified by the presence of turbulence in the oncoming flow. Experiments have shown that in most cases of interest in practice these modifications are small [12-2, 12-5]. For this reason wind tunnel tests aimed at measuring aerodynamic forces or trussed frameworks with sharp-edged members are to this day conducted in smooth flow [12-1, 12-6]. Note, however, that in some cases the effect of turbulence on the drag force can be significant. For members with rectangular cross section this effect depends upon: (a) the ratio b/h between the sides of the cross section, and (b) the turbulence in the oncoming flow. If the ratio b/h is small, no flow reattachment occurs following separation at the front corners. Depending upon its intensity, the turbulence can enhance the flow entrainment in the wake and, therefore, cause stronger suctions and larger drag (Fig. 4.5.10a). If the ratio b/h is sufficiently large, the turbulence can cause flow reattachment which would not have occurred in smooth flow—and thus result in reduced drag (Fig. 4.5.10b) [4-25, 4-26]. The dependence of the drag coefficient upon turbulence intensity is shown for two ratios b/h in Fig. 4.5.11* [4-26]. Additional studies on turbulence effects on drag and lift of sharp-edged bodies are reported in [4-27], [4-28], and [4-85]. The effect of turbulence in the case of bodies with rounded

---

*Note that for $b/h = 1$, C_D as obtained in [4-26] for smooth flow differs by about 10% from the value listed in Table 4.5.1. Differences of this order or larger are common even for results of simple wind tunnel tests.

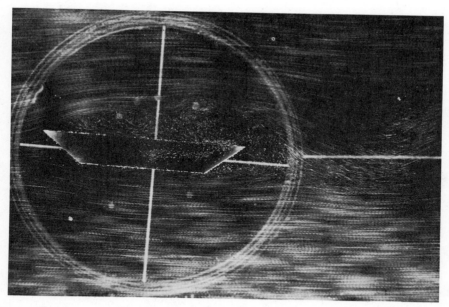

FIGURE 4.5.9a. Visualization of water flow over a model bridge deck section. Courtesy of the National Aeronautical Establishment, National Research Council of Canada.

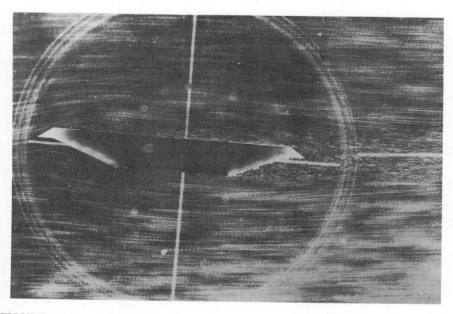

FIGURE 4.5.9b. Visualization of water flow over a partially streamlined model bridge deck section. Courtesy of the National Aeronautical Establishment, National Research Council of Canada.

TABLE 4.5.1. Two-dimensional drag and lift coefficients for structural shapes. From "Wind Forces on Structures," *Trans.* ASCE, **126** (1961), 1124–1198 and [12-2]

Profile and wind direction	C_D	C_L
	2.03	0
	1.96 – 2.01	0
	2.04	0
	1.81	0
	2.0	0.3
	1.83	2.07
	1.99	−0.09
	1.62	−0.48
	2.01	0
	1.99	−1.19
	2.19	0

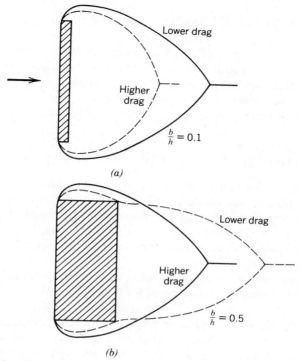

FIGURE 4.5.10. Separation layers in smooth flow (solid line) and in turbulent flow (interrupted line). After A. Laneville, I. S. Gartshore, and G. V. Parkinson, "An Explanation of Some Effects of Turbulence on Bluff Bodies," *Proceedings*, Fourth International Conference, Wind Effects on Buildings and Structures, Cambridge Univ. Press, Cambridge, U.K., 1977.

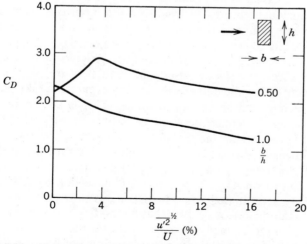

FIGURE 4.5.11. Dependence of drag coefficient upon turbulence intensity. After A. Laneville, I. S. Gartshore, and G. V. Parkinson, "An Explanation of Some Effects of Turbulence on Bluff Bodies," *Proceedings*, Fourth International Conference, Wind Effects on Buildings and Structures, Cambridge Univ. Press, Cambridge, U.K., 1977.

shapes is, essentially, to reduce the Reynolds number at which the critical region (Fig. 4.5.2) sets in. This is shown in [4-29], which includes, in addition, information on the fluctuating lift and drag forces on a rigid cylinder due to vortex shedding and to turbulence in the oncoming flow (see also [4-30]).

Reference 4-14 is a respected compendium on drag effects.

4.6 REPRESENTATIVE FLOW EFFECTS IN THREE DIMENSIONS

Most flows have a three-dimensional character, principally as a result of their contact with boundaries. For example, if a hypothetical laminar flow consisting of an air mass displaced uniformly as a single unit encounters an object, it will be diverted in several directions. Also, the passage of such a flow along a surface sets up boundary-layer velocity gradients. In addition, three-dimensionality is clearly inherent in turbulent flows.

Although the general equations for fluid flow remain available for application, few flow problems in three dimensions have been satisfactorily solved in a purely analytical fashion because of the considerable complexities involved. As a result, most three-dimensional studies rely partially or wholly upon experiment. Therefore, this section is mainly concerned with broad aspects of three-dimensional flows, with conditions of testing, and with some representative results obtained by test.

4.6.1 Cases Retaining Two-Dimensional Flow Features

The success of the two-dimensional flow models discussed in the previous section has in a few cases been considerable because some actual flows retain certain two-dimensional features, at least to a first approximation. Consider, for example, the case of a long rod of square cross section in an air flow with uniform mean velocity normal to one face. Except near the ends of the rod, the mean flow may, in this case, be considered for practical purposes as two-dimensional. However, the effects associated with flow fluctuations are not identical in different strips, the differences between events that take place at any given time increasing with separation distance. This is shown in Fig. 4.6.1 [4-16] for the pressure difference between centerlines of top and bottom faces of the rod under both laminar and turbulent approaching flow.* It is observed that the three-dimensionality of the flow manifests itself through spanwise loss of correlation R_{AB} between pressure differences (measured respectively between points A and A' at section A and points B and B' at section B), this correlation loss being strongly accentuated when turbulence is present in the oncoming flow. From this example one may infer that fluctuating phenomena, including vortex shedding, cannot normally be expected to be altogether uniform along the entire length of a cylindrical body, even if the flow has uniform mean speed and the body is geometrically uniform.

---

*The turbulence characteristics were the same as in the experiment of Fig. 4.5.8.

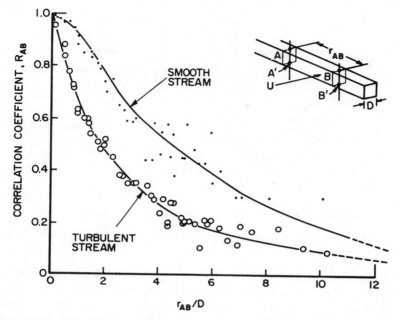

FIGURE 4.6.1. Spanwise correlation of the fluctuating pressure difference across the center line of a long square-section cylinder for flow normal to a face ($\mathscr{R}e = 10^5$). From B. J. Vickery, "Fluctuating Lift and Drag on a Long Cylinder of Square Cross-Section in a Smooth and in a Turbulent Flow," *Journal of Fluid Mechanics*, **25** (1966), Cambridge University Press, New York, pp. 481–494.

In practice, mean flow conditions upwind of tall slender structures are usually not uniform, as assumed in the simpler case discussed above; indeed, in the atmospheric boundary layer the mean flow velocity increases with height. Also, certain tall structures (e.g., stacks) are not geometrically uniform. These important features—in addition to the incident turbulence—further decrease the coherence of vortices shed in the wake of structures.

4.6.2 Structures in Three-Dimensional Flows: Case Studies

The complexities of wind flow introduced by the geometries of typical structures and by the characteristics of the terrain and obstacles upstream emphasize the need to carry out detailed studies of wind pressures experimentally using wind tunnel models and simulation. In order to give some idea of the type of results so obtained and to emphasize the important roles of the boundary layer velocity profile and of the turbulence in such results, a few examples are cited below.

Wind flows about buildings are prime examples of three-dimensional flows that cannot be described acceptably by two-dimensional models. Figure 4.6.2 [15-11] suggests such a situation. Here a tall model building in a wind flow is preceded by a lower building. This latter trips off a vortex in the space between buildings. Air descending close to the windward wall flows through openings

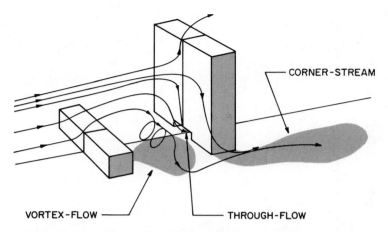

FIGURE 4.6.2.   Main features of the flow around a tall building model [15-11].

beneath the building at ground level. Regions of accelerated flow are produced around vertical and horizontal corners of the building. In the areas of vortex-flow, through-flow, and corner streams, many design problems are presented by the special characteristics of the locally accelerated flow. (See Sect. 15.3 and p. 174.)

A few examples are now shown of differences between drag or pressure coefficients measured in a uniform and in a boundary layer flow. The existence of such differences was first pointed out by Flachsbart in 1932 [4-31].

We consider first the case of a rectangular plate normal to the wind in a smooth flow. The drag coefficients depend strongly upon aspect ratio and upon whether the plate is held in midair, as in the case of a traffic sign, or stands on the ground, as in the case of a free-standing wall—see Table 4.6.1.

Note that the aerodynamic force normal to the plate is not necessarily largest when the yaw angle α (Fig. 4.6.3) is zero. For a plate with aspect ratio $\lambda = 5$, the dependence of the aerodynamic force normal to the face of the plate

TABLE 4.6.1.   Drag Coefficients for a Rectangular Plate Normal to Wind in Smooth Flow [4-10, 12-2]

								Rectangular Plate on Ground (Standing on Long Side)		
	Rectangular Plate in Normal Wind[a]									
Aspect Ratio	1.0	2.0	5.0	10.	20.	40.	∞	1.0	10.	∞
C_D	1.18	1.19	1.20	1.23	1.48	1.66	1.98	1.10	1.20	1.20

[a]The values listed in [4-10] were taken from [12-2]. Some of these values were incorrectly transcribed in [4-10] and therefore differ from those shown in this Table.

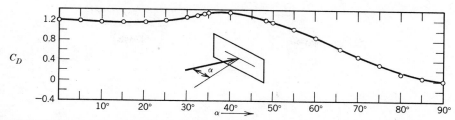

FIGURE 4.6.3. Dependence of drag coefficient for plate with aspect ratio $\lambda = 5$ upon direction of horizontal wind [12-2].

upon α is shown in Fig. 4.6.3. It is seen that for $\alpha = 40°$ the aerodynamic force is larger by about 15% than in the case $\alpha = 0°$. A similar, though somewhat smaller, increase was reported in [12-2] for a plate girder with aspect ratio $\lambda \cong 10$.

The effect of turbulence on a square plate normal to the flow was studied in [4-25], where drag coefficients were measured for both smooth flow and turbulence flow with 8.3% turbulence intensity and 7.6 cm longitudinal turbulence scale—see Table 4.6.2.

Note that the drag coefficients measured in smooth flow differ slightly among themselves and from the value of Table 4.6.1 ($C_D = 1.18$). Note also that as the ratio between the longitudinal scale of turbulence and the dimension of the plate decreases, the influence of the turbulence on the magnitude of the drag coefficient becomes smaller. These results are further discussed in Sect. 7.3.3.

Figure 4.6.4a shows a model used for measurements reported by Flachsbart in 1932 [4-31]. The measurements were conducted in both smooth and shear (boundary-layer) flow (Figs. 4.6.4b and c). The measured mean pressure coefficients C_p, referred to the free stream velocity, are shown in Fig. 4.6.4d for smooth flow and Fig. 4.6.4e for boundary-layer flow (interrupted and solid lines represent pressures and suctions, respectively). It is seen that the differences between the results obtained in the two types of flow are significant. Similar results were subsequently obtained in [4-32] and [4-33].

Figure 4.6.5a depicts mean flow patterns around a vertical wall of height-

TABLE 4.6.2. Drag Coefficients for Square Plate Normal to the Mean Flow [4-25]

Plate Size (cm)	C_D	
	Smooth	Turbulent
5.08 × 5.08	1.12	1.26
10.16 × 10.16	1.09	1.22
15.24 × 15.24	1.11	1.20
20.32 × 20.32	1.15	1.18

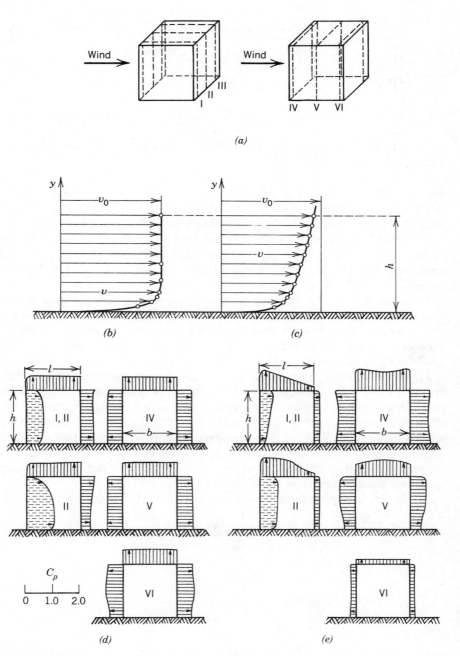

FIGURE 4.6.4. Summary of model tests in smooth and boundary-layer flow. From "Winddruck auf geschlossene und offene Gebäude," by O. Flachsbart, in *Ergebnisse der Aerodynamischen Versuchanstalt zu Göttingen*, IV. Lieferung, L. Prandtl and A. Betz (Eds.), Verlag von R. Oldenbourg, Munich and Berlin, 1932.

to-width ratio 1:1 with uniform approaching flow. Figure 4.6.5b depicts the same situation in boundary-layer flow. Figures 4.6.6a and 4.6.6b display the pressure coefficients developed on the faces of a cube resting on a horizontal surface (due to flow normal to one face) first in uniform flow, then in a boundary-layer flow. Figures 4.6.7a and 4.6.7b present similar results for a tall building. It is noted that in Figs. 4.6.5b, 4.6.6b and 4.6.7b the pressure coefficients are referred to the free stream velocity [4-20].

Loads on structural parts (e.g., cladding) are determined by the algebraic sum of the external and internal pressures acting on these parts. In the ideal case of a hermetically sealed building, the internal pressure is not affected by the external wind flow (Fig. 4.6.8a). If the building has an opening on the windward (leeward) side and is otherwise sealed, the wind flow will create a positive (negative) internal pressure, as shown in Fig. 4.6.8b (Fig. 4.6.8c).

In most cases the opening or porosity distribution over the building envelope is not known, and internal pressures could be either positive or negative (Fig. 4.6.8d). Building standards (e.g., [2-49]) specify internal pressure coefficients generally believed to be conservative for use in design. Recent investigations into the magnitude of internal pressures and of their dependence on time are reported in [4-52, 4-53, 4-54, 4-55, 4-56, 4-57].

4.7 THE RELATION OF TIME-VARYING FORCES TO WIND VELOCITY IN TURBULENT FLOW

For a given body immersed in a wind flow it is of interest to convert information on velocity fluctuations into information on pressures over the body or on resultant forces and moments. Since most real flows are sufficiently complex that analytical calculation of such results is not possible, it is usual to employ formulas featuring unknown coefficients that may be evaluated by experiment.

4.7.1 Drag Forces

The net drag force consists of the resultant over a given body surface of all components of elemental forces that are aligned with the drag, or along-wind, direction. The time-varying drag $F_D(t)$ on a body completely enveloped by a flow is conventionally given by the formula

$$F_D(t) = \tfrac{1}{2}\rho U^2(t)B^2 C_D \qquad (4.7.1)$$

where B is a typical body dimension and C_D is the usual drag coefficient.

In Eq. 4.7.1 a second term of the form $\rho B^3[dU(t)/dt]C_m$ is often included, particularly if the fluid in question is relatively dense, as for example in the case of water; C_m is an empirical "virtual mass" coefficient intended to account for effects linked to the fluid acceleration. Actually, the coefficient C_m appears to be useful in cases wherein the fluid mass involved is appreciable relative to the body mass. One may then visualize it as specifying a hypothetical mass which, given the acceleration dU/dt, accounts for the net force due to all the variously accelerated fluid elements in the entire flow around the body. In most

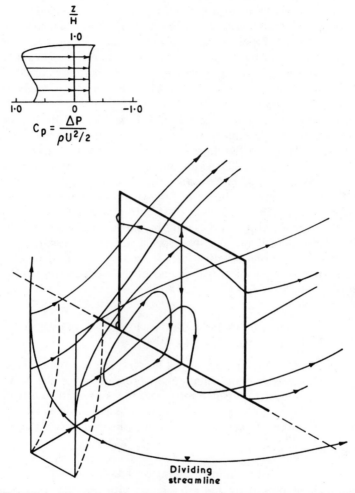

FIGURE 4.6.5a.   Flow pattern and center line pressure distribution over a wall of height-to-width ratio 1:1 in a constant velocity field. From W. D. Baines, "Effects of Velocity Distribution on Wind Loads and Flow Patterns on Buildings," *Proceedings*, Symposium No. 16, Wind Effects on Buildings and Structures, held at the National Physical Laboratory, England, in 1963, published by HMSO London in 1965.

flows of interest in wind engineering, however, the entire term containing C_m contributes only a negligible part to F_D. For this reason it is usually neglected in this context, and it is not retained in what follows.

A three-dimensional flow will have three components, $U(t)$, $V(t)$, and $W(t)$, in three mutually perpendicular directions. In the neutrally stratified flows* of strong interest to wind engineering the mean wind velocity $\bar{U}$ is horizontal,

*See Chapter 1, p. 9 and Chapter 2, p. 31.

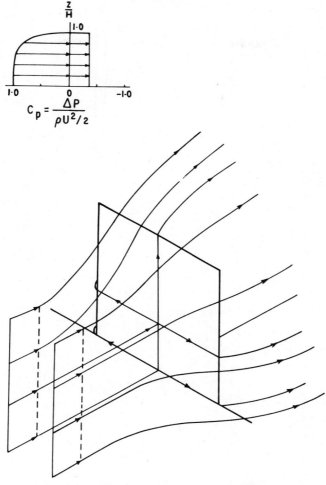

FIGURE 4.6.5b.   Flow pattern and center line pressure distribution over a wall of height-to-width ratio 1:1 in a boundary-layer velocity field. From W.D. Baines, "Effects of Velocity Distribution on Wind Loads and Flow Patterns on Buildings," *Proceedings*, Symposium No. 16, Wind Effects on Buildings and Structures, held at the National Physical Laboratory, England, in 1963, published by HMSO London in 1965.

and the wind then can be represented as the sum of mean and fluctuating components.

$$U(t) = \bar{U} + u(t)$$
$$V(t) = v(t) \qquad\qquad (4.7.2)$$
$$W(t) = w(t)$$

the means of u, v, and w being zero.

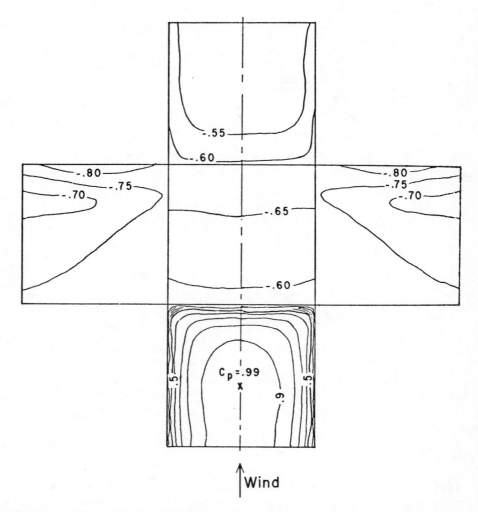

FIGURE 4.6.6a. Pressure distributions on the faces of a cube in a constant velocity field. From W. D. Baines, "Effects of Velocity Distribution on Wind Loads and Flow Patterns on Buildings," *Proceedings*, Symposium No. 16, Wind Effects on Buildings and Structures, held at the National Physical Laboratory, England, in 1963, published by HMSO London in 1965.

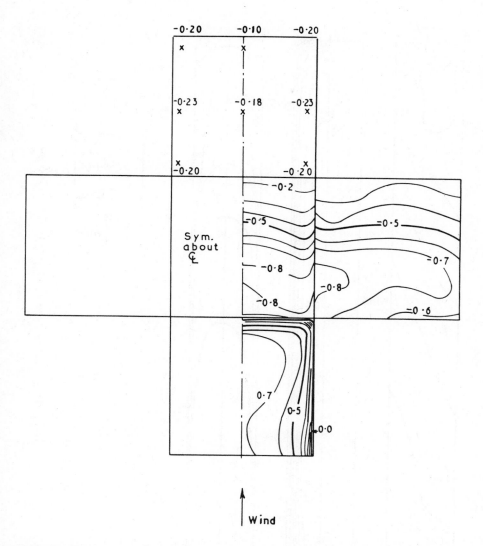

FIGURE 4.6.6b. Pressure distributions on the faces of a cube in a boundary-layer velocity field. From W.D. Baines, "Effects of Velocity Distribution on Wind Loads and Flow Patterns on Buildings," *Proceedings*, Symposium No. 16, Wind Effects on Buildings and Structures, held at the National Physical Laboratory, England, in 1963, published by HMSO London in 1965.

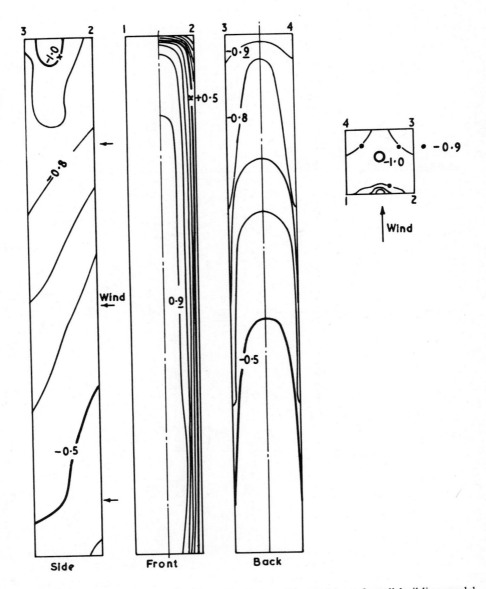

FIGURE 4.6.7a. Pressure distributions over the sides and top of a tall building model in a constant velocity field. From W. D. Baines, "Effects of Velocity Distribution on Wind Loads and Flow Patterns on Buildings," *Proceedings*, Symposium No. 16, Wind Effects on Buildings and Structures, held at the National Physical Laboratory, England, in 1963, published by HMSO London in 1965.

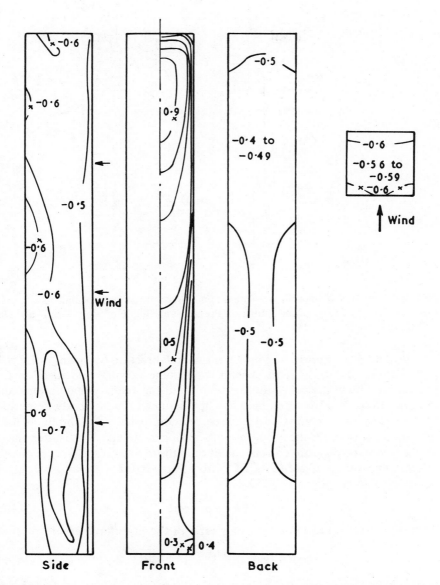

FIGURE 4.6.7b. Pressure distributions over the sides and top of a tall building model in a boundary-layer velocity field. From W. D. Baines, "Effects of Velocity Distribution on Wind Loads and Flow Patterns on Buildings," *Proceedings*, Symposium No. 16, Wind Effects on Buildings and Structures, held at the National Physical Laboratory, England, in 1963, published by HMSO London in 1965.

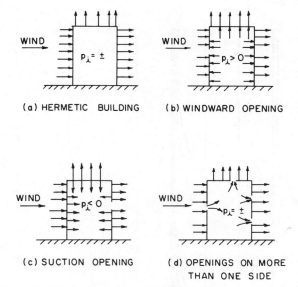

FIGURE 4.6.8.   Mean internal pressures in buildings with various opening distributions. From H. Liu and P. J. Saathoff, "Internal Pressure and Building Safety," *J. Struct. Div.*, ASCE, **108** (1982), 223–2234.

One may then express drag in the horizontal direction by means of Eq. 4.7.1 with $U(t)$ as in Eq. 4.7.2. In general, when time-varying velocities are thus introduced, the imperfect spatial correlation of the velocity fluctuations must also be considered. However, here it is first assumed that the body in question is sufficiently small compared to the correlation distances of the fluctuations u, v, and w, so that, for the purposes of the problem at hand, these latter may be considered to be perfectly correlated. Since in the high winds usually of greatest interest to wind engineering $u(t)/\bar{U}$ rarely exceeds 0.2, u^2 may generally be neglected with small error yielding

$$F_D(t) = \bar{F}_D + \rho \bar{U} u(t) B^2 C_D \qquad (4.7.3)$$

where the steady and the fluctuating parts of the drag force are, respectively

$$\bar{F}_D = \tfrac{1}{2}\rho B^2 C_D [\bar{U}^2 + \overline{u^2(t)}] \qquad (4.7.4a)$$

and

$$F_D' = \rho \bar{U} u(t) B^2 C_D \qquad (4.7.4b)$$

From Eq. 4.7.4b it is seen that $F_D'(t)$ varies directly as $u(t)$. This is true to a first approximation only, since observation of physical flows reveals that C_D may itself also vary as a function of the frequency components of $u(t)$.

In order to examine the statistical characteristics of $F_D'(t)$ it is useful to consider its spectral density. To obtain this, one may first calculate its autocovariance function (see Appendix A2, Eq. A2.21):

$$R_{F_D}(\tau) = \lim_{T \to \infty} \frac{1}{T} \int_{-T/2}^{T/2} F_D'(t)F_D'(t+\tau)\,dt = \overline{F_D'(t)F_D'(t+\tau)}$$
$$= \rho^2 \bar{U}^2 B^4 C_D^2 \overline{u(t)u(t+\tau)} \tag{4.7.5}$$

Noting further (Appendix A2, Eq. A2.20) that the spectral density $S_{F_D}(n)$ is related to $R_{F_D}(\tau)$ as follows:

$$S_{F_D}(n) = 2 \int_{-\infty}^{\infty} R_{F_D}(\tau) \cos 2\pi n\tau \, d\tau \tag{4.7.6}$$

one obtains

$$S_{F_D}(n) = \rho^2 \bar{U}^2 B^4 C_D^2 S_u(n) \tag{4.7.7}$$

Dividing this through by $(\frac{1}{2}\rho\bar{U}^2 B^2)^2$ yields the spectral density S_{C_D} of the fluctuating drag coefficient:

$$S_{C_D}(n) = 4C_D^2 \frac{S_u(n)}{\bar{U}^2} \tag{4.7.8}$$

This equation will be valid over the range of frequencies of $S_u(n)$ provided all effects remain perfectly correlated as assumed above. However, because in practice this assumption does not hold, it is usual to include an adjustment factor to preserve the validity of Eq. 4.7.8. This is done by writing (4.7.8) as

$$S_{C_D}(n) = 4C_D^2 \frac{S_u(n)}{\bar{U}^2} \chi^2(n) \tag{4.7.9}$$

where the newly-introduced factor $\chi^2(n)$ is termed the *aerodynamic admittance**
of the body in question and represents a modifying adjustment (for an actual body) of the ideal case of a body enveloped by turbulence with full spatial correlation. This modification brings the drag coefficient spectrum into alignment with actual conditions.

The aerodynamic admittance is a function of body shape and dimensions and of the characteristics of the turbulence. For a given body it is thus a frequency-dependent function. Figure 4.7.1 [4-35] suggests the manner in which $\chi^2(n)$ varies for a square flat plate placed normal to a turbulent flow with uniform mean speed. The decrease of the aerodynamic admittance with increasing frequency corresponds to the fact that the smaller turbulent eddies have shorter wavelengths; thus those eddies with higher frequencies will suffer loss of coherence more rapidly than do the larger eddies. References 4-36 and 4-37 appear to be among the earliest to have introduced and used aerodynamic admittance concepts in buffeting problems.

---

*The use of this term in wind engineering is an extension of its original use in aeronautical contexts [4-34].

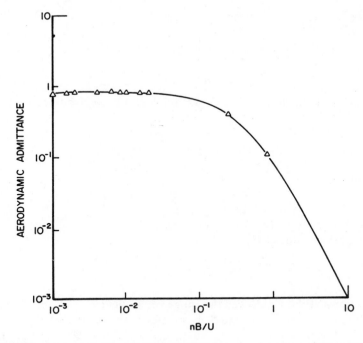

FIGURE 4.7.1. Aerodynamic admittance of a square plate in turbulent flow. After P. W. Bearman, "Wind Loads on Structures in Turbulent Flow," in *The Modern Design of Wind-Sensitive Structures*, Construction Industry Research and Information Association, London, U.K., 1971, pp. 42–48. By permission of the Director of the National Physical Laboratory, U.K., and the Director of the Construction Industry Research and Information Association, U.K.

4.7.2   Relation of Wind Pressures over Slender Buildings to Wind Velocities

The type of arguments employed in Sect. 4.7.1 in relation to total drag forces is now applied to the case of a high-rise building of rectangular plan form, with the horizontal wind blowing normal to one face. In this instance, the along-wind structural motion is dependent on the windward and leeward pressure distributions in a manner that is conceptually simple.

The pressure acting at a point Q of elevation z on the surface of such a body in a steady flow of velocity $U(z)$ may be expressed as

$$p(Q) = \tfrac{1}{2}\rho U^2(z) C_p(Q) \qquad (4.7.10)$$

where ρ is fluid density and C_p is the appropriate pressure coefficient at this point.

In the case of unsteady flow $U(z) = \bar{U}(z) + u(z, t)$ the pressure may be approximated by

$$p(Q, t) = \bar{p}(Q) + p'(Q, t) \qquad (4.7.11)$$

where $\bar{p}$ and p' have the following values:

$$\bar{p}(Q) = \tfrac{1}{2}\rho C_p(Q)\bar{U}^2(z)\left[1 + \frac{\overline{u^2(z)}}{\bar{U}^2(z)}\right] \qquad (4.7.12)$$

$$p'(Q) = \tfrac{1}{2}\rho C_p(Q)\bar{U}^2(z)\left[2\,\frac{u(z)}{\bar{U}(z)} + \frac{u^2(z) - \overline{u^2(z)}}{\bar{U}^2(z)}\right] \qquad (4.7.13)$$

where overbars indicate mean values.

A brief numerical example is in order here. In the atmosphere $\overline{u^2(z)}^{1/2} \cong 2.5u_*$ and $\bar{U}(z) = 2.5u_*\ln[(z - z_d)/z_0]$ (see Eqs. 2.2.18 and 2.3.2 and Table 2.2.1). For example, if $z_0 = 0.03$ m, $z_d \cong 0$, and $\bar{U}(10) = 40$ m/sec at $z = 50$ m:

$$\frac{\overline{u^2(z)}}{\bar{U}^2(z)} = 0.018$$

so that the error in neglecting the nonlinear term in Eq. 4.7.12 is less than 2%.

More generally and analogously to the drag results already discussed in Sect. 4.7.1, calculations reported in [4-38] indicate that the following relations are satisfactory, with insignificant error, for $\bar{p}$ and p':

$$\bar{p}(Q) = \tfrac{1}{2}\rho\bar{U}^2(z)C_p(Q) \qquad (4.7.14)$$

$$p'(Q, t) = \rho\bar{U}(z)u(z, t)C_p(Q) \qquad (4.7.15)$$

If, as in the case of many buildings, the horizontal dimensions of the body are small compared to the scale of turbulence, it is reasonable to assume that the fluctuating pressures affecting along-wind response—which consist entirely of those on the windward and leeward faces—may be given by

$$p'(Q_w) = \rho\bar{U}(z)u(z, t)C_w(Q_w) \qquad (4.7.16)$$

$$p'(Q_\ell) = \rho\bar{U}(z)u(z, t)C_\ell(Q_\ell) \qquad (4.7.17)$$

where Q_w and Q_ℓ are points on the windward and leeward faces, respectively, and where

$$C_w(Q_w) = \frac{\bar{p}(Q_w)}{\tfrac{1}{2}\rho\bar{U}^2(z)} \qquad (4.7.18)$$

$$C_\ell(Q_\ell) = \frac{\bar{p}(Q_\ell)}{\tfrac{1}{2}\rho\bar{U}^2(z)} \qquad (4.7.19)$$

where z is the elevation of point Q_w or Q . As discussed in Chapter 9, it is usual in current procedures for estimating along-wind building response to assume that Eqs. 4.7.16 and 4.7.17 are valid regardless of the ratio of building transverse dimensions to the scale of turbulence. This point is brought up again later in Sects. 4.7.3 and 4.7.4.

In calculating along-wind structural response (see Chapter 5) information is required on the spatial correlation of pressures applied at any two points Q_1 and Q_2. Such information is supplied by the co-spectra of fluctuating pressures (quadrature spectra being assumed negligible). Assuming the validity of Eqs. 4.7.16 and 4.7.17, the co-spectra take the form

$$S_{p'_1 p'_2}^C(Q_1, Q_2, n) = C_p(Q_1)C_p(Q_2)\rho^2 \bar{U}(z_1)\bar{U}(z_2)S_{u_1 u_2}^C(Q_1, Q_2, n) \quad (4.7.20)$$

that is, the co-spectra of the pressures are proportional to the co-spectra of the fluctuating longitudinal wind components in the undisturbed oncoming flow at the elevations of the two points. The pressure coefficient $C_p(Q_i)$ represents windward or leeward values depending upon whether the point Q_i is on the windward or leeward side.

The co-spectrum $S_{u_1 u_2}^C$ may be expressed in the following form:

$$S_{u_1 u_2}^C(Q_1, Q_2, n) = S_{u_1 u_2}^C(r, n)N(n) \quad (4.7.21)$$

where $S_{u_1 u_2}^C(r, n)$ is the co-spectrum of the longitudinal velocity fluctuations at points Q_1 and Q'_2 (Q'_2 being the projection of Q_2 on a plane, normal to the mean wind direction, that contains Q_1), and r is the distance between Q_1 and Q'_2. The function $N(n)$ is referred to as the along-wind cross-correlation coefficient.

If Q_1 and Q_2 are contained in the same vertical plane normal to the mean wind (i.e., if their along-wind separation is zero), then $N(n) \equiv 1$. For nonzero along-wind separation, an expression of $N(n)$ is given in Sect. 4.7.4. In the case $Q_1 \equiv Q_2$,

$$S_{p'}(Q_1, n) = C_p^2(Q_1)\rho^2 \bar{U}^2(z_1)S_u(z_1) \quad (4.7.22)$$

4.7.3   Pressure Fluctuations on the Windward Face of Bluff Bodies

A theory of turbulent flow around two-dimensional bluff bodies has recently been developed in [4-39], which has subsequently been applied in [4-40] and [4-41] to the study of surface pressures generated by turbulent velocity fluctuations. The theory is based, essentially, on the following assumptions: (a) the turbulence intensity is of the same order of magnitude as, or lower than, the turbulence intensity typical of atmospheric flows; (b) the body is sufficiently long that end effects may be neglected; (c) in the flow region upwind of the body any velocity fluctuations induced by wake flow are statistically independent of the velocity fluctuations caused by the oncoming turbulence, so that the latter can be studied separately. Fundamental to the approach of [4-39] is a generalization of "rapid-distortion theory" which allows the linearization of the equations describing the turbulent motion near the upstream face of the body. This linearization follows from the assumption that the changes in the mean flow associated with the presence of the body distort the turbulence sufficiently rapidly so that, during the distortion process, the nonlinear inertial transfer of energy between eddies of different sizes is negligible.

The equations for the turbulent nondimensionalized vorticity vector $\tilde{\omega}_i$ ($i = 1, 2, 3$) are then [4-39]

$$\frac{D\tilde{\omega}_i}{Dt} \simeq \sum_{j=1}^{3} \omega_j \frac{\partial \tilde{u}_i}{\partial x_j} \quad (i = 1, 2, 3) \quad (4.7.23)$$

where $\tilde{u}_i$ ($i=1, 2, 3$) is the nondimensionalized velocity fluctuation vector. The nondimensionalized pressure $\tilde{p}'$ is given by

$$-\frac{\partial \tilde{p}'}{\partial x_i} \simeq \frac{\partial u_i}{\partial t} + \sum_{j=1}^{3} \left(\tilde{U}_j \frac{\partial \tilde{u}_i}{\partial x_j} + \tilde{u}_j \frac{\partial \tilde{U}_i}{\partial x_j} \right) \qquad (i=1, 2, 3) \qquad (4.7.24)$$

where $\tilde{U}_i$ ($i=1, 2, 3$) is the nondimensionalized mean velocity vector, the field of which is, approximately, irrotational.* The boundary conditions are essentially the following: (a) at large distances from the body, the velocities approach their values in the undisturbed flow and (b) in the immediate vicinity of the upwind surface of the body, the velocities at each point are perpendicular to the outward normal from that surface.

Calculations carried out on the basis of the above equations and boundary conditions suggest, for example, that whenever $a/L_x < 1$ (where a is the typical horizontal dimension of body and L_x is the longitudinal turbulence scale), Eq. 4.7.22 is applicable, on the windward face, up to frequencies $n \cong 0.15 U/a$, where U is the mean speed of the undisturbed flow [4-41]. For higher frequencies, the pressure spectra decay more rapidly than the velocity spectra so that, for structural design purposes, Eq. 4.7.22 is conservative. That this is the case appears to be confirmed by experimental results reported in [4-42], [4-43], and [4-44].† Calculations also suggest that the smaller eddies are "piled up" against the upward face of the body and, therefore, that the coherence of the high frequency pressure fluctuations is somewhat greater than the coherence of the high frequency velocity fluctuations in the undisturbed flow. In structural engineering computations, this "piling up" effect can be taken into account by choosing appropriately small values of the exponential decay coefficients in Eq. 2.3.29 or 2.3.30.

4.7.4 Pressure Fluctuations on the Leeward Face of Bluff Bodies

According to Eqs. 4.7.16 through 4.7.19, the ratios of the pressures on the leeward face to the pressures on the windward face are the same for both fluctuating and mean pressures. Results of full-scale measurements suggest, however, that the pressure fluctuations on the leeward side are less strong than indicated by Eq. 4.7.17 (see, for example, Fig. 4.7.2 taken from [4-50]). It is reasonable to assume, therefore, that the use in design of Eq. 4.7.17 is conservative from a structural safety viewpoint.

Also of interest for design purposes is the question of the extent to which pressures on the windward side of a building are correlated to the pressures on the leeward side. It is intuitively clear that this correlation cannot be perfect.

---

*An irrotational flow is one in which the components

$$\omega_1 = \frac{\partial u_3}{\partial x_2} - \frac{\partial u_2}{\partial x_3}, \quad \omega_2 = \frac{\partial u_1}{\partial x_3} - \frac{\partial u_3}{\partial x_1}, \quad \omega_3 = \frac{\partial u_2}{\partial x_1} - \frac{\partial u_1}{\partial x_2}$$

are all zero.

†According to [4-67], however, at any given frequency the pressure and velocity spectra have the same slope.

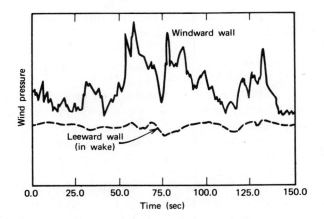

FIGURE 4.7.2. Variations with time of wind pressure on the windward and on the leeward wall of a building. After W. A. Dalgliesh, "Statistical Treatment of Peak Gust on Cladding, *J. Struct. Div.*, ASCE, **97** (1971), 2173–2187.

The correlation will be greater for eddies with large wave lengths—which can be thought of as enveloping the body in the same manner as the mean flow—and will decay as the wave lengths decrease. This dependence can be expressed by choosing an appropriate expression for the function $N(n)$ in Eq. 4.7.21. In [4-45] an expression for this function has been proposed of the form

$$N(n) = \frac{1}{\xi} - \frac{1}{2\xi^2}(1 - e^{-2\xi}) \tag{4.7.25}$$

$$\xi = \frac{15.4n\,\Delta x}{\bar{U}} \tag{4.7.26}$$

where $\bar{U}$ is the mean wind speed at elevation $(2/3)H$; Δx is the smallest of the dimensions B, H, and D; B is the width; H is the height, and D is the depth of the prismatic body. Full-scale and wind tunnel measurements reported in [4-46], [4-47], [4-48], and [4-49] suggest that this expression is adequate for practical use.

4.7.5   Peak Local Wind Loads

The adequate design of roof members, roofing, cladding, and other elements susceptible to failure due to the local action of wind (e.g., solar collectors [4-60]) is of foremost importance for reasons of both safety and economy. It is therefore desirable that wind-induced loads on such elements be ascertained as realistically as possible [4-59].

The elements potentially involved in failures due to local wind loads are usually relatively rigid so that the dynamic amplification of the response is negligible. The wind load acting on an element is then equal to the sum, over the entire area of the element, of the instantaneous pressures induced by wind.

During every storm this load reaches a peak; the element concerned, and its connections, must be designed to sustain the peak wind load attained during the $\bar{N}$-year storm, where $\bar{N}$ is the mean recurrence interval of the design wind speed specified for that element.

The total wind force acting on an element such as a roof member or a curtain wall could, in principle, be measured directly. However, the experimental set-ups required for such measurements are prohibitively expensive and impractical. For this reason forces acting over an element have recently been measured by devices that automatically add pressures occurring simultaneously at several points of the element, weighted by the respective tributary areas. In particular, such techniques have been used at the University of Western Ontario to measure wind loads on models of low-rise structures [4-71, 4-72, 4-73, 4-74, 4-75, 4-76, 4-77]. These measurements, as well as results of full scale tests, [4-51, 4-78, 4-79, 4-80] have been used to develop new design load provisions for main frames and for parts and portions of low-rise buildings that have been recently incorporated in various standards, including [2-49].

4.8 SECONDARY WIND FLOW EFFECTS

In addition to the wind loads caused by the direct action upon the structure of the wind flow, it is of interest in certain situations to examine secondary effects produced by wind, such as the blowing of roofing gravel [4-58, 9-63], and the drifting of snow. Systematic studies of these effects have been reported in [4-63, 4-64, 4-65, 4-66, 4-81, 4-82, 4-83, 4-84].

Mention is also made of wind action as a factor that influences the energy consumption of buildings by increasing air infiltration. It is shown in [4-68] that energy losses due to wind-induced air infiltration can be reduced significantly by the sheltering effect of trees acting as wind breaks; the energy savings thus achieved may in certain cases be as high as 15%. The results of [4-68] were obtained in wind tunnel tests and were subsequently confirmed by full-scale measurements [4-69].

ADDENDUM

For the sake of its historical interest, we reproduce here a note by Count Buffon describing the flow changes occurring upwind of a tower, for which it offers a charming (if scientifically no longer tenable) explanation. A translation of the note follows.

On Reflected Wind

I must report here an observation which it seems to me has escaped the attention of physicists, even though everyone is in a position to verify it; that reflected wind is stronger than direct wind, and the more so as one is closer to the obstacle that reflects it. I have experienced this a number of times, by approaching a tower that is almost 100 feet high and is situated at the north end of my garden in

Montbard; when a strong wind blows from the south, up to thirty steps from the tower one feels strongly pushed; after which there is an interval of five or six steps where one ceases to be pushed and where the wind, which is reflected by the tower, is, so to speak, in equilibrium with the direct wind; after this, the closer one approaches the tower, the more the wind reflected by it is violent, it pushes you back much more strongly than the direct wind pushed you forward. The cause of this effect which is a general one, and which can be experienced against all large buildings, against sheer cliffs, and so forth, is not difficult to find. The air in the direct wind acts only with its ordinary speed and mass; in the reflected wind, the speed is slightly lower but the mass is considerably increased by the compression that the air suffers against the obstacle which reflects it; and as the momentum of any motion is composed of the speed multiplied by the mass, this momentum is considerably larger after the compression than before. It is a mass of ordinary air that pushes you in the first case, and it is a mass of air that is once or twice as dense that pushes you back in the second case.

à l'Hiſtoire Naturelle.   15

ADDITIONS

A l'Article qui a pour titre : Des Vents réglés, *page* 224.

I.

Sur le Vent réfléchi , page 242.

Je dois rapporter ici une obſervation qui me paroît avoir échappé à l'attention des Phyſiciens , quoique tout le monde ſoit en état de la vérifier ; c'eſt que le vent réfléchi eſt plus violent que le vent direct , & d'autant plus qu'on eſt plus près de l'obſtacle qui le renvoie. J'en ai fait nombre de fois l'expérience , en approchant d'une tour qui a près de cent pieds de hauteur & qui ſe trouve ſituée au nord , à l'extrémité de mon jardin , à Montbard ; lorſqu'il ſouffle un grand vent du midi , on ſe ſent fortement pouſſé juſqu'à trente pas de la tour ; après quoi , il y a un intervalle de cinq ou ſix pas , où l'on ceſſe d'être

16   *Supplément*

pouſſé & où le vent , qui eſt réfléchi par la tour , fait , pour ainſi dire , équilibre avec le vent direct ; après cela , plus on approche de la tour & plus le vent qui en eſt réfléchi eſt violent , il vous repouſſe en arrière avec beaucoup plus de force que le vent direct ne vous pouſſoit en avant. La cauſe de cet effet qui eſt général , & dont on peut faire l'épreuve contre tous les grands bâtimens , contre les collines coupées à plomb , &c. n'eſt pas difficile à trouver. L'air dans le vent direct n'agit que par ſa vîteſſe & ſa maſſe ordinaire ; dans le vent réfléchi , la vîteſſe eſt un peu diminuée , mais la maſſe eſt conſidérablement augmentée par la compreſſion que l'air ſouffre contre l'obſtacle qui le réfléchit ; & comme la quantité de tout mouvement eſt compoſée de la vîteſſe multipliée par la maſſe , cette quantité eſt bien plus grande après la compreſſion qu'auparavant. C'eſt une maſſe d'air ordinaire , qui vous pouſſe dans le premier cas , & c'eſt une maſſe d'air une ou deux fois plus denſe , qui vous repouſſe dans le ſecond cas.

Facsimile of note on reflected wind. From *Histoire Naturelle, Générale et Particulière, Contenant les Epoques de la Nature,* Par M. le Comte de Buffon, Intendant du Jardin et du Cabinet du Roi, de l'Académie Française, de celle des Sciences, etc., Tome Treizième, A Paris, De L'Imprimerie Royale, 1778.

REFERENCES

4-1   L. M. Milne-Thompson, *Theoretical Hydrodynamics*, Macmillan, New York, 1965.

4-2   H. Rouse, *Advanced Mechanics of Fluids*, Wiley, 1958.

4-3   G. K. Batchelor, *Fluid Dynamics*, Cambridge Univ. Press, Cambridge, U.K., 1967.

4-4   V. L. Streeter, *Fluid Mechanics*, McGraw-Hill, New York, 1966.

4-5   H. Bénard, "Formation de centres de giration à l'arrière d'un obstacle en mouvement," *C. r. Acad. Sci.*, **147** (1908), 839–842.

4-6   Th. v. Kármán, "Über den Mechanismus des Widerstandes den ein bewegter Körper in einer Flüssigkeit erfährt," in *Nachrichten der Königlichen Gesellschaft der Wissenschaften*, Göttingen (1911), 509–517.

4-7   G. A. Dobrodzicki, *Flow Visualization in the National Aeronautical Establishment's Water Tunnel*, Aeronautical Report No. LR-557, National Research Council of Canada, Ottawa, 1972.

4-8   V. Strouhal, "Über eine besondere Art der Tonerregung," *Ann. Phys.*, **5** (1878), 216–250.

4-9   L. R. Wooton and C. Scruton, "Aerodynamic Stability," in *The Modern Design of Wind-Sensitive Structures*, Construction Industry Research and Information Association, London, England, 1971, pp. 65–81.

4-10   "Wind Forces on Structures," *Trans. ASCE*, **126**, Part II (1961), 1124–1198.

4-11   NASA Skylab II Photo 73-H-964, 73-HE-777, released Oct. 2, 1973.

4-12   *Weather*, **31**, 10 (Oct. 1976), 346.

4-13   A. Roshko, "On the Wake and Drag of Bluff Bodies," *J. Aeronaut. Sci.*, **22**, 2 (1955), 124–132.

4-14   S. F. Hoerner, *Fluid-Dynamic Drag*, published by the author, 148 Busteed Drive, Midland Park, N.J., 1965.

4-15   C. Scruton and E. W. E. Rogers, "Steady and Unsteady Wind Loading of Buildings and Structures," *Phil. Trans. Roy. Soc. London*, **A269** (1971), 353–383.

4-16   B. J. Vickery, "Fluctuating Lift and Drag on a Long Cylinder of Square Cross-Section in a Smooth and in a Turbulent Stream," *J. Fluid Mech.*, **25**, Part 3 (1966), 481–494.

4-17   P. Sachs, *Wind Forces in Engineering*, Pergamon, Oxford, 1972.

4-18   Swiss Code of Practice, No. 160, Article 20, 1956.

4-19   G. E. Mattingly, *An Experimental Study of the Three-Dimensionality of the Flow Around a Circular Cylinder*, Report No. BN295, Institute for Fluid Dynamics and Applied Mathematics, University of Maryland, College Park, Md., June 1972.

4-20   W. D. Baines, "Effects of Velocity Distribution on Wind Loads and Flow Patterns on Buildings," in *Proceedings of the Symposium on Wind Effects on Buildings and Structures*, Vol. 1, National Physical Laboratory, Teddington, U.K., 1963, pp. 197–223.

4-21   J. H. Gerrard, "The Mechanics of the Formation Region of Vortices Behind Bluff Bodies," *J. Fluid Mech.* **25** (1966), 401–413.

4-22   A. Roshko, "Experiments on the Flow Past a Circular Cylinder at Very High Reynolds Number," *J. Fluid Mech.* **10** (1961), 345–356.

4-23   J. Courchesne and A. Laneville, "A Comparison of Correction Methods Used in the Evaluation of Drag Coefficient Measurements for Two-Dimensional Rectangular Cylinders," Paper No. 79-WA/FE-3, Annual Meeting, Dec. 2–7, 1979, The American Society of Mechanical Engineers, New York, N.Y.

4-24   O. Güven, C. Farell, and V. C. Patel, "Surface-Roughness Effects on the Mean Flow Past Circular Cylinders," *J. Fluid Mech.* **98** (1980), 673–701.

4-25   P. W. Bearman, "An Investigation of the Forces on Flat Plates Normal to a Turbulent Flow," *J. Fluid Mech.* **46** (1971), 177–198.

4-26   A. Laneville, I. S. Gartshore, and G. V. Parkinson, "An Explanation of Some Effects of Turbulence on Bluff Bodies," *Proceedings Fourth International Conference, Wind Effects on Buildings and Structures*, Cambridge University Press, Cambridge, U.K., 1977.

4-27   J. A. Roberson, C. Y. Lin, G. S. Rutherford, and M. D. Stine, "Turbulence Effects on Drag of Sharp Edged Bodies," *J. Hydr. Div.*, ASCE, **98** (July 1972), 1187–1201.

4-28 B. E. Lee, "The Effect of Turbulence on the Surface Pressure Field of a Square Prism," *J. Fluid Mech.* **69** (1975), 263–282.

4-29 R. M. C. So and S. D. Savkar, "Buffeting Forces on Rigid Circular Cylinder in Cross-Flows," *J. Fluid Mech.* **105** (1981), 397–425.

4-30 T. M. Mulcahy, *Design Guide for Single Circular Cylinder in Turbulent Crossflow*, Technical Memorandum ANL-CT-82-7, Argonne National Laboratory, March 1982.

4-31 O. Flachsbart, "Winddruck auf geschlossene und offene Gebäude," in *Ergebnisse der Aerodynamischen Versuchanstalt zu Göttingen*, L. Prandtl and A. Betz (Eds.), Verlag von R. Oldenbourg, Munich and Berlin, 1932.

4-32 M. Jensen, *Shelter Effect Investigations*, Danish Technical Press, Copenhagen, 1954.

4-33 M. Jensen and N. Franck, *Model-Scale Tests in Turbulent Wind*, Danish Technical Press, Copenhagen, 1965.

4-34 W. R. Sears, "Some Aspects of Nonstationary Airfoil Theory and Its Applications," *J. Aero. Sci.*, **2** (1941), 104.

4-35 P. W. Bearman, "Wind Loads on Structures in Turbulent Flow," in *The Modern Design of Wind-Sensitive Structures*, Construction Industry Research and Information Association, London, 1971, pp. 42–48.

4-36 H. W. Liepmann, "On the Application of Statistical Concepts to the Buffeting Problem," *J. Aero. Sci.*, **19**, 12 (1952), 793–800, 822.

4-37 A. G. Davenport, "Buffeting of a Suspension Bridge by Storm Winds," *J. Struct. Div.*, ASCE, **88**, No ST6 (1962), 233–264.

4-38 R. Vaicaitis and E. Simiu, "Non-Linear Pressure Terms and Along-Wind Response," *J. Struct. Div.*, ASCE, **103**, No. ST4, Proc. Paper 12837 (1977), 903–906.

4-39 J. C. R. Hunt, "A Theory of Turbulent Flows Round Two-Dimensional Bodies," *J. Fluid Mech.*, **61**, Part 4 (1973), 625–706.

4-40 J. C. R. Hunt, "A Theory of Fluctuating Pressures on Bluff Bodies in Turbulent Flows," in *Proceedings of the IUTAM-IAHR Symposium on Flow-Induced Structural Vibrations*, Karlsruhe, West Germany, 1972, E. Naudascher (ed.), Springer-Verlag, Berlin, 1974, pp. 190–203.

4-41 J. C. R. Hunt, "Turbulent Velocities Near and Fluctuating Surface Pressures on Structures in Turbulent Winds," in *Proceedings of the Fourth International Conference on Wind Effects on Buildings and Structures*, London, 1975, Cambridge Univ. Press. Cambridge, U.K., 1976, pp. 309–320.

4-42 B. J. Vickery and K. H. Kao, "Drag or Along-Wind Response of Slender Structures," *J. Struct. Div.*, ASCE, **98**, No. ST1, Proc. Paper 8635 (1972), 21–36.

4-43 P. W. Bearman, "Some Measurements of the Distortion of Turbulence Approaching a Two-Dimensional Body, *J. Fluid Mech.*, **53**, Part 3 (1972), 451–467.

4-44 R. D. Marshall, *Surface Pressure Fluctuations Near an Axisymmetric Stagnation Point*, NBS Technical Note No. 563, National Bureau of Standards, Washington, D.C., 1971.

4-45 J. Vellozzi and E. Cohen, "Gust Response Factors," *J. Struct. Div.*, ASCE, **94**, No. ST6, Proc. Paper 5980 (1968), 1295–1313.

4-46 R. Lam Put, "Dynamic Response of a Tall Building to Random Wind Loads," *Proceedings of the Third International Conference on Wind Effects on Buildings and Structures*, Tokyo, Japan, 1971, Saikon, Tokyo, 1972, pp. 429–440.

4-47 H. van Koten, "The Fluctuating Wind Pressures on the Cladding and Inside a Building," in *Symposium on Full-Scale Measurements of Wind Effects on Tall Buildings*, University of Western Ontario, London, Canada, 1973.

4-48 J. D. Holmes, "Pressure Fluctuations on a Large Building and Along-Wind Structural Loading," *J. Ind. Aerodyn.*, **1**, 1 (1975), 249–278.

4-49 K. H. Kao, *Measurements of Pressure-Velocity Correlation on a Rectangular Prism in Turbulent Flow*, Report No. BLWT-20, University of Western Ontario, London, Canada, 1970.

4-50 W. A. Dalgliesh, "Statistical Treatment of Peak Gust on Cladding," *J. Struct. Div.*, ASCE, **97**, No. ST9, Proc. Paper 8356 (1971), 2173–2187.

4-51   R. D. Marshall, "A Study of Wind Pressures on a Single-Family Dwelling in Model and Full-Scale," *J. Ind. Aerodyn.*, **1**, 2 (1975), 177–199.

4-52   G. A. Euteneuer, "Druckansteig im Inneren von Gebäuden bei Windeinfall," *Der Bauingenieur*, **45** (1970), 214–216.

4-53   J. D. Holmes, "Mean and Fluctuating Internal Pressures Induced by Wind," in *Wind Engineering*, Proceedings of the Fifth International Conference, Fort Collins, Colorado, July 1979, J. E. Cermak (Ed.), Vol. 1, Pergamon Press, Oxford, 1980.

4-54   T. Stathopoulos, D. Surry, and A. G. Davenport, "Internal Pressure Characteristics of Low-Rise Buildings Due to Wind Action," in *Wind Engineering*, Proceedings of the Fifth International Conference, Fort Collins, Colorado, July 1979, J. E. Cermak (Ed.), Vol. 1, Pergamon Press, Oxford, 1980.

4-55   H. Liu and P. J. Saathoff, "Building Internal Pressure: Sudden Change," *J. Eng. Mechs. Div.*, ASCE, **107** (1981), 309–321.

4-56   H. Liu and P. J. Saathoff, "Internal Pressure and Building Safety," *J. Struct. Div.*, ASCE, **108** (1982), 2223–2234.

4-57   P. J. Saathoff and H. Liu, "Internal Pressure of Multi-Room Buildings," *J. Eng. Mechs. Div.*, ASCE, **109** (1983), 908–919.

4-58   J. E. Minor and L. W. Beason, "Window Glass Failures in Windstorms," *J. Struct. Div.*, ASCE, **102**, No. ST1, Proc. Paper 11834 (1976), 147–160.

4-59   "Hancock Glass Breakage: A Combination of Errors?" *Engineering News Record* (May 1976), 9.

4-60   R. P. McBean, "Wind Load Effects on Flat Plate Solar Collectors", *J. Struct. Eng.*, **111** (1985), 343–352.

4-61   "New Approaches to Design Against Wind Action," in A. G. Davenport (Ed.), *Course Notes*, The Boundary Layer Wind Tunnel Laboratory, University of Western Ontario, London, Canada, 1971.

4-62   J. Blessman, *Aerodinâmica das Construcoẽs*, Editora da Universidade, Porto Alegre, Brasil, 1983.

4-63   R. J. Kind and R. L. Wardlaw, *Design of Rooftops Against Gravel Blow-off*, Report No. 15544, National Research Council of Canada, Ottawa, 1976.

4-64   R. J. Kind, "A Critical Examination of the Requirements for Model Simulation of Wind-Induced Erosion/Deposition Phenomena Such as Snow Drifting," *Atmos. Environ.*, **10** (1976), 219–227.

4-65   C. Mateescu and H. Popescu, *Accumulations de neige sur les constructions, Etude expérimentale sur modèles*, Annales de l'Institut Technique du Bâtiment et des Travaux Publics, Série EM, Paris, 1974.

4-66   J. Wianecki, *Banc d'essais d'accumulation de la neige due au vent*, Annales de l'Institut Technique du Bâtiment et des Travaux Publics, Série EM, Paris, 1976.

4-67   W. Z. Sadeh and J. E. Cermak, "Turbulence Effect on Wall Pressure Fluctuations," *J. Eng. Mech. Div.*, ASCE, **98**, No. EM6, Proc. Paper 9445 (1972), 1356–1379.

4-68   G. E. Mattingly and E. F. Peters, "Wind and Trees: Air Infiltration Effects on Energy and Housing," *J. Ind. Aerodyn.*, **2** (1977), 1–9.

4-69   G. E. Mattingly (personal communication, Apr. 1977).

4-70   H. P. Pao and T. W. Kao, "On Vortex Trails Over Oceans," *Atmos. Sci.*, Meteorological Society of the Republic of China, Taiwan, **3** (1976), 28–38.

4-71   D. Surry and T. Stathopoulos, "An Experimental Approach to the Economical Measurement of Spatially Averaged Wind Loads," *J. Ind. Aerodyn.*, **2**, 4 (Jan. 1978), pp. 385–397.

4-72   A. G. Davenport, D. Surry, and T. Stathopoulos, "Wind Loads on Low Rise Buildings: Final Report of Phases I and II—Parts 1 and 2," *BWLT Report SS8-1977*, The University of Western Ontario, London, Ontario, Canada, Nov., 1977.

4-73   A. G. Davenport, D. Surry, and T. Stathopoulos, "Wind Loads on Low Rise Buildings: Final Report of Phase III—Parts 1 and 2," *BWLT—SS4—1978*. The University of Western Ontario, London, Ontario, Canada, July, 1978.

4-74   L. Apperley, D. Surry, T. Stathopoulos, and A. G. Davenport, "Comparative Measurements of Wind Pressure on a Model of a Full-Scale Experimental House at Aylesbury, England," *J. Ind. Aerodyn.*, **4** (1979), 207–228.

4-75   T. Stathopoulos, "PDF of Wind Pressures on Low-Rise Buildings," *J. Struct. Div.*, ASCE, **106** (1980), 973–990.

4-76   T. Stathopoulos, D. Surry, and A. G. Davenport, "Effective Wind Loads on Flat Roofs," *J. Struct. Div.*, ASCE, **107** (1981), 281–298.

4-77   T. Stathopoulos, "Wind Loads on Eaves of Low Buildings," *J. Struct. Div.* ASCE, **107** (1981), 1921–1934.

4-78   K. J. Eaton and J. R. Mayne, "The Measurement of Wind Pressure on Two-Story Houses at Aylesbury," *J. Ind. Aerodyn.*, **1** (1975), 67–109.

4-79   J. D. Holmes, *Wind Loads on Low Rise Buildings—A Review*, CSIRO, Division of Building Research, Highett, Victoria, Australia, 1983.

4-80   T. Stathopoulos, "Wind loads on low-rise buildings: a review of the state of the art," *Eng. Struct.*, **6** (1984), 119–135.

4-81   J. T. Templin and W. R. Schriever, "Loads Due to Drifted Snow'," *J. Struct. Div.*, ASCE, **108** (1982), 1916–1925.

4-82   J. D. Iversen, "Small-scale modeling of snow-drift phenomena," in *Wind Tunnel Modeling for Civil Engineering Applications*, T. A. Reinhold (Ed.), Cambridge Univ. Press, Cambridge, U.K., 1982.

4-83   R. J. Kind and R. L. Wardlaw, "Failure Mechanisms of Loose-Laid Roof-Insulation Systems (High-Rise Buildings)," *J. Wind Eng. Ind. Aerodyn.*, **9** (1982), 325–341.

4-84   R. J. Kind and R. L. Wardlaw, "Wind Tunnel Tests on Loose-Laid Roofing Systems for Flat Roofs" *Proceedings, Second International Symposium on Roofing Technology*, National Bureau of Standards, Gaithersburg, Md., Sept. 1985.

4-85   I. S. Gartshore, "Some Effects of Upstream Turbulence on the Unsteady Lift Forces Imposed on Prismatic Two Dimensional Bodies", *J. Fluids Eng.*, **106** (1984), 418–424.

5

Structural Dynamics

Structural dynamics is the discipline concerned with the study of structural response to time-dependent loads. It is the purpose of this chapter to review certain elementary results of structural dynamics theory and to derive on their basis expressions for the response of structures subjected to distributed stationary random loads. These results are then applied in the particular case in which the loads are induced by wind to obtain expressions for the along-wind response, including deflections and accelerations. Several of the results obtained will also be useful in other applications occurring throughout the text.

5.1 THE SINGLE-DEGREE-OF-FREEDOM LINEAR SYSTEM

Consider the system represented in Fig. 5.1.1 consisting of a single mass m concentrated at point B and of the member AB assumed to have negligible mass. The displacement $x(t)$ of the mass m is opposed by (1) a restoring force supplied by the member AB and (2) a damping force due to the internal friction that develops within the system during its motion. It is assumed that the restoring force is linear, that is, proportional to the displacement $x(t)$, and that the damping is viscous, that is, proportional to the velocity dx/dt. It follows then from Newton's second law that the motion of the mass is described by the equation

$$m\ddot{x} + c\dot{x} + kx = F(t) \tag{5.1.1}$$

where $F(t)$ is the time-dependent load acting on the mass, k is the spring constant (or the stiffness) of the member AB, c is known as the coefficient of viscous damping, and the dot denotes differentiation with respect to time. It is common to write Eq. 5.1.1 in the form

$$\ddot{x} + 2\zeta_1(2\pi n_1)\dot{x} + (2\pi n_1)^2 x = \frac{F(t)}{m} \tag{5.1.2}$$

179

FIGURE 5.1.1. Schematic representation of a single-degree-of-freedom system.

where

$$n_1 = \frac{1}{2\pi}\sqrt{\frac{k}{m}} \tag{5.1.3}$$

$$\zeta_1 = \frac{c}{2\sqrt{km}} \tag{5.1.4}$$

are known as the natural frequency and the damping ratio of the system, respectively.* The quantity $2\sqrt{km}$ is known as the critical damping coefficient and can be shown to be the value of the damping coefficient beyond which the free motion of the system is nonoscillatory. The damping ratio is expressed as a percentage of the critical damping.

5.1.1 Response to a Harmonic Load

It can be easily verified [5-1] that if

$$F(t) = F_0 \cos 2\pi n t \tag{5.1.5}$$

the steady-state solution of Eq. 5.1.2 is

$$x(t) = F_0 H(n) \cos(2\pi n t - \phi) \tag{5.1.6}$$

where

$$\phi = \tan^{-1}\frac{2\zeta_1(n/n_1)}{1 - (n/n_1)^2} \tag{5.1.7}$$

$$H(n) = \frac{1}{4\pi^2 n_1^2 m\{[1 - (n/n_1)^2]^2 + 4\zeta_1^2(n/n_1)^2\}^{1/2}} \tag{5.1.8}$$

---

*The quantity $2\pi n_1$ is referred to as the natural circular frequency and is commonly denoted by ω_1.

The quantity $H(n)$ is known as the *mechanical magnification factor* or *mechanical admittance function* of the system with parameters m, n_1, and ζ_1.

Similarly, the steady-state response to the load

$$F(t) = F_0 \sin 2\pi nt \qquad (5.1.9)$$

may be written as

$$x(t) = F_0 H(n) \sin(2\pi nt - \phi) \qquad (5.1.10)$$

5.1.2 Response to an Arbitrary Load

Let the system described by Eq. 5.1.2 be subjected to the action of a load equal to the unit impulse function $\delta(t)$ acting at time $t = 0$, that is, to a load defined as follows (see Fig. 5.1.2):

$$\left. \begin{array}{l} \delta(t) = 0 \qquad \text{for } t \neq 0 \\[2mm] \displaystyle\lim_{\Delta t \to 0} \int_0^{\Delta t} \delta(t)\, dt = 1 \end{array} \right\} \qquad (5.1.11)$$

The response of the system to the load $\delta(t)$ is a function of time and is denoted $G(t)$.

An arbitrary load $F(t)$ (Fig. 5.1.3) may be described as a sum of elemental impulses of magnitude $F(\tau')\, d\tau'$ each acting at time τ'. By virtue of the linearity of the system, the response at time t to each such impulse is $G(t - \tau')F(\tau')\, d\tau'$ and the total response at time t is

$$x(t) = \int_{-\infty}^{t} G(t - \tau')F(\tau')\, d\tau' \qquad (5.1.12)$$

where the limits of the integral indicate that all the elemental impulses that have acted before time t have been taken into account. With the change of

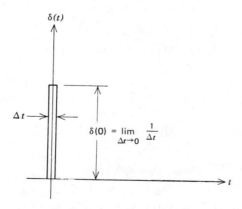

FIGURE 5.1.2.   Unit impulse function.

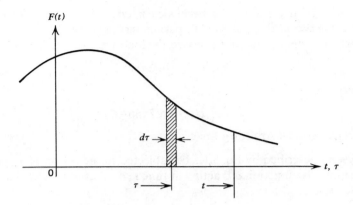

FIGURE 5.1.3.   Load $F(t)$.

variable $\tau = t - \tau'$, Eq. 5.1.12 becomes

$$x(t) = \int_0^\infty G(\tau)F(t - \tau)\, d\tau \tag{5.1.13}$$

Let $F(t) = F_0 \cos 2\pi nt$. It follows then from Eqs. 5.1.6 and 5.1.13 that

$$H(n) \cos \phi = \int_0^\infty G(\tau) \cos 2\pi n\tau\, d\tau \tag{5.1.14}$$

and

$$H(n) \sin \phi = \int_0^\infty G(\tau) \sin 2\pi n\tau\, d\tau \tag{5.1.15}$$

Using now Eqs. 5.1.14 and 5.1.15, it is possible to write

$$H^2(n) \cos^2\phi = \int_0^\infty \int_0^\infty G(\tau_1) \cos 2\pi n\tau_1 G(\tau_2) \cos 2\pi n\tau_2\, d\tau_1\, d\tau_2 \tag{5.1.16}$$

$$H^2(n) \sin^2\phi = \int_0^\infty \int_0^\infty G(\tau_1) \sin 2\pi n\tau_1 G(\tau_2) \sin 2\pi n\tau_2\, d\tau_1\, d\tau_2 \tag{5.1.17}$$

The addition of Eqs. 5.1.16 and 5.1.17 yields the following relation between $H(n)$ and $G(\tau)$:

$$H^2(n) = \int_0^\infty \int_0^\infty G(\tau_1)G(\tau_2) \cos 2\pi n(\tau_1 - \tau_2)\, d\tau_1\, d\tau_2 \tag{5.1.18}$$

5.1.3   Response to a Stationary Random Load

The case is now examined in which the load $F(t)$ is generated by a stationary process with spectral density $S_F(n)$. The expression for the spectral density of the response $S_x(n)$ can be derived using Eqs. A2.20, A2.21, and 5.1.13:

$$S_x(n) = 2 \int_{-\infty}^{\infty} R_x(\tau) \cos 2\pi n\tau \, d\tau$$

$$= 2 \int_{-\infty}^{\infty} \left[\lim_{T \to \infty} \frac{1}{T} \int_{-T/2}^{T/2} x(t)x(t+\tau) \, dt \right] \cos 2\pi n\tau \, d\tau$$

$$= 2 \int_{-\infty}^{\infty} \left\{ \lim_{T \to \infty} \frac{1}{T} \int_{-T/2}^{T/2} dt \left[\int_{0}^{\infty} G(\tau_1)F(t-\tau_1) \, d\tau_1 \right.\right.$$

$$\left.\left. \times \int_{0}^{\infty} G(\tau_2)F(t+\tau-\tau_2) \, d\tau_2 \right] \right\} \cos 2\pi n\tau \, d\tau$$

$$= 2 \int_{0}^{\infty} G(\tau_1) \left\{ \int_{0}^{\infty} G(\tau_2) \left[\int_{-\infty}^{\infty} R_F(\tau+\tau_1-\tau_2) \cos 2\pi n\tau \, d\tau \right] d\tau_2 \right\} d\tau_1$$

$$= 2 \int_{0}^{\infty} \int_{0}^{\infty} G(\tau_1)G(\tau_2) \cos 2\pi n(\tau_1-\tau_2) \, d\tau_1 \, d\tau_2$$

$$\times \int_{-\infty}^{\infty} R_F(\tau+\tau_1-\tau_2) \cos 2\pi n(\tau+\tau_1-\tau_2) \, d(\tau+\tau_1-\tau_2)$$

$$+ 2 \int_{0}^{\infty} \int_{0}^{\infty} G(\tau_1)G(\tau_2) \sin 2\pi n(\tau_1-\tau_2) \, d\tau_1 \, d\tau_2$$

$$\times \int_{-\infty}^{\infty} R_F(\tau+\tau_1-\tau_2) \sin 2\pi n(\tau+\tau_1-\tau_2) \, d(\tau+\tau_1-\tau_2) \tag{5.1.19}$$

where, in the last step, the identity

$$\cos 2\pi n\tau \equiv \cos 2\pi n[(\tau+\tau_1-\tau_2)-(\tau_1-\tau_2)] \tag{5.1.20}$$

is used.

By virtue of Eqs. A2.20, A2.23 and 5.1.18, there follows

$$S_x(n) = H^2(n)S_F(n) \tag{5.1.21}$$

This result is extremely useful in applications.

5.2   THE MULTI-DEGREE OF FREEDOM LINEAR SYSTEM

5.2.1   Natural Modes and Frequencies of a Continuously Distributed Structure

It may be regarded as an experimental fact that a continuously distributed elastic structure with low damping, when excited by a sinusoidal force, will vibrate in resonance at certain sharply-defined characteristic frequencies. Associated with each such resonant or natural frequency there will also be a characteristic, or modal, form of vibration amplitude distributed throughout the structure. Such forms are called the *normal modes* of the structure. For example, Fig. 5.2.1 depicts the first four normal modes of a vertical cantilever beam.

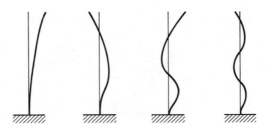

FIGURE 5.2.1.   First four normal modes of a cantilever beam.

These characteristic deflection modes and associated frequencies are properties of the structure, independent of the loads, and represent very fundamental
dynamical evidences of its internally distributed inertial and stiffness properties.* In fact, the set of normal modes may be regarded as a fundamental set of
special deflection forms by means of which *any* general deflection of the structure
may be expressed.

Thus, if z is a running coordinate (like height, for example) of a structure,
the modal deflection forms of lateral (x-direction) vibration may be written as
$x_i(z)$, where $i = 1, 2, 3, \ldots$. Then any deflection $x(z, t)$ may be expressed as the
sum:

$$x(z, t) = \sum_i x_i(z)\xi_i(t) \tag{5.2.1}$$

where the coefficients $\xi_i(t)$ indicate what fraction of each mode $x_i(z)$ enters the
given deflection pattern. The coefficients $\xi_i(t)$ are called the *generalized coordinates* of the system.

An important property of the normal modes $x_i(z)$ is their mutual orthogonality with respect to mass weighting, by which is meant that

$$\int_{\text{system}} x_i(z)x_j(z)m(z)\, dz = 0 \qquad (i \neq j) \tag{5.2.2}$$

where $m(z)$ is the mass of the structure per unit length.

Since the system is actually continuously distributed but responds at each
of its resonant frequencies like a *single* vibrating entity (or single degree of
freedom), it becomes very useful and convenient to liken continuous systems
to single-degree-of-freedom systems. It is helpful in this context to use the
energy approach. The kinetic energy of a single mass M is $\frac{1}{2}M\dot{x}^2$, where $\dot{x}$ is
its velocity of displacement. We now seek the corresponding energy for the
distributed system.

The lateral displacement being $x(z, t)$, the elemental kinetic energy at point
z is

$$\tfrac{1}{2}m(z)[\dot{x}(z, t)]^2\, dz$$

---

*Details on procedures for determining normal modes and natural frequencies may be found, for
example, in [5-1] or [5-3].

The kinetic energy of the whole system is therefore

$$\text{K.E.} = \tfrac{1}{2} \int_{\text{system}} m(z)[\dot{x}(z, t)]^2 \, dz \qquad (5.2.3)$$

If the system is vibrating in the single resonance mode $x_i(z)$, then

$$\dot{x}(z, t) = x_i(z)\dot{\xi}_i(t) \qquad (5.2.4)$$

so that the kinetic energy becomes

$$\text{K.E.} = \tfrac{1}{2} M_i \dot{\xi}_i^2 \qquad (5.2.5)$$

where

$$M_i = \int_{\text{system}} [x_i(z)]^2 m(z) \, dz \qquad (5.2.6)$$

M_i is known as the *generalized mass* of the system in the ith normal mode.

In this sense a continuous system vibrating in any one of its normal modes may be viewed as though it were simply a single-degree-of-freedom system with a mass M_i and velocity $\dot{\xi}_i$.

5.2.2 General Expression of the Response

Consider a structure for which it may be assumed that the displacement in the direction x is the same for all points in the structure that have the same coordinate z (Fig. 5.2.2). It can be shown [5-2] that if the damping ratio is small the generalized coordinates $\xi_i(t)$ satisfy the equations

$$\ddot{\xi}_i(t) + 2\zeta_i(2\pi n_i)\dot{\xi}_i(t) + (2\pi n_i)^2 \xi_i(t) = \frac{Q_i(t)}{M_i} \qquad (i = 1, 2, 3, \ldots) \qquad (5.2.7)$$

where ζ_i, n_i, and M_i are the damping ratio, the natural frequency, and the generalized mass (Eq. 5.2.6) in the ith mode. The quantity $Q_i(t)$ is known as

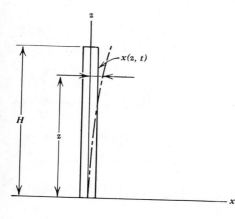

FIGURE 5.2.2. Schematic representation of a slender structure.

the *generalized force* in the ith mode and has the expression

$$Q_i(t) = \int_0^H p(z, t) x_i(z) \, dz \tag{5.2.8}$$

where H is the height of the structure, and $p(z, t)$ is the time-dependent load per unit of length acting on the system. It is seen that each of the equations 5.2.7 is of exactly the same form as the equation of motion of the single-degree-of-freedom system Eq. 5.1.2.

If the load $p(z, t)$ is such that

$$p(z, t) = F(t)\delta(z - z_1) \tag{5.2.9}$$

where $\delta(z - z_1)$ is defined in a manner similar to Eqs. 5.1.11, that is, if the structure is subjected to a concentrated force $F(t)$ acting at a point of coordinate z_1, the generalized force $Q_i(t)$ will be

$$Q_i(t) = \lim_{\Delta z \to 0} \int_{z_1}^{z_1 + \Delta z} p(z, t) x_i(z) \, dz$$
$$= x_i(z_1) F(t) \tag{5.2.10}$$

5.2.3   Response to a Harmonic Load

If a concentrated load

$$F(t) = F_0 \cos 2\pi n t \tag{5.2.11}$$

is acting on the structure at a point of coordinate z_1, by virtue of Eq. 5.2.10 the generalized force in the ith mode will be

$$Q_i(t) = F_0 x_i(z_1) \cos 2\pi n t \tag{5.2.12}$$

and the steady-state solutions of Eqs. 5.2.7 will be similar to the solution 5.1.6 of Eq. 5.1.5, i.e.,

$$\xi_i(t) = F_0 x_i(z_1) H_i(n) \cos(2\pi n t - \phi_i) \tag{5.2.13}$$

where

$$H_i(n) = \frac{1}{4\pi^2 n_i^2 M_i \{[1 - (n/n_i)^2]^2 + 4\zeta_i^2 (n/n_i)^2\}^{1/2}} \tag{5.2.14}$$

$$\phi_i = \tan^{-1} \frac{2\zeta_i(n/n_i)}{1 - (n/n_i)^2} \tag{5.2.15}$$

It follows from Eq. 5.2.1 that the response of the structure at a point of coordinate z is

$$x(z, t) = F_0 \sum_i x_i(z) x_i(z_1) H_i(n) \cos(2\pi n t - \phi_i) \tag{5.2.16}$$

It is convenient to write Eq. 5.2.16 in the form

$$x(z, t) = F_0 H(z, z_1, n) \cos[2\pi n t - \phi(z, z_1, n)] \tag{5.2.17}$$

where, as follows immediately from Eqs. A2.4a and A2.4b,

$$H(z, z_1, n) = \left\{ \left[\sum_i x_i(z)x_i(z_1)H_i(n) \cos \phi_i \right]^2 \right.$$
$$\left. + \left[\sum_i x_i(z)x_i(z_1)H_i(n) \sin \phi_i \right]^2 \right\}^{1/2} \tag{5.2.18}$$

$$\phi(z, z_1, n) = \tan^{-1} \frac{\sum_i x_i(z)x_i(z_1)H_i(n) \sin \phi_i}{\sum_i x_i(z)x_i(z_1)H_i(n) \cos \phi_i} \tag{5.2.19}$$

Similarly, the steady-state response at a point of coordinate z to a concentrated load

$$F(t) = F_0 \sin 2\pi nt \tag{5.2.20}$$

acting on the structure at a point of coordinate z_1 can be written as

$$x(z, t) = F_0 H(z, z_1, n) \sin[2\pi nt - \phi(z, z_1, n)] \tag{5.2.21}$$

5.2.4 Response to a Concentrated Stationary Random Load

Let the response at a point of coordinate z to a concentrated unit impulsive load $\delta(t)$ acting at time $t = 0$ at a point of coordinate z_1 be denoted $G(z, z_1, t)$. Following the same reasoning that led to Eq. 5.1.13, the response $x(z, t)$ of the structure at a point of coordinate z to an arbitrary load $F(t)$ acting at a point of coordinate z_1 can be expressed as

$$x(z, t) = \int_0^\infty G(z, z_1, t)F(t - \tau) \, d\tau \tag{5.2.22}$$

Note the complete similarity of Eqs. 5.2.17, 5.2.21, and 5.2.22 to Eqs. 5.1.6, 5.1.10, and 5.1.13, respectively. Therefore, by following the same steps that led to Eq. 5.1.21 there results

$$S_x(z, z_1, n) = H^2(z, z_1, n)S_F(n) \tag{5.2.23}$$

where $S_x(z, z_1, n)$ is the spectral density of the displacement $x(z, t)$, the mechanical admittance function $H(z, z_1, n)$ is given by Eq. 5.2.18, and $S_F(n)$ is the spectral density of the force $F(t)$.

5.2.5 Response to Two Concentrated Stationary Random Loads

Let $x(z, t)$ now denote the response of the structure at a point of coordinate z to the action of two stationary random loads $F_1(t)$ and $F_2(t)$ acting at points of coordinates z_1 and z_2, respectively. The autocovariance of the response can be written as

$$R_x(z, \tau) = \lim_{T \to \infty} \frac{1}{T} \int_{-T/2}^{T/2} x(z, t)x(z, t+\tau)\, dt$$

$$= \lim_{T \to \infty} \frac{1}{T} \int_{-T/2}^{T/2} \left[\int_0^\infty G(z, z_1, \tau_1)F_1(t-\tau_1)\, d\tau_1 \right.$$

$$+ \int_0^\infty G(z, z_2, \tau_1)F_2(t-\tau_1)\, d\tau_1 \Bigg]$$

$$\times \left[\int_0^\infty G(z, z_1, \tau_2)F_1(t+\tau-\tau_2)\, d\tau_2 \right.$$

$$+ \int_0^\infty G(z, z_2, \tau_2)F_2(t+\tau-\tau_2)\, d\tau_2 \Bigg]\, dt$$

$$= \int_0^\infty G(z, z_1, \tau_1) \left[\int_0^\infty G(z, z_1, \tau_2)R_{F_1}(\tau+\tau_1-\tau_2)\, d\tau_2 \right] d\tau_1$$

$$+ \int_0^\infty G(z, z_2, \tau_2) \left[\int_0^\infty G(z, z_2, \tau_2)R_{F_2}(\tau+\tau_1-\tau_2)\, d\tau_2 \right] d\tau_1$$

$$+ \int_0^\infty G(z, z_1, \tau_1) \left[\int_0^\infty G(z, z_2, \tau_2)R_{F_1 F_2}(\tau+\tau_1-\tau_2)\, d\tau_2 \right] d\tau_1$$

$$+ \int_0^\infty G(z, z_2, \tau_1) \left[\int_0^\infty G(z, z_1, \tau_2)R_{F_2 F_1}(\tau+\tau_1-\tau_2)\, d\tau_2 \right] d\tau_1$$

$$\text{(5.2.24)}$$

where the definition of the cross-covariance function (Eq. A2.29) was used.

The spectral density of the displacement $x(z, t)$ is

$$S_x(z, n) = 2 \int_{-\infty}^\infty R_x(z, \tau) \cos 2\pi n\tau\, d\tau$$

$$= 2 \int_{-\infty}^\infty R_x(z, \tau) \cos 2\pi n[(\tau+\tau_1-\tau_2)-(\tau_1-\tau_2)]\, d(\tau+\tau_1-\tau_2) \quad \text{(5.2.25)}$$

Let Eq. 5.2.24 be substituted into Eq. 5.2.25. Using the relations

$$H(z, z_i, n) \cos \phi(z, z_i, n) = \int_0^\infty G(z, z_i, \tau) \cos 2\pi n\tau\, d\tau \quad \text{(5.2.26)}$$

$$H(z, z_i, n) \sin \phi(z, z_i, n) = \int_0^\infty G(z, z_i, \tau) \sin 2\pi n\tau\, d\tau \quad \text{(5.2.27)}$$

(which are similar to Eqs. 5.1.14 and 5.1.15) and

$$H(z, z_1, n)H(z, z_2, n) \cos[\phi(z, z_1, n) - \phi(z, z_2, n)]$$

$$= \int_0^\infty \int_0^\infty G(z, z_1, \tau_1)G(z, z_2, \tau_2) \cos 2\pi n(\tau_1 - \tau_2)\, d\tau_1\, d\tau_2 \quad \text{(5.2.28)}$$

$$H(z, z_1, n)H(z, z_2, n) \sin[\phi(z, z_1, n) - \phi(z, z_2, n)]$$

$$= \int_0^\infty \int_0^\infty G(z, z_1, \tau_1)G(z, z_2, \tau_2) \sin 2\pi n(\tau_1 - \tau_2) \, d\tau_1 \, d\tau_2 \qquad (5.2.29)$$

which can be derived immediately from Eqs. 5.2.26 and 5.2.27, and following the steps that led to Eq. 5.1.21, there results

$$S_x(z, n) = H^2(z, z_1, n)S_{F_1}(n) + H^2(z, z_2, n)S_{F_2}(n)$$

$$+ 2H(z, z_1, n)H(z, z_2, n)\{S^C_{F_1 F_2}(n) \cos[\phi(z, z_1, n) - \phi(z, z_2, n)]$$

$$+ S^Q_{F_1 F_2}(n) \sin[\phi(z, z_1, n) - \phi(z, z_2, n)]\} \qquad (5.2.30)$$

in which $S_x(z, n)$ is the spectral density of the displacement at a point of co-ordinate z, $H(z, z_i, n)$ are the mechanical admittance functions defined as in Eq. 5.2.18, $S_{F_i}(n)$ is the spectral density of the force $F_i(t)$, and $S^C_{F_1 F_2}(n)$, $S^Q_{F_1 F_2}$ are the co-spectrum and the quadrature spectrum of the forces $F_1(t)$ and $F_2(t)$ defined as in Eqs. A2.33 and A2.34, respectively.

5.2.6   Effect of the Cross-Correlation of the Loads Upon the Magnitude of the Response

Consider two random stationary loads $F_1(t)$ and $F_2(t)$ acting at points of co-ordinates z_1 and z_2, respectively, and such that $F_1(t) = F_2(t)$ at all times. By definition, in this case the cross-correlation equals the autocorrelation, $S^C_{F_1 F_2} = S_{F_1}(n)$, and $S^Q_{F_1 F_2} = 0$ (Eqs. A2.21 and A2.29, A2.20 and A2.33, A2.23 and A2.34). The loads $F_1(t)$ and $F_2(t)$ are said to be *perfectly correlated*. The spectral density of the response to the two loads can then be written as (Eq. 5.2.30)

$$S_x(z, n) = \{H^2(z, z_1, n) + H^2(z, z_2, n)$$

$$+ 2H(z, z_1, n)H(z, z_2, n) \cos[\phi(z, z_1, n) - \phi(z, z_2, n)]\}S_{F_1}(n) \qquad (5.2.31)$$

In the particular case in which $z_1 = z_2$

$$S_x(z, n) = 4H^2(z, z_1, n)S_{F_1}(n) \qquad (5.2.32)$$

Consider now the case in which the loads $F_1(t)$ and $F_2(t)$ are such that their cross-covariance $R_{F_1 F_2}(\tau) = 0$. Then, by virtue of Eqs. A2.33 and A2.34,

$$S^C_{F_1 F_2}(n) = S^Q_{F_1 F_2}(n) = 0 \qquad (5.2.33)$$

and, if the statistical properties of the loads are the same, that is, if $S_{F_1}(n) \equiv S_{F_2}(n)$,

$$S_x(z, n) = [H^2(z, z_1, n) + H^2(z, z_2, n)]S_{F_1}(n) \qquad (5.2.34)$$

or, if $z_1 = z_2$,

$$S_x(z, n) = 2H^2(z, z_1, n)S_{F_1}(n) \qquad (5.2.35)$$

The spectrum of the structural response to the action of the uncorrelated loads is thus seen to be only one half as large as in the case of the perfectly correlated loads.

5.2.7 Distributed Stationary Random Loads

The spectral density of the response to a distributed stationary random load may be obtained by generalizing Eq. 5.2.30 to the case where an infinite number of elemental loads rather than two concentrated loads are acting on the structure. Thus, if the load is distributed over an area A, and if it is noted that in the absence of torsion the mechanical admittance functions are independent of the across-wind coordinate y, the spectral density of the along-wind fluctuating deflection may be written as

$$
\begin{aligned}
S_x(z, n) = \int_A \int_A & H(z, z_1, n) H(z, z_2, n) \\
& \times \{ S^C_{p'_1 p'_2}(n) \cos[\phi(z, z_1, n) - \phi(z, z_2, n)] \\
& + S^Q_{p'_1 p'_2}(n) \sin[\phi(z, z_1, n) - \phi(z, z_2, n)] \} \, dA_1 \, dA_2
\end{aligned} \tag{5.2.36}
$$

where p'_1 and p'_2 denote pressures acting at points of coordinates y_1, z_1 and y_2, z_2, respectively.

It can be verified that from Eq. 5.2.36 there follows[*]

$$
\begin{aligned}
S_x(z, n) = \frac{1}{16\pi^4} \sum_i \sum_j & \frac{x_i(z) x_j(z)}{n_i^2 n_j^2 M_i M_j} \\
& \times \frac{1}{\{[1 - (n/n_i)^2]^2 + 4\zeta_i^2(n/n_i)^2\}\{1 - (n/n_j)^2]^2 + 4\zeta_j^2(n/n_j)^2\}} \\
& \times \left[\left\{ \left[1 - \left(\frac{n}{n_i}\right)^2 \right]\left[1 - \left(\frac{n}{n_j}\right)^2 \right] + 4\zeta_i\zeta_j \frac{n}{n_i}\frac{n}{n_j} \right\} \int_A \int_A x_i(z_1) x_j(z_2) \right. \\
& \times S^C_{p'_1 p'_2}(n) \, dA_1 \, dA_2 + \left\{ 2\zeta_j \frac{n}{n_j}\left[1 - \left(\frac{n}{n_i}\right)^2 \right] - 2\zeta_i \frac{n}{n_i}\left[1 - \left(\frac{n}{n_j}\right)^2 \right] \right\} \\
& \left. \times \int_A \int_A x_i(z_1) x_j(z_2) S^Q_{p'_1 p'_2}(n) \, dA_1 \, dA_2 \right]
\end{aligned} \tag{5.2.37}
$$

If the damping is small and the resonant peaks are well separated, the cross-terms in Eq. 5.2.37 become negligible and

$$
S_x(z, n) = \sum_i \frac{x_i^2(z) \displaystyle\int_A \int_A x_i(z_1) x_i(z_2) S^C_{p'_1 p'_2}(n) \, dA_1 \, dA_2}{16\pi^4 n_i^4 M_i^2 \{[1 - (n/n_i)^2]^2 + 4\zeta_i^2(n/n_i)^2\}} \tag{5.2.38}
$$

5.3 EXAMPLE: ALONG-WIND RESPONSE

To illustrate the application of the material presented in this chapter, the case of the along-wind response of tall buildings will be dealt with below.

---

[*]By using Eqs. 5.2.18 and 5.2.19, 5.2.14 and 5.2.15, A2.4a and A2.4b. For a derivation of Eq. 5.2.37 in terms of complex variables, see [5-2, p. 91].

5.3.1 Mean Response

If in Eq. 5.2.8 the load p per unit of length is independent of time, the corresponding along-wind deflection, which will be denoted by $\bar{x}(z)$, results immediately from Eqs. 5.2.1 and 5.2.7:

$$\bar{x}(z) = \sum_i \frac{\int_0^H \bar{p}(z)x_i(z)\,dz}{4\pi^2 n_i^2 M_i} x_i(z) \qquad (5.3.1)$$

where

$$M_i = \int_0^H x_i^2(z)m(z)\,dz \qquad (5.3.2)$$

and $\bar{p}$ denotes the time-invariant load.

As indicated in Chapter 4, the mean wind load acting on a building of width B (Fig. 5.3.1) may be written as

$$\bar{p}(z) = \tfrac{1}{2}\rho(C_w + C_l)BU^2(z) \qquad (5.3.3)$$

where ρ is the air density, C_w and C_l are the values, averaged over the building width, of the mean pressure coefficient on the windward face and suction coefficient on the leeward face, respectively, and $U(z)$ is the mean speed at

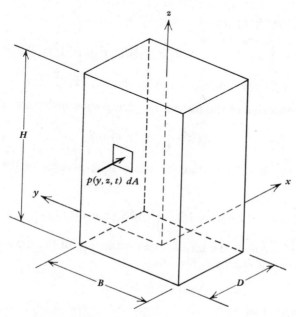

FIGURE 5.3.1. Schematic view of a building.

elevation z in the undisturbed oncoming flow. Equation 5.3.1 then becomes

$$\bar{x}(z) = \tfrac{1}{2}\rho(C_w + C_l)B \sum_i \frac{\displaystyle\int_0^H U^2(z)x_i(z)\, dz}{4\pi^2 n_i^2 M_i} x_i(z) \tag{5.3.4}$$

5.3.2   Fluctuating Response to Wind: Deflections and Accelerations

As indicated in Chapter 4, the co-spectrum of the pressures at point M_1, M_2 of coordinates (y_1, z_1), (y_2, z_2), respectively, may be written as

$$S^C_{p_1' p_2'}(n) = S_{p'}^{1/2}(z_1, n)S_{p'}^{1/2}(z_2, n)\,\mathrm{Coh}(y_1, y_2, z_1, z_2, n)N(n) \tag{5.3.5}$$

where $S_{p'}^{1/2}(z_1, n)$ is the spectral density of the pressures at point $M_i\,(i=1, 2)$ and $\mathrm{Coh}(y_1, y_2, z_1, z_2, n)$ and $N(n)$ are the across-wind and the along-wind cross-correlation coefficient, respectively. By definition, if both M_1 and M_2 are on the same—windward or leeward—face of the structure, $N(n) \equiv 1$. The expression for $S_{p'}(z_i, n)$ is, approximately,

$$S_{p'}(z_i, n) = \rho^2 C^2 U^2(z)S_u(z_i, n) \tag{5.3.6}$$

where $C = C_w$ or $C = C_l$ according as M_i is on the windward or on the leeward face, and $S_u(z_i, n)$ is the spectral density of the longitudinal velocity fluctuations at elevation z_i in the undisturbed oncoming flow $(i=1, 2)$. Equation 5.2.38 thus becomes

$$\begin{aligned}
S_x(z, n) = \frac{\rho^2}{16\pi^4} &\sum_i \frac{x_i^2(z)[C_w^2 + 2C_w C_l N(n) + C_l^2]}{n_i^4 M_i^2\{[1-(n/n_i)^2]^2 + 4\zeta_i^2(n/n_i)^2\}} \\
&\times \int_0^B \int_0^B \int_0^H \int_0^H x_i(z_1)x_i(z_2)U(z_1)U(z_2)S_u^{1/2}(z_1)S_u^{1/2}(z_2) \\
&\times \mathrm{Coh}(y_1, y_2, z_1, z_2, n)\, dy_1\, dy_2\, dz_1\, dz_2
\end{aligned} \tag{5.3.7}$$

The mean square value of the fluctuating along-wind deflection is (Eq. A2.15)

$$\sigma_x^2(z) = \int_0^\infty S_x(z, n)\, dn \tag{5.3.8}$$

From Eq. A2.16b it follows that the mean square value of the along-wind acceleration is

$$\sigma_{\ddot{x}}^2(z) = 16\pi^4 \int_0^\infty n^4 S_x(z, n)\, dn \tag{5.3.9}$$

The expected value of the largest peak occurring in the time interval T is, in the case of the fluctuating deflection,

$$x_{\max}(z) = K_x(z)\sigma_x(z) \tag{5.3.10}$$

where, as indicated in Appendix A2 (Eqs. A2.38 and A2.43), the peak factor $K_x(z)$ is, approximately,

$$K_x(z)=[2 \ln v_x(z)T]^{1/2}+\frac{0.577}{[2 \ln v_x(z)T]^{1/2}} \tag{5.3.11}$$

and

$$v_x(z)=\left[\frac{\displaystyle\int_0^\infty n^2 S_x(z,\,n)\,dn}{\displaystyle\int_0^\infty S_x(z,\,n)\,dn}\right]^{1/2} \tag{5.3.12}$$

Similarly, the largest peak of the along-wind acceleration is, approximately,

$$\ddot{x}_{\max}(z)=K_{\ddot{x}}(z)\sigma_{\ddot{x}}(z) \tag{5.3.13}$$

where

$$K_{\ddot{x}}(z)=[2 \ln v_{\ddot{x}}(z)T]^{1/2}+\frac{0.577}{[2 \ln v_{\ddot{x}}(z)T]^{1/2}} \tag{5.3.14}$$

and

$$v_{\ddot{x}}(z)=\left[\frac{\displaystyle\int_0^\infty n^6 S_x(z,\,n)\,dn}{\displaystyle\int_0^\infty n^4 S_x(z,\,n)\,dn}\right]^{1/2} \tag{5.3.15}$$

It is convenient for computational purposes to rewrite the fluctuating response in terms of nondimensional quantities in the form

$$\frac{\sigma_x(z)}{H}=\frac{\rho BH}{m_0}\frac{J_0(z)}{4\pi^2} \tag{5.3.16}$$

$$\frac{v_x(z)H}{u_*}=\frac{J_1(z)}{J_2(z)} \tag{5.3.17}$$

$$\frac{\sigma_{\ddot{x}}(z)H}{u_*^2}=\frac{\rho BH}{m_0}J_2(z) \tag{5.3.18}$$

$$\frac{v_{\ddot{x}}(z)H}{u_*}=\frac{J_3(z)}{J_2(z)} \tag{5.3.19}$$

where m_0 is the mass of the building per unit height at some specified elevation, u_* is the friction velocity (see Chapter 2)—or any suitably chosen reference velocity—and

$$J_L(z)=\left[\sum_r x_r^2(z)I_{rrL}/\tilde{f}_r^4\tilde{M}_r^2\right]^{1/2} \quad (L=0,\,1,\,2,\,3) \tag{5.3.20}$$

$$\tilde{M}_r=\int_0^1 x_r^2(Z)\frac{m(Z)}{m_0}\,dZ \tag{5.3.21}$$

$$Z=\frac{z}{H}; \quad Y=\frac{y}{H} \tag{5.3.22}$$

$$\tilde{f}_r = \frac{n_r H}{u_*} \tag{5.3.23}$$

$$I_{rrL} = \int_0^\infty \tilde{f}^{2L} \phi_{rr}^*(\tilde{f}) Y_{rr}(\tilde{f})\, d\tilde{f} \tag{5.3.24}$$

$$\tilde{f} = \frac{nH}{u_*} \tag{5.3.25}$$

$$\phi_{rr}^*(\tilde{f}) = \frac{1}{[1 - (\tilde{f}/\tilde{f}_r)^2]^2 + [2\zeta_r(\tilde{f}/\tilde{f}_r)]^2} \tag{5.3.26}$$

$$Y_{rr}(\tilde{f}) = \int_0^1 \int_0^1 \int_0^1 \int_0^1 [C_w^2 + 2 C_w C_l N(u_* \tilde{f}/H) + C_l^2]$$
$$\times x_r(Z_1) x_r(Z_2) \tilde{U}(Z_1) \tilde{U}(Z_2) \left[\frac{n S_u(Z_1, n)}{u_*^2 \tilde{f}} \frac{n S_u(Z_2, n)}{u_*^2 \tilde{f}} \right]^{1/2}$$
$$\times \text{Coh}(Y_1, Y_2, Z_1, Z_2, n)\, dY_1\, dY_2\, dZ_1\, dZ_2 \tag{5.3.27}$$

$$\tilde{U}(Z) = \frac{U(Z)}{u_*} \tag{5.3.28}$$

5.3.3 Total Fluctuating Response to Wind as a Sum of Background and Resonant Contributions

Consider a single-degree-of-freedom linearly elastic system with mass m, natural frequency n_1, and damping ratio ζ_1. Let this system be subjected to the action of a forcing function with a spectrum $S(n)$ such that (Fig. 5.3.2a)

$$S(n) \equiv S_0 \qquad (n \geqslant 0) \tag{5.3.29}$$

where S_0 is a constant. The mean square value of the response can be written as

$$\sigma_x^2 = S_0 \int_0^\infty |H(n)|^2\, dn \tag{5.3.30}$$

where

$$|H(n)|^2 = \frac{1}{16\pi^4 n_1^4 m^2} \frac{1}{[1 - (n/n_1)^2]^2 + [2\zeta_1(n/n_1)]^2} \tag{5.3.31}$$

The quantity $|H(n)|^2$, which is represented in Fig. 5.3.2b, is an analytic function; therefore, the integral in Eq. 5.3.30 can be evaluated by means of complex integration or integral tables to yield (see [5-3], p. 501)

$$\sigma_x^2 = \frac{1}{16\pi^4 n_1^4 m^2} \frac{\pi n_1}{4\zeta_1} S_0 \tag{5.3.32}$$

If the damping ratio ζ_1 is small, the bulk of the contributions to the total value σ_x^2 is due to the forcing function components of frequencies $n_1 - \Delta n_1 < n < n_1 + \Delta n_1$, where $\Delta n_1/n_1$ is small (hatched area in Fig. 5.3.2a). If, as in the

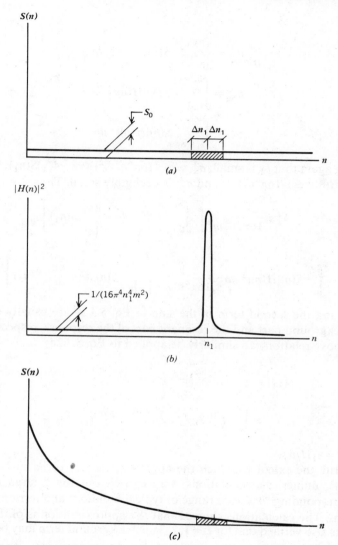

FIGURE 5.3.2. (*a*) White noise spectrum; (*b*) square of modulus of mechanical admittance; (*c*) decaying curve $S(n)$.

case of atmospheric turbulence or of the structural response it generates, the quantity $S(n)$ is represented by a decaying curve (Fig. 5.3.2*c*), the integral

$$\sigma_x^2 = \int_0^\infty S(n)|H(n)|^2 \, dn \tag{5.3.33}$$

may be written as the sum of three contributions:

$$\sigma_x^2 = \sigma_{x_1}^2 + \sigma_{x_2}^2 + \sigma_{x_3}^2 \tag{5.3.34}$$

where

$$\sigma_{x_1}^2 = \int_0^{n_1 - \Delta_1} S(n)|H(n)|^2 \, dn \tag{5.3.35}$$

$$\sigma_{x_2}^2 = \int_{n_1 - \Delta_1}^{n_1 + \Delta_1} S(n)|H(n)|^2 \, dn \tag{5.3.36}$$

$$\sigma_{x_3}^2 = \int_{n_1 + \Delta_1}^{\infty} S(n)|H(n)|^2 \, dn \tag{5.3.37}$$

Assuming again that ζ_1 is small, $\sigma_{x_2}^2 \simeq [1/(16\pi^4 n_1^4 m^2)](\pi n_1/4\zeta_1)S(n_1)$; for $0 < n < n_1 - \Delta_1$, $|H(n)|^2 \simeq 1/16\pi^4 n_1^4 m^2$; and $\sigma_{x_3}^2$ is negligibly small. Thus,

$$\sigma_x^2 \simeq \frac{1}{16\pi^4 n_1^4 m^2} \left[\int_0^{n_1 - \Delta} S(n) \, dn + \frac{\pi n_1}{4\zeta_1} S(n_1) \right] \tag{5.3.38}$$

or

$$\int_0^{\infty} S(n)|H(n)|^2 \, dn \simeq \frac{1}{16\pi^4 n_1^4 m^2} \left[\int_0^{\infty} S(n) \, dn + \frac{\pi n_1}{4\zeta_1} S(n_1) \right] \tag{5.3.39}$$

The first and the second term of the sum in Eq. 5.3.39 are usually referred to as the background part and the resonant part of the response, respectively.

The above relation can similarly be applied to Eq. 5.3.24:

$$I_{11L} = \int_0^{\infty} \phi_{11}^*(\tilde{f})\tilde{f}^{2L} Y_{11}(\tilde{f}) \, d\tilde{f}$$

$$\simeq \int_0^{\infty} \tilde{f}^{2L} Y_{11}(\tilde{f}) \, d\tilde{f} + \frac{\pi \tilde{f}_1}{4\zeta_1} \tilde{f}_1^{2L} Y_{11}(\tilde{f}_1) \tag{5.3.40}$$

where $\tilde{f}_1 = n_1 H/u_*$.

To verify the extent to which the approximation involved in Eq. 5.3.40 is acceptable, numerical calculations were carried out for a large number of cases corresponding to a wide range of typical buildings and terrain roughness conditions. The calculations showed that the approximation is of the order of 1%. It was also verified that for $L = 1, 2, 3$ the background term may be neglected and, therefore,

$$I_{11L} \simeq \frac{\pi \tilde{f}_1}{4\zeta_1} \tilde{f}_1^{2L} Y_{11}(\tilde{f}_1) \qquad (L = 1, 2, 3) \tag{5.3.41}$$

It is convenient to define the quantities $\mathscr{B}$ and $\mathscr{R}$ as follows:

$$\mathscr{B} = \frac{1}{C_D^2} \int_0^{\infty} Y_{11}(\tilde{f}) \, d\tilde{f} \tag{5.3.42}$$

$$\mathscr{R} = \frac{1}{C_D^2} \frac{\pi \tilde{f}_1}{4\zeta_1} Y_{11}(\tilde{f}_1) \tag{5.3.43}$$

If the notation

$$\tilde{Y}_{11}(\tilde{f}_1) = \frac{1}{C_w^2 + 2C_w C_l N(n_1) + C_l^2} Y_{11}(\tilde{f}_1) \tag{5.3.44}$$

is used, the quantity $\mathscr{R}$ may be written as

$$\mathscr{R} = \frac{C_w^2 + 2C_w C_l N(n_1) + C_l^2}{(C_w + C_l)^2} \frac{\pi \tilde{f}_1}{4\zeta_1} \tilde{Y}_{11}(\tilde{f}) \tag{5.3.45}$$

Let the mean speed U in Eq. 5.3.28 be represented by the logarithmic law. The zero plane displacement z_d will then be a parameter in the expression for $\mathscr{B}$. If the quantity $\tilde{\mathscr{B}}$ is defined as

$$\tilde{\mathscr{B}} = \frac{1}{C_D^2} \int_0^\infty Y_{11}(\tilde{f})|_{z_d=0} \, d\tilde{f} \tag{5.3.46}$$

it can be verified that, approximately,

$$\mathscr{B} \simeq \left(1 - \frac{z_d}{H}\right) \tilde{\mathscr{B}} \tag{5.3.47}$$

Equations 5.3.42 through 5.3.47 may be used for the computation of along-wind response.

Finally, recall that expressions for $S_u(z, n)$ in Eq. 5.3.27 are found in Chapter 2 and that, as indicated in Chapters 2 and 4, it is reasonable to assume

$$\text{Coh}(y_1, y_2, z_1, z_2, n) = \exp\left\{-\frac{n[C_z^2(z_1 - z_2)^2 + C_y^2(y_1 - y_2)^2]^{1/2}}{\frac{1}{2}[U(z_1) + U(z_2)]}\right\} \tag{5.3.48}$$

$$N(n) = \frac{1}{\tilde{\xi}} - \frac{1}{2\tilde{\xi}^2}(1 - e^{-2\tilde{\xi}}) \tag{5.3.49}$$

where C_y and C_z are known as exponential decay parameters,

$$\tilde{\xi} = \frac{15.4n\,\Delta x}{\bar{U}} \tag{5.3.50}$$

$\bar{U} = U(\frac{2}{3}H)$, and Δx is the smallest of the dimensions B, H, and D.

REFERENCES

5-1   W. C. Hurty and M. F. Rubinstein, *Dynamics of Structures*, Prentice-Hall, Englewood Cliffs, N.J., 1964.

5-2   J. D. Robson, *An Introduction to Random Vibration*, Elsevier, New York, 1964.

5-3   L. Meirovitch, *Analytical Methods in Vibrations*, MacMillan, Collier-MacMillan Canada, Ltd., Toronto, 1967.

6

Aeroelastic Phenomena

A body immersed in a flow is subjected to surface stresses induced by that flow. If there is turbulence in the incident flow, this will be the source of time-dependent surface stresses. Such stresses are also caused by flow fluctuations initiated by the body itself.

Further, if the body moves or deforms appreciably under the induced surface forces, these deflections, changing as they do the boundary conditions of the flow, will affect the fluid forces that in turn will influence the deflections. *Aeroelasticity* is the discipline concerned with the study of phenomena wherein aerodynamic forces and structural motions interact significantly.

An *aerodynamic instability* can be a phenomenon occurring wholly within the flow alone, as when a trail of vortices or a rapidly diverging wake is shed from a fixed body. But if a body in a fluid flow deflects under some force and the initial deflection gives rise to succeeding deflections of oscillatory and/or divergent character, an *aeroelastic instability* is said to be produced. A purely aerodynamic instability such as vortex shedding may occasion structural deflection as well, initiating a phenomenon having aeroelastic character. All aeroelastic instabilities involve aerodynamic forces that act upon the body as a consequence of its motion. Such forces are termed *self-excited*.

The purpose of this chapter is to discuss fundamental aspects of aeroelastic phenomena that need to be taken into account in the design of certain structural members, towers, stacks, tall buildings, suspended-span bridges, cable roofs, piping systems, and power lines. Not all of these phenomena are presently completely understood. Indeed, only a few theoretical formulations from first principles exist for modeling aerodynamic forces on oscillating bodies. In most investigations, empirical models are set up in which the essence of the aero-dynamics must be contributed by experiment. The corresponding analytical models usually include just enough parameters to match the strongest observed features of the phenomena. Such models are thus minimally descriptive, but not explanatory in the sense of revealing basic physical causes; subtle but important details of the actual fluid-structure interaction may in certain cases be left unattended to.

Empirical models may only be used for the prediction of aeroelastic effects if the ranges of the governing nondimensional parameters in the model are close to those of the prototype. Most commonly, it is the Reynolds number of the prototype that is not realized in the model. As a result, uncertainties subsist in interpretation of model test results. (See also Chapter 7.)

Most of the empirical models described in this chapter apply to situations that may be considered, at least approximately, as two-dimensional. In practice, three-dimensional effects are present, owing to any number of factors such as: flow adjustments near the ends of finite cylinders; spanwise variations, either of the body cross section (as in the case of tapered stacks) or of the body deformation; nonuniform mean flows; or imperfect spatial coherence of the incident turbulence or of the vorticity shed in the wake of the body. Information on three-dimensional effects is in most cases scarce and must be obtained from wind tunnel experiments.

The topics dealt with in this chapter include vortex shedding and the associated lock-in phenomena, across-wind galloping, wake galloping, torsional divergence, flutter, and buffeting response in the presence of self-excited forces.

6.1 VORTEX SHEDDING AND THE LOCK-IN PHENOMENON

It was seen in Sect. 4.4 that under certain conditions a fixed bluff body sheds alternating vortices whose primary frequency N_s is, according to the Strouhal relation,

$$\frac{N_s D}{U} = \mathcal{S} \tag{6.1.1}$$

where $\mathcal{S}$ depends upon body geometry and the Reynolds number, D is the across-wind dimension of the body, and U is the mean velocity of the uniform flow in which the body is immersed. The frequency N_s is also that of the net primary forces acting transversely to the direction of U while the primary frequency of net forces acting in the flow direction will be $2N_s$. Actually the net force vector defined by the integral of instantaneous pressures over a given bluff body will vary in magnitude and direction with time in a fairly complex manner depending upon detailed body geometry and Reynolds number of the flow. Only the frequencies of its principal harmonics are given by N_s and $2N_s$.

If the body that instigates the vortex shedding is elastically supported or if it is subject to local contour deformation, it will deflect wholly or locally and, by this action, influence the local flow. Not many of the full range of possibilities latent in this situation have been studied in detail. Deformable steel shells have given rise to so-called ovalling oscillations [6-1] under these conditions. Many examples of cross-wind rigid-contour oscillations have been noted; and in water flows important along-flow deflections have been observed [6-2, 6-3, 6-4].

Unless otherwise noted, it will be assumed in this section that: the structure is a cylinder with a rigid surface; the oncoming flow has uniform mean velocity;

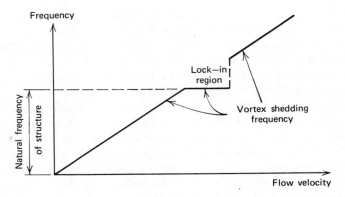

FIGURE 6.1.1. Evolution of vortex shedding frequency with wind velocity over elastic structure.

the deflections of the body are the same throughout its length; the body is elastically sprung and possesses mechanical damping in the across-wind direction; and it is rigidly constrained in the along-wind direction. Under the action of the vortices shed in its wake the cylinder will be driven periodically, but this driving will elicit only small response unless the Strouhal frequency of alternating pressures approaches the natural across-flow mechanical frequency of the cylinder. Near this frequency greater body movement is elicited and the body begins to interact strongly with the flow. It is experimentally observed at this point that the body mechanical frequency controls the vortex-shedding phenomenon even when variations in flow velocity displace the nominal Strouhal frequency away from the natural mechanical frequency by a few percent. This control of the phenomenon by the mechanical forces is commonly known as *lock-in*. Observations show that during lock-in the amplitude of the oscillations attains some fraction, rarely exceeding half, of the across-wind dimension of the body. The effect of lock-in upon vortex shedding is represented in Fig. 6.1.1, which shows that in the lock-in region the vortex-shedding frequency is constant rather than being a linear function of wind velocity, as suggested by Eq. 6.1.1 (and as it in fact is outside the lock-in region).

No completely successful analytical method has yet been developed, starting from basic flow principles, to represent the full range of response behavior of a bluff elastic body under the action of vortex shedding. It has, instead, been found reasonably fruitful to build empirical models and match their performance to reality by a judicious choice of parameters. References 6-1 to 6-33 and 6-87 provide an overview of the recent literature in this area.

6.1.1 Analytical Models of Vortex-Induced Response

Assume first that the circular cylinder dealt with above is fixed not only in the along-wind direction but in the across-wind direction as well. In this case, a reasonable first approximation to the across-wind force per unit span acting on the cylinder is

$$F = \tfrac{1}{2}\rho U^2 D C_{LS} \sin \omega_s t \qquad (6.1.2)$$

where $\omega_s = 2\pi N_s$, N_s satisfies the Strouhal relation (Eq. 6.1.1), and C_{LS} is the lift coefficient. (For a circular cylinder and Reynolds numbers $40 < \mathcal{R}e < 3 \times 10^5$, in a uniform smooth flow $C_{LS} \simeq 0.6$ [6-4, p. 72].) One important feature of this across-wind force is that it is imperfectly correlated along the cylinder span.

When the cylinder is allowed to oscillate, however, this simple expression for the forcing function F is inadequate for two reasons. First, the across-wind force increases with oscillation amplitude until a limiting amplitude is reached. Second, the spanwise correlation of the across-wind force also increases, as indicated in Fig. 6.1.2, in which η denotes the ratio of the maximum across-wind displacement to the diameter. Let y denote the across-wind displacement of a cylinder of unit length for which the effect of the imperfect spanwise force correlation is not explicitly accounted for.* The equation of motion of the cylinder can be written as

$$m\ddot{y} + c\dot{y} + ky = \mathscr{F}(y, \dot{y}, \ddot{y}, t) \qquad (6.1.3)$$

where m is the cylinder mass, c its mechanical damping constant, k its spring stiffness, and $\mathscr{F}$ its fluid-induced forcing function per unit span, which may be dependent on displacement y and its time derivatives $\dot{y}$ and $\ddot{y}$ as well as on time.

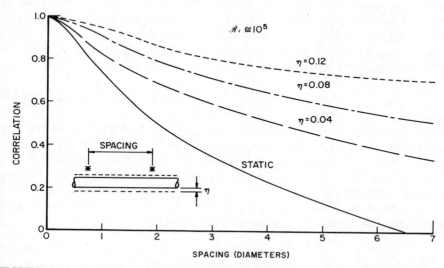

FIGURE 6.1.2. The effect of increasing the amplitude of oscillation of a circular cylinder on the correlation between the flow at two points [6-15, 6-38].

*This effect is accounted for empirically in [10-2] (see also Chapter 10, Eqs. 10.2.6, 10.2.7, and 10.2.16).

Much effort has been spent on finding by empirical means a suitable expression for $\mathscr{F}$ in Eq. 6.1.3 that fits the experimentally observed facts. The complexity of such an expression will depend on the detail and completeness with which the experimental facts are observed, on the one hand, and on the needs to be met by the subsequent predictions from the model, on the other.

A Lift Oscillator Model.   An empirical model that has achieved a certain degree of realism and acceptance was presented in [6-7] and [6-13]. This was later modified in [6-16], [6-17], [6-20], and [6-26]. The main elements of this model, as summarized in [6-33], are presented here. Let D be the across-wind projected dimension of the cylinder. Then, with the notations $\omega_1^2 = k/m$, $c/m = 2\zeta\omega_1$, ζ being the ratio of damping c to its critical value (see Sect. 5.1), Eq. 6.1.3 may be rewritten as

$$\frac{\ddot{y}}{D} + 2\zeta\omega_1 \frac{\dot{y}}{D} + \omega_1^2 \frac{y}{D} = \frac{\rho U^2}{2m} C_L \qquad (6.1.4)$$

where C_L is a time-varying lift coefficient. In particular, C_L is assumed to satisfy an oscillation equation with the characteristics of a Van der Pol oscillator [6-34] that is linked to the velocity of motion as follows:

$$\ddot{C}_L + a_1 \dot{C}_L + a_2 \dot{C}_L^3 + a_3 C_L = a_4 \dot{y} \qquad (6.1.5)$$

where $a_1, \ldots, a_4$ are constants to be evaluated from experiment. This assumption is suggested by similarities between the behavior of the system being studied and that of the Van der Pol oscillator, which exhibits low damping at low amplitudes and strong damping at high amplitudes.

Under the particular conditions of the problem, it is found expedient [6-33] to modify the oscillator equation to the following special form:

$$\ddot{C}_L + \omega_s^2 C_L - \left[C_{L0}^2 - C_L^2 - \left(\frac{\dot{C}_L}{\omega_s} \right)^2 \right] (\omega_s G \dot{C}_L - \omega_s^2 H C_L) = \omega_s F \frac{\dot{y}}{D} \qquad (6.1.6)$$

where $\omega_s = 2\pi N_s$ is the Strouhal circular oscillation frequency defined by Eq. 6.1.1, and C_{L0}, G, H, and F are parameters to be fitted according to experimental results. Note that when $\dot{y} = 0$ (stationary cylinder), the solution for C_L is approximated by $C_L = C_{L0} \sin \omega_s t$. Equations 6.1.4 and 6.1.6 constitute a coupled nonlinear pair.

Notwithstanding the complexity implied by this, the solutions in first approximation are taken to be sinusoidal, that is, in the forms

$$\frac{y}{D} = A \sin \omega t \qquad (6.1.7)$$

$$\frac{C_L}{C_{L0}} = B \sin(\omega t + \phi) \qquad (6.1.8)$$

with $\omega/\omega_1 \cong 1$ and $\omega/\omega_s \cong 1$ implied. A and B are amplitude factors of fluctuating displacement and lift, respectively, and ϕ is a phase angle.

Use of these solution forms in Eqs. 6.1.4 and 6.1.6 leads to the following results:

$$A = \frac{BC_{L0}/S_G}{(\delta^2 + 4)^{1/2}} \qquad (6.1.9a)$$

$$B^2 = 1 - \frac{F}{GS_G C_{L0}^2}\left[\frac{\delta}{\delta^2 + 4}\right] \qquad (6.1.9b)$$

$$\phi = \arctan\left(-\frac{2}{\delta}\right) \qquad (6.1.9c)$$

where

$$S_G = \frac{\zeta}{\mu} \qquad (6.1.10a)$$

$$\mu = \frac{\rho D^2}{8\pi^2 \mathscr{S}^2 m} \qquad (6.1.10b)$$

and δ is required to satisfy the cubic equation

$$\delta^3 - \Delta\delta^2 + \left(4 - \frac{HF}{\zeta GS_G}\right)\delta - 4\left(\Delta - \frac{F}{2\zeta S_G}\right) = 0 \qquad (6.1.11)$$

The parameters δ and Δ, called detunings, are given by

$$\delta = \frac{2}{\zeta}\left(\frac{\omega}{\omega_1} - 1\right) \qquad (6.1.12)$$

$$\Delta = \frac{2}{\zeta}\left(\frac{\omega_s}{\omega_1} - 1\right) \qquad (6.1.13)$$

For details, the reader is referred to [6-26] and related references.

It has been shown [6-33] that Eqs. 6.1.9 and 6.1.11 yield good representations of the entrained resonant response of spring-mounted rigid cylinders provided that the empirical parameters G, H, and F are appropriately selected. Results given for G, H, and F in [6-33] for the case of the circular-section cylinder are reproduced below. These have been based upon experimental data:

$$\log_{10} G = 0.25 - 0.21 S_G \qquad (6.1.13a)$$

$$H = \zeta h \qquad (6.1.13b)$$

$$\log_{10} h S_G^2 = -0.83 + 0.98 S_G \qquad (6.1.13c)$$

$$F = \frac{4GS_G}{h} \qquad (6.1.13d)$$

Figure 6.1.3 [6-33] presents a comparison between experimental and theoretical values of the amplitude A for one case of application of the lift oscillator model. In this case the parameters employed were $S_G = 0.5$, $D = 6$ mm, and $\omega_1 = 321.7$ rad/sec (see also [6-91]).

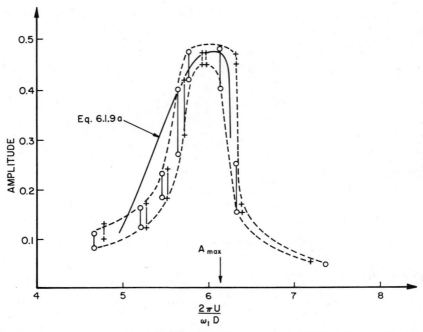

FIGURE 6.1.3. Comparison of theoretical and experimental results, vortex lock-in, lift oscillator model [6-33].

It is clear from what has been briefly presented above that the lift oscillator model is relatively accurate and that it requires considerable experimental data and analytic care.

An Empirical Linear Model.   A simpler model may be employed if the aims it is intended to serve are modest. In wind engineering a model simply identifying the maximum deflections of an experimental structure may often suffice. Regarding the response to vortex shedding, there is some incentive then to retain a linear model that can reproduce some of the principal observed experimental results near lock-in. The fact that, even in the case of the nonlinear lift oscillator model treated above only pure sinusoidal response is finally considered, reinforces the idea that a linear model may retain some usefulness in this context.

To set up such a model it is assumed that aerodynamic excitation, aerodynamic damping, and aerodynamic stiffness* must be provided to a linear mechanical oscillator. Since lock-in implies that the natural frequency of the mechanical oscillator controls the whole mechanical-aerodynamic system, the model should be driven at this frequency, that is, at $\omega = \omega_1$.

---

*Aerodynamic stiffness and damping are defined as aerodynamic forces expressible as products of flow-dependent constants by the displacement y and its time derivative $\dot{y}$, respectively.

With these observations in view, the following model is postulated:*

$$m[\ddot{y} + 2\zeta\omega_1\dot{y} + \omega_1^2 y] = \tfrac{1}{2}\rho U^2 (2D)\left[Y_1(K_1)\frac{\dot{y}}{U} + Y_2(K_1)\frac{y}{D}\right.$$

$$\left. + \tfrac{1}{2}C_L(K_1)\sin(\omega_1 t + \phi)\right] \qquad (6.1.14)$$

where $K_1 = D\omega_1/U$ and Y_1, Y_2, C_L, and ϕ are parameters to be fitted.

The form of Eq. 6.1.14 is simplified if the following notations are introduced: Let $\eta = y/D$, $s = Ut/D$, $\eta' = d\eta/ds$, etc.; then Eq. 6.1.14 becomes

$$\eta'' + 2\zeta K_1\eta' + K_1^2\eta = \frac{\rho D^2}{m}\left[Y_1\eta' + Y_2\eta + \tfrac{1}{2}C_L\sin(K_1 s + \phi)\right] \qquad (6.1.15)$$

By defining

$$K_0^2 = K_1^2 - \frac{\rho D^2}{m}Y_2(K_1) \qquad (6.1.16a)$$

$$\gamma = \frac{1}{2K_0}\left[2\zeta K_1 - \frac{\rho D^2}{m}Y_1(K_1)\right] \qquad (6.1.16b)$$

Equation 6.1.15 simplifies further to

$$\eta'' + 2\gamma K_0\eta' + K_0^2\eta = \frac{\rho D^2}{2m}C_L\sin(K_1 s + \phi) \qquad (6.1.17)$$

which describes an oscillator of natural dimensionless circular frequency K_0 and damping ratio γ, acted upon by a force of amplitude $\rho D^2 C_L/2m$ at dimensionless circular frequency K_1.

The means of the experimental data of Fig. 6.1.3 [6-33] are replotted in Fig. 6.1.4 as functions of K_1. The resemblance of these experimental results to a typical resonance curve is evident. The steady-state solution to Eq. 6.1.17 is

$$\eta = \frac{\rho D^2 C_L/2m}{\sqrt{(K_0^2 - K_1^2)^2 + (2\gamma K_0 K_1)^2}}\sin(K_1 s - \theta) \qquad (6.1.18)$$

where

$$\theta = \text{arc tan}\frac{2\gamma K_0 K_1}{K_0^2 - K_1^2} \qquad (6.1.19)$$

The response given by Eq. 6.1.18 is matched to Fig. 6.1.4 in the following manner. The value of K_1 at the resonant peak of Fig. 6.1.4 is identified as $K_1 \cong K_0$. The maximum amplitude η_{max} at this point is equal to

$$\eta_{max} = \frac{\rho D^2 C_L}{4\gamma m K_0^2} \qquad (6.1.20)$$

---

*This model borrows its form, in part, from models used to describe flutter (see Sect. 6.5).

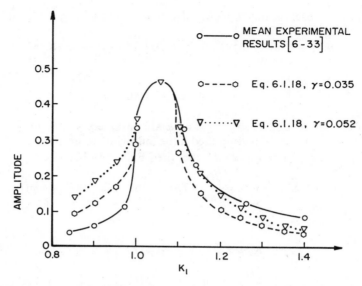

FIGURE 6.1.4.   Comparison of theory [Eq.6.1.18] with replotted experimental results of [6-33].

The width ΔK_1 of the resonance peak at the height $\eta = (1/\sqrt{2})\eta_{max}$, for example, permits evaluation of γ as

$$\gamma = \frac{\Delta K_1}{2K_0} \qquad (6.1.21a)$$

or, if ΔK_1 is measured at the height $1/R$ of the peak amplitude η_{max}, then γ is given by

$$\gamma = \frac{\Delta K_1}{2K_0\sqrt{R^2 - 1}} \qquad (6.1.21b)$$

Thus, Eqs. 6.1.16, 6.1.19, 6.1.20, and 6.1.21 permit determination of all the parameters Y_1, Y_2, and C_L needed in this model to represent the response of the oscillator that at velocity U has the natural dimensionless circular frequency K_0U/B.

The data of Fig. 6.1.4 illustrate the application of this model. On the same graph as the mean experimental data are plotted the amplitudes of Eq. 6.1.18 for values of $\gamma = 0.052$ and $\gamma = 0.035$, with $K_0 = 1.05$. The fit is good in the near vicinity of the peak response. Use of Eqs. 6.1.16 then permits evaluation of Y_1 and Y_2 as functions of K_1. If desired, a more detailed fitting of the model to test results can also be made adjusting $Y_1(K_1)$ point by point to fit the graph of response amplitude.

The present model can also be used to fit data corresponding to oscillations in which "decay-to-resonance" and "growth-to-resonance" effects occur. If the initial deflection of a model started from rest is greater than the amplitude of the steady-state oscillation caused by vortex shedding, the model will initially exhibit a decaying oscillation, as in Fig. 6.1.5a. The oscillation that corresponds to an initial deflection lower than the steady-state amplitude is depicted in Fig. 6.1.5b. These physical situations may be described fairly well by an expression representing the sum of a transient and a steady-state response. Assuming that the transient and steady frequencies differ little, the expression can be given approximately by

$$\eta = [\eta_1 e^{-\gamma\omega_0 t} + \eta_0] \sin(\omega_0 t + \psi) \qquad (6.1.22)$$

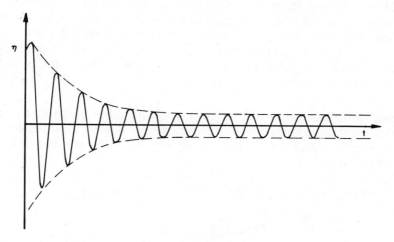

FIGURE 6.1.5a. Decaying oscillation to steady state of bluff, elastically sprung model under vortex lock-in.

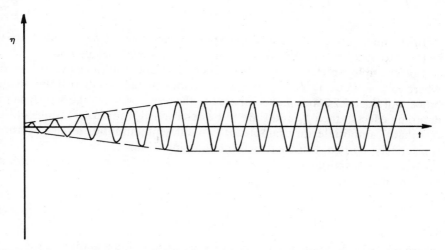

FIGURE 6.1.5b. Growth of oscillation to steady state of bluff, elastically sprung model under vortex lock-in.

where ω_0 is the observed circular frequency of oscillation and η_1, η_0, and ψ are chosen to match initial conditions (at $t=0$, the amplitude is $(\eta_1 + \eta_0)\sin\psi$ and at steady-state conditions, $t \to \infty$, the amplitude is η_0). Note that in the case of amplitude *growth* $\eta_1 < 0$. The response given by Eq. 6.1.22 may be rewritten in terms of dimensionless time s as

$$\eta = [\eta_1 e^{-\gamma K_0 s} + \eta_0] \sin(K_0 s + \psi) \qquad (6.1.23)$$

Again, the important parameters γ and K_0 can be recognized by matching the solution to experimental observations. This method is cited in [6-35] and [6-36]. In a given situation it is not certain a priori that decay-to-resonance and growth-to-resonance tests will provide identical values for γ in this model. However, some tests [6-37] suggest that the differences are not great.

It is again emphasized that the above linear model is also an empirical one that is intended to be retrofitted to experimental results for basically nonlinear phenomena not yet fully understood.

An Empirical Nonlinear Model. Use may also be made of the idea of a Van der Pol oscillator in another manner which simply extends the foregoing linear model by addition of a nonlinear (cubic) aerodynamic term. Thus Eq. 6.1.14 may be modified to the form:

$$m[\ddot{y} + 2\zeta\omega_1\dot{y} + \omega_1^2 y] = \tfrac{1}{2}\rho U^2(2D)\left[Y_1(K)\left(1 - \varepsilon\frac{y^2}{D^2}\right)\frac{\dot{y}}{U} + Y_2(K)\frac{y}{D} \right.$$

$$\left. + \tfrac{1}{2}C_L(K)\sin(\omega t + \phi) \right] \qquad (6.1.24)$$

where $K = D\omega/U$ and ω satisfies the Strouhal relation

$$\omega D/U = 2\pi\mathscr{S} \qquad (6.1.25)$$

In this model Y_1, ε, Y_2, and C_L are parameters, functions of K, that are to be fitted to observations. Various exploitations of this model may occur.

For example, an elementary illustration of its use is in the case, at lock-in, where $\omega \cong \omega_1$ (and $Y_2 \cong 0$, $C_L \cong 0$, since at lock-in the last two terms in the bracket of Eq. 6.1.24 are small compared to the term reflecting the aerodynamic damping effects); then Y_1 and ε are to be determined from observations of free "resonant" amplitude, for example for two different amplitude levels governed, respectively, by two distinct structural damping values ζ.

At steady amplitudes the average energy dissipation per cycle is zero, so that one may write

$$\int_0^T \left[2m\zeta\omega - \rho UDY_1\left(1 - \varepsilon\frac{y^2}{D^2}\right) \right]\dot{y}^2\, dt = 0 \qquad (6.1.26)$$

where $\omega T = 2\pi$. Assuming that y behaves practically sinusoidally:

$$y = y_0 \cos \omega t \qquad (6.1.27)$$

leads to the results

$$\int_0^T \dot{y}^2 \, dt = \omega y_0^2 \pi \tag{6.1.28}$$

$$\int_0^T y^2 \dot{y}^2 \, dt = \omega y_0^4 \frac{\pi}{4} \tag{6.1.29}$$

so that (6.1.26) yields the condition

$$4\pi m \mathscr{S} \zeta - \rho D^2 Y_1 + \rho Y_1 \varepsilon y_0^2 = 0 \tag{6.1.30}$$

Solution of (6.1.30) with two different pairs of values (ζ, y_0), obtained from two steady-state experiments, determines Y_1 and ε. This model, with Y_1 and ε now prescribed, can be employed to predict amplitudes y_0 in a similar context, from the equation

$$y_0 = \left[\frac{4\pi m \mathscr{S} \zeta - \rho D^2 Y_1}{-\rho \varepsilon Y_1} \right]^{1/2} \tag{6.1.31}$$

The model is useful in predicting prototype action from the behavior of laboratory tests. It can be exploited in a number of other ways that will not be pursued here.

6.1.2 An Empirical Model Developed for the Estimation of the Response of the Chimneys and Towers

A model derived in effect from Eq. 6.1.24 was developed in [6-88] for application to the design of chimneys and towers with circular cross-section. It is noted in [6-88] that the product $\rho U^2 Y_2(K)$ of Eq. 6.1.24 is considerably less than $m \omega_1^2$, so that in practice the term $Y_2(K) y/D$ may be ignored. It is also noted in [6-88] that in the case of a random motion, the term $\varepsilon y^2/D^2$ of Eq. 6.1.24 may be replaced by the ratio $\overline{y^2}/(\lambda D)^2$, where λ is a coefficient whose physical significance is discussed subsequently. The term

$$\tfrac{1}{2}\rho U^2 (2D) Y_1(K) \left(1 - \varepsilon \frac{y^2}{D^2}\right) \frac{\dot{y}}{U}$$

of Eq. 6.1.24 is written in [6-88] in the form

$$2\omega_1 \rho D^2 K_{a0} \left(\frac{U}{U_{cr}}\right) \left[1 - \frac{\overline{y^2}}{(\lambda D)^2}\right] \dot{y}$$

where $K_{a0}(U/U_{cr})$ is an aerodynamic coefficient, and $U_{cr} = \omega_1 D/(2\pi\mathscr{S})$. The above term is equated to the product $-2m\zeta_a \omega_1$, where ζ_a is defined as the aerodynamic damping ratio, which may thus be written as

$$\zeta_a = -\frac{\rho D^2}{m} K_{a0} \left(\frac{U}{U_{cr}}\right) \left[1 - \frac{\overline{y^2}}{(\lambda D)^2}\right] \tag{6.1.32}$$

(For $\overline{y^2}^{1/2} = \lambda D$, the aerodynamic damping vanishes, so that the structure no longer experiences any aeroelastic effects causing the response to increase. The coefficient λ may thus be interpreted as the ratio between the limiting rms value of the aeroelastic response and the diameter D.) The total damping ratio of the system is then

$$\zeta_t = \zeta + \zeta_a \tag{6.1.33}$$

where ζ is the structural damping ratio. The aeroelastic effects are, in effect, introduced in the equation of motion simply by substituting into that equation the total damping ratio ζ_t for the structural damping ratio ζ.

The validity of this simple approach was verified in [6-88] by numerical studies and by comparisons with experimental results reported in [6-39]. Figure 6.1.6 shows the dependence of the measured response $\eta_{rms} = \overline{y^2}^{1/2}/D$ upon the reduced wind speed $2\pi U/\omega_1 D$ for various structural damping ratios ζ. Figure 6.1.7 shows calculated versus measured ratios $\overline{y^2}^{1/2}_{max}/D$ for various values of the parameter $K_s = m\zeta/\rho D^2$, where $\overline{y^2}^{1/2}_{max}$ is the rms response corresponding to the most unfavorable reduced wind speed. Three regimes are noted in Fig. 6.1.7, corresponding, respectively to: (1) vibrations whose character is largely due to the random nature of the forces associated with vortex shedding (forced vibration regime); (2) a transition zone; and (3) self-induced vibrations (lock-in regime). Vibrations typical of these three regimes are shown in Fig. 6.1.8. Note

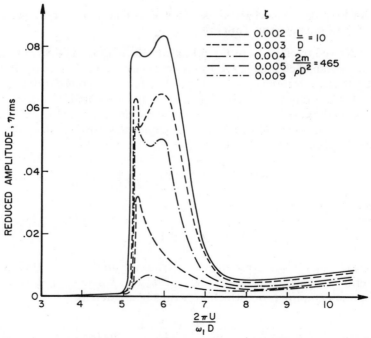

FIGURE 6.1.6. The response of a model stack of circular section for different values of structural damping ($\mathcal{R}e$ subcritical). From L. R. Wooton, "The Oscillations of Large Circular Stacks in Wind," *Proc. Inst. Civ. Eng.*, **43** (1969), 573–598.

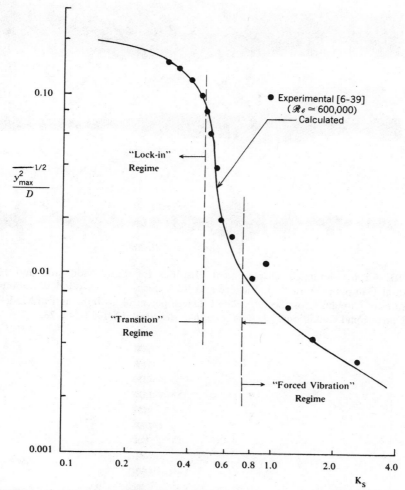

FIGURE 6.1.7.   Measured and estimated response in smooth flow. From B. J. Vickery and R. I. Basu, "Across-Wind Vibrations of Structures of Circular Cross-Section. Part I. Development of a Mathematical Model for Two-Dimensional Conditions," *J. Wind Eng. Ind. Aerodyn.* **12** (1983), 49–73.

that the ratios of peak to rms response are about 4.0 in the forced vibration regime, and about $\sqrt{2}$ in the lock-in regime.

Based on inferences from experimental data available in the literature, [6-88] proposed curves representing: (a) the dependence of K_{a0max} upon the Reynolds number $\mathscr{R}e = UD/v$, where K_{a0max} denotes the maximum value of $K_{a0}(U/U_{cr})$ in smooth flow (Fig. 6.1.9), and (b) the dependence of the ratio $K_{a0}(U/U_{cr})/K_{a0max}$ upon U/U_{cr} for smooth flow and flows with various turbulence intensities $\overline{u^2}^{1/2}/U$ (Fig. 6.1.10).

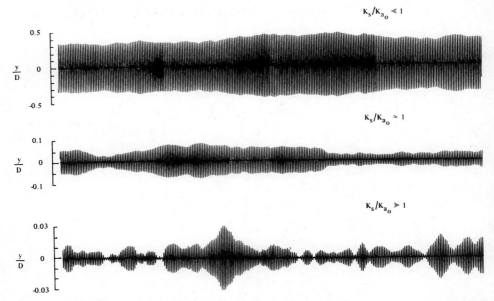

FIGURE 6.1.8. Simulated Displacement Histories for Low, Moderate, and High Structural Damping. From B. J. Vickery and R. I. Basu, "Across-Wind Vibrations of Structures of Circular Cross-Section. Part I. Development of a Mathematical Model for Two-Dimensional Conditions," *J. Wind Eng. Ind. Aerodyn.*, **12** (1983), 49–73.

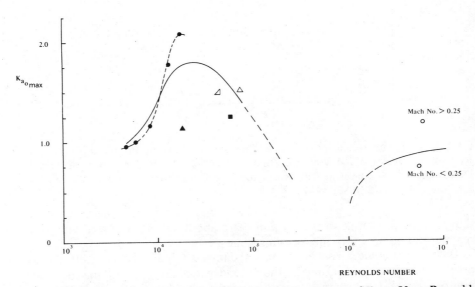

FIGURE 6.1.9. Experimental Data, and Suggested Dependence of K_{a0max} Upon Reynolds Number. From B. J. Vickery and R. I. Basu, "Across-Wind Vibrations of Structures of Circular Cross-Section. Part I. Development of a Mathematical Model for Two-Dimensional Conditions," *J. Wind Eng. Ind. Aerodyn.*, **12** (1983), 49–73.

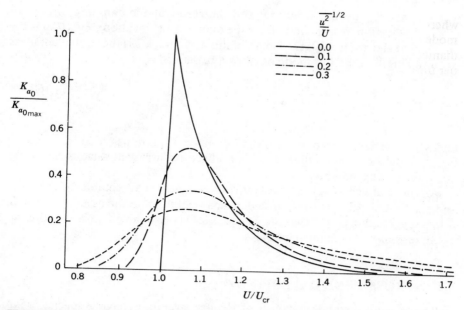

FIGURE 6.1.10. Dependence of Ratio K_{a0}/K_{a0max} Upon Ratio U/U_{cr} for Various Turbulence Intensities. From B. J. Vickery and R. I. Basu, "Across-Wind Vibrations of Structures of Circular Cross-Section. Part I. Development of a Mathematical Model for Two-Dimensional Conditions," *J. Wind Eng. Ind. Aerodyn.*, **12** (1983), 49–73.

For a vertical structure experiencing random motions described by the relation

$$\overline{y^2(z)} = \sum_i \overline{\xi_i^2}\, y_i^2(z) \tag{6.1.34}$$

[6-89] proposes the following expression for the total damping in the i-th mode:

$$\zeta_{ti} = \zeta_i + \zeta_{ai} \tag{6.1.35}$$

$$\xi_{ai} = -\frac{\rho D_0^2}{m_{ei}} \left[K_{1i} - K_{2i}\frac{\overline{\xi_i^2}}{D_0^2} \right] \tag{6.1.36}$$

$$K_{1i} = \frac{\displaystyle\int_0^h K_{a0}(z)\left[\frac{D(z)}{D_0}\right]^2 y_i^2(z)\, dz}{\displaystyle\int_0^h y_i^2(z)\, dz} \tag{6.1.37}$$

$$K_{2i} = \frac{\displaystyle\int_0^h K_{a0}(z)y_i^4(z)\, dz}{\lambda^2 \displaystyle\int_0^h y_i^2(z)\, dz} \tag{6.1.38}$$

where ζ_i and ζ_{ai} are the structural and the aerodynamic damping in the i-th mode of vibration, respectively, D_0 is the diameter at elevation $z=0$, $D(z)$ is the diameter at elevation z, h is the height of the structure, m_{ei} is the equivalent mass per unit length in the i-th mode of vibration, defined as

$$m_{ei} = \frac{M_i}{\displaystyle\int_0^h y_i^2(z)\, dz} \tag{6.1.39}$$

and M_i is the generalized mass in the i-th mode. Equations 6.1.35 to 6.1.38 are based on the assumption that aeroelastic effects occurring at various elevations are linearly superposable.

For the relatively small values of the response that are acceptable for chimneys and stacks, the estimated response depends weakly upon the assumed value of λ. It is suggested in [6-89] that the value $\lambda \simeq 0.4$ is reasonable for use in estimates of the response of concrete chimneys.

6.2   ACROSS-WIND GALLOPING

Galloping is an instability typical of slender structures having special cross-sectional shapes such as, for example, rectangular or "D" sections or the effective sections of some ice-coated power line cables. Under certain conditions that are defined later herein, these structures can exhibit large-amplitude oscillations in the direction normal to the flow (one to ten or even many more across-wind dimensions of the section) at frequencies that are much lower than those of vortex shedding from the same section. A classical example of this type of instability is the across-wind large-amplitude galloping of power line conductor cables that have received a coating of ice under conditions of freezing rain.

Early and clarifying analyses of the galloping problem appeared in [6-40], [6-41], and [6-42]. References 6-43 to 6-50 have dealt with the problem as a nonlinear phenomenon. In across-wind galloping the relative angle of attack of the wind to the structural cross section depends directly on the across-wind velocity of the structure. Experience has proved that knowledge of the mean lift and drag coefficients of the cross section obtained under *static* conditions as functions of angle of attack suffices as a basis upon which to build a satisfactory analytical description of the galloping phenomenon. Galloping is thus governed essentially by quasi-steady forces.

As in the case of the vortex-induced oscillation, the phenomenon will be conceived of, and dealt with analytically, as two-dimensional in nature. Further questions related to galloping response are discussed in [6-46] to [6-50].

6.2.1   Analytical Formulation of the Galloping Problem

Before presenting the basic analytical formulation, it is of interest to note some of the recognized literature that treats the galloping phenomenon. Reference 6-48 reviews the state of the art and presents a compact analysis of the problem.

It also points out the early and basic contributions of Glauert [6-40] and Den Hartog [6-41, 6-42]. References 6-43 to 6-49 constitute important contributions particularly toward clarifying the nonlinear questions related to the aerodynamics. Reference 6-50 offers a critical discussion of existing analytical models of galloping.

Consider a section of a prismatic body in a smooth oncoming flow (Fig. 6.2.1). Assume that the body is *fixed* (i.e., experiences no motion, oscillatory or otherwise) and that the angle of attack of the flow velocity U_r is α. Below is obtained an expression for the force coefficient in the y direction. First, the component of the mean drag (mean force in the direction of U_r) can be written as

$$D(\alpha) = \tfrac{1}{2}\rho U_r^2 B C_D(\alpha) \tag{6.2.1}$$

while the mean lift (mean force in the direction normal to U_r) is

$$L(\alpha) = \tfrac{1}{2}\rho U_r^2 B C_L(\alpha) \tag{6.2.2}$$

The projection of these components on the direction y is then

$$F_y(\alpha) = -D(\alpha)\sin\alpha - L(\alpha)\cos\alpha \tag{6.2.3}$$

If $F_y(\alpha)$ is written in the alternative form

$$F_y(\alpha) = \tfrac{1}{2}\rho U^2 B C_{F_y}(\alpha) \tag{6.2.4}$$

where

$$U = U_r \cos\alpha \tag{6.2.5}$$

it follows from Eqs. 6.2.3 and 6.2.4 that

$$C_{F_y}(\alpha) = -[C_L(\alpha) + C_D(\alpha)\tan\alpha]\sec\alpha \tag{6.2.6}$$

The case is now considered in which the same body *oscillates* in the across-wind direction y in a flow with velocity U (Fig. 6.2.2). The magnitude of the relative velocity of the flow with respect to the moving body is denoted by U_r and can be written as

$$U_r = (U^2 + \dot{y}^2)^{1/2} \tag{6.2.7}$$

The angle of attack, denoted by α, is

$$\alpha = \arctan\frac{\dot{y}}{U} \tag{6.2.8}$$

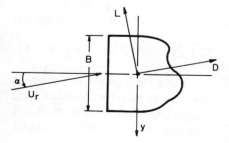

FIGURE 6.2.1. Lift and drag on a fixed bluff object.

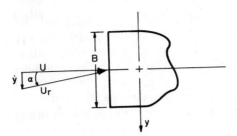

FIGURE 6.2.2. Effective angle of attack on an oscillating bluff object.

If the body has mass m per unit length, is elastically sprung, and has linear mechanical damping, its equation of motion can be written in the usual form

$$m[\ddot{y}+2\zeta\omega_1\dot{y}+\omega_1^2y]=F_y \qquad (6.2.9)$$

where ζ is the damping ratio and ω_1 the natural circular frequency, and where F_y denotes the aerodynamic force acting on the body. It is assumed that the mean aerodynamic lift and drag coefficients $C_L(\alpha)$ and $C_D(\alpha)$ for the oscillating body and for the fixed body are the same so that $F_y(\alpha)$ is given by Eq. 6.2.4 where $C_{F_y}(\alpha)$ is given by Eq. 6.2.6.

Let us first consider the case of incipient (small) motion, that is, the condition in the vicinity of $\dot{y}=0$ wherein

$$\alpha \cong \frac{\dot{y}}{U} \cong 0$$

For this condition

$$F_y \cong \frac{\partial F_y}{\partial \alpha}\bigg|_{\alpha=0} \alpha \qquad (6.2.10)$$

which leads to examination of the factor $dC_{F_y}/d\alpha$ found upon differentiation of Eq. 6.2.6 to have the value at $\alpha=0$

$$\frac{dC_{F_y}}{d\alpha}\bigg|_{\alpha=0} = -\left(\frac{dC_L}{d\alpha}+C_D\right)_0 \qquad (6.2.11)$$

Thus, for small motion the equation of motion takes the form

$$m[\ddot{y}+2\zeta\omega_1\dot{y}+\omega_1^2y]=-\tfrac{1}{2}\rho U^2B\left(\frac{dC_L}{d\alpha}+C_D\right)_0\frac{\dot{y}}{U} \qquad (6.2.12)$$

Considering the aerodynamic (right-hand) side of the equation as a contribution to overall system damping, the net damping coefficient of the system is

$$2m\zeta\omega_1+\tfrac{1}{2}\rho UB\left(\frac{dC_L}{d\alpha}+C_D\right)_0=d \qquad (6.2.13)$$

where, by analogy to the first term of the left-hand side, which is known as mechanical damping, the second term is referred to as aerodynamic damping. From the well-known theory of the linear single-degree-of-freedom oscillator with viscous damping it follows that the system tends toward oscillatory stability if $d > 0$ and toward instability if $d < 0$. Since ζ, the mechanical damping ratio, is usually positive, instability will occur only if

$$\left(\frac{dC_L}{d\alpha} + C_D\right)_0 < 0 \qquad (6.2.14)$$

This is the well-known Glauert–Den Hartog criterion, a *necessary* condition for incipient galloping instability (a sufficient one being $d < 0$). It is clear from Eq. 6.2.14 that circular cylinders, for which $dC_L/d\alpha \equiv 0$ because of their symmetry, cannot gallop.

To summarize the problem to this point: The initial tendency of a slender, prismatic structure toward galloping instability can be assessed by evaluating its time-averaged section lift and drag coefficients and assessing the sign of the expression $dC_L/d\alpha + C_D$ at $\alpha = 0$.

For many problems of wind engineering this initial assessment suffices to describe possibilities of incipient instability relative to galloping. For example, Fig. 6.2.3 [6-51, 6-52] depicts the lift and drag coefficients for an octagonal post structure having a region of wind approach angle $(-5° < \alpha < 5°)$ where the structure is susceptible to galloping according to the Den Hartog criterion. To pursue the problem farther, however, and describe the galloping action in detail requires full development of C_{F_y} in powers of $\dot{y}/U$. Reference 6-48 suggests an abbreviated power series with several odd powers of $\dot{y}/U$ and with an

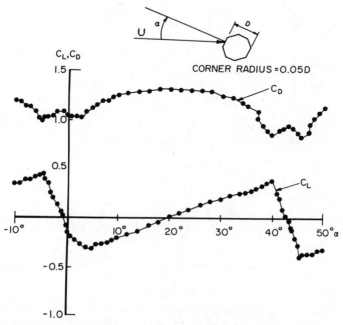

FIGURE 6.2.3. Force coefficients on an octagonal cylinder $(\mathscr{R}e = 1.2 \times 10^6)$ [6-52].

appropriately-signed second-power term to smooth the fit:

$$C_{F_y} = A_1 \left(\frac{\dot{y}}{U}\right) - A_2 \left(\frac{\dot{y}}{U}\right)^2 \frac{\dot{y}}{|\dot{y}|} - A_3 \left(\frac{\dot{y}}{U}\right)^3 + A_5 \left(\frac{\dot{y}}{U}\right)^5 - A_7 \left(\frac{\dot{y}}{U}\right)^7 \quad (6.2.15)$$

If the dependence of C_L and C_D upon α is known, the coefficients A_1 through A_7 can be evaluated as follows. First, C_{F_y} is plotted against $\tan \alpha$. Since $\tan \alpha = \dot{y}/U$, C_{F_y} can then be approximated by the above polynomial using either a least squares fit or some other technique as desired. Reference 6-48 applies the method of Kryloff and Bogoliuboff [6-53] to the solution of the resulting non-linear equation, postulating as a first response approximation:

$$y = a \cos(\omega_1 t + \phi) \qquad (6.2.16a)$$

$$\dot{y} = -a\omega_1 \sin(\omega_1 t + \phi) \qquad (6.2.16b)$$

where a and ϕ are considered to be slowly varying functions of time. Three basic types of curve C_{F_y} vs. α and the corresponding galloping response amplitudes a as functions of reduced velocity $U/D\omega_1$ are identified (see Fig. 6.2.4).

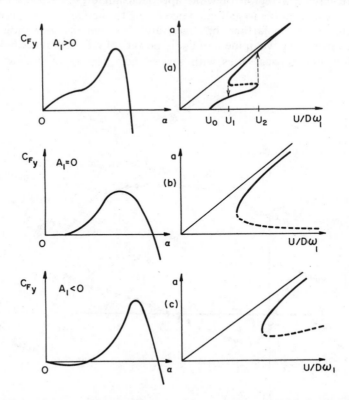

FIGURE 6.2.4.   Three basic types of lateral force coefficients and the corresponding galloping response amplitudes a. From M. Novak, "Galloping Oscillations of Prismatic Structures," *J. Eng. Mech. Div.*, ASCE, **98** (1972), 27–46.

It is noted that the only possible oscillatory motions are those with amplitudes a traced in full lines in Fig. 6.2.4. If the speed increases from U_0 to U_2 (Fig. 6.2.4a), the amplitude of the response is likely to jump from the lower to the upper branch of the solid curve. If the speed decreases from U_2 to U_0 the jump occurs from the upper to the lower curve.

Reference 6-49 discusses the response of elongated three-dimensional bodies by use of the sectional theory outlined above and mentions the effect of flow turbulence upon the galloping. It is noted that turbulence can transform steady oscillations into unsteady ones, reduce the magnitude of the aerodynamic damping, and, in certain cases, depending upon its scale and intensity, destroy the necessary conditions for galloping. Under certain conditions of an initial triggering disturbance larger than the steady-state amplitude, certain sections can experience galloping at much lower velocities than those required in smooth flow. Finally, it is noted in [6-49] that galloping oscillations also depend upon the extent to which the mean angle of attack varies as a function of the magnitude of the wind drag.

The closely similar problem of a long flexible beam free to deflect in both along-wind and across-wind directions is analyzed in [6-54]. Reference 6-55 discusses the effect of incident wind skewed to the long axis of a galloping body. For information on galloping tendencies of stranded cables, see [6-90].

6.3 WAKE GALLOPING

The case is now considered of two cylinders, one of which is located upstream of the other. Under certain conditions, the downstream cylinder may be subjected to galloping oscillations induced by the turbulent wake of the upstream cylinder. This has proven to be the case, for example, for power transmission line cables grouped in so-called bundles, that is, for groups of conductors consisting of two, four, six, eight, or more parallel cables separated by mechanical spacers in the direction transverse to their span. (Figure 6.3.1 depicts a spacer in a 4-cable bundle of a power line.) With the spacers in place, it is the cable region between them that is most susceptible to wake galloping conditions since cable freedom of motion is greatest there.

Wake galloping may occur only under conditions where the frequencies of response of the downstream cylinder are low compared to its vortex-shedding frequencies and to those of the cylinder located upstream. Just as with the phenomenon treated in Sect. 6.2, wake galloping is governed by parameters that describe mean (rather than instantaneous) aerodynamic phenomena and can be measured when the body is fixed.

The wake of the upstream cylinder may be pictured as suggested in Fig. 6.3.2. Investigating this wake with a "probe" consisting of the downstream cylinder itself reveals a distribution of along- and across-wind forces (Fig. 6.3.3) acting on this cylinder as a consequence of its particular locations in the wake. One important finding is that the across-wind wake forces have a tendency to center the downstream cylinder, that is, draw it toward the wake centerline, contrary

FIGURE 6.3.1. Spacer in four-bundle power line.

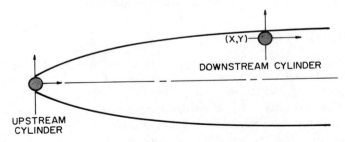

FIGURE 6.3.2. Sectional geometry, cylinders in wake galloping phenomenon.

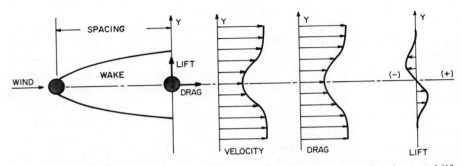

FIGURE 6.3.3. Qualitative sketch of the distributions of mean velocity, drag, and lift on a circular cylinder in the wake of another.

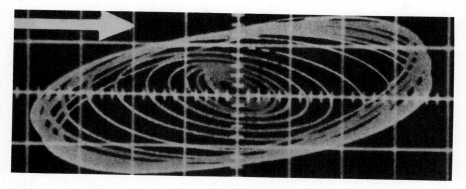

FIGURE 6.3.4.   Amplitude trace of a wake galloping orbit [6-52]. Courtesy of the National Aeronautical Establishment, National Research Council of Canada.

to the possible intuitive expectation that since the outer flow beyond the wake edges is faster it should, by Bernoulli's principle, tend to pull the downstream cylinder outward, away from the wake center.

An explanation has been sought for this apparent anomaly, which may tentatively be ascribed to numerous criss-crossings of the flow field inside the wake by time-varying local jets of fluid that have strong components directed inward toward the center. These jets, or local fluid velocities, would tend to create repetitive drag forces directed, on the average, toward the wake center. This view of the phenomenon has been supported to some degree by flow visualization studies in a water tunnel [6-56]. As indicated in Fig. 6.3.3, the centering lift is strongest at about a quarter of the total wake width outward from the centerline.

When the downstream cylinder located a few diameters of the upstream body behind this latter is displaced—for any reason—into approximately the outer quarter of the wake (see Fig. 6.3.2), it enters a region of galloping instability. In this region a galloping motion will begin, growing in amplitude until an apparent limit cycle is reached. This motion consists of large oscillations in an elliptical orbit with the long ellipse axis oriented approximately along the main flow direction. The direction of the elliptical orbit is such that the cylinder moves downstream near the outer edges of the wake and upstream nearer the center of the wake, or clockwise above the centerline in Fig. 6.3.3 and counterclockwise below it. These directions coincide with the intuitive assessment that net drag forces will be higher in the outer, faster portion of the wake and lower in its interior. References 6-56 to 6-65 cover various aspects of the wake galloping phenomenon. An oscilloscope trace of a developing wake galloping orbit is shown in Fig. 6.3.4 [6-63].

6.3.1   Analysis of the Wake Galloping Phenomenon

The phenomenon is analyzed as if its basic ingredients were two-dimensional, as was done in the preceding sections. Consider two cylinders (Fig. 6.3.5), one

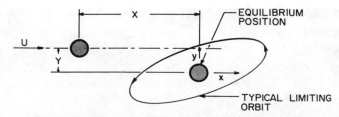

FIGURE 6.3.5. Coordinates for wake galloping analysis.

windward, producing a wake, and one leeward, within that wake. The leeward cylinder will be assumed to be elastically sprung in both horizontal and vertical directions about some position (X, Y), where X, Y are along-wind and across-wind coordinates conveniently centered on the windward cylinder.

The equations of motion for the leeward cylinder may be stated in terms of the excursions (x, y) of that cylinder away from (X, Y):

$$m\ddot{x} + d_x\dot{x} + K_{xx}x + K_{xy}y = F_x \qquad (6.3.1a)$$

$$m\ddot{y} + d_y\dot{y} + K_{yx}x + K_{yy}y = F_y \qquad (6.3.1b)$$

where m is the mass per unit span (normal to the figure) of the leeward cylinder; d_x, d_y are respective damping constants; $K_{rs}(r, s = x, y)$ are direct and cross-coupling spring constants restraining the motion of the leeward cylinder; and F_x, F_y are the net X- and Y-force components.

Next, if C_x and C_y are defined as the steady average force coefficients referred to free-stream dynamic pressure $\frac{1}{2}\rho U^2$ that apply to the cylinder located at point (X, Y), then it can be shown that the incipient forces in x and y directions may be expressed as [6-65]

$$F_x = \frac{1}{2}\rho U^2 D \left\{ \left(\frac{\partial C_x}{\partial x} x + \frac{\partial C_x}{\partial y} y \right) + C_y \frac{\dot{y}}{U_w} - 2C_x \frac{\dot{x}}{U_w} \right\} \qquad (6.3.2a)$$

$$F_y = \frac{1}{2}\rho U^2 D \left\{ \left(\frac{\partial C_y}{\partial x} x + \frac{\partial C_y}{\partial y} y \right) - C_x \frac{\dot{y}}{U_w} - 2C_y \frac{\dot{x}}{U_w} \right\} \qquad (6.3.2b)$$

where U is the free upstream velocity and U_w is the average wake velocity in the U direction at (X, Y), and D is the projected across-wind dimension of the cylinder. Expressions similar to Eqs. 6.3.2 were first developed in [6-58] and [6-59]. Values of C_x, C_y, and their derivatives are obtained by direct measurement of time-averaged values in wind tunnel model studies. Cases of interest have concerned smooth circular cylinders and the rougher surfaces of stranded wire cables.

Analytical solution of the problem, in which the forces given by Eqs. 6.3.2 are clearly self-excited only, proceeds by assigning values

$$x = x_0 e^{\lambda t} \qquad (6.3.3a)$$

$$y = y_0 e^{\lambda t} \qquad (6.3.3b)$$

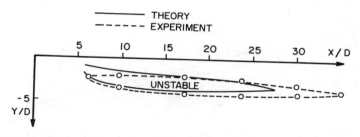

FIGURE 6.3.6. Measured and predicted stability boundaries for wake galloping [6-56].

to x and y in Eqs. 6.3.1 and 6.3.2 and setting the determinant of coefficients of Eqs. 6.3.1 equal to zero. It follows from Eqs. 6.3.3 that the solutions λ are unstable if $\lambda_1 \geqslant 0$ in the calculated value of form

$$\lambda = \lambda_1 + i\lambda_2$$

(where $i = \sqrt{-1}$), since they then contain a diverging exponential factor. Such solutions are then sought for the parameters associated with a number of points X, Y.

The agreement between the theory and experiment has been found to be satisfactory, as seen in Fig. 6.3.6 [6-56], where the curves indicate points at which marginally unstable solutions (i.e., where $\lambda = i\lambda_2$) are found. For these solutions, the orbit $[x(t), y(t)]$ may be calculated, if desired, by using Eqs. 6.3.1 and 6.3.2.

As in other aeroelastic phenomena, the structural parameters exert strong control over the characteristics of wake galloping. In particular, in carrying out model studies the values of the spring constants $K_{rs}(r, s = x, y)$ require particular attention. This is especially true in the representation of the action of cables, a subject that has received much attention [6-60, 6-62, 6-65] but is outside the scope of the present discussion.

6.4 TORSIONAL DIVERGENCE

The phenomenon of torsional divergence was at first most closely associated with aircraft wings and their susceptibility to twisting off at some excessive air speed. To form a conceptual picture of what occurs in such a situation, consider a thin airfoil, or any other analogous structure, such as a bridge deck (Fig. 6.4.1). Under the effect of wind, the structure will be subjected to, and will act to resist, a drag force, a lift force, and a twisting moment. As the wind velocity increases, the twisting moment in particular increases also. This in turn twists the structure farther; but this condition may also, by increasing the effective angle of attack of the wind relative to the structure, further increase the twisting moment, which then demands additional reactive moment from the structure. Finally, a velocity is reached at which the magnitude of the wind-induced moment, together with the tendency for twist to demand additional structural

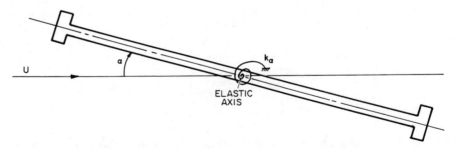

FIGURE 6.4.1. Geometry and parameters for torsional divergence problem.

reaction, creates an unstable condition and the structure twists to destruction. The problem is one of stability, quite analogous in a structural sense to column buckling. Just as column buckling occurs when a critical column load is reached, torsional divergence occurs at some critical divergence velocity of the wind. The phenomenon depends upon structural flexibility and the manner in which the aerodynamic moments develop with twist; it does not depend upon ultimate structural strength.

In the case of thin airfoils, the aerodynamic twisting moment increases with increased angle of attack. In other, more complex structures, it may be that the aerodynamic twisting moment does not follow this simple tendency. As a result, such structures may not follow the pattern described above; in fact, depending upon the relation between aerodynamic moment and angle of attack, some structures may be immune to torsional divergence. Finally, it should be noted that in most cases of practical interest in civil engineering the critical divergence velocities are extremely high, well beyond the range of velocities normally considered in design.

6.4.1 Analytical Modeling of Torsional Divergence

To analyze the torsional divergence phenomenon, consider, as in Fig. 6.4.1, the section of a structure that can rotate against a torsional spring about some pivot point (or *elastic center*). Let the spring constant and the angle of rotation be denoted by k_α and α, respectively.

Assuming that the mean wind velocity is U and that the deck width is B, the aerodynamic moment per unit span can be written as

$$M_\alpha = \tfrac{1}{2}\rho U^2 B^2 C_M(\alpha) \qquad (6.4.1)$$

where $C_M(\alpha)$ is the aerodynamic moment coefficient about the twist axis. An example of the dependence of C_M upon α in the case of an open truss bridge deck is shown in Fig. 6.4.2.

At zero angle of attack the value of this moment is

$$M_\alpha(0) = \tfrac{1}{2}\rho U^2 B^2 C_{M0} \qquad (6.4.2)$$

where $C_{M0} = C_M(0)$. For a small change in α away from $\alpha = 0$, M_α may be given

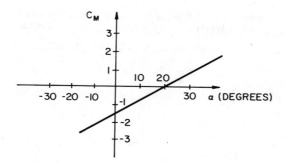

FIGURE 6.4.2. Moment coefficient for a bluff structure as a function of angle of attack.

to first approximation by

$$M_\alpha = \tfrac{1}{2}\rho U^2 B^2 \left[C_{M0} + \frac{dC_M}{d\alpha}\bigg|_{\alpha=0} \alpha \right] \tag{6.4.3}$$

Equating the aerodynamic to the internal structural moment leads to the equation

$$\tfrac{1}{2}\rho U^2 B^2 [C_{M0} + C'_{M0}\alpha] = k_\alpha \alpha \tag{6.4.4}$$

where

$$C'_{M0} = \frac{dC_M}{d\alpha}\bigg|_{\alpha=0} \tag{6.4.5}$$

The divergence problem is summarized (in this two-dimensional description) by Eq. 6.4.4. We now examine its solution.

Define $\lambda \equiv \tfrac{1}{2}\rho U^2 B^2$. Equation 6.4.4 then becomes

$$(k_\alpha - \lambda C'_{M0})\alpha = \lambda C_{M0}$$

or

$$\alpha = \frac{\lambda C_{M0}}{k_\alpha - \lambda C'_{M0}} \tag{6.4.6}$$

The solution of Eq. 6.4.6 for α approaches infinity (diverges) for the value

$$\lambda = \frac{k_\alpha}{C'_{M0}} \tag{6.4.7}$$

This therefore defines the critical divergence velocity:

$$U_c = \sqrt{\frac{2k_\alpha}{\rho B^2 C'_{M0}}} \tag{6.4.8}$$

The problem may readily be generalized to three dimensions, but this is reserved for a specific application in Chapter 13 (Sect. 13.1.2). It should also be noted that the problem considered here is that of incipient instability only. If more complex structural action with increasing velocity occurs (due to a more complex curve of C_M vs. α, for example, than that shown in Fig. 6.4.2), the divergence problem can be solved by a systematic solution of the relation

$$\tfrac{1}{2}\rho U^2 B^2 C_M(\alpha) = k_\alpha \alpha \qquad (6.4.9)$$

for any range of velocities desired. The pursuit of this problem is beyond the aim of this section.

6.5   FLUTTER

One of the earliest aeroelastic oscillations to be recognized was the flutter of airfoils. The term "flutter" has been variously used; recently, however, this use has become more restricted. The most common present uses of the term employ additional qualifying terms, for example, classical flutter, stall flutter, single-degree-of-freedom flutter, and panel flutter. All of these terms were originally employed in aerospace applications, but some have carried over to wind engineering.

Classical flutter originally applied to thin airfoils. The term also finds application today to suspended-span bridge decks. It implies an aeroelastic phenomenon in which two degrees of freedom of a structure, rotation and vertical translation, couple together in a flow-driven, unstable oscillation. Coupling of the two degrees of freedom—indispensable for thin airfoil flutter under normal structural circumstances—has come to be the identifying sign for classical flutter.

Stall flutter is a single-degree-of-freedom oscillation of airfoils in torsion driven by the nonlinear characteristics of the lift in the vicinity of the stall, or loss-of-lift condition. This phenomenon can also occur with structures having broad surfaces that can stall depending on the angle of approaching wind. So-called "stop-sign-flutter," the torsional oscillation of traffic stop-signs about torsionally weak posts, is an example in a nonaeronautical area.

Single-degree-of-freedom flutter may include stall flutter, but may simply be associated with systems undergoing strongly separated flows. Bluff, un-streamlined bodies are typical examples. Prominent among these are the decks of suspended-span bridges, which can in various instances exhibit single-degree torsional instability. These are discussed in more detail in Chapter 8.

Panel flutter is a sustained oscillation of panels—typically the sides of large rockets—caused by the high-speed passage of air *along* the panel. The most prominent cases have been in supersonic flow regimes and so have not appeared in the usual wind engineering context. Flutter of taut canvas covers and flag flutter are, however, phenomena related to panel flutter.

It is likely that, in its detail, flutter in practically all cases involves nonlinear aerodynamics. It has been possible in a number of instances, however, to treat the problem successfully by linear analytical approaches. The main reasons for this are two: First, the supporting structure is usually treatable as linearly elastic and its actions dominate the *form* of the response, which is usually an exponentially modified sinusoidal oscillation. Second, it is the incipient or starting condition, which may be treated as having only small amplitude, that separates the stable and unstable regimes. These two main features enable a flutter analysis to be based on the standard stability considerations of linear elastic systems.

It is characteristic of flutter as a typical self-excited oscillation that a structural system by means of its deflections and their time derivatives taps off energy from the wind flow. If the system is given an initial disturbance, its motion will either decay or diverge (i.e., its oscillations will be damped or will grow indefinitely) according to whether the energy of motion extracted from the flow is less than or exceeds the energy dissipated by the system through mechanical damping. The theoretical dividing line between the decaying and divergent cases, namely, sustained sinusoidal oscillation, is then recognized as the critical flutter condition.

In the treatment of flutter, in the present wind engineering context, only classical flutter and single-degree-of-freedom flutter will be discussed.

6.5.1 Equation of Motion for an Airfoil or a Bridge Deck

Consider a section of an airfoil or a bridge deck (Fig. 6.5.1) subjected to the action of a smooth oncoming flow. The section is assumed to have two degrees of freedom: bending displacement and twist denoted by h and α, respectively. A unit span of the system has mass m, mass moment of inertia I, static unbalance S (equal to the product of mass m and a distance, a, which separates the center of mass from the elastic center),* vertical and torsional restoring forces characterized by spring constants C_h and C_α, and coefficients of viscous damping

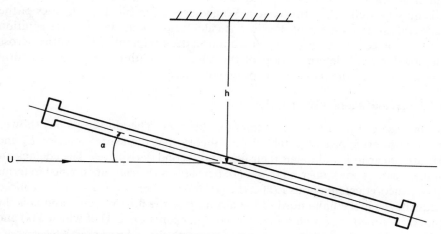

FIGURE 6.5.1. Degrees of freedom of section of a line-like structure.

---

*Note that with a fixed sign convention S may be positive or negative depending on the location (forward or aft) of the center of mass with respect to the elastic center.

c_h and c_α. With these definitions the equations of motion can be written [6-66, 6-67]

$$m\ddot{h} + S\ddot{\alpha} + c_h \dot{h} + C_h h = L_h \tag{6.5.1a}$$

$$S\ddot{h} + I\ddot{\alpha} + c_\alpha \dot{\alpha} + C_\alpha \alpha = M_\alpha \tag{6.5.1b}$$

where L_h and M_α are, respectively, the self-excited aerodynamic lift and moment about the rotation axis per unit span. Designating by r_g the radius of gyration of the body about the center of rotation and using notations similar to those of Sect. 5.1, Eqs. 6.5.1 become

$$m[\ddot{h} + a\ddot{\alpha} + 2\zeta_h \omega_h \dot{h} + \omega_h^2 h] = L_h \tag{6.5.2a}$$

$$I\left[\frac{a}{r_g^2}\ddot{h} + \ddot{\alpha} + 2\zeta_\alpha \omega_\alpha \dot{\alpha} + \omega_\alpha^2 \alpha\right] = M_\alpha \tag{6.5.2b}$$

where ζ_h, ζ_α are damping ratios-to-critical, and ω_h, ω_α are the natural circular frequencies in h and α degrees of freedom, respectively, defined by

$$\omega_h^2 = \frac{C_h}{m} \tag{6.5.3a}$$

$$\omega_\alpha^2 = \frac{C_\alpha}{I} \tag{6.5.3b}$$

In the case of bridge decks that are symmetrical, the center of mass lies in the vertical plane of the centerline. In this case $a = 0$. Usually the rotation axis lies in this plane also, though it may be at some vertical distance from the center of mass. In the case of bridges with arched decks the effective rotation axis may lie well below this center. When accounting for the dynamics of the deck, the mass moment of inertia I is calculated about the effective rotation axis and hence is typically, even for a uniform deck, a quantity that varies across the span. Actual determination of the effective rotation axis is a structural problem outside the scope of the present discussion.

6.5.2   Aerodynamic Lift and Moment

In the case of thin airfoils in incompressible flow, Theodorsen [6-66] showed from basic principles of potential flow theory that the expressions for L_h and M_α are linear in h and α and their first and second derivatives. The coefficients in these expressions, referred to as *aerodynamic coefficients*, are defined in terms of two theoretical functions $F(k)$ and $G(k)$ [6-66], where $k = b\omega/U$ is the reduced frequency, b is the half-chord of the airfoil, U is the flow velocity, and ω is the circular frequency of oscillation. The complex function $C(k)$ of which $F(k)$ and $G(k)$ are the real and imaginary parts, respectively, is known as *Theodorsen's circulation function* (Fig. 6.5.2). For aircraft flight regimes in all velocity ranges, wide research has developed further analytical expressions for all necessary aerodynamic coefficients. There exists a vast literature on the subject, to which [6-67] to [6-70] are useful introductions.

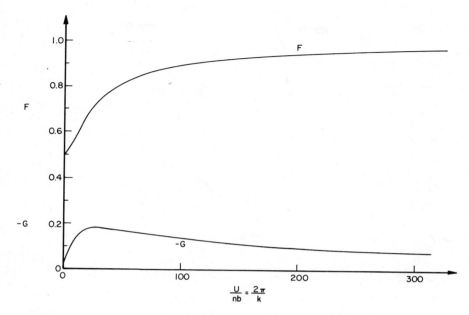

FIGURE 6.5.2. Real and imaginary parts of the Theodorsen circulatory function $C(K) = F(K) + iG(K)$.

For bluff objects of wind engineering applications, it has not to date been possible to develop expressions for the aerodynamic coefficients starting from basic fluid-flow principles. However, it has been shown in [6-71] that for small oscillations the self-excited lift and moment on a bluff body may be treated as linear in the structural displacement and rotation and their first two derivatives, and that it is possible to measure the aerodynamic coefficients by means of specially designed wind tunnel tests. Such tests indicate that just as in the case of the airfoil the aerodynamic coefficients of a bluff body are functions of the reduced velocity.

Various forms for the linear expressions for L_h and M_α have been employed. The classical theoretical (and some experimental) work has used complex number forms based on the representation of the flutter oscillation as having the complex form $e^{i\omega t}$. However, in the wind engineering practice developed to date in the United States real forms have been employed. Below are stated commonly used linearized forms of this type [6-71]:

$$L_h = \tfrac{1}{2}\rho U^2 (2B) \left[KH_1^*(K)\frac{\dot{h}}{U} + KH_2^*(K)\frac{B\dot{\alpha}}{U} + K^2 H_3^*(K)\alpha \right] \quad (6.5.4a)$$

$$M_\alpha = \tfrac{1}{2}\rho U^2 (2B^2) \left[KA_1^*(K)\frac{\dot{h}}{U} + KA_2^*(K)\frac{B\dot{\alpha}}{U} + K^2 A_3^*(K)\alpha \right] \quad (6.5.4b)$$

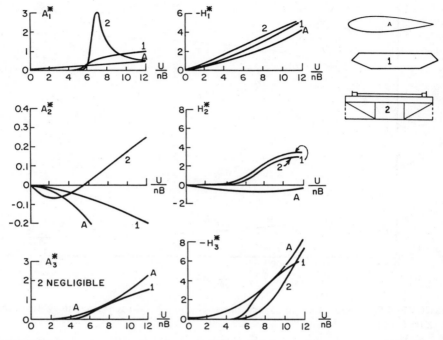

FIGURE 6.5.3.   Aerodynamic coefficients H_i^* and A_i^* for a thin airfoil, a partly stream-lined box, and an open-truss deck section. After R. H. Scanlan and J. J. Tomko, "Airfoil and Bridge Deck Flutter Derivatives," *J. Eng. Mech. Div.*, ASCE, **97** (1971), 1717–1737.

where the reduced frequency K is defined as†

$$K = \frac{B\omega}{U} = \frac{B(2\pi n)}{U} \tag{6.5.5}$$

B is the chord, deck width, or along-wind dimension of the structure, U is the uniform approach velocity of the wind, and ω is the circular frequency of oscillation (n is the frequency of oscillation). In Eqs. 6.5.4 terms in $\ddot{h}$, $\ddot{\alpha}$, and h have been omitted as being of negligible importance in wind engineering. (In aeronautical practice terms in $\ddot{h}$ and $\ddot{\alpha}$ but not h are retained.) The coefficients H_i^* and A_i^* ($i = 1, 2, 3$) are nondimensional functions of K. The quantities α, $\dot{h}/U$, and $B\dot{\alpha}/U$ are effective angles of attack and therefore also nondimensional. The typical term in Eqs. 6.5.4 can be viewed as following the classical pattern of expressions for aerodynamic lift force per unit span, such as

$$L = \tfrac{1}{2}\rho U^2 (2B) C_L \cong \tfrac{1}{2}\rho U^2 (2B) \frac{dC_L}{d\alpha}\alpha$$

†The reduced frequencies k, used in aeronautical practice, and K, used in wind engineering, differ in that k is defined in terms of the half-chord $b = B/2$, whereas for reasons of convenience K is defined in terms of the full chord B, as in Eq. 6.5.5.

for small angle of attack α. Formally, terms such as KH_1^* or $K^2A_3^*$ are thus analogous to lift coefficient derivatives $dC_L/d\alpha$. These terms should be referred to as *motional* derivatives, however, and they go over into steady-state derivatives, such as $dC_L/d\alpha$, only for $K \to 0$ (zero frequency). From an experimental point of view this means that the aerodynamic coefficients of Eqs. 6.5.4 can be measured only if the body is in an oscillatory state, whereas $dC_L/d\alpha$ is obtained under static conditions (i.e., with the body fixed; see Sect. 6.2). The factors K or K^2 preceding H_i^* and A_i^* could just as well be included with these latter in a total coefficient of some other designation if desired, but the evolution of the theory [6-71] has identified them as nondimensional factors. References 6-71 through 6-77 discuss various experimental techniques used in the United States, Japan, and France for obtaining the nonstationary aerodynamic (flutter) derivatives.

Sample experimental values of the coefficients H_i^* and A_i^* for a partly streamlined box and for an open-truss deck are shown in Fig. 6.5.3 [6-71], where for purposes of comparison the analogous coefficients H_i^* and A_i^* for a thin airfoil are also given.*

6.5.3 Solution of the Flutter Equations

Because of the dependence of the aerodynamic terms upon K, the analytical solution of the flutter problem becomes more involved than the comparable stability solutions where quasi-steady aerodynamics holds. Under K-dependent conditions, a typical solution method is as follows. A value of K is chosen and the values of H_i^* and A_i^* corresponding to that value are obtained from plots of these experimental functions. It is then assumed that h and α have solutions proportional to $e^{i\omega t}$ which are inserted into Eqs. 6.5.2 and 6.5.4. The determinant of coefficients of the amplitudes of h and α is then set equal to zero as the basic stability condition. This constitutes in fact a complex quartic equation in the

---

*In aeronautical practice it is usual to write the equations for L_h and M_α as given in Eqs. 6.5.4, except that B is replaced by b and K by k. In this case H_i^* and A_i^* must be replaced by $\tilde{H}_i^*$ and $\tilde{A}_i^*$, where

$$\tilde{H}_1^* = 4H_1^* \qquad \tilde{A}_1^* = 8A_1^*$$
$$\tilde{H}_2^* = 8H_2^* \qquad \tilde{A}_2^* = 16A_2^*$$
$$\tilde{H}_3^* = 8H_3^* \qquad \tilde{A}_3^* = 16A_3^*$$

It can be shown [6-71] that the following theoretical relations hold:

$$k\tilde{H}_1^*(k) = -2\pi F(k) \qquad\qquad k\tilde{A}_1^*(k) = \pi F(k)$$
$$k\tilde{H}_2^*(k) = -\pi\left[1 + F(k) + \frac{2G(k)}{k}\right] \qquad k\tilde{A}_2^*(k) = -\frac{\pi}{2}\left[1 - F(k) - \frac{2G(k)}{k}\right]$$
$$k^2\tilde{H}_3^*(k) = 2\pi\left[F(k) - \frac{kG(k)}{2}\right] \qquad k^2\tilde{A}_3^*(k) = \pi\left[F(k) - \frac{kG(k)}{2}\right]$$

As previously indicated, the functions $F(k)$ and $G(k)$ hold for thin airfoils only and are incorrect for other geometric forms.

unknown flutter frequency ω, which must then be solved. The solution obtained will, in general, be of the form $\omega = \omega_1 + i\omega_2$ with $\omega_2 \neq 0$, and will therefore represent either a decaying ($\omega_2 > 0$) or a divergent ($\omega_2 < 0$) oscillation. A new value of K is then chosen and the procedure is repeated until the solution is purely (or very nearly) imaginary, that is, until $\omega_2 \simeq 0$, so that $\omega \simeq \omega_1$. To that solution there corresponds the flutter condition at real frequency ω_1. Let K_c be the value of K for which $\omega \simeq \omega_1$. The critical flutter velocity is then

$$U_c = \frac{B\omega_1}{K_c} \tag{6.5.6}$$

Because of its interest in applications, a useful variant on the solution outlined above is now presented in somewhat more detail. Let

$$s = \frac{Ut}{B}$$

be a nondimensional time (or distance). Noting that

$$(\dot{\ }) = \frac{d(\ )}{dt} = \frac{d(\ )}{ds}\frac{ds}{dt} = (\ )' \frac{U}{B}$$

Equations 6.5.2 and 6.5.4 can be reduced to

$$\frac{h''}{B} + 2\zeta_h K_h \frac{h'}{B} + K_h^2 \frac{h}{B} = \frac{\rho B^2}{m}\left[KH_1^* \frac{h'}{B} + KH_2^*\alpha' + K^2 H_3^*\alpha \right] \tag{6.5.7a}$$

$$\alpha'' + 2\zeta_\alpha K_\alpha \alpha' + K_\alpha^2\alpha = \frac{\rho B^4}{I}\left[KA_1^* \frac{h'}{B} + KA_2^*\alpha' + K^2 A_3^*\alpha \right] \tag{6.5.7b}$$

where $K_h = B\omega_h/U$, $K_\alpha = B\omega_\alpha/U$.

Posing now the solution forms

$$\frac{h}{B} = \frac{h_0}{B} e^{i\omega t} = \frac{h_0}{B} e^{iKs}$$

$$\alpha = \tilde{\alpha}_0 e^{i(\omega t + \phi)} = \alpha_0 e^{i\omega t} = \alpha_0 e^{iKs}$$

Equations 6.5.7 take the form

$$\left[-K^2 + 2i\zeta_h K_h K + K_h^2 - \frac{\rho B^2}{m} iK^2 H_1^* \right]\frac{h_0}{B}$$

$$- \left[\frac{\rho B^2}{m} iK^2 H_2^* + \frac{\rho B^2}{m} K^2 H_3^* \right]\alpha_0 = 0 \tag{6.5.8a}$$

$$\left[-\frac{\rho B^4}{I} iK^2 A_1^* \right]\frac{h_0}{B}$$

$$+ \left[-K^2 + 2i\zeta_\alpha KK_\alpha + K_\alpha^2 - \frac{\rho B^4}{I} iK^2 A_2^* - \frac{\rho B^4}{I} K^2 A_3^* \right]\alpha_0 = 0 \tag{6.5.8b}$$

Defining an unknown X as

$$X = \frac{\omega}{\omega_h}$$

and setting the determinant of Eqs. 6.5.8 equal to zero results in a complex polynomial in X of degree four. This breaks down into the following two equations, assuming that X is always real at the flutter condition:

From the real part:

$$X^4 \left(1 + \frac{\rho B^4}{I} A_3^* - \frac{\rho B^2}{m} \frac{\rho B^4}{I} A_2^* H_1^* + \frac{\rho B^2}{m} \frac{\rho B^4}{I} A_1^* H_2^* \right)$$

$$+ X^3 \left(2\zeta_\alpha \frac{\omega_\alpha}{\omega_h} \frac{\rho B^2}{m} H_1^* + 2\zeta_h \frac{\rho B^4}{I} A_2^* \right)$$

$$+ X^2 \left(-\frac{\omega_\alpha^2}{\omega_h^2} - 4\zeta_h\zeta_\alpha \frac{\omega_\alpha}{\omega_h} - 1 - \frac{\rho B^4}{I} A_3^* \right)$$

$$+ X \cdot 0$$

$$+ \left(\frac{\omega_\alpha}{\omega_h} \right)^2 = 0 \tag{6.5.9a}$$

From the imaginary part:

$$X^3 \left(\frac{\rho B^4}{I} A_2^* + \frac{\rho B^2}{m} H_1^* + \frac{\rho B^2}{m} \frac{\rho B^4}{I} H_1^* A_3^* - \frac{\rho B^2}{m} \frac{\rho B^4}{I} A_1^* H_3^* \right)$$

$$+ X^2 \left(-2\zeta_\alpha \frac{\omega_\alpha}{\omega_h} - 2\zeta_h - 2\zeta_h \frac{\rho B^4}{I} A_3^* \right)$$

$$+ X \left(-\frac{\rho B^2}{m} H_1^* \frac{\omega_\alpha^2}{\omega_h^2} - \frac{\rho B^4}{I} A_2^* \right)$$

$$+ \left(2\zeta_h \frac{\omega_\alpha^2}{\omega_h^2} + 2\zeta_\alpha \frac{\omega_\alpha}{\omega_h} \right) = 0 \tag{6.5.9b}$$

These two real equations are solved successively for different assumed values of K and their roots X are plotted vs. K. At the point (X_c, K_c) where the two plots cross, the flutter condition is identified [6-66, 6-67].

The flutter problem as treated above is seen to be a semi-inverse problem since the aerodynamic coefficients are functions of the solution frequency, and a range of frequency parameters K must therefore be used to survey the solution region.

Alternate methods are also available, though they are beyond the scope of the present discussion. One of the more important of these approaches involves the use of so-called *aerodynamic indicial functions* [6-67 to 6-70 and 6-78, 6-79]. Such functions, derivable from the coefficients H_i^* and A_i^*, represent the response of the bluff section to a step change in angle of attack. They also permit representation of transient response problems. Reference 6-80 makes use of individual response functions in predicting bridge response under natural wind (see also Sect. 6.6). In general, the use of such functions gives rise to more involved calculations than the stability determinant method sketched above. Avoidance of the more general indicial function approach is justified in those cases where structural frequencies and natural modes are not greatly altered by the aerodynamic forces.

A few basic qualitative considerations are in order regarding the solution of the flutter equations and the nature of the flutter phenomenon in the case of bridges as opposed to that of airfoils. In the flutter of airfoils under normal structural conditions (center of mass not excessively far aft of the rotation point) it is impossible for single-degree-of-freedom flutter to occur since both degrees h and α are individually positively damped† (i.e., H_1^* and A_2^* are negative for all values of K). This is the basic reason why classical airfoil flutter, if and when it occurs, must involve coupled freedoms; that is, it must be a condition in which it is mainly the coupling (not the damping) terms that govern the response.

On the other hand, as seen in Fig. 6.5.3 [6-71], certain types of structure (in this case an open-truss suspension bridge deck model) exhibit A_2^* (torsional damping) coefficients that change sign from negative to positive with advancing values of reduced wind velocity U/nB (where $n = \omega/2\pi$). As a result, whether or not coupling coefficients exist, single-degree torsional motion becomes unstable and drives a self-excited flutter due to its net negative damping. Thus, purely single-degree flutter, or "single-degree-driven" flutter, can exist for cases where A_2^* evolves as described above.

The flutter of three-dimensional structures is essentially based on the two-dimensional theory presented above and is discussed in Chapter 13.

6.6 BUFFETING RESPONSE IN THE PRESENCE OF AEROELASTIC PHENOMENA

Buffeting is defined as the unsteady loading of a structure by velocity fluctuations in the oncoming flow. If these velocity fluctuations are clearly associated with the turbulence shed in the wake of an upstream body, the unsteady loading is referred to as wake buffeting.‡ Effective analytical models of the wake buffeting phenomenon do not currently exist in the wind engineering field. On the other hand, notable contributions [6-82 to 6-85] have been made to the problem of

---

†Because of the formal similarity between the mechanical damping terms in the left-hand sides of Eqs. 6.5.2 and the terms containing the coefficients H_1^* and A_2^* in Eqs. 6.5.4, the latter are referred to as aerodynamic damping terms. The differences $2\zeta_h \omega_h m - \frac{1}{2}\rho U^2(2B)KH_1^*$ and $2\zeta_\alpha \omega_\alpha I - \frac{1}{2}\rho U^2(2B^2)KA_2^*(B/U)$ are referred to as net (or total) damping in the translational and the rotational mode, respectively. (See also Sect. 6.2.1.)

‡An example of oscillations attributed to wake buffeting is provided in [6-81], which reports across-wind motions of up to 0.3 m in welded steel towers (diameter: 3 m; height: 70 m; separation distance: 2.5 diameters) in moderate winds.

the buffeting of line-like structures by atmospheric turbulence. Many of the ideas employed below can be traced to origins in these references.

The problem dealt with in this section is that of buffeting by incident turbulence that develops in an atmospheric flow over relatively homogeneous terrain—open, suburban, or urban. For such turbulence it is possible, in certain cases, to model the response to buffeting forces for both those structures that do not and those that do exhibit aeroelastic interaction with the wind forces. Section 5.3 deals with aerodynamic loadings that are independent of structural motion. However, structures like slender towers or the decks of suspended-span bridges, which exhibit aeroelastic effects, are also of considerable interest in practical applications. The present section is concerned principally with the response of such line-like structures.

6.6.1   Aerodynamic Forces on Line-Like Structures

Consider a line-like structure, with spanwise coordinate x, that is being buffeted by atmospheric turbulence. If the oscillations of the structure in each responding mode are small, it may be assumed that the aerodynamic behavior of the structure is linear, so that the aerodynamic forces consist of a superposition of (a) self-excited forces of the type dealt with in Section 6.5 and (b) buffeting forces induced by the incident turbulence.

Buffeting Forces.   For turbulence intensities typical of winds in the atmospheric boundary layer, and for turbulence components with frequencies that are of interest in practice, it may be assumed that the squares and products of the velocity fluctuations u, v, and w are negligible with respect to the square of the mean velocity U, and that the force coefficients C_D, C_L, and C_M are independent of frequency in the range considered. As a result, expressions for the buffeting forces based on quasi-steady theory are acceptable, so that for section x of the span the buffeting drag, lift, and aerodynamic moment (see Fig. 6.6.1) can be

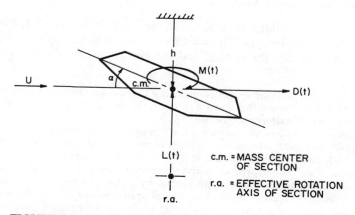

FIGURE 6.6.1.   Buffeting forces on section of a line-like structure.

written as

$$\frac{D(t)}{\frac{1}{2}\rho U^2 B} = C_D(\alpha_0)\frac{A}{B}\left[1+2\frac{u(x,\,t)}{U}\right] \tag{6.6.1a}$$

$$\frac{-L(t)}{\frac{1}{2}\rho U^2 B} = C_L(\alpha_0)\left[1+2\frac{u(x,\,t)}{U}\right]+\left[\frac{dC_L}{d\alpha}\bigg|_{\alpha=\alpha_0}+\frac{A}{B}C_D(\alpha_0)\right]\frac{w(x,\,t)}{U} \tag{6.6.1b}$$

$$\frac{M(t)}{\frac{1}{2}\rho U^2 B^2} = \left[C_M(\alpha_0)+C_D(\alpha_0)\frac{Ar}{B^2}\right]\left[1+2\frac{u(x,\,t)}{U}\right]+\frac{dC_M}{d\alpha}\bigg|_{\alpha=\alpha_0}\frac{w(x,\,t)}{U} \tag{6.6.1c}$$

where B is a typical body dimension such as deck width, A is the across-wind area per unit length projected on the plane normal to the mean wind speed U, r is the distance of the deck mass center to the effective rotation axis, $U+u(t)$ and $w(t)$ are the wind speed components in the along-wind and the vertical directions, respectively,† and α_0 is the mean angle of attack under wind action. In Eqs. 6.6.1b and 6.6.1c the dimensionless ratio $w(t)/U$ represents an angular fluctuation from the mean angle α_0. In Eqs. 6.6.1, the quantity $[1+2u(t)/U]$ is obtained by squaring the sum $[1+u(t)/U]$ and neglecting the square of its second term, as shown in Sect. 4.7.

Self-Excited Forces. It was indicated in Sect. 6.5 that for a body oscillating with circular frequency ω in both the vertical displacement and the torsional modes, the self-excited lift and moment L_h and M_α may be expressed as in Eqs. 6.5.4.

Since the random buffeting load action on a structure may be viewed as a superposition of elemental harmonic loads (see Appendix A2), the vibrations of that structure may, correspondingly, be viewed as a superposition of harmonic responses induced by these loads. Each such oscillation induces, in turn, an elemental self-excited load expressible by Eqs. 6.5.4.‡

6.6.2 Buffeting Response of a Suspension Bridge Without Aerodynamic Coupling Between Modes

For many types of bridge deck sections the coefficients H_2^*, H_3^*, and A_1^* in Eqs. 6.5.4 may be disregarded in first approximation as having minor or negligible influence, so that the vertical and torsional motions of a straight bridge

---

†Equations 6.6.1b and 6.6.1c are written assuming that the line-like structure is horizontal (e.g., a bridge). In the case of a vertical structure (e.g., a tower), the vertical velocity component $w(t)$ must be replaced by the lateral velocity component $v(t)$.

‡An equivalent—and alternative—formulation is to employ the aerodynamic indicial function approach [6-78, 6-79, 6-80] wherein the frequency dependent information contained in the self-excitation aerodynamic coefficients H_i^* and A_i^* is first converted into time-dependent indicial aerodynamic functions and the aerodynamic forces are then expressed in terms of an integral over the product of an indicial function and the structural motion. This approach, typically employed in gust response studies for aircraft, usually leads to explicit time-history calculations, but these are avoided in the present context. Here, time-dependent formulations will be transformed into spectral, or frequency-dependent descriptions of response amplitudes.

may be taken as uncoupled. The above aerodynamic coupling coefficients are of secondary importance particularly in those cases of common occurrence wherein single-degree torsional instability is manifest (i.e., where A_2^* changes sign with increasing U/nB).

Expressions for the bridge response will now be sought following a procedure closely parallel to that employed in Chapter 5 to study along-wind response. Here, however, the effect of aerodynamic self-excitation terms will be taken into account in addition to the aerodynamic buffeting forces.

Torsion will be dealt with first. Consider a full bridge for which the torsional response at any spanwise section x is $\alpha(x, t)$. The response can be written in terms of generalized coordinates as

$$\alpha(x, t) = \sum_i \alpha_i(x) p_i(t) \tag{6.6.2}$$

where $p_i(t)$ are the corresponding time-dependent generalized coordinates of the problem and $\alpha_i(x)$ are the torsional vibration modes. The equation of motion of the deck section x is

$$I(x)\ddot{\alpha}(x, t) + c_\alpha(x)\dot{\alpha}(x, t) + k_\alpha(x)\alpha(x, t) = \mathcal{M}(x, t) \tag{6.6.3}$$

where $I(x)$ is the local mass moment of inertia of the deck about the effective rotation axis and $c_\alpha(x)$ and $k_\alpha(x)$ are, respectively, the effective structural damping and stiffness of the section. To bring the generalized coordinates into the problem, Eq. 6.6.2 is used for $\alpha(x, t)$ in Eq. 6.6.3. The result is then multiplied through by $\alpha_i(x)$ and integrated over the full span L, yielding

$$I_i[\ddot{p}_i(t) + 2\zeta_{\alpha_i}(2\pi n_{\alpha_i})\dot{p}_i(t) + (2\pi n_{\alpha_i})^2 p_i(t)] = M_{\alpha_i}(t) \tag{6.6.4}$$

where I_i is the generalized inertia:

$$I_i = \int_0^L I(x)\alpha_i^2(x)\, dx \tag{6.6.5}$$

ζ_{α_i} and n_{α_i} are, respectively, the damping ratio and the natural frequency (Hz) in the ith torsion mode and M_{α_i} is the generalized force. Implicit use has been made of the orthogonality relation

$$\int_0^L I(x)\alpha_i(x)\alpha_j(x)\, dx = 0 \qquad (i \neq j) \tag{6.6.6}$$

The generalized force $M_{\alpha_i}(t)$ has the form

$$M_{\alpha_i} = \int_0^L \mathcal{M}(x, t)\alpha_i(x)\, dx \tag{6.6.7}$$

The attention of the reader is drawn at this point to the similarity between Eqs. 6.6.2–6.6.5 and Eqs. 5.2.1, 5.2.6, 5.2.7, and 5.2.8. Both sets depict the usual modal approach to a dynamics problem in a continuous structure.

In the present context the distributed moment per unit span will have both self-excited and active, time-dependent components, the former associated with the motion and the latter a function of the gust velocity components in the atmospheric flow passing over the structure. The self-excited components (see Eq. 6.5.4b) will be assumed to take the form* (with $A_1^* \cong 0$)

$$M_\alpha(K) = \tfrac{1}{2}\rho U^2 (2B^2) \left[K A_2^*(K) \frac{B\dot{\alpha}}{U} + K^2 A_3^*(K)\alpha \right] \qquad (6.6.8)$$

where $K = 2\pi n B/U$ while the time-dependent (gust contribution will be random (see Eq. 6.6.1c).

Before applying the full random gust moment, let a single sinusoidal component of amplitude M_0 and frequency n be applied at spanwise section $x = x_1$. Then the applied distribution moment is

$$\mathscr{M}(x, t) = M_\alpha(K) + M_0 \delta(x - x_1) \cos 2\pi nt \qquad (6.6.9)$$

where $\delta(x - x_1)$ is the Dirac delta function (see Eq. 5.1.11) so that the generalized force, Eq. 6.6.5, becomes

$$M_{\alpha_i} = \int_0^L [M_\alpha(K) + M_0 \delta(x - x_1) \cos 2\pi nt] \alpha_i(x)\, dx \qquad (6.6.10)$$

Use of $\alpha(x, t)$ from Eq. 6.6.2 in Eqs. 6.6.8 and 6.6.10 implies that calculation will be required of factors having the form

$$\sum_j p_j(t) \int_0^L \alpha_j(x)\alpha_i(x)\, dx = \sum_j G_{ij} p_j(t) \qquad (6.6.11)$$

where $G_{ij} = \int_0^L \alpha_i \alpha_j\, dx$ and

$$\int_0^L \delta(x - x_1)\alpha_i(x)\, dx = \alpha_i(x_1) \qquad (6.6.12)$$

The first occurs in $M_\alpha(K)$ and the second occurs in the single sinusoidal component.

Since the modes $\alpha_i(x)$ are dimensionless and of arbitrary scale, it is convenient to normalize them arbitrarily, for example, setting

$$\frac{1}{L} \int_0^L \alpha_i^2(x)\, dx = 1 \qquad (6.6.13)$$

One may then note that $G_{ii} = L$, but that for $i \neq j$ the values of G_{ij} are much

---

*Wind tunnel tests performed by the writers have tended to indicate that the destabilizing effect of the self-excited forces acting on a suspension bridge deck is somewhat reduced by the presence of turbulence in the incident flow. The use in calculations of aerodynamic coefficients H_i^* and A_i^* obtained under smooth flow conditions is therefore thought here to be conservative. Recent model experiments [6-86] employing techniques of random analysis have shed further light on the effect of turbulence upon the values of H_i^* and A_i^*. Full explorations of the effect of appropriately scaled turbulence on the flutter derivatives of bridge decks remain to be carried out.

less than L, in general. It will be assumed here that G_{ij} $(i \neq j)$ is negligible, which is reasonable for bridges in which $I(x)$ is approximately constant across the span, as can be seen from Eq. 6.6.6. The net value of the generalized force M_{α_i} then is

$$M_{\alpha_i} \cong \rho U^2 B^2 L \left[K A_2^*(K) \frac{B\dot{p}_i}{U} + K^2 A_3^*(K) p_i \right] + M_0 \alpha_i(x_1) \cos 2\pi nt \quad (6.6.14)$$

Equation 6.6.4, which describes the motion of the ith mode, may then be written with use of Eq. 6.6.14 as

$$I_i[\ddot{p}_i(t) + 2\tilde{\zeta}_{\alpha_i}(2\pi\tilde{n}_{\alpha_i})\dot{p}_i(t) + (2\pi\tilde{n}_{\alpha_i})^2 p_i(t)] = M_0 \alpha_i(x_1) \cos 2\pi nt \quad (6.6.15)$$

where new effective frequency $\tilde{n}_{\alpha_i}$ and damping $\tilde{\zeta}_{\alpha_i}$ have been introduced such that

$$\tilde{n}_{a_i}^2 = n_{\alpha_i}^2 - n^2 \left[\frac{\rho B^4 L}{I_i} A_3^*(K) \right] \quad (6.6.16)$$

$$\tilde{\zeta}_{\alpha_i} = \frac{1}{\tilde{n}_{\alpha_i}} \left[\zeta_{\alpha_i} n_{\alpha_i} - \frac{\rho B^4 L}{2I_i} A_2^*(K) n \right] \quad (6.6.17)$$

Equations 6.6.16 and 6.6.17 introduce the effect of the aerodynamic self-excited forces into the response at frequency n.

Equation 6.6.15 $(i = 1, 2, 3, \ldots)$ is similar in form to Eq. 5.2.7 for which the generalized force is given by Eq. 5.2.12. In Chapter 5 the system defined by Eq. 5.2.7 is analyzed under distributed random loading, leading to Eq. 5.2.38. Completely analogous steps hold here, yielding the following result for the spectrum of torsional response:

$$S_\alpha(x, n) \cong \sum_i \frac{\alpha_i^2(x) \int_0^L \int_0^L \alpha_i(x_1)\alpha_i(x_2) S_{M_1 M_2}^C(n) \, dx_1 \, dx_2}{16\pi^4 \tilde{n}_{\alpha_i}^4 I_i^2 \{[1 - (n/\tilde{n}_{\alpha_i})^2]^2 + 4\tilde{\zeta}_{\alpha_i}^2 (n/\tilde{n}_{\alpha_i})^2\}} \quad (6.6.18)$$

where $S_{M_1 M_2}^C(n)$ is the co-spectrum of the buffeting moments M_1 and M_2 per unit span, which act respectively at the coordinates x_1 and x_2.

Equation 6.6.1c describes the applied aerodynamic moment per unit span due to steady wind and gust components. In this equation, the moment and drag coefficients C_M and C_D are functions of the mean twist angle $\alpha_0(x)$ at the spanwise section x, and the velocity components u and v are also functions of x and time. For convenience the following notation is introduced:

$$C_{ME}[\alpha_0(x)] \equiv C_M[\alpha_0(x)] + C_D[\alpha_0(x)] \frac{Ar}{B^2} \quad (6.6.19)$$

and

$$\left. \frac{dC_M}{d\alpha} \right|_{\alpha = \alpha_0(x)} = C_M'[\alpha_0(x)] \quad (6.6.20)$$

Thus, referring to Eqs. A2.29 and A2.33 (Appendix A2), the moment cospectrum between sections x_1 and x_2 may be written

$$S^C_{M_1 M_2}(n) = [\tfrac{1}{2}\rho U^2 B^2]^2 \left\{ 4C_{ME}[\alpha_0(x_1)]C_{ME}[\alpha_0(x_2)] \frac{S^C_{u_1 u_2}(n)}{U^2} \right.$$

$$+ 2C_{ME}[\alpha_0(x_1)]C'_M[\alpha_0(x_2)] \frac{S^C_{u_1 w_2}(n)}{U^2}$$

$$+ 2C_{ME}[\alpha_0(x_2)]C'_M[\alpha_0(x_1)] \frac{S^C_{u_1 w_2}(n)}{U^2}$$

$$\left. + C'_M[\alpha_0(x_1)]C'_M[\alpha_0(x_2)] \frac{S^C_{w_1 w_2}(n)}{U^2} \right\} \tag{6.6.21}$$

Evaluation of C_{ME} and C'_M at values x_1 and x_2 requires knowledge of the mean deflection distribution $\alpha_0(x)$ over the span. This can be obtained by a static study of the type discussed in Sect. 6.4.1 or may be described in terms of the torsional vibration modes by the expression

$$\alpha_0(x) = \sum_i \frac{\int_0^L \tfrac{1}{2}\rho U^2 B^2 C_{ME}[\alpha_0(x_1)]\alpha_i(x_1)\,dx_1}{4\pi^2 n_{\alpha_i}^2 I_i} \alpha_i(x) \tag{6.6.22}$$

which is a result derived from Eq. 6.6.4 by neglecting all time-dependent terms. The solution of Eq. 6.6.22 for a given wind velocity U requires an iterative approach, starting conveniently with $\alpha_0 = 0$.

In Eq. 6.6.21 the co-spectra $S^C_{u_1 w_2}(n)$ and $S^C_{u_2 w_1}(n)$ are negative in value and appreciably smaller in magnitude than $S^C_{u_1 u_2}(n)$ and $S^C_{u_1 w_1}(n)$; they may conservatively be neglected.

The root mean square of the fluctuating torsional response at section x is

$$\overline{\alpha^2(x)} = \int_0^\infty S_\alpha(x, n)\,dn \tag{6.6.23}$$

Peak values of the fluctuating torsional response may be obtained by following steps similar to those of Sect. 5.3. Methods of calculation relative to the quantities mentioned above are discussed in Chapter 13.

If the vertical (bending) response of the bridge is written as

$$h(x, t) = \sum_i h_i(x)q_i(t) \qquad (i = 1, 2, \ldots) \tag{6.6.24}$$

where $h_i(x)$ are the vertical bending modes of vibration and q_i are the generalized coordinates for these modes, then, by a process completely analogous to that described above for torsion, the spectrum of the vertical response can be shown to be

$$S_h(x, n) \cong \sum_i \frac{h_i^2(x) \int_0^L \int_0^L h_i(x_1)h_i(x_2)S^C_{L_1 L_2}(n)\,dx_1\,dx_2}{16\pi^4 n_{h_i}^4 M_i^2 \{[1 - (n/n_{h_i})^2]^2 + 4\tilde{\zeta}_{n_i}^2(n/n_{h_i})^2\}} \tag{6.6.25}$$

where

$$M_i = \int_0^L m(x)h_i^2(x)\,dx$$

is the generalized inertia, $m(x)$ being the deck section mass per unit length, n_{h_i} the natural frequency† in the ith mode, and $\tilde{\zeta}_{h_i}$ the aerodynamically-influenced system damping defined by

$$\tilde{\zeta}_{h_i} = \zeta_{h_i} - \frac{\rho B^2 L}{2M_i} H_1^*(K) \frac{n}{n_{h_i}} \tag{6.6.26}$$

where $K = 2\pi Bn/U$ and ζ_{h_i} is the mechanical damping ratio in the ith mode. The co-spectrum of the time-dependent lift forces L_1 and L_2 per unit length of span, which act respectively at span points x_1 and x_2, is (from Eq. 6.6.1b)

$$S_{L_1 L_2}^C(n) = [\tfrac{1}{2}\rho U^2 B]^2 \left\{ 4C_L[\alpha_0(x_1)]C_L[\alpha_0(x_2)] \frac{S_{u_1 u_2}^C(n)}{U^2} \right.$$

$$+ 2C_L[\alpha_0(x_1)]C_{LE}'[\alpha_0(x_2)] \frac{S_{u_1 w_2}^C(n)}{U^2}$$

$$+ 2C_L[\alpha_0(x_2)]C_{LE}'[\alpha_0(x_1)] \frac{S_{u_2 w_1}^C(n)}{U^2}$$

$$\left. + C_{LE}'[\alpha_0(x_1)]C_{LE}'[\alpha_0(x_2)] \frac{S_{w_1 w_2}^C(n)}{U^2} \right\} \tag{6.6.27}$$

where

$$C_{LE}'[\alpha_0(x)] = \frac{dC_L}{d\alpha}\bigg|_{\alpha = \alpha_0(x)} + \frac{A}{B} C_D[\alpha_0(x)] \tag{6.6.28}$$

The mean, mean square and peak vertical responses can then be calculated, as was indicated above for torsional response, by following steps similar to those of Sect. 5.3.

To calculate the *along-wind* response, completely analogous procedures to those above are used, the basic forcing function being the drag as given by Eq. 6.6.1a; a knowledge of along-wind vibration modes is also required.

6.6.3 Outline of the General Buffeting Response Problem of Line-Like Structures

Let the across-wind bending and torsional modes of a symmetrical line-like structure‡ be represented by $h_i(x)$ and $\alpha_i(x)$ as in Sect. 6.6.2, so that sectional deflections h and α (Fig. 6.5.1) under dynamic excitation are

$$h(x, t) = \sum_i h_i(x)q_i(t) \tag{6.6.29a}$$

$$\alpha(x, t) = \sum_i \alpha_i(x)p_i(t) \tag{6.6.29b}$$

---

†There is no aerodynamic influence in this case upon the natural frequency, owing to the absence in the basic model (Eq. 6.5.4) of terms in h.

‡For unsymmetrical structures ($S \neq 0$, Eq. 6.5.1), the treatment is analogous, with aerodynamic force and moment referred to the elastic axis.

Analogously to previous formulations (Sect. 6.6.2) the equations of motion (mechanically uncoupled about the centerline) become

$$M_i[\ddot{q}_i + 2\zeta_{h_i}(2\pi n_{h_i})\dot{q}_i + (2\pi n_{h_i})^2 q_i] = \int_0^L \mathscr{L}(x, t)h_i(x)\,dx \qquad (6.6.30a)$$

$$I_i[\ddot{p}_i + 2\zeta_{\alpha_i}(2\pi n_{\alpha_i})\dot{p}_i + (2\pi n_{\alpha_i})^2 p_i] = \int_0^L \mathscr{M}(x, t)\alpha_i(x)\,dx \qquad (6.6.30b)$$

where $\mathscr{L}(x, t)$ and $\mathscr{M}(x, t)$ are, respectively, the lift and moment per unit span at section x of the span.

In order to obtain the necessary system admittance functions, $\mathscr{L}(x, t)$ and $\mathscr{M}(x, t)$ are alternately specified in the following manners. For lift-associated admittances:

$$\mathscr{L}(x, t) = L_h + L_0 e^{i2\pi nt}\delta(x - x_1)$$

$$\mathscr{M}(x, t) = M_\alpha$$

For moment-associated admittances:

$$\mathscr{L}(x, t) = L_h$$

$$\mathscr{M}(x, t) = M_\alpha + M_0 e^{i2\pi nt}\delta(x - x_1)$$

where L_h and M_α are the self-excited aerodynamic lift and moment per unit span given by Eqs. 6.5.4.

Modified equations of motion (6.6.30) can then be written that are similar to Eq. 6.6.15 though now coupled by the presence of the full set of unsteady motion derivatives proportional to H_i^* and A_i^*. From these equations, aerodynamically modified mechanical admittances can be calculated analogously to previous results, but now for two coupled equations. The results, representing, respectively, (1) the across-wind deflection due to a concentrated harmonic lift at section x_1, (2) the torsional deflection due to a concentrated harmonic lift at x_1, (3) the across-wind deflection due to a concentrated harmonic moment at x_1, and (4) the torsional deflection due to a concentrated harmonic moment at x_1, may be designated $H_{hL}(x, x_1, n)$, $H_{\alpha L}(x, x_1, n)$, $H_{hM}(x, x_1, n)$, and $H_{\alpha M}(x, x_1, n)$.

Assuming now that the structure is subjected to a distributed buffeting lift $L(x, t)$ and moment $M(x, t)$ as defined by Eqs. 6.6.1b and 6.6.1c, the spectra of across-wind bending and torsional response can be calculated by integrating elemental effects. Designating by $S_{L_1 L_2}$, $S_{L_1 M_2}$, $S_{M_1 L_2}$, and $S_{M_1 M_2}$ the cross-spectra corresponding respectively to the lifts and moments at x_1 and x_2 as suggested by their subscripts, the following typical expression for vertical response spectrum $S_h(x, n)$ is obtained:

$$S_h(x, n) = \int_0^L \int_0^L [H_{hL}^*(x, x_1, n)H_{hL}(x, x_2, n)S_{L_1 L_2}(n)$$

$$+ H_{hL}^*(x, x_1, n)H_{hM}(x, x_2, n)S_{L_1 M_2}(n)$$

$$+ H_{hM}^*(x, x_1, n)H_{hL}(x, x_2, n)S_{M_1 L_2}(n)$$

$$+ H_{hM}^*(x, x_1, n)H_{hM}(x, x_2, n)S_{M_1 M_2}(n)] \, dx_1 \, dx_2$$

where H^* denotes the complex conjugate of H.

It should be remarked that both the mean speed of the flow and the values of lift and moment may, in the above expressions, be a function of x. In that case modal orthogonality relations can no longer be used (as was done, for example, in Eq. 6.6.14), and the expressions for the modified admittances become more elaborate; however, the attendant calculations can be conveniently programmed for electronic computers.

Possible applications of the expressions for the response of line-like structures dealt with here include the calculation of the responses of tall prismoidal buildings with strong torsional motions, and those of tall towers and suspended-span bridges.

REFERENCES

6-1   D. J. Johns et al., "On Wind-Induced Instabilities of Open-Ended Circular Cylindrical Shells," in *Proceedings of the Conference on Tower Shaped Structures*, The Hague, 1969, A. M. Haas and H. van Koten (Eds.), Insititue TNO for Building Materials and Structures, Delft, The Netherlands, pp. 185–212.

6-2   L. R. Wootton, M. H. Warner, R. N. Sainsbury, and D. H. Cooper, *Oscillation of Piles in Marine Structures*, Construction Industry Research and Information Association, London, Report No. 41, 1972.

6-3   R. King, M. Prosser, and D. J. Johns, "On Vortex Excitation of Model Piles in Water," *J. Sound Vib.*, **29** (1973), 169–188.

6-4   R. E. D. Bishop and A. Y. Hassan, "The Lift and Drag Forces on a Circular Cylinder Oscillating in a Flowing Fluid," *Proc. Roy. Soc., London, Series A*, **277** (1964), 51–74.

6-5   G. V. Parkinson, G. Feng, and N. Ferguson, "Mechanisms of Vortex-Excited Oscillations of Bluff Cylinders," *Proceedings of the Symposium on Wind Effects on Buildings and Structures*, Loughborough University of Technology, Leicestershire, U.K., 1966.

6-6   N. Ferguson and G. V. Parkinson, "Surface and Wake Phenomena of Vortex-Excited Oscillations of Bluff Cylinders," *J. Eng. Ind.*, ASME, **89** (1967), 831–838.

6-7   R. T. Hartlen, W. D. Baines, and I. G. Currie, *Vortex-Excited Oscillations of a Circular Cylinder*, University of Toronto Technical Report No. 6809, Toronto, 1968.

6-8   G. H. Toebes, "The Unsteady Flow and Wake Near an Oscillating Cylinder," *Trans. ASME, J, Basic Eng.*, **91** (1969), 493–505.

6-9   V. C. Mei and I. G. Currie, "Flow Separation on a Vibrating Cylinder," *Phys. Fluids*, **12**, (1969), 2248–2254.

6-10   G. H. Koopman, *Wind-Induced Vibrations of Skewed Circular Cylinders*, Civil and Mechanical Engineering Department Report No. 70-11, Catholic University of America, Washington, D.C., 1970.

6-11   R. A. Skop, S. E. Ramberg, and K. M. Ferer, *Added Mass and Damping Forces on Circular Cylinders*, Naval Research Laboratory Formal Report No. 7970, Washington, D.C., 1970.

6-12   R. J. Glass, *A Study of the Hydroelastic Vibrations of Spring Supported Cylinders in a Steady Fluid Stream Due to Vortex Shedding*, ONR Project N0014-69-C-0148 Final Report, Washington, D.C., 1970.

6-13   R. T. Hartlen and I. G. Currie, "Lift-Oscillator Model of Vortex-Induced Vibration," *J. Eng. Mech. Div.*, ASCE, **96** (1970), 577–591.

6-14   R. Sainsbury and D. King, "The Flow-Induced Oscillations of Marine Structures," *Proc. Inst. Civ. Eng.*, **49** (1971), 269–302.

6-15   L. R. Wooton and C. Scruton, "Aerodynamic Stability," in *The Modern Design of Wind-Sensitive Structures*, Construction Industry Research and Information Association, London, 1971.

6-16   O. M. Griffin, R. A. Skop, and G. H. Koopmann, "The Vortex-Excited Resonant Vibrations of Circular Cylinders," *J. Sound Viv.*, **31** (1973), 235–249.

6-17   R. A. Skop and O. M. Griffin, "A Model for the Vortex-Excited Response of Bluff Cylinders," *J. Sound Vib.*, **27** (1973), 225–233.

6-18   O. M. Griffin and S. E. Ramberg, "The Vortex-Street Wakes of Vibrating Cylinders," *J. Fluid Mech.*, **66** (1974), 553–578.

6-19   W. D. Iwan and R. D. Blevins, "A Model for the Vortex-Induced Oscillation of Structures," *J. Appl. Mech.*, ASME, **41** (1974), 581–585.

6-20   R. A. Skop, *On Modeling Vortex-Excited Oscillations*, Naval Research Laboratory Memorandum Report No. 2927, Washington, D.C., 1974.

6-21   F. Angrilli, G. DiSilvio, and A. Zanardo, "Hydroelasticity Study of Circular Cylinders in a Water Stream," in *Proceedings of the IUTAM-IAHR Symposium on Flow-Induced Structural Vibrations*, Karlsruhe, West Germany, 1972, E. Naudascher (Ed.), Springer-Verlag, Berlin, 1974, pp. 504–512.

6-22   R. King and M. J. Prosser, "Criteria for Flow-Induced Oscillations of a Cantilevered Cylinder in Water," in *Proceedings of the IUTAM-IAHR Symposium on Flow-Induced Structural Vibrations*, Karlsruhe, West Germany, 1972, E. Naudascher (ed.), Springer-Verlag, Berlin, 1974, pp. 488–503.

6-23   O. M. Griffin and S. E. Ramberg, "On Vortex Strength and Drag in Bluff Body Wakes," *J. Fluid Mech.*, **69** (1975), 721–729.

6-24   R. King, *Vortex Excited Oscillations of Inclined (Yawed) Cylinders in Flowing Water*, British Hydro-Mechanics Research Association, Bedford, U.K., Report No. RR 1214, 1975.

6-25   R. D. Blevins and T. E. Burton, "Fluid Forces Induced by Vortex-Shedding," *J. Fluids Eng.*, **95** (1976), 19–24.

6-25   R. A. Skop and O. M. Griffin, "On a Theory for the Vortex-Excited Oscillations of Flexible Cylindrical Structures," *J. Sound Vib.*, **41** (1975), 263–274.

6-27   W. D. Iwan, *The Vortex Induced Oscillation of Elastic Structural Elements*, ASME paper 75-DET-28, 1975.

6-28   W. K. Blake, "Periodic and Random Excitation of Streamline Structures by Trailing-Edge Flows," in *Proceedings of the Symposium on Turbulence in Liquids*, University of Missouri, Rolla, Mo., 1975.

6-29   S. E. Ramberg, O. M. Griffin, and R. A. Skop, "Some Resonant Transverse Vibration Properties of Marine Cables With Application to the Prediction of Vortex-Induced Structural Vibrations," in *Ocean Eng. Mech.*, N. Monney (Ed.), ASME, New York, 1975, 29–42.

6-30   R. King, "An Investigation of the Criteria Controlling Sustained Self-Excited Oscillations of Cylinders in Flowing Water," in *Proceedings of the Symposium on Turbulence in Liquids*, University of Missouri, Rolla, Mo., 1975.

6-31   S. E. Ramberg and O. M. Griffin, "Velocity Correlation and Vortex Spacing in the Wake of a Vibrating Cable," *J. Fluids Eng.*, **98** (1976), 10–18.

6-32   S. E. Ramberg and O. M. Griffin, *The Effects of Vortex Coherence, Spacing and Circulation on the Flow-Induced Forces on Vibrating Cables and Bluff Structures*, Naval Research Laboratory Formal Report No. 7945, Washington, D.C., 1976.

6-33   O. M. Griffin, R. A. Skop, and S. E. Ramberg, "Modeling of the Vortex-Induced Oscillations of Cables and Bluff Structures," paper delivered to Society for Experimental Stress Analysis, Silver Spring, Md., 1976.

6-34   N. Minorsky, *Nonlinear Oscillations*, Van Nostrand, New York, 1962.

6-35   R. H. Scanlan, "Theory of the Wind Analysis of Long-Span Bridges Based on Data Obtainable from Section Model Tests," in *Proceedings of the Fourth International Conference on*

Wind Effects, London, 1975, Cambridge Univ. Press, Cambridge, U.K., 1976, pp. 259–269.

6-36   R. H. Gade, H. R. Bosch, and W. Podolny, Jr., "Recent Aerodynamic Studies of Long-Span Bridges," *J. Struct. Div.*, ASCE, **102**, No. ST7 (July 1976), 1299–1315.

6-37   R. H. Gade and H. R. Bosch, *Interim Report, Wind Tunnel Studies on the Luling, La. Cable-Stayed Bridge*, F.H.W.A., Fairbank Lab, McLean, Va., U.S. Department of Transportation, 1975.

6-38   G. H. Toebes, "Fluidelastic Features of Flow Around Cylinders," in *Proceedings of the International Research Seminar on Wind Effects on Buildings and Structures*, Ottawa, Canada, 1967, Vol. 2, University of Toronto Press, Toronto, 1968, pp. 323–334.

6-39   L. R. Wooton, "The Oscillations of Large Circular Stacks in Wind," *Proc. Inst. Civ. Eng.*, **43** (1969), 573–598.

6-40   H. Glauert, *Rotation of an Airfoil About a Fixed Axis*, Aeronautical Research Committee, R & M 595, Great Britain, 1919.

6-41   J. P. Den Hartog, "Transmission Line Vibration Due to Sleet," *Trans. AIEE*, **51** (1932), 1074–1076.

6-42   J. P. Den Hartog, *Mechanical Vibrations*, 4th ed., McGraw-Hill, New York, 1956.

6-43   G. V. Parkinson and N. P. H. Brooks, "On the Aeroelastic Instability of Bluff Cylinders," *Trans. ASME, J. Appl. Mech.*, **83** (1961), 252–258.

6-44   G. V. Parkinson and J. D. Smith, "An Aeroelastic Oscillator with Two Stable Limit Cycles," *Trans. ASME, J. Appl. Mech.*, **84** (1962), 444–445.

6-45   G. V. Parkinson, "Aeroelastic Galloping in One Degree of Freedom," in *Proceedings of the Symposium on Wind Effects on Buildings and Structures*, Vol. 1, National Physical Laboratory, Teddington, U.K., 1963, pp. 581–609.

6-46   G. V. Parkinson and J. D. Smith, "The Square Prism as an Aeroelastic Nonlinear Oscillator," *Quart. J. Mech. Appl. Math.*, **17**, Pt. 2 (1964), 225–239.

6-47   G. V. Parkinson and T. V. Santosham, "Cylinders of Rectangular Section as Aeroelastic Nonlinear Oscillators," Vibrations Conference, ASME, Boston, Mass., 1967.

6-48   M. Novak, "Aeroelastic Galloping of Prismatic Bodies," *J. Eng. Mech. Div.*, ASCE, **95**, No. EM1 (Feb. 1969), 115–142.

6-49   M. Novak, "Galloping Oscillations of Prismatic Structures," *J. Eng. Mech. Div.*, ASCE, **98**, No. EM1 (Feb. 1972), 27–46.

6-50   G. V. Parkinson, "Mathematical Models of Flow-Induced Vibrations of Bluff Bodies," in *Proceedings of the IUTAM-IAHR Symposium on Flow-Induced Structural Vibrations*, Karlsruhe, West Germany, 1972, E. Naudascher (Ed.), Springer-Verlag, Berlin, 1974, pp. 81–127.

6-51   R. L. Wardlaw, "Wind Tunnel Investigations in Industrial Aerodynamics," *Can. Aeronaut. Space J.*, **18**, No. 3 (March 1972).

6-52   R. H. Scanlan and R. L. Wardlaw, "Reduction of Flow-Induced Structural Vibrations," in *Isolation of Mechanical Vibration, Impact, and Noise*, AMD Vol. 1, Sect. 2 ASME, New York, 1973, pp. 35–63.

6-53   N. Kryloff and N. Bogoliuboff, *Introduction to Nonlinear Mechanics*, trans. S. Lefschetz, Annals of Mathematics Studies, No. 11, Princeton Univ. Press, Princeton, N.J., 1947.

6.54   V. Mukhopadhyay and J. Dugundji, "Wind Excited Vibration of a Square Section Cantilever Beam in Smooth Flow," *J. Sound Vib.*, **45**, No. 3 (1976), 329–339.

6-55   R. Skarecky, "Yaw Effects on Galloping Instability," *J. Eng. Mech. Div.*, ASCE, **101**, No. EM6 (Dec. 1975), 739–754.

6-56   R. L. Wardlaw, K. R. Cooper, and R. H. Scanlan, "Observations on the Problem of Subspan Oscillation of Bundled Power Conductors," *DME/NAE Quarterly Bulletin* No. 1973 (1), National Research Council, Ottawa, Canada, 1973 (reprint), pp. 1–20.

6-57   K. R. Cooper and R. L. Wardlaw, *Preliminary Wind Tunnel Investigation of Twin Bundle Sub-Conductor Oscillations*, Report No. LTR-LA-41 NAE, NRC, Ottawa, Canada, 1970.

6-58   A. Simpson, "Stability of Subconductors of Smooth Cross-Section," *Proc. Inst. Electr. Eng.*, **117**, 4 (1970), 741–750.

6-59   A. Simpson, "On the Flutter of a Smooth Cylinder in a Wake," *Aeronaut. Q.* (Feb. 1971), 25–41.

6-60   A. Simpson, "Wake-Induced Flutter of Circular Cylinders: Mechanical Aspects," *Aeronaut. Q.* (May 1971), 101–118.

6-61   K. R. Cooper and R. L. Wardlaw, "Aeroelastic Instabilities in Wakes," in *Proceedings of the Third International Conference on Wind Effects on Buildings and Structures*, Tokyo, 1971, Saikon, Tokyo, 1972, pp. 647–655.

6-62   A. Simpson, "Determination of the Natural Frequencies of Multi-Conductor Overhead Transmission Lines," *J. Sound Vib.*, **20**, 4 (1972), 417–449.

6-63   K. R. Cooper, *A Wind Tunnel Investigation of Twin Bundled Power Conductors*, Report No. LTR-LA-96, NAE, NRC, Ottawa, Canada, 1972.

6-64   J. A. Watts, K. R. Cooper, and R. L. Wardlaw, *Proposed Wind Tunnel Tests Programs for Bundled Conductor Subspan Oscillations*, Report No. LTR-LA-99, NAE, NRC, Ottawa, Canada, 1972.

6-65   R. H. Scanlan, *A Wind Tunnel Investigation of Bundled Power-Line Conductors, Part VI. Observations on the Problem*, Report No. LTR-LA-121, NAE, NRC, Ottawa, Canada, 1972.

6-66   T. Theodorsen, *General Theory of Aerodynamic Instability and the Mechanism of Flutter*, NACA Report No. 496, 1935.

6-67   R. H. Scanlan and R. Rosenbaum, *Aircraft Vibration and Flutter*, Macmillan, New York, 1951 (reprint, Dover, 1968).

6-68   Y. C. Fung, *The Theory of Aeroelasticity*, Wiley, New York, 1955 (reprint, Dover, 1969).

6-69   R. L. Bisplinghoff, H. Ashley, and R. L. Halfman, *Aeroelasticity*, Addison-Wesley, Cambridge, Mass., 1955.

6-70   R. L. Bisplinghoff and H. Ashley, *Principles of Aeroelasticity*, Wiley, New York, 1962.

6-71   R. H. Scanlan and J. J. Tomko, "Airfoil and Bridge Deck Flutter Derivatives," *J. Eng. Mech. Div.*, ASCE, **97**, No. EM6, Proc. Paper 8609 (Dec. 1971), 1717–1737.

6-72   R. H. Scanlan and A. Sabzevari, "Experimental Aerodynamic Coefficients in the Analytical Study of Suspension Bridge Flutter," *J. Mech. Eng. Sci.*, **11**, 3 (1969), 234–242.

6-73   N. Ukeguchi, H. Sakata, and H. Nishitani, "An Investigation of Aeroelastic Instability of Suspension Bridges," in *Proceedings of the International Symposium on Suspension Bridges*, Laboratorio Nacional de Engenharia Civil, Lisbon, 1966, pp. 273–284.

6-74   T. Okubo and N. Narita, "A Comparative Study on Aerodynamic Forces Acting on Cable-Stayed Bridge Girders," in *Proceedings of the Second U.S.–Japan Research Seminar on Wind Effects on Structures*, Kyoto, 1974, Univ. of Tokyo Press, Tokyo, 1976, pp. 271–283.

6-75   T. Okubo and K. Yokoyama, "Some Approaches for Improving Wind Stability of Cable-Stayed Girder Bridges," in *Proceedings of the Fourth International Conference on Wind Effects on Buildings and Structures*, London, 1975, Cambridge Univ. Press, Cambridge, U.K., 1976, pp. 241–249.

6-76   Y. Otsuki, K. Washizu, H. Tomizawa, and A. Ohya, "A Note on the Aeroelastic Instability of a Prismatic Bar with Square Section," *J. Sound Vib.*, **34**, 2 (1974), 233–248.

6-77   H. Loiseau and E. Szechenyi, "Étude du comportement aéroélastique du tablier d'un pont à haubans," T.P. 1975-75, Office National d'Études et de Recherches Aérospatiales, Châtillon, France.

6-78   R. H. Scanlan and K. S. Budlong, "Flutter and Aerodynamic Response Considerations for Bluff Objects in a Smooth Flow," in *Proceedings of the IUTAM-IAHR Symposium on Flow-Induced Vibrations*, Karlsruhe, West Germany, 1972, E. Naudascher (Ed.), Springer-Verlag, Berlin, 1974, pp. 339–354.

6-79   R. H. Scanlan, J.-G. Béliveau, and K. S. Budlong, "Indical Aerodynamic Functions for Bridge Decks," *J. Eng. Mech. Div.*, ASCE, **100**, No. EM4 (Aug. 1974), 657–672.

6-80   J.-G. Béliveau, R. Vaicaitis, and M. Shinozuka, "Motion of a Suspension Bridge Subject to Wind Loads," *J. Struct. Div.*, ASCE, **103**, No. ST6 (1977), 1189–1205.

6-81   K. R. Cooper and R. L. Wardlaw, "Aeroelastic Instabilities in Wakes," in *Proceedings of the Third International Conference on Wind Effects on Buildings and Structures*, Tokyo, 1971, Saikon, Tokyo, 1972, pp. 647–655.

6-82 H. W. Liepmann, "On the Application of Statistical Concepts to the Buffeting Problem," *J. Aeronaut. Sci.*, **19**, 12 (Dec. 1952), 793–800, 822.

6-83 A. G. Davenport, "The Application of Statistical Concepts to the Wind Loading of Structures," *Proc. Inst. Civ. Eng.*, **19** (1961), 449–472.

6-84 A. G. Davenport, "The Response of Slender, Line-Like Structures to a Gusty Wind," *Proc. Inst. Civ. Eng.*, **23** (1962), 389–407.

6-85 A. G. Davenport, "The Action of Wind on Suspension Bridges," in *Proceedings of the International Symposium on Suspension Bridges*, Laboratorio Nacional de Engenharia Civil, Lisbon, 1966, pp. 79–100.

6-86 W.-H. Lin, "Forced and Self-Excited Responses of a Bluff Structure in a Turbulent Wind," doctoral dissertation, Department of Civil Engineering, Princeton University, 1977.

6-87 R. D. Blevins, *Flow-Induced Vibration*, Van Nostrand Reinhold, New York, 1977.

6-88 B. J. Vickery, and R. I. Basu, "Across-Wind Vibrations of Structures of Circular Cross-Section, Part I, Development of a Two-Dimensional Model for Two-Dimensional Conditions," *J. Wind Eng. Ind. Aerodyn.*, **12** (1983), 49–73.

6-89 R. I. Basu, and B. J. Vickery, "Across-Wind Vibrations of Structures of Circular Cross-Section, Part 2, Development of a Mathematical Model for Full Scale Application," *J. Wind Eng. Ind. Aerodyn.*, **12** (1983), 75–97.

6-90 D. J. B. Richards, "Aerodynamic Properties of the Severn Crossing Conductor," *Proceedings of the Symposium on Wind Effects on Buildings and Structures*, Vol. II, National Physical Laboratory, Teddington, U.K., Her Majesty's Stationery Office, London 1965, pp. 688–765.

6-91 O. M. Griffin and R. A. Skop, "The Vortex-Induced Oscillations of Structures," *J. Sound Vib.*, **44** (1976), 303–305.

7

Wind Tunnels

Although the science of theoretical fluid mechanics is well developed and computational methods are experiencing rapid growth in the area, it remains necessary to perform physical experiments to gain needed insights into many complex effects associated with fluid flow. This is the case in the well-established field of aeronautics, for which wind tunnels were first developed, and, to an even greater extent, in the practical study of buildings, structures, and machines that stand in the earth's near-surface atmospheric layer.

For the most part such structures have been designed for other purposes than providing minimal resistance to the air moving about them. They have therefore, in recent decades, been the focus of what is termed bluff-body aerodynamics. In such aerodynamics there is much emphasis on flows around sharp corners, on separated flows, and so forth. These situations are among the most recondite when it comes to both theoretical and computational methods. The wind tunnel is thus naturally resorted to as an investigative tool in this context.

Typically, the full scale bluff body is immersed in a turbulent atmospheric flow. Flachsbart determined as early as 1932 (see Sect. 4.6.2 and Fig. 4.6.4) that simulations of the aerodynamic behavior of buildings should be conducted in wind tunnel flows with characteristics similar to those of the natural wind. Currently, the vast majority of tests are carried out in wind tunnels that simulate atmospheric flows. (In some instances tests in smooth flow are still accepted, for example, in the case of trussed frameworks—see Sect. 4.5 and Chapter 12— or for preliminary investigations of the geometric shape of bridge deck section models. However, these instances are the exception rather than the rule). There is therefore a strong interest in gaining a knowledge—for later reproduction in the wind tunnel—of the nature of wind flows in the earth's boundary layer; "target" characteristics to be duplicated in the wind tunnel are acquired from meteorological investigations of the atmospheric boundary layer (see Chapter 2 and [7-1] to [7-4]).

Simulation occurs at reduced geometric scale for obvious reasons of economy and convenience. The question of scale then opens up the whole area of physical similitude and the necessary underlying theory, which places emphasis on a set of dimensionless numbers and/or *similarity* criteria applicable to both flow and test models of structures placed in it. With characteristics of the target flow and scale factors for similitude established, it soon becomes apparent that certain of the model criteria established for similarity cannot in fact be satisfied under typical, everyday test conditions. The wind tunnel modeler is thus launched upon a series of inevitable compromises that render his task complex, revealing it as an art of both performance and interpretation rather than an exact science.

A basic discussion of similarity criteria is presented in Sect. 7.1. Wind tunnels used in civil engineering applications are briefly described in Sect. 7.2, which also includes comments on some difficulties in achieving similarity between wind tunnel and atmospheric flows. Section 7.3 is devoted to scaling problems, insofar as they affect the aerodynamic and aeroelastic behavior of the models to be tested, and to the question of wind tunnel blockage. Section 7.4 reviews some attempts to validate results of wind tunnel tests by comparisons with full-scale measurements. Information on general wind tunnel testing techniques is provided in [7-5] to [7-10]. Reference 7-11 is a useful compendium on wind tunnel modeling for civil engineering applications and includes, in particular, up to date information on modern wind tunnel instrumentation.

7.1 BASIC SIMILARITY REQUIREMENTS

In analyzing any physical problem—more particularly one that is expected to be studied experimentally—it is usual to identify a set of governing dimensionless parameters. These parameters are in certain cases obtained by first writing the partial differential equations that describe the physical system at hand. These equations are then rendered dimensionless by dividing each of the key variables by a reference value having corresponding dimension. When the process is completed, a number of dimensionless groups emerge as factors governing the physical behavior of the system. Maintaining the values of such groups intact from one situation (prototype) to another (model) will automatically ensure similarity. In the case of fluid flow, this process involves the conservation equations for mass, momentum, and energy, together with the equation of state of the fluid. These are written and converted to dimensionless form in the manner described. In the present chapter, however, an alternative and simpler method for arriving at the dimensionless groups will suffice. This is a dimensional analysis based on a set of physical parameters assumed *a priori* to affect the wind tunnel flow.

7.1.1 Dimensional Analysis

For concreteness, let it be assumed that the force F developed somewhere on a body immersed in a flowing fluid is a function only of the following six parameters: density ρ, flow velocity U, some typical dimension D, some frequency n, fluid viscosity μ, and gravitational acceleration g. One writes

$$F \stackrel{\mathrm{d}}{=} \rho^\alpha U^\beta D^\gamma n^\delta \mu^\varepsilon g^\zeta \tag{7.1.1}$$

where $\alpha, \ldots, \zeta$ are exponents to be determined. There are three basic quantities: mass M, length L, and time T, to which all of the above parameters are dimensionally related. Writing the dimensional equivalent of each of the quantities in Eq. 7.1.1 results in the following dimensional equality:

$$\frac{ML}{T^2} \stackrel{\mathrm{d}}{=} \left(\frac{M}{L^3}\right)^\alpha \left(\frac{L}{T}\right)^\beta (L)^\gamma \left(\frac{1}{T}\right)^\delta \left(\frac{M}{LT}\right)^\varepsilon \left(\frac{L}{T^2}\right)^\zeta \tag{7.1.2}$$

from which the following three independent equations are obtained by equating corresponding exponents:

$$M: \quad 1 = \alpha + \varepsilon$$

$$L: \quad 1 = -3\alpha + \beta + \gamma - \varepsilon + \zeta$$

$$T: -2 = -\beta - \delta - \varepsilon - 2\zeta \tag{7.1.3}$$

These equations may now be solved for any three of the exponents in terms of the remaining three; for example,

$$\alpha = 1 - \varepsilon$$

$$\beta = 2 - \varepsilon - \delta - 2\zeta$$

$$\gamma = 2 - \varepsilon + \delta + \zeta \tag{7.1.4}$$

whence it is seen that

$$F \stackrel{\mathrm{d}}{=} \rho^{1-\varepsilon} U^{2-\varepsilon-\delta-2\zeta} D^{2-\varepsilon+\delta+\zeta} n^\delta \mu^\varepsilon g^\zeta \tag{7.1.5}$$

or

$$F \stackrel{\mathrm{d}}{=} \rho U^2 D^2 \left(\frac{Dn}{U}\right)^\delta \left(\frac{\mu}{\rho UD}\right)^\varepsilon \left(\frac{Dg}{U^2}\right)^\zeta \tag{7.1.6}$$

From this it follows that the dimensionless force coefficient* $F/\rho U^2 D^2$ is a function of the dimensionless numbers Dn/U, $\mu/\rho UD$, and Dg/U^2.

The dimensionless numbers mentioned are of course already well known in fluid mechanics. For example, when n is the frequency n_s of vortex shedding from a bluff objective of cross-sectional dimension D, then

$$\frac{Dn_s}{U} = \mathscr{S} \tag{7.1.7}$$

---

*Typically, coefficients of lift force F_L and drag force F_D are written

$$C_L = \frac{F_L}{\frac{1}{2}\rho U^2 D^2} \qquad C_D = \frac{F_D}{\frac{1}{2}\rho U^2 D^2}$$

where $\frac{1}{2}\rho U^2$ is recognized as the dynamic pressure from the Bernoulli equation (see Eq. 4.1.20).

is the well-known *Strouhal number*. When n is n_m, a characteristic mechanical frequency associated with a structure, then Dn_m/U is termed the *reduced frequency* relative to a steady flow past the structure of velocity U; its reciprocal U/n_mD is the associated *reduced velocity*. The group nz/U, where z is height above ground, n represents a frequency associated with a component of variable wind velocity, and U is mean wind velocity, is a dimensionless frequency f (called the Monin coordinate) often used as abscissa in depicting wind velocity spectra (see Eq. 2.3.17). Further, if n is replaced by the circular frequency $f_c = 2\omega \sin \phi$, which is the Coriolis parameter (where ω is the rotational speed of the earth in radians/sec and ϕ is the latitude—see Eq. 1.2.3), then the quantity

$$\frac{U}{Df_c} = \mathscr{R}o \tag{7.1.8}$$

is called the *Rossby* number.

The group $\mu/\rho UD$ is the reciprocal of the well-known *Reynolds* number

$$\mathscr{R}e = \frac{\rho UD}{\mu} = \frac{UD}{v} \tag{7.1.9}$$

which is sometimes more specifically called the *molecular* Reynolds number when $v = \mu/\rho$ is the kinematic molecular viscosity of the fluid. In some applications (see Sects. 2.1.2 and 4.4), a turbulent Reynolds number may be employed in which v is replaced by v_{turb}, an "eddy" or "turbulence" kinematic viscosity. It is tentatively suggested in [7-12] that in the atmosphere such a viscosity has an order of magnitude near the ground given by

$$v_{\text{turb}} \cong u_* z_0 \tag{7.1.10}$$

where u_* is the shear, or friction, velocity, and z_0 is the surface roughness length (see Table 2.2.1). Note that Eq. 7.1.10 yields considerably lower values than those suggested (also tentatively) in [2-117].

Finally, the reciprocal of the group Dg/U^2 is called the *Froude number*:

$$\mathscr{F}i = U^2/Dg \tag{7.1.11}$$

Thus, simple analysis reveals the several dimensionless groups that play key roles in wind tunnel similitude, particularly in aiding the transfer of results from experimental model to full-scale prototype.

Though it is not directly pertinent to the present discussion, it is worth pointing out here that, were thermal effects to be included in the above analysis, three additional commonly occurring dimensionless numbers would emerge, namely:

Prandtl number:

$$\mathscr{P}i = \frac{\mu C_p}{K} \tag{7.1.12}$$

Eckert number:

$$\mathscr{E}c = \frac{U^2}{C_p\theta} \tag{7.1.13}$$

and

Richardson number:

$$\mathscr{R}i = \frac{\Delta\theta}{\theta}\left(\frac{Dg}{U^2}\right) \tag{7.1.14}$$

where C_p is specific heat at constant pressure, K is thermal conductivity, and θ is absolute temperature. Note that the Richardson number consists of a dimensionless temperature divided by a Froude number; $\mathscr{R}i$ plays an important role in thermally induced convection in the atmosphere. Because of the examination in this chapter principally of mechanical effects, the last three numbers are not emphasized in what follows.

7.1.2 Basic Scaling Considerations

It will be recognized in the consideration of dimensionless numbers above that no distinction is made as to source or origin of a given parameter: it can be fluid, structural, or other. For example, a length, frequency, density, or velocity may be associated with *any* characteristic of the fluid or structure immersed in it. This implies that *ratios* among such quantities must be maintained constant from prototype to model. For example, if ρ_s and ρ_f are the density of the structure and of the fluid, respectively, then

$$(\rho_s/\rho_f)_m = (\rho_s/\rho_f)_p \tag{7.1.15}$$

where the subscripts m and p refer respectively to model and prototype. Since this holds as well for geometric ratios and geometric shapes in general, it implies that all model shapes must be geometrically similar to prototype shapes and that, for example, vibrational modal shapes of prototype structure must be maintained in the corresponding model. Likewise, frequencies from all sources must bear the same ratios to each other in model as in prototype. Further, since oscillatory deflections must maintain proper proportionality from prototype to model, dimensionless damping ratios that affect such deflections must remain the same in prototype and model.

There now may be examined a typical set of scaling factors together with the process by which they are set. Three such factors may be arbitrarily chosen. The first might be an arbitrary *length scale*:

$$\lambda_L = \frac{D_m}{D_p} \tag{7.1.16}$$

set, for example, by comparison of model size to prototype size. (It will be seen subsequently what particular considerations enter into the setting of a length scale when turbulence is involved.) A second choice might be a convenient *velocity scale*:

$$\lambda_V = \frac{U_m}{U_p} \tag{7.1.17}$$

set perhaps by available wind tunnel speeds compared to expected natural wind speeds; and a third might be a *density scale*:

$$\lambda_\rho = \frac{\rho_m}{\rho_p} \tag{7.1.18}$$

usually forced upon the experimentalist by fixed circumstances (such as testing in air of the same density as that surrounding the prototype, whence $\lambda_\rho = 1$).

Given the fundamental exigencies of mass, length, and time, the three fixed scale choices, once made, condition all others in consequence of the requirement that the dimensionless groups maintain their constancy from prototype to model and vice versa. Thus, for example, the reduced frequency requirement

$$\left(\frac{Dn}{U}\right)_m = \left(\frac{Dn}{U}\right)_p \tag{7.1.19}$$

sets the *frequency scale* λ_n for *all* pertinent test frequencies:

$$\lambda_n = \frac{\lambda_V}{\lambda_L} \tag{7.1.20}$$

the reciprocal of which is the time scale λ_T.

It may be emphasized at this point in this illustrative discussion that λ_L, λ_V, and λ_ρ have been fixed either arbitrarily or in consequence of some unavoidable circumstance. We now inquire as to the consequence of invoking Froude number similitude, requiring

$$\left(\frac{U^2}{Dg}\right)_m = \left(\frac{U^2}{Dg}\right)_p \tag{7.1.21}$$

or

$$\frac{\lambda_V^2}{\lambda_L \lambda_g} = 1 \tag{7.1.22}$$

where λ_g is the gravitational scale factor. In most instances gravitational effects must be considered to be the same in model and prototype, so that $\lambda_g = 1$, whence, when Froude scaling is respected

$$\lambda_V = \sqrt{\lambda_L} \tag{7.1.23}$$

which may contradict an original choice for λ_V. In numerous (though by no means all) cases, it proves convenient to accede to Froude number scaling, adjusting λ_V accordingly, whence frequency scaling takes the value

$$\lambda_n = 1/\sqrt{\lambda_L} \tag{7.1.24}$$

Attention to gravitational effects may be required for certain structures (suspension bridges, for example) or for certain cases where convective air motions are important. As noted above, the latter are disregarded in the present discussion.

We now may examine the effect of invoking Reynolds number scaling, that is,

$$\left(\frac{\rho U D}{\mu}\right)_m = \left(\frac{\rho U D}{\mu}\right)_p \tag{7.1.25}$$

If prototype and model are both in air under atmospheric conditions, Reynolds number scaling requires simply that $\lambda_V \lambda_L = 1$ or

$$\lambda_V = 1/\lambda_L \tag{7.1.26}$$

which is, in general, in sharp conflict with other requirements set above, for example with:

$$\lambda_V = \sqrt{\lambda_L} \tag{7.1.27}$$

in the case of Froude scaling. Thus Reynolds number scaling is seen to be incompatible with the prior setting of length and velocity scales unless testing is undertaken at full scale $\lambda_L = 1$.

Another view of the same effect is that, for example, under Froude scaling, Reynolds number scaling is hugely distorted:

$$\lambda_{\mathscr{R}e} = \frac{(\mathscr{R}e)_m}{(\mathscr{R}e)_p} = \lambda_V \lambda_L = \lambda_L^{3/2} \tag{7.1.28}$$

To illustrate: if $\lambda_L = 1/300$, then

$$\lambda_{\mathscr{R}e} = \left(\frac{1}{300}\right)^{3/2} = \frac{1}{5196} \tag{7.1.29}$$

indicating a test Reynolds number less than one five-thousandth of $\mathscr{R}e$ for the prototype. It is noted that some aeronautical testing achieves Reynolds numbers closer to prototype values by using rarefield or compressed fluids, or fluids with lower kinematic viscosity than air, such as freon. A further recent stage involves use of gases at cryogenic temperatures [7-13].

Rossby number scaling also proves to be intractable under most circumstances, since an equivalent Coriolis acceleration effect (as represented by f_c) cannot practically be realized to the frequency scale λ_n mentioned above. Such an effect would require some means for imparting lateral acceleration to the flow, which is not easily achieved ([2-28], [2-29], [7-14]).

Thus, normal wind tunnel testing in air under standard gravity and atmospheric conditions typically entails fundamental scale violations of the Reynolds and of the Rossby number.

7.2 WIND TUNNEL SIMULATIONS OF ATMOSPHERIC FLOWS

To achieve similarity between the model and the prototype, it is desirable to reproduce at the requisite scale the characteristics of the atmospheric flows expected to affect the structure of concern (see Sects. 4.6 and 4.7). These characteristics have been outlined in Chapter 2. They include: (1) the variation of the mean wind speed with height; (2) the variation of turbulence intensities and integral scales with height; and (3) the spectra and cross-spectra of turbulence in the along-wind, across-wind, and vertical directions.

Attempts to simulate atmospheric flows have so far been confined to flows of the boundary-layer type; few, if any, laboratory investigations have been devoted to the simulation of downslope winds, hurricane eyewalls, tornadoes, and thunderstorms. (Note, however, the tentative simulation of tornado-induced forces in [7-15].)

Tunnels used for civil engineering purposes have cross sections that rarely exceed 3 m × 3 m. (A notable exception is the 9 m × 9 m tunnel of the National Research Council, Ottawa, Canada.) Three types of wind tunnels have been used for simulating atmospheric flows. They are referred to as long tunnels, short tunnels, and tunnels with active devices, and are described in Sects. 7.2.1, 7.2.2, and 7.2.3, respectively. Sections 7.2.4 and 7.2.5 comment on the possible effects of violating the Reynolds and Rossby number similarity requirements upon the simulation of flow turbulence.

7.2.1 Long Wind Tunnels

In long wind tunnels ([7-16, 7-17]) a boundary layer with a typical depth of 0.5 m to 1 m develops naturally over a rough floor of the order of 20 m to 30 m in length (Figs. 7.2.1 [7-54], 7.2.2, 7.2.3). The depth of the boundary layer can be increased by placing at the test section entrance passive devices of the types described in Sect. 7.2.2. Such an artificial increase may be necessary, particularly in simulations of flow over the ocean or over terrain with low or moderate roughness. The height of most tunnels may be adjusted to increase slightly with position downstream. The purpose of such an adjustment is to achieve a zero pressure gradient streamwise, which would otherwise not obtain, owing to energy losses associated with flow friction at the walls and with internal friction due to turbulence.

Atmospheric turbulence simulations in long wind tunnels are probably the best that can be achieved in the present state of the art. However, even when passive devices such as spires are not used, similitude between the turbulence in the laboratory flow and in the atmosphere is generally not achieved (see Sects. 7.2.5 and 7.3.1). The lack of similitude becomes stronger if, for example, spires are employed (see Sect. 7.2.2).

7.2.2 Short Wind Tunnels

Tunnels used for aeronautical purposes are usually designed for testing in smooth flow and therefore need not have long test sections. Many such tunnels have been converted for use in civil engineering applications by adding, at the test section entrance, passive devices, such as grids, barriers, fences, and spires, that generate a thick boundary layer. The floor of the test section, which is usually on the order of 5 m long, is covered with roughness elements (Fig. 7.2.4). Various types, shapes, and combinations of passive devices have been suggested and commented upon in [7-16] to [7-25].

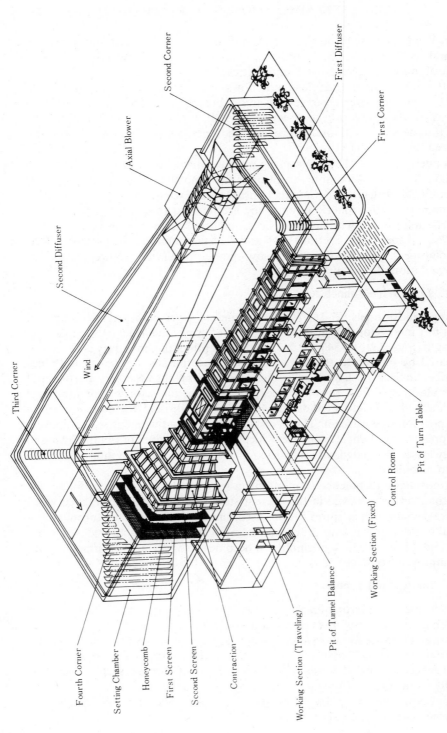

FIGURE 7.2.1. Wind tunnel operated by Kawasaki Heavy Industries, Ltd., Japan, at its Akashi Technical Institute. (Wind speed range: 0.2 to 25 m/sec; test section dimensions: 2.5 m × 3 m × 20 m). From R. D. Marshall, "Wind Tunnels Applied to Wind Engineering in Japan," *Journal of Structural Engineering*, 110 (June 1984), 1203–1221.

The following labels appear in the figure:

Second Corner
Axial Blower
First Diffuser
First Corner
Second Diffuser
Third Corner
Wind
Fourth Corner
Setting Chamber
Honeycomb
First Screen
Second Screen
Contraction
Working Section (Traveling)
Pit of Tunnel Balance
Working Section (Fixed)
Control Room
Pit of Turn Table

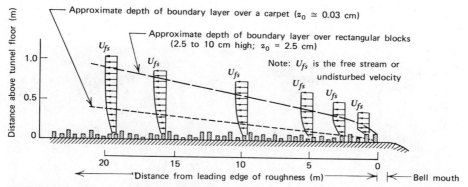

FIGURE 7.2.2. Development of boundary layer in a long wind tunnel. After A. G. Davenport and N. Isyumov. "The Application of the Boundary-Layer Wind Tunnel to the Prediction of Wind Loading," in *Proceedings*. International Research Seminar on Wind Effects on Buildings and Structures, Vol. 1, p. 221. Copyright, Canada, 1968, University of Toronto Press.

FIGURE 7.2.3. Upstream view of a long wind tunnel (courtesy Boundary-Layer Wind Tunnel Laboratory, The University of Western Ontario).

FIGURE 7.2.4. Spire and roughness arrays in a short wind tunnel (courtesy of the National Aeronautical Establishment, National Research Council of Canada).

Reference 7-26 proposes the following procedure for the design of spires with the configuration of Fig. 7.2.5:*

1. Select the desired boundary-layer depth, δ.

2. Select the desired shape of mean velocity profile defined by the power law exponent, α (Eq. 2.2.26).

3. Obtain the height h of the spires from the relation

$$h = 1.39\delta/(1 + \alpha/2) \tag{7.2.1}$$

4. Obtain the width of the spire base from Fig. 7.2.6, in which H is the height of the tunnel test section.

The desired mean wind profile occurs at a distance $6h$ downstream from the spires. According to [7-26], the wind tunnel floor downwind of the spires should be covered with roughness elements, for example, cubes with height k such that ([7-26] to [7-28]):

$$k/\delta = \exp\{(2/3)\ln(D/\delta) - 0.1161[(2/C_f) + 2.05]^{1/2}\} \tag{7.2.2}$$

---

*The base length of the triangular splitter plate in Fig. 7.2.5 is $h/4$. The lateral spacing between the spires is $h/2$. In practice the width of the tunnel need not be an integral multiple of $h/2$.

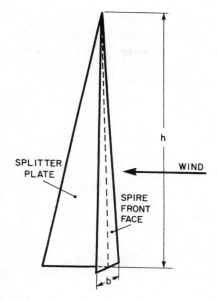

FIGURE 7.2.5. A proposed spire configuration. From H. P. A. H. Irwin, "The Design of Spires for Wind Simulation," *J. Wind Eng. Ind. Aerodyn.*, **7** (1981), 361–366.

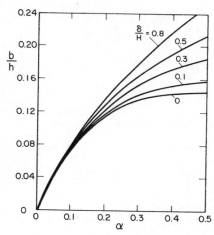

FIGURE 7.2.6. Graph for obtaining spire base width. From H. P. A. H. Irwin, "The Design of Spires for Wind Simulation," *J. Wind Eng. Ind. Aerodyn.*, **7** (1981), 361–366.

where D is the spacing of the roughness elements and

$$C_f = 0.136[\alpha/(1+\alpha)]^2 \tag{7.2.3}$$

Equation 7.2.2 is valid in the range $30 < \delta D^2/k^3 < 2000$.

A study of the dependence of flow features upon the type of passive devices being used was recently presented in [7-17]. Figure 7.2.7 [7-17] shows the mean velocity, longitudinal turbulence intensity, and vertical turbulence intensity profiles at (a) 6.1 m and (b) 18.3 m downwind of the test section entrance, for flows obtained by using three different types of spires, the wind tunnel floor being covered by staggered 1.27 cm cubes spaced 5.08 cm apart. In Fig. 7.2.7 the boundary-layer thickness δ, the mean wind speed U at elevation δ, and the power law exponent α (Eq. 2.2.26) are denoted by *delta*, *Uinf*, and *EXP*, respectively. It may be assumed that the mean flow with exponent $\alpha = 0.16$ at station $x = 6.1$ m, and the mean flow with exponent $\alpha = 0.29$ at station $x = 18.3$ m, are

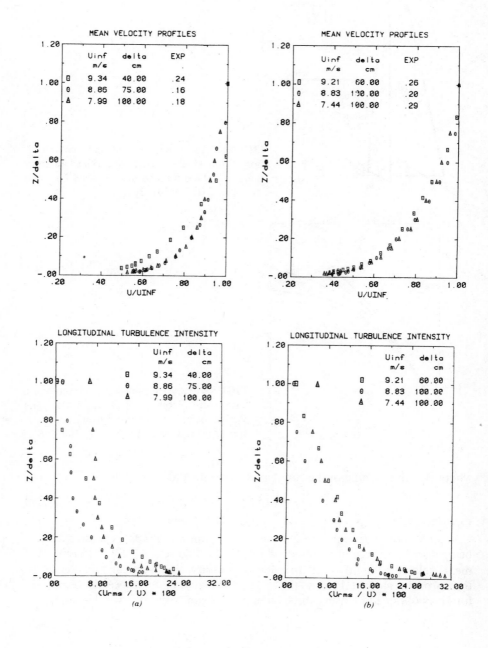

(a)   (b)

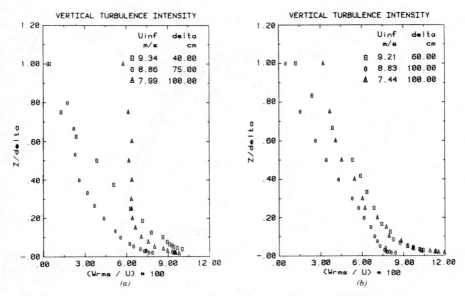

FIGURE 7.2.7. Wind tunnel flow features at (*a*) 6.1 m and (*b*) 18.3 m downwind of spires, obtained by using three types of spire configurations. Reprinted with permission from J. E. Cermak, "Physical Modeling of the Atmospheric Boundary Layer (ABL) in Long Boundary-Layer Wind Tunnels (BLWT)," in *Wind Tunnel Modeling for Civil Engineering Applications*, T. A. Reinhold (Ed.), Cambridge University Press, Cambridge, U.K., 1982.

approximately representative of open terrain and suburban terrain conditions, respectively (see Table 2.2.2).

Some modelers adopt a geometric scale equal to the ratio between the boundary-layer thickness measured in the laboratory and the value δ of Table 2.2.2, even though the latter is nominal, rather than physically significant (see Eq. 2.2.15 and Sect. 2.2.4). If this geometric scaling criterion is used for the simulations of Fig. 7.2.7, the geometric scales are found to be $0.75/275 = 1/367$ for the flow with $\alpha = 0.16$, and $1/400$ for the flow with $\alpha = 0.29$. The respective longitudinal turbulence intensities at 50 m above ground are about 0.07 and 0.15, versus about 0.15 and 0.225, as obtained from Eqs. 2.2.18, 2.3.1, 2.3.2, and Table 2.3.1. As expected, the discrepancy between the longitudinal turbulence intensity in the wind tunnel and the "target" value in the atmosphere is more severe at the station $x = 6.1$ m, which would correspond to the fetch available in a short tunnel.

Figure 7.2.8 [7-17] shows spectra of the longitudinal velocity fluctuations measured at station $x = 18.3$ m and elevation $z/\delta = 0.05$ in the three flows described in Fig. 7.2.7*b*. For the flow with $\alpha = 0.29$, it is seen in Fig. 7.2.8 that at the nondimensional frequency $nz/U(z) = 0.8$, $nS(n)/\overline{u^2} \simeq 0.05$, versus 0.06, as obtained from Eqs. 2.3.2, 2.3.16, and Table 2.3.1. Unlike the turbulence intensity,

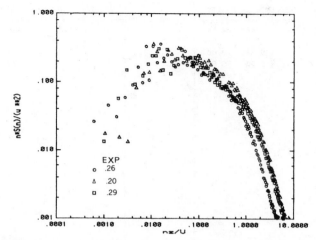

FIGURE 7.2.8. Spectra of longitudinal velocity fluctuations measured at 18.3 m downwind of spires. Reprinted with permission from J. E. Cermak, "Physical Modeling of the Atmospheric Boundary Layer (ABL) in Long Boundary-Layer Wind Tunnels (BLWT)," in *Wind Tunnel Modeling for Engineering Applications*, T. A. Reinhold (Ed.), Cambridge University Press, Cambridge, U. K., 1982.

the higher frequency spectrum measured in the wind tunnel is in this instance relatively close to the "target" value.*

The results of [7-17] and of other studies (e.g., [7-24], [7-53]) indicate that, regardless of the type of passive devices being used, simulations in short wind tunnels generally do not achieve similitude between the turbulence in the laboratory and in the atmospheric flow.

7.2.3 Tunnels with Active Devices

In tunnels equipped with jets (Fig. 7.2.9) it is possible, within certain limits, to vary the mean velocity profile and the flow turbulence independently of each other [7-29, 7-30]. Such tunnels are relatively expensive and do not necessarily result in superior flow simulations. However, they may be useful for basic studies in which the effect of varying some flow characteristics independently of the others can be studied in detail.

Active cascades of moving airfoils (Fig. 7.2.10) have been recently designed with a view to creating, and simulating effects of, large-scale turbulence over bridge deck section models [7-31, 7-32].

7.2.4 Reynolds Number and Turbulent Flow Simulation

It is suggested in [7-33, p. 204, 7-34, p. 266, and 7-35, p. 290], that Reynolds numbers of turbulent flows obtained in the laboratory downwind of square mesh grids may in some cases be too small to give rise to a turbulence spectrum

---

*Information on integral scales for the wind tunnel flows is not reported in [7-17].

FIGURE 7.2.9. Upstream view of the test section and jets of the 1.20 m × 1.70 m closed-circuit jet tunnel. University of Toronto Institute for Aerospace Studies (courtesy Dr. H. W. Teunissen).

FIGURE 7.2.10. Mechanically driven airfoil cascade for low-frequency turbulence simulation [7-32].

having an inertial subrange. It is further suggested [7-35] that the Reynolds number based on eddy size should be the order of 10^5 to assure existence of this subrange. Applying analogous reasoning to a developed turbulent boundary layer of depth, say, 0.5 m, in which the integral scale length L_u^x (a measure of typical eddy size) is about 0.125 m, a Reynolds number at a velocity of 12 m/sec may be calculated:

$$\mathscr{R}e = \frac{UL_u^x}{v} \cong \frac{12(0.125)}{1.5 \times 10^{-5}} = 10^5$$

Thus, typical boundary-layer simulations of the kind discussed may be expected to develop velocity spectra with satisfactory inertial subranges, though at lower velocities and turbulence integral scales they may be borderline.

7.2.5   Rossby Numbers and Turbulent Flow Simulation

Failure of Rossby number equivalence in typical test circumstances is due to the difficulty of scaling the Coriolis parameter f_c above its automatically achieved full-scale value. Rotating wind tunnels ([2-28, 2-29]), or tunnels with porous walls and across-wind suction imparting lateral acceleration to the flow [7-17] are currently not used in civil engineering applications. An investigation into the effect of the Rossby number on boundary-layer flow is therefore in order.

It was shown in Sect. 2.2 that the approximate depth of the atmospheric boundary layer may be expressed as

$$\delta \simeq 0.25 \frac{u_*}{f_c} \tag{7.2.4}$$

where $u_* = U(h)/\{2.5 \ln(h/z_0)\}$, $U(h)$ is the mean speed at a reference height h, z_0 is the roughness length, and f_c is the Coriolis parameter. Equation 7.2.4 can also be written in a form that emphasizes the dependence of the atmospheric boundary-layer depth δ upon Rossby number:

$$\delta = c\mathscr{R}o \tag{7.2.5}$$

where $c \simeq 0.25h/\{2.5\ln(h/z_0)\}$ and $\mathscr{R}o = U(h)/hf_c$. The boundary-layer depth δ is seen to be an increasing function of wind speed. For high wind speeds such as are of interest to the structural designer, it follows from Eq. 7.2.5 that δ is of the order of several kilometers. For example, if $z_0 = 0.05$ m (open terrain), $U(10) = 25$ m/sec, and $f_c = 10^{-4}$ sec^{-1} (corresponding to an angle of latitude $\phi \simeq 45°$, see Table 1.2.1), then $\delta \simeq 5$ km. The region of interest to the structural designer, that is, the lowest few hundred meters of the atmospheric boundary layer, is thus seen to amount to about one-tenth or less of the full atmospheric boundary-layer depth.

As noted in [7-36], both in the atmosphere and in the laboratory the mean velocity profile is very nearly logarithmic over the region consisting of the lower one-tenth of the boundary layer or so (see Sect. 2.2 and Fig. 7.2.11). Moreover, measurements suggest that in this region the turbulent energy

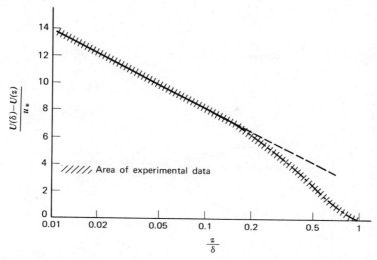

FIGURE 7.2.11. Logarithmic plot of velocity distributions in turbulent boundary layers over plates. After F. H. Clauser. "The Turbulent Boundary Layer," *Advances in Applied Mechanics*, **4** (1956), Academic Press, New York, p. 9.

production is approximately balanced by the energy dissipation (Fig. 7.2.12 and [7-37]) so that the longitudinal velocity spectrum in the inertial subrange may be expressed in nondimensional form as

$$\frac{nS(z, n)}{u_*^2} = 0.26 f^{-2/3} \tag{2.3.16}$$

where n is the frequency, z is the height, and $f = nz/U(z)$ (see Sect. 2.3). Equation 2.3.16 is not valid in the upper region of the boundary layer where the energy production differs significantly from the energy dissipation (see Fig. 7.2.12).

Consider now a long wind tunnel in which the boundary layer develops naturally over a rough floor and in which the boundary-layer depth is of the order of 1 m (Fig. 7.2.2). Assume that the height of the building being tested is 200 m and that the model scale is 1/400. Since, as was shown above, the region of the atmospheric boundary layer over which the logarithmic law holds is (under strong wind conditions) a few hundred meters high, it is reasonable to assume that Eq. 2.3.16 is valid throughout the building height. However, in the laboratory similarity theory suggests that Eq. 2.3.16 can only be applied up to a height of approximately 0.1 m from the wind tunnel floor, to which there would correspond a full-scale height of just 40 m above ground.

A schematic representation of the situation just described is given in Fig. 7.2.13, which shows the boundary layer that develops in a long wind tunnel (full line), and the atmospheric boundary layer reduced to model scale (broken line). The lower one-tenth and the outer nine-tenths of the boundary layer are denoted by L_{wt} and O_{wt}, respectively, for the wind tunnel flow, and by L_a and

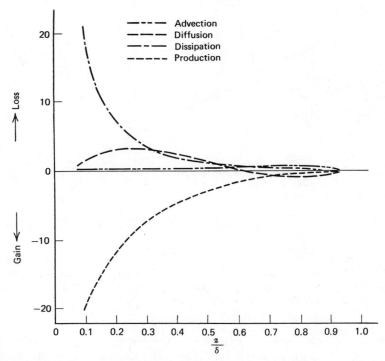

FIGURE 7.2.12.   Energy balance in a turbulent boundary layer. After A. A. Townsend, *The Structure of Turbulent Shear Flow*, Cambridge University Press, New York, 1956, p. 234.

O_a, respectively, for the atmospheric flow. It can be seen in Fig. 7.2.13 that over most of the wind tunnel boundary-layer depth the atmospheric flow in the lower layer L_a is simulated by the flow in the outer region O_{wt} in which, according to similarity theory, Eq. 2.3.16 would not be expected to hold.

This situation does not necessarily imply that the actual flow characteristics

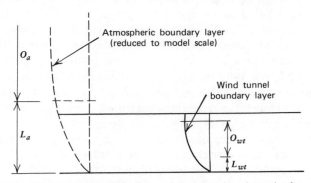

FIGURE 7.2.13.   Lower and outer regions of the boundary layer in the wind tunnel and in the atmosphere.

developed in the wind tunnel are unacceptable for testing purposes. However, if the objective of a laboratory test is to predict the along-wind response of a tall building surrounded by homogeneously rough terrain, then the test results must be interpreted with care: adjustments of these results may be required to account for possible differences between wind tunnel and full-scale spectra in the inertial subrange.

7.3 WIND TUNNEL SIMULATION OF AERODYNAMIC AND AEROELASTIC BEHAVIOR OF BLUFF BODIES

The purpose of this section is to discuss some practical aspects of the dependence of the aerodynamic and aeroelastic response of wind tunnel models upon the turbulence characteristics and the Reynolds number of the flow. Also included here is a brief discussion of aerodynamic distortions due to blockage effects.

7.3.1 Effect of Turbulence Characteristics of the Flow

The details of the dependence of the aerodynamic and aeroelastic behavior of bodies upon the turbulence characteristics of the flow are not fully understood. However, it is clear that in order for the effects of turbulence on the model to be similar to those on the prototype (i.e., in order for the turbulent eddies to envelop or otherwise affect the body or part thereof in a similar way in the atmosphere and in the laboratory), it is necessary that the ratio between some typical length characterizing the turbulence and some characteristic dimension of the body be the same in both situations.

It is convenient to adopt the integral scale L_u^x (see Sect. 2.3.2) as the characteristic length of turbulence. The geometric scale factor of the simulation, $D_L = D_m/D_p$, should then be given by

$$D_L = \frac{(L_u^x)_m}{(L_u^x)_p} \tag{7.3.1}$$

where $(L_u^x)_p$ and $(L_u^x)_m$ are, respectively, an estimate of the integral scale that obtains in the atmosphere at some representative elevation (see Sect. 2.3.2), and the integral scale measured in the wind tunnel flow at the corresponding elevation above the tunnel floor. The application of Eq. 7.3.1 is discussed in [7-38].

Equation 7.3.1 is violated in many instances because of the difficulty of achieving sufficiently large integral scales in the laboratory, particularly in short wind tunnels (see Sect. 7.2.2). However, even if Eq. 7.3.1 is nominally satisfied, it should be recalled that integral scales are poorly known and can vary from measurement to measurement by a factor of five or even ten (see Sect. 2.3.2); thus, the assumed value of the ratio $(L_u^x)_m/(L_u^x)_p$ can differ significantly from its actual value.

An attempt to assess errors due to the imperfect simulation of the integral scale of turbulence is reported in [7-39] for the pressures at various points of the building represented schematically in Fig. 7.3.1. In the investigation of [7-39] the integral scale was not varied independently of the other flow features. Rather, the wind tunnel boundary-layer flow was kept unchanged while the dimensions of the model were increased. It was estimated that the integral scale L_u^x in the wind tunnel was equal to about 1/500 times a nominal integral scale judged to be typical of atmospheric flows. Measurements were made on 1/500, 1/250, and 1/100 models of the same building. Ratios between the peak, mean, and rms pressures measured at several points on the 1/100 and 1/250 models, and the corresponding pressures on the 1/500 model, are listed in Table 7.3.1.

It is seen that in some instances the influence of the model size upon the test results is significant (e.g., for the peak pressures at tap 29, 1/100 scale, or tap 111, 1/100 scale and 1/250 scale). Note also that the pattern of variation of the ratios of Table 7.3.1 is irregular. This may be due, at least in part, to the fact that by changing the height of the model by the factors 2.5 and 5, the turbulence intensities at the elevation of the points under consideration also changed.

The effect of turbulence features upon the modeling of aerodynamic behavior is discussed in [7-40] to [7-42].

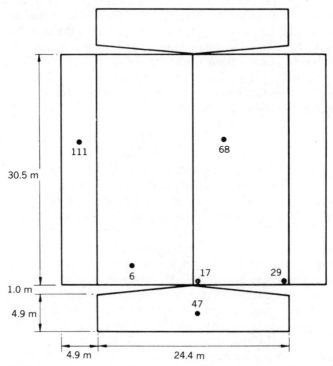

FIGURE 7.3.1.   Schematic view of building with pressure taps (After [7-39]).

TABLE 7.3.1. Ratios of Peak, Mean, and RMS Pressures on 1/100 and 1/250 Models to Corresponding Pressures on 1/500 Model[a]

TAP	1/100			1/250		
	Peak	Mean	rms	Peak	Mean	rms
29[b]	1.34(0.93)	1.65(0.19)	1.09(0.53)	0.90(0.72)	1.26(2.40)	1.03(0.88)
17[b]	0.90(0.97)	1.16(1.48)	0.62(0.95)	0.78(1.01)	0.98(1.07)	0.62(0.88)
6[b]	1.00(1.02)	1.67(1.90)	1.13(1.90)	0.84(0.81)	1.20(0.81)	1.04(0.78)
111[c]	0.69(0.75)	1.00(0.43)	0.60(0.67)	0.63(0.78)	1.00(0.75)	0.53(0.57)
47[c]	0.84(0.83)	0.96(0.93)	0.83(0.91)	0.90(0.95)	0.81(0.79)	0.83(0.91)
68[b]	1.05(1.07)	1.40(1.40)	0.80(0.67)	0.83(0.90)	1.07(0.97)	0.73(0.81)

[a]Numbers not between parentheses correspond to open exposure. Numbers between parentheses correspond to built-up exposure.
[b]Suctions.
[c]Pressures.

7.3.2 Reynolds Number Effects

Sharp corners tend to cause immediate flow separation, independently of the Reynolds number of the flow. For this reason it is generally assumed that if the flow is adequately simulated, pressures on rectangular and other sharp-cornered structures are adequately reproduced in the wind tunnel. However, bluff bodies with long afterbody extensions downstream may exhibit flow re-attachment, which does depend on the Reynolds number. Such circumstances may affect the values of the across-wind forces experienced by the body. Few full-scale supporting data on this topic are available to date. Note also that if the details of a scale model require extremely small dimensions (as, for example, in modeling the members of a truss structure at a scale of 1/500 or below) it may be that the drag coefficient applicable to such a member can be unduly influenced (raised) by Reynolds number effects. Figure 4.5.6 bears out this tendency.

In the case of bodies with curved surfaces, Reynolds number deficiencies can have significant effects. This is simply illustrated by the evolution of both mean drag coefficient and Strouhal number for a circular cylindrical section as a function of Reynolds number (see Figs. 4.4.4 and 4.5.2).

As indicated in Chapter 4, the aerodynamic behavior of such bodies depends on whether the boundary layers on the curved surfaces are laminar or (partially or fully) turbulent. Since boundary layers occurring at high Reynolds numbers are turbulent, it is logical to attempt the reproduction of full-scale flows around smooth cylinders by changing laminar boundary layers into turbulent ones. This can be done by providing the surface with roughness elements (see [4-15], [7-43] to [7-46], and Fig. 4.5.5). It is suggested in [7-44] that the thickness e of the roughness elements should satisfy the relations

$$\frac{Ue}{v} > 400$$

$$\frac{e}{D} < 10^{-2}$$

where U is the mean wind speed, v is the kinematic viscosity ($v \simeq 1.5 \times 10^{-5}$ m^2/sec in air), and D is the characteristic transverse dimension of the model.

For example, in the case of the DMA tower (Fig. 15.3.22), the roughness was achieved by fixing onto the surface of the 1/200 model thirty-two equidistant vertical wires. Three sets of experiments are reported in [7-44] in which the surface of the cylinder was (1) smooth, (2) provided with 0.6 mm wires ($e/D \simeq 7 \times 10^{-3}$), and (3) provided with 1 mm wires, respectively. It was found that the highest mean and peak pressures were more than twice as high on the smooth model than on the models provided with wires. The differences between pressures on the model with 0.6 mm and the model with 1 mm wires were small. The influence of the roughness on the magnitude of the mean pressures at 20 m (full-scale) below the top of the building is shown in Fig. 7.3.2 in which the mean pressure coefficient $\bar{C}_p$ is defined as follows:

$$\bar{C}_p = \frac{\bar{p} - p_r}{\frac{1}{2}\rho U_r^2}$$

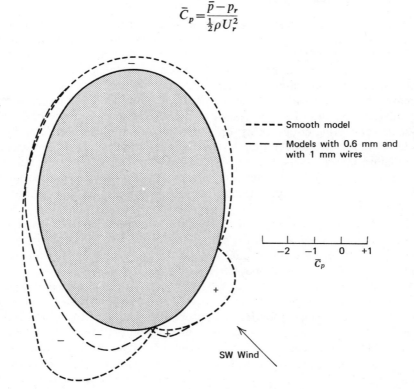

FIGURE 7.3.2.   Influence of model surface roughness on pressure distribution [7-44].

where $\bar{p}$ is the measured mean pressure, p_r is the static reference pressure, U_r is the mean speed at top of the building, and ρ is the air density.

Approaches of the type described above were found to yield acceptable results in cases not involving aeroelastic motions. However, if aeroelastic effects are present, wind tunnel tests in which such approaches are used can provide an utterly misleading picture of the behavior of the prototype (see Chapter 10).

7.3.3 Wind Tunnel Blockage

A body placed in a wind tunnel will partially obstruct the passage of air, causing the flow to accelerate. This effect is referred to as blockage. If the blockage is substantial, the flow around the model, and the model's aerodynamic behavior, are no longer representative of prototype conditions.

Corrections for blockage depend upon the body shape, the nature of the aerodynamic effect of concern (i.e., whether drag, lift, Strouhal number, and so forth), the characteristics of the wind tunnel flow, and the relative body/wind tunnel dimensions. Basic studies on blockage are summarized in [7-47] to [7-49] and in [7-50], which also contains a bibliography on this topic.

It is concluded in [7-50] that, in the case of drag, the following approximate relation may be used for the great majority of model configurations in all flows, including boundary layer flows:

$$C_{D_c} = \frac{C_D}{1 + K \dfrac{S}{C}} \tag{7.3.2}$$

where C_{D_c} is the corrected drag coefficient, C_D is the drag coefficient measured in the wind tunnel, S is the reference area for the drag coefficients C_{D_c} and C_D, and C is the wind tunnel cross-sectional area. The ratio S/C is referred to as blockage ratio.

The coefficient K has been determined only for a limited number of situations. For example, in the case of a bar with a rectangular cross-section spanning the entire height of a wind tunnel with nominally smooth flow, K was determined to depend upon the ratio a/b as shown in Fig. 7.3.3 (a and b are the dimensions of the along-wind and across-wind sides of the rectangular cross-section, respectively).

The effect of turbulence on blockage by flat plates was studied in [7-48] for flows with uniform mean speed, and was found to be negligible in most situations. On the other hand, it is stated in [7-50] that this effect can be significant. Thus, according to [7-50], turbulence does not increase the drag on a square plate, as concluded in [7-25] (see Table 4.6.2); rather, the increased drag reported in [4-25] was only apparent, and it was the blockage effects that were affected differently by various turbulence levels.

In spite of such ongoing debates and of various continuing uncertainties, it may be assumed that for blockage ratios of 2% the blockage corrections are likely to be about 5%, and that the magnitude of the blockage correction is proportional to the blockage ratio [7-50].

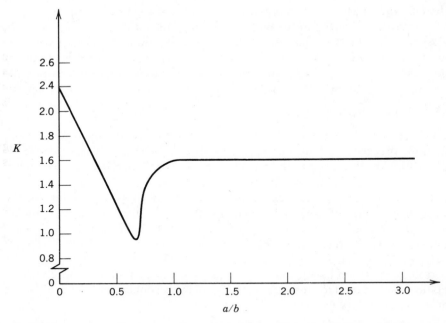

FIGURE 7.3.3.   Blockage correction factor K for two-dimensional prism with along-wind dimension a and across-wind dimension b in nominally smooth flow [7-49].

7.4   VALIDATION OF WIND TUNNEL TESTING

In spite of the numerous full-scale measurements reported in the literature, the number of dependable comparisons between model and prototype results remains relatively small. Figures 7.4.1 and 7.4.2 show a comparison between wind tunnel and full-scale measurements of pressures on the Commerce Court tower (Fig. 15.3.17). The wind tunnel values were provided at the design stage and are represented by open circles. The solid lines join average values of estimates derived from actual observations of pressure differences on the building; the shaded areas indicate the standard deviation of the full-scale estimates [7-51]. It is seen that the agreement between model and full-scale measurements of the mean pressures is satisfactory.* However, it appears from Figs. 7.4.1 and 7.4.2 that local fluctuating pressures attributable to vortex shedding (fluctuating lift) differ at some points significantly in the wind tunnel from the pressures on the prototype.

Further data for this building are available in [7-51] and [7-52]. Figure 7.4.3 shows acceleration spectra obtained from full-scale measurements and from tests on a model of the building with seven lumped mass levels. It is seen that

---

*In Figs. 7.4.1 and 7.4.2 the abbreviation RMSM denotes root mean square value about the mean [7-51].

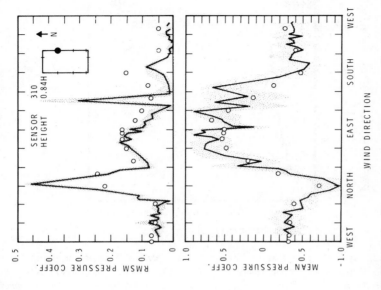

FIGURE 7.4.1. Pressures measured on west wall at 20.6 m from NW corner, 41st floor, Commerce Court Tower. From W. A. Dalgliesh, "Comparison of Model Full-Scale Wind Pressures on a High-Rise building," *J. Ind. Aerodyn.,* **1** (1975), 55–66.

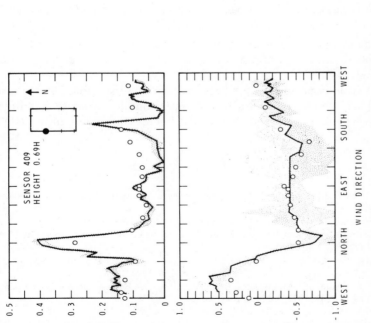

FIGURE 7.4.2. Pressures measured on east wall at 20.6 m from NE corner, 50th floor, Commerce Court Tower. From W. A. Dalgliesh, "Comparison of Model Full-Scale Wind Pressures on a High-Rise Building," *J. Ind. Aerodyn.,* **1** (1975), 55–66.

273

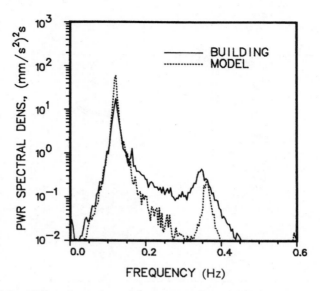

FIGURE 7.4.3. Full-scale and model north-south acceleration spectra, Commerce Court Building. Reprinted with permission from E. A. Dalgliesh, "Comparison of Model and Full-Scale Tests of the Commerce Court Building in Toronto," in *Wind Tunnel Modeling for Civil Engineering Applications*, T. A. Reinhold (Ed.), Cambridge, U.K., 1982.

in this case the model tests tended to underestimate the response in the intermediate frequency range but appear to be adequate at the low and high ends of the spectrum.

Model/full scale comparisons for pressures on low-rise buildings are also reported in [7-53] and [4-74]. Further comparisons for low-rise buildings, and for high-rise buildings, bridges, towers, and stacks, are reported in [7-11].

REFERENCES

7-1   D. A. Haugen (Ed.), *Workshop on Micrometeorology*, American Meteorological Society, Boston, MA, 1973.

7-2   *Characteristics of Windspeed in the Lower Layers of the Atmosphere Near the Ground: Strong Winds (Neutral Atmosphere)*, ESDU Data Item No. 72026, Engineering Sciences Data Unit, London, 1972.

7-3   *Characteristics of Atmospheric Turbulence Near the Ground*, ESDU Data Items Nos. 74030, 74031, 75001, Engineering Sciences Data Unit, London, 1974, 1975.

7-4   J. Counihan, "Adiabatic Atmospheric Boundary Layers: A Review and Analysis of Data from the Period 1880–1972," *Atmos. Environ.*, **9** (1975), 871–905.

7-5   A. Pope and J. J. Harper, *Low-Speed Wind Tunnel Testing*, Wiley, New York, 1966.

7-6   S. M. Gorlin and I. I. Slezinger, *Wind Tunnels and Their Instrumentation*, Israel Program for Scientific Translations, Jerusalem, 1966.

7-7   R. C. Pankhurst and D. W. Holder, *Wind Tunnel Technique*, Putnam, London, 1968.

7-8   E. Ower and R. C. Pankhurst, *The Measurement of Air Flow*, 4th ed., Pergamon, Oxford, 1969.

7-9   P. Bradshaw, *An Introduction to Turbulence and Its Measurement*, Pergamon, Oxford, 1971.

7-10 W. Merzkirch, *Flow Visualization*, Academic, New York, 1974.

7-11 T. A. Reinhold (Ed.), *Wind Tunnel Modeling for Civil Engineering Applications*, Proceedings of International Workshop, Gaithersburg, Maryland, April 1982, Cambridge University Press, New York, 1982.

7-12 R. Britter, "Modeling Flow over Complex Terrain and Implications for Determining the Extent of Adjacent Terrain to be Modeled," *Wind Tunnel Modeling for Civil Engineering Applications*, T. A. Reinhold (Ed.), Cambridge University Press, pp. 186–196. 186–196.

7-13 *High Reynolds Number Research*, D. D. Baals (Ed.), NASA CP-2009 (1977) Proceedings of Workshop, Langley Research Center, Hampton, Va., Oct. 1976.

7-14 D. R. Caldwell and C. W. Van Atta, "Ekman Boundary Layer Instabilities," *J. Fluid Mech.*, **44**, Part I (Oct. 1970), 79–95.

7-15 M. C. Jischke and B. D. Light, "Laboratory Simulation of Tornadic Wind Loads on a Cylindrical Structure," in *Wind Engineering*, Proceedings of the Fifth International Conference, Fort Collins, Colorado, July 1979, J. E. Cermak (Ed.), Vol. 2, pp. 1049–1059, Pergamon Press, Oxford and New York, 1980.

7-16 A. G. Davenport and N. Isyumov, "The Application of the Boundary-Layer Wind Tunnel to the Prediction of Wind Loading," in *Proceedings of the International Research Seminar on Wind Effects on Buildings and Structures*, University of Toronto Press, Toronto, 1968, pp. 201–230.

7-17 J. E. Cermak, "Physical Modeling of the Atmospheric Boundary Layer in Long Boundary-Layer Wind Tunnels," in *Wind Tunnel Modeling for Civil Engineering Applications*, T. A. Reinhold (Ed.), Cambridge University Press, 1982, pp. 97–125.

7-18 J. Counihan, "An Improved Method of Simulating an Atmospheric Boundary Layer in a Wind Tunnel," *Atmos. Environ.*, **3** (1969), 197–214.

7-19 J. Counihan, "Simulation of an Adiabatic Urban Boundary Layer in a Wind Tunnel," *Atmos. Environ.*, **7** (1973), 673–689.

7-20 N. J. Cook, "On Simulating the Lower Third of the Urban Adiabatic Boundary Layer in a Wind Tunnel," *Atmos. Environ.*, **7** (1973), 691–705.

7-21 N. J. Cook, "A Boundary-Layer Wind Tunnel for Building Aerodynamics," *J. Ind. Aerodyn.*, **1** (1975), 3–12.

7-22 N. J. Cook, "Wind-Tunnel Simulation of the Adiabatic Atmospheric Boundary Layer by Roughness, Barrier, and Mixing-Device Methods," *J. Ind. Aerodyn.*, **3** (1978), 157–176.

7-23 N. M. Standen, *A Spire Array for Generating Thick Turbulent Shear Layers for Natural Wind Simulation in Wind Tunnels*, Report No. LTR-LA-94, National Aeronautical Establishment, National Research Council, Ottawa, Canada, 1972.

7-24 J. A. Peterka and J. E. Cermak, *Simulation of Atmospheric Flows in Short Wind Tunnel Test Sections*, Fluid Mechanics Program Research Report, Colorado State University, 1974.

7-25 J. C. R. Hunt and H. Fernholz, "Wind Tunnel Simulation of the Atmospheric Boundary Layer: A Report on Euromech 50," *J. Fluid Mech.*, **70**, Part 3 (Aug. 1975), 543–559.

7-26 H. P. A. H. Irwin, "The Design of Spires for Wind Simulation," *J. Wind Eng. Ind. Aerodyn.* **7** (1981), 361–366.

7-27 I. S. Gartshore, *A relationship between roughness geometry and velocity profile shape for turbulent boundary layers*, National Research Council of Canada, NAE Rep. LTR-LA-140 (Oct. 1973).

7-28 R. A. Wooding, E. F. Bradley and J. K. Marshall, "Drag due to regular arrays of roughness elements of varying geometry," *Bound. Layer Meteorol.*, **5** (1973), 285–308.

7-29 H. W. Teunissen, "Simulation of the Planetary Boundary Layer in a Multiple-Jet Wind Tunnel," *Atmos. Environ.*, **9** (1975), 145–174.

7-30 H. M. Nagib, M. V. Morkovin, J. T. Yung, and J. Tan-atichat, "On Modeling of Atmospheric Surface Layers by the Counter-Jet Technique," *AIAA Journal*, **14**, No. 2 (1976), 185–190.

7-31 B. Bienkiewicz, J. E. Cermak, J. Peterka, and R. H. Scanlan, "Active Modeling of Large-Scale Turbulence," *J. Wind Eng. Ind. Aerodyn.* **13** (1983), 465–476.

7-32   J. E. Cermak, B. Bienkiewicz, and J. Peterka, *Active Modeling of Turbulence for Wind Tunnel Studies of Bridge Models*, Report No. FHWA/RD-82/148 Federal Highway Administration, McLean, Va., February 1983.

7-33   J. O. Hinze, *Turbulence*, McGraw-Hill, New York, 1959.

7-34   H. Rouse (Ed.), *Advanced Mechanics of Fluids*, Wiley, New York, 1965.

7-35   H. Tennekes and J. L. Lumley, *A First Course in Turbulence*, MIT Press, Cambridge, MA, 1972.

7-36   H. Tennekes, "The Logarithmic Wind Profile," *J. Atmos. Sci.*, **30** (1973), 234–238.

7-37   J. L. Lumley and H. A. Panofsky, *The Structure of Atmospheric Turbulence*, Wiley, New York, 1964.

7-38   N. J. Cook, "Determination of the Model Scale Factor in Wind-Tunnel Simulation of the Adiabatic Atmospheric Boundary Layer," *J. Ind. Aerodyn.*, **2** (1977–78), 311–321.

7-39   A. G. Davenport, D. Surry, T. Stathopoulos, *Wind Loads on Low Rise Buildings*, Final Report of Phases I and II, BLWT-SS8-1977, University of Western Ontario, London, Ont., Canada, 1977.

7-40   H. P. A. H. Irwin, *Wind Tunnel and Analytical Investigation of the Response of the Lions' Gate Bridge to a Turbulent Wind*, Report No. LTR-LA-210, NAE, National Research Council, Ottawa, Canada.

7-41   H. W. Tieleman, T. A. Reinhold and R. D. Marshall, "On the Wind-Tunnel Simulation of the Atmospheric Surface Layer for the Study of Wind Loads on Low-Rise Buildings," *J. Ind. Aerodyn.*, **3** (1978), 21–38.

7-42   D. Surry, "Consequences of Distortions in the Flow Including Mismatching Scales and Intensities of Turbulence," in *Wind Tunnel Modeling for Civil Engineering Applications*, T. A. Reinhold (Ed.), Cambridge University Press, 1982, pp. 137–185.

7-43   E. Szechenyi, "Supercritical Reynolds Number Simulation for Two-Dimensional Flow Over Circular Cylinders," *J. Fluid Mech.*, **70**, Part 3 (August 1975), 529–542.

7-44   J. Gandemer, G. Barnaud, and J. Biétry, *Etude de la tour D.M.A. Partie 1, Étude des efforts dûs au vent sur les façades*, Centre Scientifique et Technique du Bâtiment, Nantes, France, 1975.

7-45   B. J. Vickery, "The Aeroelastic Modeling of Chimneys and Towers," in *Wind Tunnel Modeling for Civil Engineering Applications*, T. A. Reinhold (Ed.), Cambridge University Press, 1982, pp. 408–428.

7-46   O. Güven, C. Farell, and V. C. Patel, "Surface-Roughness Effects on the Mean Flow Past Circular Cylinders," *J. Fluid Mech.*, **98** (1980), 673–701.

7-47   P. Sachs, *Wind Forces in Engineering*, Pergamon Press, Oxford and New York, 1972.

7-48   V. J. Modi and S. El-Sherbiny, "Wall Confinement Effects on Bluff Bodies in Turbulent Flows," *Proc. 4th International Conference on Wind Effects on Buildings and Structures*, Heathrow, U.K. (1975), pp. 121–130.

7-49   J. Courchesne and A. Laneville, "A Comparison of Correction Methods Used in the Evaluation of Drag Coefficient Measurements for Two-dimensional Rectangular Cylinders," ASME Winter Meeting, Paper No. 79-WA/FE3 (1979).

7-50   W. H. Melbourne, "Wind Tunnel Blockage Effects and Correlations," in *Wind Tunnel Modeling for Civil Engineering Applications*, T. A. Reinhold (Ed.), Cambridge University Press, 1982, pp. 197–216.

7-51   W. A. Dalgliesh, "Comparison of Model/Full Scale Wind Pressures on a High-Rise Building," *J. Ind. Aerodyn.*, **1** (1975), 55–66.

7-52   W. A. Dalgliesh, "Comparison of Model and Full-Scale Tests of the Commerce Court Building in Toronto," in *Wind Tunnel Modeling for Civil Engineering Applications*, T. A. Reinhold (Ed.), Cambridge University Press, 1982, pp. 575–589.

7-53   R. D. Marshall, "A Study of Wind Pressures on a Single-Family Dwelling in Model and Full Scale," *J. Ind. Aerodyn.*, **1**, 2 (Oct. 1975), 177–199.

7-54   R. D. Marshall, "Wind Tunnels Applied to Wind Engineering in Japan," *J. Struct. Eng.*, **110** (1984), 1203–1221.

8

Wind Directionality Effects

Wind effects on structural members depend upon direction for climatological, aerodynamic, and structural reasons. The extreme wind climate at any one site is, in general, nonuniform with respect to direction owing to basic atmospheric circulation patterns and/or the presence of local obstructions. Aerodynamic behavior depends upon direction for most structural members; examples range from cladding to bridges. The dependence upon direction of the structural response of a member subjected to a given aerodynamic load can be simply illustrated in the case of a circular flagpole in horizontally homogeneous terrain, anchored to its foundation by four bolts located at the corners of a square base plate. For any given wind speed, the uplift force on the anchor bolts is greater by a factor of $\sqrt{2}$ when the wind direction is parallel to the diagonal of the base plate than when it is parallel to one of the sides.

This chapter describes procedures for estimating probability distributions of largest yearly wind effects which account for the dependence upon direction of the extreme wind climate and of the aerodynamic and structural behavior of the member. Also described in this chapter are procedures for estimating probabilities of failure and safety indices for members sensitive to wind directionality effects.

8.1 PROCEDURES FOR ESTIMATING PROBABILITY DISTRIBUTIONS OF LARGEST YEARLY WIND EFFECTS

Three such procedures are currently available. The first procedure is based on the theory of stationary random processes [8-1, 8-2, 8-3]. It is shown in Sect. 8.1.1 that in the present state of the art this procedure is not suited for practical use. A second procedure, discussed in Sect. 8.1.2, is appropriate for the design of cladding and other members not subjected to significant dynamic amplification or aeroelastic effects. It utilizes (1) extreme wind speed data recorded or estimated for each of the eight (or sixteen) principal compass directions, and

(2) aerodynamic data, based on wind tunnel tests, on the dependence upon direction of the wind effect being considered. The wind speed and aerodynamic data are used to create time series of extreme wind effects, from which it is possible to estimate a univariate probability distribution of the largest wind effect, as well as the requisite design loading (e.g., the wind load with a 50-year mean recurrence interval). The practical application of this procedure—particularly for cladding design—is simple and straightforward. The third procedure, discussed in Sect. 8.1.3, utilizes the eight univariate probability distributions of the largest yearly wind speeds recorded for the principal compass directions [8-7, 8-9], and the fact that the time series of the largest yearly winds blowing from different directions have as a rule weak mutual correlations (see Sect. 3.4). This procedure can be applied to any type of structure or structural member, including structures or members subjected to wind-induced aerodynamic amplification or aeroelastic effects.

8.1.1 Procedure Based on Theory of Random Processes

In this procedure the mean wind velocity is regarded as a stationary two-dimensional random vector process, $U(t)$, with speed $U(t)$ and direction $\theta(t)$. Failure is assumed to occur if $U(\theta) \geqslant g(\theta)$ (that is, the curve $g(\theta)$ is the failure boundary in the velocity space—see Sect. A3.1.2). For example, if the relation between the wind effect $Q(\theta)$ and the wind speed U blowing from direction θ is

$$Q(\theta) = h(\theta)U^2(\theta) \qquad (8.1.1)$$

then the boundary $g(\theta)$ has the expression

$$g(\theta) = \left[\frac{R}{h(\theta)}\right] \qquad (8.1.2)$$

where R is the limit state (e.g., the wind pressure causing failure of a cladding panel).

The mean rate at which the vector $U(t)$ crosses the boundary $g(\theta)$ in the outward direction is denoted by v_D and may be referred to as the mean failure rate. If the values of v_D are small, failure is a rare event and its probability may be assumed to be of the Poisson type. The probability that in the time interval T no failure will occur (i.e., the probability that the velocity vector will not cross the failure boundary $g(\theta)$ in the outward direction) can be written as

$$p(0, T) = e^{-v_D T} \qquad (8.1.3)$$

(Eqs. A1.34, A2.39) so that the probability of failure during time T is

$$P_f = 1 - e^{-v_D T} \qquad (8.1.4)$$

The fact that there is no failure during the time interval T means that the largest wind effect Q occurring during that interval is less than R. Thus, Eq. 8.1.3 can be interpreted as representing the ordinate of the cumulative probability distribution of Q corresponding to the value $Q = R$, that is

$$F_Q(Q < R) = e^{-v_D T} \qquad (8.1.5)$$

where v_D is a function of R. In particular, if $T=1$ year, Eq. 8.1.5 represents the cumulative distribution function of the largest yearly wind effect, that is, the probability that the largest actual wind effect in any one year is less than a specified wind effect R.

The mean outcrossing rate v_D may be obtained by using Eq. A2.47, which is valid under the assumption that the random process is stationary. If polar coordinates are used, it follows from Eq. A2.47 that

$$v_D = \int_{-\pi}^{\pi} E_0^{\infty}[\dot{U}_n(\theta)|U(\theta)=g(\theta)]f_{U,\theta}[g(\theta),\theta]\left[1+\frac{g'^2(\theta)}{g^2(\theta)}\right]^{1/2}g(\theta)d\theta \quad (8.1.6)$$

[8-1, 8-2], where $\dot{U}_n$ = derivative with respect to time of the projection of the velocity vector $U(\theta)$ on the normal to the boundary $g(\theta)$, $U(\theta)$ = wind speed, $f_{U,\theta}$ = joint probability density function of wind speed and direction, and E_0^{∞} = average of the positive values of $\dot{U}_n$ given that $U(\theta)=g(\theta)$.

Attempts to evaluate the mean rate v_D have been reported in [8-1] and [8-3], where in addition to the assumption of stationarity of the wind velocity process, the assumption that $\dot{U}_n$ and U are statistically independent was used, so that

$$E_0^{\infty}[\dot{U}_n(\theta)|U(\theta)=g(\theta)] = E_0^{\infty}[\dot{U}_n(\theta)] \quad (8.1.7)$$

The quantity $E_0^{\infty}[\dot{U}_n(\theta)]$ can be estimated from spectra of wind velocities on the basis of Eq. A2.16a. Under the assumption of stationarity, these spectra and the probability density $f_{U,\theta}$ of Eq. 8.1.6 can be estimated from continuous wind velocity records or from comparable types of records, such as wind velocities recorded at one- or three-hour intervals [8-1, 8-3]. Once $E_0^{\infty}[\dot{U}_n(\theta)]$ and $f_{U,\theta}$ are obtained, Eqs. 8.1.5 and 8.1.6 can be used to estimate the cumulative distribution function of the largest yearly wind effect.

In [8-3] largest yearly wind loads estimated by the procedure described in this section were compared in a number of cases with largest yearly loads obtained on the basis of actual measurements. The discrepancies between the respective cumulative distribution functions were found to be unacceptably large. These discrepancies can be attributed to the use of wind speed data associated to a large extent with meteorological phenomena (e.g., morning breezes) that are unrelated to the strong winds of interest in structural design. As noted in Sect. 3.2.3, such records can provide a misleading basis for statistical inferences concerning extreme wind speeds. For this reason, until reliable estimates of the terms $E_0^{\infty}[\dot{U}_n|U(\theta)=g(\theta)]$ and $f_{U,\theta}[u,\theta]$ can be made on the basis of data pertaining to strong winds, the method reviewed in this section cannot be used with confidence for structural design purposes.

8.1.2 Procedure Based on the Time Series of the Largest Yearly Wind Effects

This procedure is applicable if the wind effect $Q(\theta)$ can be described by an expression of the form

$$Q(\theta) = \tfrac{1}{2}\rho C(\theta)C_p(\theta)U^2(h,\theta) \quad (8.1.8)$$

where ρ = air density, $C(\theta)$ = coefficient transforming wind load into wind effects (if $Q(\theta)$ is a wind pressure or suction, $C(\theta)\equiv 1$), $C_p(\theta)$ = aerodynamic coefficient corresponding to wind blowing from direction θ, and $U(h, \theta)=$mean wind speed corresponding to the direction θ at the reference height h above ground. It is assumed that the influence coefficient $C(\theta)$ is independent of $U(h, \theta)$. It is also assumed that the coefficient $C_p(\theta)$ is independent of or only weakly dependent upon $U(h, \theta)$. These assumptions exclude from consideration members subjected to significant dynamic amplification or aeroelastic effects.

The details of the procedure discussed in this section differ according to whether the region being considered has a well-behaved wind climate or can experience hurricane winds. The two cases are therefore treated separately.

Structures in Well-Behaved Wind Climates. Let $U_i(h, \theta)$ denote the largest value of $U(h, \theta)$ during year i. The largest wind effect Q_i during that year is the largest of the values $Q_i(\theta)$ obtained by substituting $U_i(h, \theta)$ for $U(h, \theta)$ in Eq. 8.1.8, that is

$$Q_i = \tfrac{1}{2}\rho \max_{\theta} [C(\theta)C_p(\theta)U_i^2(h, \theta)] \tag{8.1.9}$$

Note that in conventional engineering practice, wind directionality effects are not taken into account; that is, the largest yearly wind effect calculated for design purposes, Q_i^{nom}, is assumed to be given by

$$Q_i^{nom} = \tfrac{1}{2}\rho \max_{\theta} [C(\theta)C_p(\theta)] \max_{\theta} [U_i^2(h, \theta)] \tag{8.1.10}$$

$$= \tfrac{1}{2}\rho C_{tmax}U_i^2(h) \tag{8.1.11}$$

where $U_i(h) = \max_{\theta} [U_i(h, \theta)]$ denotes the largest annual wind speed regardless of direction, and

$$C_{tmax} = \max_{\theta} [C(\theta)C_p(\theta)] \tag{8.1.12}$$

Example

Consider a 100 m tall building located in an urban environment, for which it was estimated in [8-5] that $U(h, \theta)/U_f(\theta) \simeq 1.39$, where $U_f(\theta)=$fastest-mile wind speed at 10 m above ground in open terrain, and $h=318$ m. The largest yearly fastest-mile wind speeds $U_{f,1}(\theta)$ recorded during a given year in the region being considered are listed in line 1 of Table 8.1.1. The measured peak

TABLE 8.1.1. Largest Yearly Wind Speeds, Suction Coefficients, and Largest Yearly Suctions

	Direction	N	NE	E	SE	S	SW	W	NW
1	$U_{f,1}(\theta)$ (m/sec)	12.5	8.9	10.3	22.3	10.3	22.3	22.8	31.3
2	$C_p(\theta)$	0.07	1.06	3.33	0.51	0.67	0.66	1.12	0.24
3	$Q_1(\theta)$ (Pa)	13	101	426	306	86	296	703	284

suction coefficients $C_p(\theta)$ reported in [8-5] for a cladding panel located at 94 m elevation near a corner of the building are listed in line 2 of Table 8.1.1. The corresponding suctions, $Q_1(\theta)$, calculated by Eq. 8.1.9 in which $C(\theta) \equiv 1$ and $\rho = 1.25 \text{ kg/m}^3$, are listed in line 3 of Table 8.1.1, and represent the largest suctions induced in the cladding panel by winds blowing from the eight directions of the compass during the year being considered.

It is seen in Table 8.1.1 that the largest suction induced during the year of concern by winds blowing from any direction is

$$Q_1 = \max_{\theta} [Q_1(\theta)]$$

$$= 703 \text{ Pa} \qquad (8.1.13)$$

If wind directionality effects were not taken into account, it would follow from Eq. 8.1.11 that $Q_1^{\text{nom}} = \frac{1}{2} \times 1.25 \times 3.33 \times (1.39 \times 31.3)^2 = 3939$ Pa. It is seen that in this example the value obtained by ignoring wind directionality effects is considerably higher than the actual value of the largest yearly wind suction, $Q_1 = 703$ Pa. Note, however, that this would not have been the case had the directional distribution of the wind speeds and/or of the suction coefficients been relatively uniform, or had the directions corresponding to the maximum values of $C_p(\theta)$ and $U_{f,1}(\theta)$ coincided.

If extreme wind speed data $U_i(h, \theta_j)(j = 1, 2, \ldots, 8$ or $1, 2, \ldots, 16)$ are available for a sufficient number m of consecutive years (for example, $m \geqslant 20$), a set of largest yearly load data, $Q_i, (i = 1, 2, \ldots, m)$ can be calculated by using Eq. 8.1.9. From these data it is possible to estimate the best fitting distribution of Q and various statistics that may be used for design purposes. One such statistic is the wind load $Q_{\bar{N}}$ corresponding to any mean recurrence interval $\bar{N}$.

It is convenient for computational purposes to define the quantity, referred to as equivalent wind speed,

$$U_{\text{eq}} = \left\{ \max_{\theta} [Q(\theta)]/(\tfrac{1}{2}\rho C_{t\text{max}}) \right\}^{1/2} \qquad (8.1.14)$$

where $C_{t\text{max}}$ is defined as in Eq. 8.1.12. The largest yearly equivalent wind speed during year i is

$$U_{\text{eq},i} = [Q_i/(\tfrac{1}{2}\rho C_{t\text{max}})]^{1/2} \qquad (8.1.15)$$

Statistical analyses of sets of data, $U_{\text{eq},i}$, reported in [8-6] have revealed that the probability distributions of the largest annual equivalent wind speeds may be assumed for design purposes to be Extreme Value Type I. This assumption is usually conservative. Note that if the equivalent wind speeds have an Extreme Value Type I distribution, the distribution of Q is not Extreme Value Type I.

Let $U_{\text{eq}\bar{N}}$ denote the equivalent wind speed corresponding to a $\bar{N}$-year mean recurrence interval. Assuming that the distribution of U_{eq} is Extreme Value Type I, it is possible to write (see Eq. 3.2.1):

$$U_{\text{eq}\bar{N}} \simeq a_{\text{eq}} + b_{\text{eq}} \ln \bar{N} \qquad (8.1.16)$$

where

$$a_{eq} = \bar{X}_{eq} - 0.45 s_{eq} \qquad (8.1.17)$$

$$b_{eq} = 0.78 s_{eq} \qquad (8.1.18)$$

and $\bar{X}_{eq}$ and s_{eq} = sample mean and sample standard deviation of $U_{eq,i}$. The wind effect $Q_{\bar{N}}$ corresponding to the mean recurrence interval $\bar{N}$ can then be written (Eqs. 8.1.15 and 8.1.16) as

$$Q_{\bar{N}} \simeq \tfrac{1}{2}\rho C_{t\max}(a_{eq} + b_{eq}\ln \bar{N})^2 \qquad (8.1.19)$$

It is clear that failure to take wind directionality into account (i.e., the use of largest yearly wind effects estimated by Eq. 8.1.10, rather than by Eq. 8.1.9) would result in some cases in unrealistically inflated estimates of the wind load corresponding to an $\bar{N}$-year mean recurrence interval. This is shown in the following example, presented in detail on account of the practical importance of such calculations in cladding glass design.

Example

The largest yearly fastest-mile wind speeds at 10 m above ground in open terrain, U_f, obtained from records taken at Sheridan, Wyoming, in the period 1958–1977 are listed in mph in Table 8.1.2. (Summary statistics for these data are shown in Fig. 3.4.1.) We seek the 50-year wind-induced suction, $Q_{\bar{N}=50}$, on the cladding panel of the previous example, for which the aerodynamic coefficients are given in line 2 of Table 8.1.1, and the estimated ratio $U(h, \theta)/U_f(\theta)$ is approximately equal to 1.39. From Eqs. 8.1.9 (in which $C(\theta)=1$) and 8.1.15 it follows that the largest yearly equivalent wind speeds during the period 1958–1977 have the values shown in Table 8.1.2, where the corresponding sample mean $\bar{X}_{eq}$ and sample standard deviation s_{eq} are also shown. From Eqs. 8.1.16 through 8.1.19, $Q_{\bar{N}=50}=974$ Pa (20.3 psf).

If the load is calculated without taking wind directionality into account, the nominal 50-year load is (Eq. 8.1.11):

$$Q_{\bar{N}=50}^{\mathrm{nom}} = \tfrac{1}{2} \times 1.25 \times 3.33 \times (1.39 U_{f50})^2 \qquad (8.1.20)$$

where U_{f50} = 50-year fastest-mile wind speed (in m/sec) estimated from the set of largest yearly speeds regardless of direction. From the data of table 8.1.2, $U_{f50}=74.25$ mph (33.2 m/sec), and $Q_{\bar{N}=50}^{\mathrm{nom}}=4430$ Pa (92.5 psf), versus the actual 50-year load, $Q_{\bar{N}=50}=974$ Pa (20.3 psf).

Directional largest yearly fastest-mile wind speed data at a number of weather stations in the United States are available in [8-7]. Similar data that may be used for design purposes can also be obtained from monthly Local Climatological Data summaries published by the National Oceanic and Atmospheric Administration (see Sect. 3.4).

Structures in Hurricane-Prone Regions. In hurricane-prone regions the load data used for inferences concerning design loads are not yearly maxima.

TABLE 8.1.2. Largest Yearly Fastest-Mile Wind Speeds, U_f, Loads, Q_i, and Equivalent Speeds, $U_{eq,i}$

Year	\multicolumn{8}{c}{Largest Annual Fastest-Mile Wind Speed at 10 m above Ground in Open Terrain (miles per hour)[a,b]}	Q_i(Pascals)	$U_{eq,i}$(m/sec)

Year	N	NE	E	SE	S	SW	W	NW	Q_i(Pascals)	$U_{eq,i}$(m/sec)
1958	28	20	23	50	23	50	51	70	703	18.4
1959	41	25	19	29	25	40	38	60	399	13.9
1960	36	21	16	34	26	43	45	60	553	16.3
1961	25	18	30	36	27	47	38	60	718	18.6
1962	22	23	22	36	16	47	52	60	728	18.7
1963	31	14	23	33	36	63	48	57	643	17.6
1964	22	15	18	34	19	54	54	60	782	19.4
1965	33	31	20	33	17	66	43	55	708	18.5
1966	36	21	19	34	14	51	39	61	419	14.2
1967	44	14	16	40	36	51	41	62	464	15.0
1968	36	19	19	35	21	39	40	47	438	14.5
1969	28	16	15	36	22	53	34	66	459	14.9
1970	28	13	20	35	37	61	37	53	598	17.0
1971	33	15	22	31	22	49	31	47	384	13.5
1972	23	19	26	36	37	55	44	47	553	16.3
1973	28	23	19	32	15	46	39	64	419	14.2
1974	24	28	19	37	25	57	49	56	653	17.7
1975	22	22	19	27	28	39	33	51	289	11.8
1976	31	24	28	33	38	47	33	47	648	17.6
1977	44	20	19	40	36	34	44	56	513	15.7

$$\bar{X}_{eq} = 16.17 \text{ m/sec}$$
$$s_{eq} = 2.11 \text{ m/sec}$$

[a]1 mph $\simeq 0.447$ m/sec.
[b]Values in italics are largest yearly wind speeds from all directions.

Rather, they are associated with hurricanes, which occur at irregular intervals. The approach used in this case is the following. A large number, m, of hurricanes is generated by Monte Carlo simulation on the basis of climatological information on hurricane storms, as shown in Sect. 3.3. For each hurricane, the load Q_i and the corresponding equivalent wind speed $U_{eq,i}$ (see Eqs. 8.1.9 and 8.1.15) are then obtained. Following exactly the same steps used in Sect. 3.3.2, the cumulative distribution functions of the largest load and of the largest equivalent wind speed occurring in any one year are found to be

$$F_Q(Q < Q_i) = F_{U_{eq}}(U_{eq} < U_{eq,i})$$

$$= e^{-\lambda\left(1 - \frac{i}{m+1}\right)} \tag{8.1.21}$$

where λ is the annual occurrence rate of hurricanes in the area of interest for the site being considered. Continuous probability distribution curves, $F(Q < q)$ and $F(U_{eq} < u_{eq})$, that best fit Eq. 8.1.21 (e.g., Weibull or Extreme Value Type I distributions) can be estimated by using standard statistical techniques. Note that the mean recurrence interval of the load Q_i and of the equivalent wind speed $U_{eq,i}$ is

$$\bar{N} = \frac{1}{1 - e^{-\lambda\left(1 - \frac{i}{m+1}\right)}} \tag{8.1.22}$$

A similar approach is reported in [8-4].

A computer program for estimating hurricane-induced wind loads in accordance with the procedure outlined here is described briefly in [8-8], and is available on tape in [8-9]. Stored in the program are hurricane wind speeds corresponding to the 16 compass directions at 56 mileposts located at distances of 50 nautical miles along the Gulf and Atlantic coasts. These speeds were obtained from 999 hurricane wind fields generated by Monte Carlo simulation at each milepost, as described in Sect. 3.3.2, and were used in [8-10] and in the development of the wind speed map included in the American National Standard A58.1-1982 [8-11].

In cases where it is judged that the probability distribution of the largest yearly loads, $F_Q(Q < q)$, may be affected by both hurricane and non-hurricane winds, the following expression should be used:

$$F_Q(Q < q) = F_{Q_H}(Q_H < q) F_{Q_{NH}}(Q_{NH} < q) \tag{8.1.23}$$

where $F_{Q_H}(Q_H < q) =$ cumulative distribution function of hurricane-induced wind loads, Q_H, estimated as shown in this section, and $F_{Q_{NH}}(Q_{NH} < q) =$ cumulative distribution function of loads induced by non-hurricane winds, estimated as shown in the previous section (see also Sect. 3.3.2).

8.1.3 Procedure Based on the Univariate Probability Distributions of the Largest Yearly Wind Speeds Recorded for Each of the Principal Compass Directions

A simple procedure is now presented that may be applied to any type of structure, including structures subjected to aerodynamic amplification or aeroelastic effects.

It was pointed out in Section 3.4 that the correlation between extreme wind speeds occurring in any two directions is generally weak. As shown in Appendix A1 (Eqs. A1.64), two uncorrelated variables having a joint Extreme Value Type I distribution are statistically independent. It can be shown that statistical independence also holds for any number of uncorrelated variables whose joint distribution is of an extreme value type [A1-24]. In practice it can therefore be assumed that the largest yearly winds blowing from the eight principal compass directions are statistically independent. The cumulative probability distributions of the largest yearly wind effect may thus be written as

$$F_Q(Q<R)=\text{Prob}(v_1<v_1^l, v_2<v_2^l, \ldots, v_8<v_8^l) \qquad (8.1.24a)$$

$$\simeq \text{Prob}(v_1<v_1^l)\,\text{Prob}(v_2<v_2^l)\ldots\text{Prob}(v_8<v_8^l) \qquad (8.1.24b)$$

where v_i^l is the wind speed from direction i causing the occurrence of the wind effect R [8-14]. Note that if the wind speeds occurring in all directions were perfectly correlated, then

$$F_Q(Q<R)=\text{Prob}(v_k<v_k^l) \qquad (8.1.25)$$

where $1\leqslant k\leqslant 8$. Equation 8.1.25 indicates that Eq. 8.1.24b is conservative from a structural design viewpoint.

The assumption of statistical independence used in Eq. 8.1.24b is also acceptable in hurricane-prone regions. In this case lower and upper bounds for $F_Q(Q<R)$ can be estimated as in [8-15].

8.2 ESTIMATION OF FAILURE PROBABILITIES AND SAFETY INDICES FOR MEMBERS SENSITIVE TO WIND DIRECTIONALITY EFFECTS

To determine whether a member sensitive to wind directionality effects is acceptable from a safety point of view, its nominal failure probability (or its safety index) is compared to that of a member judged to be acceptable. The member is then redesigned as needed until the result of this comparison is satisfactory. An application of this reliability-based approach to the design of glass cladding for a tall building is presented in Chapter 9. This section describes procedures for estimating nominal failure probabilities and safety indices required for the application of this approach.

8.2.1 Estimation of Failure Probabilities

Consider a member whose resistance is R, and denote by Q the largest load effect acting on the member during any one year. Failure occurs for any pair of values R, Q such that $R-Q<0$. In most applications R and Q may be assumed to be independent, so that the probability of failure in any one year can be written as

$$P_f = \int_0^\infty F_R(q)f_Q(q)\,dq \qquad (8.2.1)$$

(Eq. A1.21), where F_R = cumulative distribution function of R, and f_Q = probability density function of Q. The function f_Q is related by Eq. A1.11 to the cumulative distribution function, F_Q, estimated as shown in Sects. 8.1.2 and 8.1.3. The probability of failure during the n-year lifetime of the structure can be obtained from Eqs. 8.2.1 and A1.31. The probability of failure so obtained is conditional upon a given set of values of the random parameters that determine the functions R, Q, F_R, and f_Q. Conditional probabilities of failure can be useful in certain applications in which the objective is to assess qualitatively the relative reliabilities of various members.

Unconditional failure probabilities can be estimated by using an expression similar to Eq. A3.1, provided that (1) the probability distributions of the various random parameters that determine R, Q, F_R, and f_Q are known, and (2) such estimates are not computationally prohibitive. In a number of situations of practical interest it is in principle possible to use reliability-based design methods that employ the safety index as a measure of structural reliability. Nevertheless, difficulties pertaining to the choice of the target safety index for at least some situations remain unsolved.

8.2.2   Estimation of Safety Indices

The most commonly used safety index on which reliability calculations are based has the expression

$$\beta = \frac{\ln \bar{R} - \ln \bar{Q}_n}{(V_R^2 + V_{Q_n}^2)^{1/2}} \tag{8.2.2}$$

(see Eq. A3.29), where $\bar{R}$ and V_R = mean value and coefficient of variation of the limit state, and $\bar{Q}_n$ and V_{Q_n} = mean value and coefficient of variation of the largest lifetime load effect.

We consider here only members that do not experience significant dynamic amplification or aeroelastic motions [8-13]. Load effects for such members can be described by Eq. 8.1.8. Expressions for $\bar{Q}_n$ and V_{Q_n} are first developed for the case of structures with a specified orientation angle, α (Fig. 8.2.1). We also consider the case where gravity loads, in addition to wind loads, are present. We then treat the case of structures with unknown orientation, which is of interest for the development of building code provisions on wind loads.

In general

$$R = R(X_1, X_2, \ldots, X_m) \tag{8.2.3}$$

$$Q_n = Q_n(X_{m+1}, X_{m+2}, \ldots, X_n) \tag{8.2.4}$$

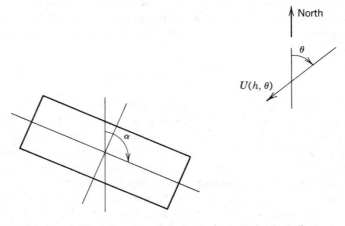

FIGURE 8.2.1.   Wind direction, θ, and angle of orientation of structure, α.

where, for example, the random variables $X_i (i = 1, 2, \ldots, m)$ may denote member dimensions and material strength, and $X_j (j = m+1, m+2, \ldots, n)$ may represent aerodynamic and micrometerological parameters.* Equations 8.2.3 and 8.2.4 can be expanded in Taylor series; approximate expressions for the mean values and coefficients of variation of R and Q_n can then be obtained in terms of means, coefficients of variation, and correlation coefficients of the random variables $X_i (i = 1, 2, \ldots, n)$ and $X_j (j = m+1, m+2, \ldots, n)$, higher order terms in these expressions being neglected. (For examples of such calculations, see Eqs. A3.26–A3.28 and [8-12]).† In most design situations that involve wind action, $\bar{R}$ and V_R can be estimated independently of $\bar{Q}_n$ and V_{Q_n}.‡

Structures with Specified Orientation. The cumulative distribution function $F_{U_{eqn}}$ of the largest lifetime equivalent wind speed U_{eqn} is

$$F_{U_{eqn}}(u_{eq}) = [F_{U_{eq}}(u_{eq})]^n \tag{8.2.5}$$

(Eq. A3.2), where $n =$ lifetime in years and $F_{U_{eq}} =$ cumulative distribution function of the largest yearly equivalent wind speed U_{eq} obtained as shown in Sect. 8.1.2. From the distribution $F_{U_{eqn}}$ it is possible to estimate the mean $\bar{U}_{eqn}$ and the standard deviation $s(u_{eqn})$.

Associated with the largest lifetime equivalent wind speed U_{eqn} is the largest lifetime wind effect Q_n, that is (see Eq. 8.1.14)

$$Q_n = \bar{Q}_n + Q_n'$$
$$= \tfrac{1}{2}(\bar{\rho} + \rho')(\bar{C}_{t\max} + C_{t\max}')(\bar{U}_{eqn} + U_{eqn}')^2 \tag{8.2.6}$$

where the bars and primes indicate mean values and deviations from the mean, respectively. We assume that the values ρ and $C_{t\max}$ used in calculating $U_{eq,i}$ (Eq. 8.1.15) are the mean values of these two variables. The following approximate relations follow from Eq. 8.2.6:

$$\bar{Q}_n \simeq \tfrac{1}{2}\bar{\rho}\bar{C}_{t\max}\bar{U}_{eqn}^2(1 + V_{U_{eqn}}^2) \tag{8.2.7}$$

$$V_{Q_n}^2 \simeq V_\rho^2 + V_{C_{t\max}}^2 + 4V_{U_{eqn}}^2 \tag{8.2.8}$$

The coefficient of variation, $V_{C_{t\max}}$, reflects the variabilities of the influence coefficients $C(\theta)$ and of the aerodynamic coefficients $C_p(\theta)$. For example, the influence coefficients transforming pressure on cladding into maximum tensile stress in the panel depend upon the panel thickness, which, for the same nominal thickness, may actually vary somewhat from panel to panel. Aerodynamic

---

*The expressions for R and Q_n may contain a number of common variables, that is, $X_i \equiv X_j$ for some values $i \leqslant m$ and $j \geqslant m+1$.
†Simpler manipulations are possible in certain instances—see Eq. 8.2.6 and subsequent derivations.
‡This is true even in the case of glass cladding, which experiences fatigue under wind loading and whose strength is influenced by the nature of the wind pressure fluctuations. As shown in Chapter 9, this influence can be incorporated in the expression of the load Q, so that values of R and V_R can be used in Eq. 8.2.2 that correspond to conventional loading patterns independent of the time history of the actual wind loads (for example, to constant loads of 60-sec duration).

coefficients for any given structure can—and usually do—vary from experiment to experiment. It is possible to write

$$V_{C_{t\max}}^2 \simeq V_C^2 + V_{C_p}^2 \tag{8.2.9}$$

where V_C and V_{C_p} are the coefficients of variation of $C(\theta)$ and $C_p(\theta)$.*

We now derive expressions for $\bar{U}_{eqn}$ and $V_{U_{eqn}}$ for the case where it may be assumed that the data $U_{eq,i}$ (Eq. 8.1.15) are best fitted by an Extreme Value Type I distribution. As mentioned previously, this assumption is generally conservative. From Eqs. A3.8–A3.9 it follows that

$$\bar{U}_{eqn} = \bar{X}_{eq} + 0.78 s_{eq} \ln n \tag{8.2.10}$$

$$s(u_{eqn}) = s_{eq} \tag{8.2.11}$$

where $\bar{X}_{eq}$ and s_{eq} are defined following Eq. 8.1.18 and $n=$ lifetime of structure in years.

To estimate $V_{U_{eqn}}$ we consider the relation

$$\bar{U}_{eqn} = c_1 c_2 c_3 c_4 c_s \bar{U}_{eqn}^{true} \tag{8.2.12}$$

where $\bar{U}_{eqn}$ and $\bar{U}_{eqn}^{true} =$ estimated and true (but unknown) mean value of the largest lifetime equivalent wind speed, and c_1, c_2, c_3, c_4, $c_s =$ coefficients with mean equal to unity which reflect, respectively, (1) errors in the measurement of the fastest-mile wind speeds over open terrain, (2) errors in the transformation of the fastest-mile wind speeds over open terrain into mean wind speeds at 10 m above ground in open terrain, $U_0(10)$, (3) errors in the transformation of $U_0(10)$ into mean wind speeds at 10 m above ground near the building site, $U(10)$, (4) errors in the transformation of $U(10)$ into mean wind speeds at the elevation, h, near the building site, $U(h)$, and (5) sampling errors in the estimation of $\bar{U}_{eqn}$ due to the limited size m of the sample of data $U_{eq,i}(i=1, 2, \ldots, m)$. The coefficient of variation $V_{U_{eqn}}$ can therefore be written as

$$V_{U_{eqn}}^2 \simeq V_{c_1}^2 + V_{c_2}^2 + V_{c_3}^2 + V_{c_4}^2 + \frac{s_s^2}{\bar{U}_{eqn}^2} + \frac{s^2(u_{eqn})}{\bar{U}_{eqn}^2} \tag{8.2.13}$$

where $V_{c_j} (j=1, 2, 3, 4)$ are the coefficients of variation of c_j, and $s_s =$ standard deviation of sampling error in the estimation of U_{eqn}.

Approximate estimates of the sampling errors s_s can be obtained by noting from Eqs. 8.2.10 and 8.1.16–8.1.18 that, for $n=50$ years, $\bar{U}_{eqn=50yr} \simeq U_{eq\bar{N}=86yr}$, so that the respective sampling errors are approximately the same for these two quantities. From Eq. 3.2.2 it then follows that

$$s_s \simeq \frac{4}{m^{1/2}} s_{eq} \tag{8.2.14}$$

where $m=$ size of data sample $U_{eq,i}$ ($i=1, 2, \ldots, m$).

---

*Equations 8.2.7 and 8.2.8 are approximate because (1) terms of order higher than two are neglected, and (2) it is assumed that the variabilities of the factors that determine $C_{t\max}$ and U_{eqn} depend negligibly upon direction.

Example

Estimate the safety index, β (Eq. 8.2.2), for $\bar{R} = 75$ psf (3590 Pa), $V_R = 0.22$, $\bar{\rho} = 1.25$ kg/m^3, $V_\rho = 0.05$, $\bar{C}_{tmax} = 3.33$, $V_C = 0$, $V_{C_\rho} = 0.1$, $\bar{X}_{eq} = 16.17$ m/sec, $s_{eq} = 2.11$ m/sec, and $m = 20$ (see Table 8.1.2), $V_{c_1} = V_{c_2} = V_{c_3} = V_{c_4} = 0.05$, $n = 50$ years. From Eqs. 8.2.11, 8.2.10, 8.2.14, 8.2.13, 8.2.9, 8.2.8, and 8.2.7, $s(u_{eqn}) = 2.11$ m/sec, $\bar{U}_{eqn} = 22.61$ m/sec, $s_s = 1.89$ m/sec, $V_{U_{eqn}} = 0.161$, $V_{C_{tmax}} = 0.1$, $V_{Q_n} = 0.340$, $\bar{Q}_n = 1092$ Pa. From Eq. 8.2.2, $\beta = 2.94$.

Case Where Gravity Loads are Present. Equations 8.2.7 and 8.2.8 are applicable if the gravity loads acting on the member may be neglected (as in the case of cladding panels subjected to wind loads). However, if (1) the effect of the gravity load is significant, and (2) the most unfavorable load combination occurs when the wind load reaches its largest lifetime value while the gravity load has an "arbitrary-point-in-time," rather than an extreme, value,* then

$$\bar{Q}_n = \bar{Q}_{n|G=0} + \bar{G} \qquad (8.2.15)$$

$$V_{Q_n}^2 = \frac{V_{Q_n|G=0}^2}{\left(1 + \dfrac{\bar{G}}{\bar{Q}_{n|G=0}}\right)^2} + \frac{V_G^2}{\left(1 + \dfrac{\bar{Q}_{n|G=0}}{\bar{G}}\right)^2} \qquad (8.2.16)$$

where $\bar{Q}_{n|G=0}$ and $V_{Q_n|G=0}$ = mean value and coefficient of variation of largest lifetime wind load estimated by Eq. 8.2.7 and 8.2.8, respectively, and $\bar{G}$ and V_G = mean value and coefficient of variation of "arbitrary-point-in-time" gravity load, respectively.

Structures with Unknown Orientation. A procedure is now presented for estimating safety indices for members of structures whose orientation is not known. Such a procedure can be useful for the development of building code provisions on wind loads.

The unknown orientation of the structure can be considered as one among several uncertain factors that determine member reliability (such as member resistance, aerodynamic coefficients, influence coefficients, and so forth). To the extent that structure orientation is included as a random variable in a properly conducted reliability analysis, the reliability of members in a structure sampled at random will be acceptable regardless of structure orientation, just as the reliability of any properly designed steel member will be acceptable even though, owing to the variability of the steel strength, actual yield stresses might be lower in some cases than the average yield stress.

The mean value and the variance of the largest lifetime loads acting on the member under consideration, averaged over all possible angles of orientation

---

*See discussion in Appendix A3 following Eq. A3.9.

α of the structure to which the member belongs (Fig. 8.2.1) are

$$\bar{Q}_n = \int_0^{2\pi} \overline{Q_n(\alpha)} f(\alpha)\, d\alpha \qquad (8.2.17)$$

$$\overline{(Q_n - \bar{Q}_n)^2} = \int_0^{2\pi} [Q_n(\alpha) - \overline{Q_n(\alpha)}]^2 f(\alpha)\, d\alpha \qquad (8.2.18)$$

where $Q_n(\alpha) = $ largest lifetime wind load acting on the member given that the angle of orientation of the structure is α, and $f(\alpha) = $ probability density function of structure orientation in the region being considered. It is reasonable to assume that α is uniformly distributed, that is, $f(\alpha) = 1/2\pi$. Other assumptions can be made, as necessary, if predominant structure orientations are known to exist.

In addition, it is assumed that wind speed data are available from 8 compass directions. (If data are available from 16 directions, the number 8 in the equations that follow must be changed into 16.) Using Eqs. 8.2.17 and 8.2.18 it can be shown after some algebra that

$$\bar{Q}_n \simeq \tfrac{1}{2}\rho C_{t\max}\left\{ \frac{1}{8} \sum_{r=1}^{8} \overline{U_{\mathrm{eqn}}(\alpha_r)}^2 \,[1 + V_{U_{\mathrm{eqn}}}^2(\alpha_r)] \right\} \qquad (8.2.19)$$

$$V_{Q_n} \simeq \frac{\left\{ (1 + V_\rho^2 + V_{C_{t\max}}^2) \dfrac{1}{8} \displaystyle\sum_{r=1}^{8} \overline{U_{\mathrm{eqn}}(\alpha_r)}^4 [1 + 6V_{U_{\mathrm{eqn}}}^2(\alpha_r)] - \left[\dfrac{1}{8} \displaystyle\sum_{r=1}^{8} \overline{U_{\mathrm{eqn}}(\alpha_r)}^2 [1 + V_{U_{\mathrm{eqn}}}^2(\alpha_r)] \right]^2 \right\}^{1/2}}{\dfrac{1}{8} \displaystyle\sum_{r=1}^{8} \overline{U_{\mathrm{eqn}}(\alpha_r)}^2 [1 + V_{U_{\mathrm{eqn}}}^2(\alpha_r)]}$$

$$(8.2.20)$$

[8-13]. In Eqs. 8.2.19 and 8.2.20, $\overline{U_{\mathrm{eqn}}(\alpha_r)}$ and $V_{U_{\mathrm{eqn}}}(\alpha_r)$ are the mean value and the coefficient of variation of the largest lifetime equivalent wind speed $U_{\mathrm{eqn}}(\alpha_r)$, estimated for the structure with angle of orientation α_r as in Eqs. 8.2.10 and 8.2.13. Use of Eqs. 8.2.19 and 8.2.20 in Eq. 8.2.2 yields the safety index of the member being considered in the case where the orientation of the structure is unknown. The case where gravity loads are present is treated in a manner entirely similar to that shown for structures with specified orientation.

REFERENCES

8-1   A. G. Davenport, "The Prediction of Risk Under Wind Loading," *Proceedings 2nd International Conference on Structural Safety and Reliability*, Munich, West Germany, September 1977, pp. 511–538.

8-2   Y. K. Wen, "Wind Direction and Structural Reliability," *J. Struct. Eng.*, **109** (April 1983), 1028–1041.

8-3   Y. K. Wen, "Wind Direction and Structural Reliability: II," *J. Struct. Eng.*, **110** (1984), 1253–1264.

8-4   B. V. Tryggvason, D. Surry, and A. G. Davenport, "Predicting Wind-Induced Response in Hurricane Zones," *J. Struct. Div.*, ASCE, **102** (Dec. 1976), 2333–2350.

8-5   J. A. Peterka and J. E. Cermak, *Wind Tunnel Study of Atlanta Office Building*, Fluid Mechanics and Wind Engineering Program, College of Engineering, Colorado State University, Ft. Collins, Colorado, Nov. 1978.

8-6   E. Simiu and J. J. Filliben, "Wind Direction Effects on Cladding and Structural Loads," *Eng. Struct.*, **3** (July 1981), 181–186.

8-7   M. E. Changery, E. Dumitriu-Valcea, and E. Simiu, *Directional Extreme Wind Speed Data for the Design of Buildings and Other Structures*, Building Science Series BSS 160, National Bureau of Standards, Washington, D.C., March 1984.

8-8   E. Simiu and M. E. Batts, "Wind-Induced Cladding Loads in Hurricane-Prone Regions," *J. Struct. Eng.*, **109** (Jan. 1983), 262–266.

8-9   *Hurricane-Induced Wind Loads*, Computer Program, Accession No. PB 82132259, National Technical Information Service, Springfield, VA, 1982.

8-10   M. E. Batts, L. R. Russell, and E. Simiu, "Hurricane Wind Speeds in the United States," *J. Struct. Div.*, ASCE, **106** (Oct. 1980), 2001–2016.

8-11   *American National Standard A58.1*, Building Code Requirements for Minimum Design Loads, American National Standards Institute, New York, 1982.

8-12   K. Rojiani and Y. K. Wen, "Reliability of Steel Buildings Under Winds," *J. Struct. Div.*, ASCE, **107** (Jan. 1981), 203–221.

8-13   E. Simiu, "Aerodynamic Coefficients and Risk-Consistent Design," *J. Struct. Eng.*, **109** (May 1983), 1278–1289.

8-14   E. Simiu, E. Hendrickson, W. Nolan, I. Olkin, and C. Spiegelman, "Multivariate Distributions of Directional Wind Speeds," *J. Struct. Eng.*, **111** (April 1985), 939–943.

8-15   O. Ditlevsen, "Narrow Reliability Bounds for Structural Systems," *J. Struct. Mech.*, **7** (1979), 453–472.

WIND LOADS AND THEIR EFFECTS ON STRUCTURES

II APPLICATIONS TO DESIGN

9

Tall Buildings:
Structural Response
and Cladding Design

The design of tall buildings is based on estimates of: (1) overall wind effects, which must be taken into account in the design of the structure, and (2) local wind effects, which govern the design of cladding. In general, the aerodynamic information needed to estimate overall as well as local wind effects cannot be determined from first principles, and must be obtained from wind tunnel tests. However, for a number of common situations the aerodynamic information is already available, and procedures for estimating structural response which incorporate that information may be employed. This is the case for buildings that (1) are not subjected to strong interference effects caused by the presence of neighboring structures, and (2) have geometric shapes that are not unusually unfavorable aerodynamically or structurally.

The resultant of the aerodynamic forces experienced by a structure subjected to wind action can be resolved into a drag (along-wind) force acting in the direction of the mean wind and a lift (across-wind) force acting perpendicularly to that direction. In general, the point of application of the total wind force, the elastic center, and the mass center of the structure do not coincide. For this reason structures are also subjected to torsional moments. This is true even for a symmetric building immersed in a symmetric mean flow, since the instantaneous flow will in general be asymmetric on account of the randomness of the flow fluctuations.

In Sects. 9.1, 9.2, and 9.3 procedures are presented for estimating the along-wind, the across-wind, and the torsional response of tall buildings whose behavior under wind loads is not significantly affected by the presence of neighboring tall buildings. As a very approximate guide, it may be assumed that if the distance between two tall buildings exceeds say, six to eight times the average of the horizontal dimensions of the buildings, mutual interference effects will be negligible for practical purposes. More refined guidelines can be obtained, for various specific situations, from studies reported in [9-1] to [9-5]. However, in the present state of the art it is not possible to provide general criteria for buildings subjected to interference effects. Nevertheless, it was noted in [9-1] that a square building located in urban terrain near a building with similar geometry and dimensions will perform satisfactorily, regardless of the relative position of the two buildings, if it is designed to withstand the loads (including the across-wind loads) it would experience in the absence of the neighboring structure.

The magnitude of wind-induced oscillations can be of concern not only from a structural point of view, but from the point of view of occupant comfort as well (see Sect. 15.1). Two types of device used to mitigate wind-induced oscillations are the tuned mass damper (TMD) and the viscoelastic damper. Information pertaining to these devices is presented in Sect. 9.4. The design of cladding for wind loads is discussed in Sect. 9.5.

9.1   ALONG-WIND RESPONSE

Until recently, drag forces used in structural design calculations were in all cases specified on the basis of climatological, meteorological, and aerodynamic considerations alone, independently of the mechanical properties of the structure, that is, of its mass distribution, flexibility, and damping. However, it has been recognized that in the case of modern tall structures—which are more flexible, lower in damping, and lighter in weight than their predecessors—the natural frequencies of vibration may be in the same range as the average frequencies of occurrence of powerful gusts and that, therefore, large resonant motions induced by wind may occur, which must be taken into account in design.

The resonant amplification of structural response to forces induced by atmospheric turbulence was first studied by Liepmann in a classic paper on the buffeting problem published in 1952 [9-6]. The application of Liepmann's concepts to civil engineering structures required the development of models representing the turbulent wind flow near the ground. Such models were proposed in 1961 by Davenport [9-7], who developed on their basis a procedure for estimating along-wind tall building response [9-8]. Vellozzi and Cohen developed a modified procedure, in which, in contrast to [9-8], it is recognized that the fluctuating pressures on the windward face of a building are not perfectly correlated to those acting on the leeward face [9-9]. This imperfect correlation is accounted for in [9-9] by a reduction factor. However, it has been shown that owing to the way in which this factor is applied, the procedure of [9-9] underestimates the resonant amplification effect [9-10], [9-11]. A procedure for estimating along-wind response based essentially on [9-8] has been included in the Canadian Structural Design Manual [9-12]. Vickery subsequently developed a procedure similar to that of [9-8] that allows, however, for more flexibility with respect to the choice of certain meteorological parameters [9-13]. An alternative approach is used in [9-43], which utilizes equations of equilibrium among horizontal forces at each floor.

In the procedures of [9-12] and [9-13] it is assumed that the characteristics of the turbulence do not vary with height above ground. Actually, according to the results of modern meteorological research, the energy of the turbulent fluctuations that cause resonant oscillations in tall buildings decreases significantly at higher elevations (see Sect. 2.3.3). Computer programs for calculating along-wind response, in which this decrease is taken into account and which allow therefore more economical designs, have been developed independently in [9-14] to [9-16].

On the basis of [9-14] and [9-16], respectively, simple procedures were developed in [9-17] and [9-18], which account for the dependence of turbulent fluctuations on height, and on the basis of which rapid manual calculations of the along-wind response can be performed. Elements of the procedure of [9-17] were incorporated in the Appendix to the American National Standard ANSI A58.1 [9-19].* However, the procedure developed in [9-18] is more practical for routine use and will therefore be reproduced in this chapter.

All the procedures mentioned above are based on the assumption that, around the structure, the terrain is approximately horizontal and that its roughness is reasonably uniform over a sufficiently large fetch. In practice it may be necessary to adjust the results obtained on the basis of this assumption by taking into account the effect upon the flow of changes in the terrain roughness upwind of the structure (see Sect. 2.4.1). However, if the topography of the surrounding terrain is unusual, or if the building is strongly affected by the flow in the wake of large neighboring buildings, analytical procedures become inapplicable and wind tunnel testing is necessary.

Another assumption common to all the above-mentioned procedures is that the mean wind is normal to the building face under consideration. Wind tunnel tests suggest that, in cases commonly encountered in tall-building design practice, to this assumption there correspond the highest values of the along-wind response [9-1, 9-20]. In the case of a square building, the peak along-wind response decreases continuously as a function of mean wind direction, from a maximum value which corresponds to the case where the direction is normal to a building face to about 0.8 times that value when the direction is parallel to a diagonal [9-1].

The general framework of the along-wind response problem is presented in Sect. 9.1.1. Section 9.1.2 describes the procedure developed in [9-18] for estimating the along-wind response of prismatic, or almost prismatic, structures

---

*However, for the sake of continuity with the earlier (1972) version of the ANSI A58.1 Standard, the Appendix to [9-19] retains from that version (1) the method for estimating the nonresonant part of the fluctuating response, and (2) a ratio of peak to rms fluctuating response $K_x = 3.0$, even though the latter is demonstrably lower (by about 15%) than the actual ratios K_x (see Sect. 9.1.2). (Note that the Appendix to [9-19] is *not* part of the ANSI A58.1-1982 Standard.)

for which it may be assumed that: (1) the fundamental vibration mode shape is approximately a straight line, and (2) the contribution to the response of the second and higher vibration modes is negligible. Also described in Sect. 9.1.2 is a procedure for estimating the along-wind response of point structures, that is, structures that may be viewed as consisting of a single mass concentrated at a height H (e.g., water towers) [9-18]. In the procedures described in Sect. 9.1.2, referred to here as simplified, all computations can be carried out manually. If the shape of the fundamental vibration mode deviates strongly from a straight line, or if the contribution of higher vibration modes is significant, the use of a computer program is required as indicated in Sect. 9.1.3. In Sect. 9.1.4, results of numerical calculations are used to discuss some of the approximations and errors inherent in the models being used.

9.1.1   Basic Relations. Equivalent Static Wind Loads

The total along-wind deflection may be viewed as a sum of two parts: the mean deflection, induced by the mean wind, and the fluctuating deflection, induced by the wind gustiness. The maximum along-wind deflection of the structure at elevation z may thus be written as

$$X_{\max}(z) = \bar{x}(z) + x_{\max}(z) \tag{9.1.1}$$

where $\bar{x}(z)$ is the mean deflection, and $x_{\max}(z)$ is the maximum fluctuating deflection in the direction of the mean wind. It is convenient to express $x_{\max}(z)$ in the form

$$x_{\max}(z) = K_x(z)\sigma_x(z) \tag{9.1.2}$$

where $\sigma_x(z)$ is the root mean square value of the fluctuating deflection and $K_x(z)$ is the peak factor, the value of which is usually about 3 to 4. Similarly, the maximum along-wind acceleration may be expressed as

$$\ddot{x}_{\max}(z) = K_{\ddot{x}}(z)\sigma_{\ddot{x}}(z) \tag{9.1.3}$$

where $\sigma_{\ddot{x}}(z)$ is the root mean square value of the along-wind accelerations and $K_{\ddot{x}}(z)$ is a peak factor, the value of which is usually about 4.

The gust response factor is defined as

$$G(z) = 1 + \frac{x_{\max}(z)}{\bar{x}(z)} \tag{9.1.4}$$

The maximum along-wind deflection can then be written as

$$X_{\max}(z) = G(z)\bar{x}(z) \tag{9.1.5}$$

It is convenient to define an equivalent static wind load that would induce in the structure along-wind deflections equal to those caused by the gusty wind. It follows from Eq. 9.1.5 and the assumed linearity of the structure that the equivalent static wind load is equal to the product of the gust response factor by the mean wind load.

The general expression for the mean deflection $\bar{x}(z)$ is given by Eq. 5.3.1.

The fluctuating deflections and accelerations as well as the respective peak factors (Eqs. 9.1.2 and 9.1.3) are obtained from Eqs. 5.3.8 through 5.3.15, in which the general expression for the quantity $S_x(z, n)$ (the spectral density of the along-wind fluctuating deflections) is given by Eq. 5.2.37. It follows from these equations that the calculated deflections and accelerations depend upon the properties of the structure, that is, its dimensions, mass distribution, natural frequencies, damping ratios, and modal shapes, and upon the assumed mean and fluctuating wind loads.

9.1.2 A Simplified Procedure for Estimating Along-Wind Response

Following [9-17], a procedure for calculating along-wind response is now presented, applicable to prismatic, or almost prismatic, structures for which it may be assumed that (1) the shape of the fundamental mode of vibration is linear, and (2) the response to wind loading is dominated by the fundamental mode. The first of these assumptions is acceptable in a large number of situations of practical interest such as in the case of typical multistory framed structures (see, for example, [9-21, p. 428] or [9-22, pp. 60 and 242]). The second assumption will generally hold if the ratios of natural frequencies in the second and higher modes to the fundamental frequency are sufficiently large (see Sect. 9.1.4). Also given in this section is a procedure for estimating the along-wind response of point structures, that is, structures that may be viewed approximately as consisting of a simple mass M concentrated at a height H.

Basic Assumptions. The procedure presented in this section is based on the following assumptions:

1. The behavior of the structure is linearly elastic.

2. The fundamental mode of vibration is a linear function of height above ground, that is, $x_1(z) = z/H$.

3. The contribution of the second and higher vibration modes to the response is negligible.

4. The mean velocity profile is described by the relation

$$U(z) = 2.5 u_* \ln \frac{z - z_d}{z_0} \qquad z \geqslant z_d + 10 \qquad (9.1.6)$$

$$U(z) = 2.5 u_* \ln \frac{10}{z_0} \qquad z \leqslant z_d + 10 \qquad (9.1.7)$$

(In Eqs. 9.1.6 and 9.1.7, z, z_0, and z_d are expressed in meters.)

The use of the logarithmic profile above elevation $(z_d + 10)$ meters implies the assumption of horizontal homogeneity of the flow. This assumption may not hold over regions near a change in surface roughness, as indicated in Sect. 2.4. However, in such regions Eq. 9.16—with suitable values of the parameters u_*, z_0, and z_d—may be used to obtain reasonable upper and lower bounds for the value of the response. Equation 9.1.7 is used, conservatively, on account of the uncertainty with regard to the actual nature of the flow near a building for $z < z_d + 10$ or so.

5.   The mean velocity $U(z)$ in Eqs. 9.1.6 and 9.1.7 is averaged over a period of one hour.

6.   The longitudinal velocity fluctuations are described by Eq. 2.3.2 and Table 2.3.1, and by Eq. 2.3.16.

7.   The mean and the fluctuating pressures are described by Eqs. 5.3.3 and 5.3.6, respectively. The expressions for the mean response are therefore given by Eqs. 5.3.4 and 5.3.2, and those for the fluctuating response by Eqs. 5.3.7 through 5.3.15 (or the equivalent expressions in nondimensional form, Eqs. 5.3.16 through 5.3.28).

8.   The spatial cross-correlations of the fluctuating pressures in the across-wind and along-wind directions are described by Eqs. 5.3.48 and 5.3.49, respectively.

Response Parameters.   A brief discussion is now presented of some of the structural, micrometeorological, and aerodynamic parameters involved in the estimation of along-wind response with a view to assisting the structural designer in their interpretation and selection.

a. Damping Ratio, ζ_1.   Suggested values for mechanical damping ratios of steel and reinforced concrete frames are 0.01 and 0.02, respectively [9-12, 9-19]. Lower values of the mechanical damping may have to be used, for example, in the case of welded steel stacks, of certain prestressed structures, or of structures of the framed tube type [9-12, 9-23]. In addition to the mechanical damping, the aerodynamic damping may, in principle, also be taken into account. The aerodynamic damping, which may help reduce the magnitude of the resonant oscillations, is associated with changes in the relative velocity of the air with respect to the building as the latter oscillates about its mean deformed position. It appears that it may be unconservative to rely upon the effect of the aerodynamic damping; for this reason, the latter is neglected in [9-12] and [9-19] and is also neglected herein.

b. Terrain Roughness Parameters, z_0, z_d.   The variation of mean wind speed with height is determined by two parameters, the roughness length z_0 and the zero plane displacement z_d (Eq. 9.1.6). The roughness length may be interpreted physically as a measure of the turbulent eddy size at the ground level. Values of z_0 suggested for structural design purposes are given in Table 9.1.1 (see Sect. 2.2.4).

In densely built-up cities (or in forests) the buildings (or trees) obstruct the flow near the ground; the mean flow thus begins to develop above an elevation which is referred to as the zero plane displacement and is slightly lower than the average height of the surrounding buildings (or trees). For design purposes, the zero plane displacement may be assumed to be zero in coastal and open terrain, and if the values of z_0 of Table 9.1.1 are used, in built-up terrains as well.

TABLE 9.1.1. Suggested Values of Roughness Lengths z_0 for Various Types of Terrain

Type of Terrain	Coastal[a,b]	Open[b]	Sparsely Built-up Suburbs[b]	Towns, Densely Built-up Suburbs[b]	Centers of Large Cities[b]
$z_0(m)$	0.005–0.01	0.03–0.10	0.20–0.40	0.80–1.20	2.00–3.00

[a]Applicable to structures directly exposed to winds blowing from open water.
[b]Values of z_0 to be used in conjunction with the assumption $z_d = 0$.

c. *Exponential Decay Parameters, C_y, C_z.* The narrow-band spatial cross-correlation of the fluctuating pressures in the across-wind direction (Eq. 5.3.48) is a measure of the extent to which pressures applied at different points of the same building face act coherently or at cross-purposes. The smaller the values of the parameters C_y and C_z in the expression for the cross-correlation the more coherent will be the action of such pressures and, therefore, the larger the response.

On the basis of wind tunnel tests, it has been suggested that it is reasonable to assume $C_y = 16$ and $C_z = 10$ [9-13]. The procedures presented in this section are based on these values. However, as indicated in Sect. 2.3.4, full-scale measurements do not always confirm this assumption. As shown in Sect. 9.1.4, the effect upon the total along-wind response of changes in the values of C_y and C_z of as much as 30% to 40% is, in general, relatively small (of the order of 5%–10%). However, the effect of such changes upon the accelerations may be considerable.

d. *Friction Velocity, u_*.* The friction (or shear) velocity u_* is a measure of the wind intensity over terrain of given roughness. If the mean wind at a specified reference height above ground z_R is known, u_* can be obtained by using Eq. 9.1.6:

$$u_* = \frac{U(z_R)}{2.5 \ln[(z_R - z_d)/z_0]} \qquad (9.1.8)$$

In meteorological work, the reference height most commonly used is $z_R = 10$ m.

In designing tall buildings it is reasonable to use mean wind speeds averaged over a period of one hour. In this chapter, therefore, the symbol U will denote hourly mean speeds. If mean wind speeds U^t are specified that are averaged over periods t different from one hour, the mean winds averaged over one hour can be obtained by using Fig. 2.3.10. For convenience, the information included in Fig. 2.3.10 is summarized in Table 9.1.2.

For values of t not included in Table 9.1.2, linear interpolation is permissible. If the wind speeds are given in terms of fastest-miles U_f as in [9-19], the averaging time in seconds is given by

$$t = 3600/U_f. \qquad (9.1.9)$$

As indicated in Chapter 2, the retardation of the flow due to increased terrain roughness causes the mean speeds over built-up terrain to be lower—

TABLE 9.1.2. Approximate Ratios of Probable Maximum Speed Averaged over Period t to That Averaged over One Hour (at 10 m above Ground in Open Terrain)

t (sec)	2	5	10	30	60	100	200	500	1000	3600
U^t/U	1.53	1.47	1.42	1.28	1.24	1.18	1.13	1.07	1.03	1.00

for any given large-scale storm—than the mean speeds at equal elevations over open terrain. Since wind climatological information is commonly provided in terms of wind speeds measured over open terrain (generally at airport weather stations), the problem arises of converting this information into wind speeds applicable to a built-up environment. In Sect. 2.2.5 this problem was shown to be solved as follows. Let u_{*1}, z_{01} denote the friction velocity and roughness length over open terrain, and let u_* denote the friction velocity over terrain with roughness length z_0. For the surface roughness categories of Table 9.1.1, approximate ratios u_*/u_{*1} can be obtained from Table 9.1.3. Once u_* is known, $U(z)$ can be calculated by using Eq. 9.1.6.

e. Duration of Storm, T. This parameter appears in Eqs. 5.3.11 and 5.3.14, which indicate in effect that the expected peak values of the fluctuations will be higher if the duration of the storm increases. The assumed storm duration is implicit in the use of design mean speeds averaged over one hour, that is, $T = 3600$ sec.

f. Mean Pressure and Suction Coefficients, C_w, C_l. The mean pressure and suction coefficients are functions of the shape of the structure (see Chapter 4). In the case of tall buildings with a rectangular shape in plan, it may be assumed $C_w = 0.8$, $C_l = 0.5$, and $C_D = C_w + C_l = 1.3$.

g. Mean Square Value of Turbulent Velocity Fluctuations. The ratio, β, between the mean square value of the longitudinal velocity fluctuations, $\overline{u^2}$, and the square of the friction velocity, u_*^2 (Eq. 2.3.2) depends upon surface roughness, as shown in Table 9.1.4.

TABLE 9.1.3. Ratios u_*/u_{*1} For Various Surface Roughness Categories

Type of Terrain	Coastal	Open	Sparsely Built-up Suburbs	Towns, Densely Built-up Suburbs	Centers of Large Cities
u_*/u_{*1}	0.85	1.00	1.15	1.33	1.45

TABLE 9.1.4. Approximate Ratio $\beta = \overline{u^2}/u_*^2$ for Various Surface Roughness Categories

Type of Terrain	Coastal	Open	Sparsely Built-up Suburbs	Densely Built-up Suburbs	Centers of Large Cities
β	6.50	6.00	5.25	4.85	4.00

Expressions for the Along-Wind Response. Using the basic assumptions listed earlier in this section and relations given in Sect. 5.3, results of numerical integrations were closely fitted in [9-18] by simple functions, and expressions for the along-wind response were developed that are listed in Table 9.1.5 for buildings with a nearly linear fundamental mode shape (Fig. 9.1.1), and in Table 9.1.6 for point structures (Fig. 9.1.2).

TABLE 9.1.5. Equations for Estimating the Along-Wind Response of Buildings with an Approximately Linear Fundamental Modal Shape [9-18]

(1) $Q = 2\left(1 - \dfrac{z_d^2}{h^2}\right) \ln \dfrac{h - z_d}{z_0} - 1$

(2) $J = 0.78 Q^2$

(3) $\mathcal{B} = \dfrac{6.71 Q^2}{1 + 0.26 \dfrac{b}{h}}$

(4) $\tilde{f}_1 = \dfrac{n_1 h}{u_*}$

(5) $C(x) = \dfrac{1}{x} - \dfrac{1}{2x^2}(1 - e^{-2x})$

(6) $x_1 = 12.32 \dfrac{\tilde{f}_1}{Q} \dfrac{d}{h}$

(7) $N(\tilde{f}_1) = C(x_1)$

(8) $C_{Df}^2(\tilde{f}_1) = C_w^2 + 2 C_w C_l N(\tilde{f}_1) + C_l^2$

(9) $x_2 = 3.55 \dfrac{\tilde{f}_1}{Q}$

(10) $M(z) = b d \rho_b(z)$

(11) $\mathcal{R} = 0.59 \dfrac{Q^2}{\zeta_1} \left(\dfrac{Q}{\tilde{f}_1}\right)^{2/3} \dfrac{C_{Df}^2(\tilde{f}_1)}{C_D^2} C(x_2) \dfrac{1}{1 + 3.95 \dfrac{\tilde{f}_1}{Q} \dfrac{b}{h}}$

(12) $M_1 = \dfrac{1}{h^2} \displaystyle\int_0^h M(z) z^2 \, dz$

(13) $q_* = \tfrac{1}{2} \rho u_*^2$

(14) $\bar{x} = \dfrac{C_D b h q_*}{M_1 (2\pi n_1)^2} J$

(15) $\sigma_x = \dfrac{C_D b h q_*}{M_1 (2\pi n_1)^2} \left(\dfrac{\beta \mathcal{B}}{6} + \mathcal{R}\right)^{1/2}$

(16) $v_x = n_1 \left(\dfrac{\mathcal{R}}{\dfrac{\beta \mathcal{B}}{6} + \mathcal{R}}\right)^{1/2}$

(17) $K_x = [1.175 + 2 \ln(v_x T)]^{1/2}$

(18) $G = 1 + K_x \dfrac{\sigma_x}{\bar{x}}$

(19) $X_{\max} = G \bar{x}$

(20) $\sigma_{\ddot{x}} = \dfrac{C_D b h q_*}{M_1} \mathcal{R}^{1/2}$

(21) $K_{\ddot{x}} = [1.175 + 2 \ln(n_1 T)]^{1/2}$

(22) $\ddot{X}_{\max} = K_{\ddot{x}} \sigma_{\ddot{x}}$

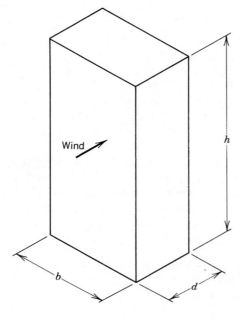

FIGURE 9.1.1. Schematic representation of tall building with rectangular cross section.

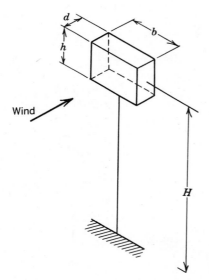

FIGURE 9.1.2. Schematic representation of point structure.

In Tables 9.1.5 and 9.1.6, h and H are the vertical dimensions shown in Figs. 9.1.1 and 9.1.2, b = across-wind dimension of structure, d = along-wind dimension of structure, z_0 = roughness length (see Table 9.1.1), z_d = zero plane displacement (for practical calculations it may be assumed $z_d = 0$), n_1 = natural frequency of vibration in fundamental mode of vibration, u_* = friction velocity, C_D = drag coefficient ($C_D = C_w + C_l$), C_w and C_l = average pressure coefficient

TABLE 9.1.6. Equations for Estimating the Along-Wind Response of Point Structures [9–18]

(1) $P = \ln \dfrac{H - z_d}{z_0}$

(2) $J = 6.25 P^2$

(3) $\mathscr{B} = \dfrac{150 P^2}{1.06 + 0.4 \dfrac{b}{H} + 0.25 \dfrac{b}{H} - 0.23 \dfrac{b}{H}\dfrac{h}{H}}$

(4) $\tilde{f}_1 = \dfrac{n_1 H}{u_*}$

(5) $C(x) = \dfrac{1}{x} - \dfrac{1}{2x^2}(1 - e^{-2x})$

(6) $x_1 = 6.16 \dfrac{\tilde{f}_1}{P} \dfrac{d}{H}$

(7) $N(\tilde{f}_1) = C(x_1)$

(8) $C_{Df}^2(\tilde{f}_1) = C_w^2 + 2 C_w C_l N(\tilde{f}_1) + C_l^2$

(9) $x_2 = 2.56 \dfrac{\tilde{f}_1}{P} \dfrac{b}{H}$

(10) $x_3 = 1.60 \dfrac{\tilde{f}_1}{P} \dfrac{h}{H}$

(11) $\mathscr{R} = 9.40 \dfrac{P^2}{\zeta_1}\left(\dfrac{P}{\tilde{f}_1}\right)^{2/3} \dfrac{C_{Df}^2(\tilde{f}_1)}{C_D^2} C(x_2)C(x_3)$

(12) $M_1 = M$

(13) $q_* = \frac{1}{2}\rho u_*^2$

(14) $\bar{x} = \dfrac{C_D bh q_*}{M_1 (2\pi n_1)^2} J$

(15) $\sigma_x = \dfrac{C_D bh q_*}{M_1 (2\pi n_1)^2}\left(\dfrac{\beta \mathscr{B}}{6} + \mathscr{R}\right)^{1/2}$

(16) $v_x = n_1 \left(\dfrac{\mathscr{R}}{\dfrac{\beta \mathscr{B}}{6} + \mathscr{R}}\right)^{1/2}$

(17) $K_x = [1.175 + 2\ln(v_x T)]^{1/2}$

(18) $G = 1 + K_x \dfrac{\sigma_x}{\bar{x}}$

(19) $X_{\max} = G \bar{x}$

(20) $\sigma_{\ddot{x}} = \dfrac{C_D bh q_*}{M_1} \mathscr{R}^{1/2}$

(21) $K_{\ddot{x}} = [1.175 + 2\ln(n_1 T)]^{1/2}$

(22) $\ddot{X}_{\max} = K_{\ddot{x}} \sigma_{\ddot{x}}$

of windward and leeward face of building, respectively, $M =$ total mass of structure with dimensions b, h, and d in Fig. 9.1.2, $z =$ height above ground, $M(z) =$ mass of building per unit height, $\rho_b(z) =$ bulk mass of building per unit volume, $\zeta_1 =$ damping ratio, $\rho =$ mass of air per unit volume, $\beta =$ coefficient given in Table 9.1.4, $T =$ duration of storm ($T \simeq 3600$ sec), $\bar{x} =$ mean displacement at top of structure, $G =$ gust response factor, $X_{\max} =$ peak displacement at top of structure, $\sigma_{\ddot{x}} =$ rms acceleration at top of structure, and $X_{\max} =$ peak acceleration at top of structure. Tables 9.1.5 and 9.1.6 are in principle applicable only if $n_1 h / U(h) \geqslant 0.1$ and $n_1 H / U(H) \geqslant 0.1$, respectively, which is the case for most structures. In practice, they may be used even if these conditions are not met, in which case the results obtained will be slightly conservative.

Numerical Example

Consider a building with $h = 200$ m; $b = 35$ m; $d = 35$ m; $n_1 = 0.175$ Hz; $\zeta_1 = 0.01$; $\rho_b = 200$ kg/m³; $C_w = 0.8$; $C_l = 0.5$; and $C_D = 1.3$. The building is located in a town† ($z_0 \simeq 1$ m, see Table 9.1.1). It is assumed $\rho = 1.25$ kg/m³, and the fastest-mile wind speed at 10 m above ground in open terrain ($z_0 = 0.07$ m, see Table 9.1.1) is $U_f(10) = 78$ mph.

From Eq. 9.1.9, the averaging time for the fastest-mile wind speed is $t \simeq 46$ sec, and from Table 9.1.2 the ratio $U^{46}/U \simeq 1.25$, that is, the hourly wind speed at 10 m above ground in open terrain is $U_1(10) \simeq (78/1.25)$ mph $\simeq 27.8$ m/sec and $u_{*1} = 27.8/[2.5 \ln(10/0.07)] = 2.24$ m/sec (Eq. 9.1.8). From Table 9.1.3, $u_*/u_{*1} = 1.33$, and $u_* = 2.98$ m/sec. Then, referring to the equations of Table 9.1.5, $Q = 9.60$ (Eq. 1); $J = 71.83$ (Eq. 2); $\mathcal{B} = 591$ (Eq. 3); $\tilde{f}_1 = 11.74$ (Eq. 4); $x_1 = 2.63$ (Eq. 6); $N(\tilde{f}_1) = 0.31$ (Eqs. 7 and 5); $C_{Df}^2(\tilde{f}_1) = 1.14$ (Eq. 8); $x_2 = 4.34$ (Eq. 9); $M(z) = 245,000$ kg (Eq. 10); $\mathcal{R} = 353$ (Eq. 11); $M_1 = 16,333,300$ kg (Eq. 12); $q_* = 5.55$ kg/m/sec² (Eq. 13); $\bar{x} = 0.184$ m (Eq. 14); $\sigma_x = 0.074$ m (Eq. 15); $v_x = 0.114$ sec⁻¹ (Eq. 16); $K_x = 3.63$ (Eq. 17); $G = 2.46$ (Eq. 18); $X_{max} = 0.452$ m (Eq. 19); $\sigma_{\ddot{x}} = 0.058$ m/sec² (Eq. 20); $K_{\ddot{x}} = 3.75$ (Eq. 21); and $\ddot{X}_{max} = 0.218$ m/sec² (Eq. 22).

9.1.3 Computer Programs for Estimating Along-Wind Response

The procedure of Table 9.1.5 is applicable to structures for which the response to wind is dominated by the fundamental mode and for which the fundamental modal shape may be represented approximately by a straight line. Also, the procedure is based upon assumptions with regard to the structure of the atmospheric flow and the relation between wind pressures and wind speeds that were defined in Sect. 9.1.2.

For certain structures the assumption that the fundamental mode of vibration is a straight line or that the contribution to the response of the higher modes can be neglected may not be realistic. Also, it may be of interest in certain situations to employ micrometeorological and aerodynamic models different from those incorporated in the procedure of Table 9.1.5. In such cases, in lieu of that procedure, a computer program must be used to estimate the along-wind response. The computation of the response amounts essentially to the evaluation of the integrals in Eqs. 5.3.1, 5.3.2, and 5.3.7 through 5.3.15. Computer programs have been developed in which suitable numerical integration schemes are used and in which the specified structural, micrometeorological, and aerodynamic information is incorporated as input or in specialized subroutines. A computer program developed by the National Bureau of Standards is available on tape in [9-14].

9.1.4 Approximations and Errors in Estimation of the Along-Wind Response

In this section, estimates based on numerical calculations are presented of errors associated with uncertainties regarding certain features and parameter values of the models employed. The calculations were carried out for three typical buildings selected as case studies and described in Table 9.1.7. The

---

†It is assumed that the terrain roughness is homogeneous over a distance upwind of at least 12.5h (see Sect. 2.4.1).

TABLE 9.1.7. Description of Buildings Selected as Case Studies

Building	H (m)	B	D	n_1 (Hz)	ζ_1	ρ_b kg/m³
1	365	60	45	0.10	0.01	150
2	150	60	45	0.20	0.01	150
3	45	45	45	1.00	0.01	150

wind speed at 10 m above ground in open terrain ($z_0 = 0.07$ m) was assumed to be $U_f = 75$ mph, where U_f is the fastest-mile of wind.

Contribution of the Higher Vibration Modes to the Response. The root mean square of the fluctuating deflections and accelerations were calculated for buildings 1 and 2 in open and town exposure. The assumed modal shapes in the first three modes are similar to those represented in Fig. 5.2.1. The damping ratios were assumed to be $\zeta_1 = \zeta_2 = \zeta_3 = 0.01$. Calculations were carried out separately for the cases $n_2/n_1 = 1.2$, $n_3/n_1 = 1.5$ and $n_2/n_1 = 2.5$, $n_3/n_1 = 5$. The contributions of the higher (i.e., of the second and third) modes of vibration to the response are listed in Table 9.1.8. The contribution of the cross-mode product was also included in Table 9.1.8. This contribution represented about half of the amounts shown in columns (1) and (5) and was altogether negligible in all other cases.

Influence upon Calculated Response of the Deviation from a Straight Line of Fundamental Modal Shape. A convenient means for estimating the influence upon response of the fundamental modal shape is provided by the expression

$$\frac{\sigma_x}{\bar{x}} = \frac{1 + \gamma + 2\alpha}{1 + \gamma + \alpha} Q \tag{9.1.10}$$

derived by Vickery [9-13] on the basis of the assumptions that the power law (Eq. 2.2.26) holds and that the fundamental modal shape is described as follows:

$$x_1(z) = \left(\frac{z}{H}\right)^{\gamma} \tag{9.1.11}$$

where γ is a constant. In Eq. 9.1.10, σ_x is the rms of the fluctuating deflections, $\bar{x}$ is the mean deflection, Q is a function of geometrical, mechanical, and environmental parameters, independent of γ. It may be assumed, roughly, that α can vary between 0.10 for open exposure and 0.40 for centers of large cities. It follows then from Eq. 9.1.10 that for $\alpha = 0.10$ the calculated ratios $\sigma_x/\bar{x}$ calculated assuming $\gamma = 0.5$ and $\gamma = 1.5$ differ by about 1% from that calculated assuming $\gamma = 1$ (i.e., a linear fundamental modal shape). For $\alpha = 0.4$, the corresponding differences are about 3%. It is thus seen that moderate deviations

TABLE 9.1.8. Percent Contribution of Higher Modes of Vibration to Root Mean Square of Fluctuating Response

Building	Open Exposure				Town			
	$n_2/n_1 = 1.2$; $n_3/n_1 = 1.5$		$n_2/n_1 = 2.5$; $n_3/n_1 = 5$		$n_2/n_1 = 1.2$; $n_3/n_1 = 1.5$		$n_2/n_1 = 2.5$; $n_3/n_1 = 5$	
	Defl. (1)	Accel. (2)	Defl. (3)	Accel. (4)	Defl. (5)	Accel. (6)	Defl. (7)	Accel. (8)
1	5	14	0.1	10	7	8	0.1	7
2	8	20	0.1	10	14	9	2	8

from a straight line of the fundamental modal shape have an insignificant effect upon the calculated ratio $\sigma_x/\bar{x}$.

Influence upon Calculated Response of Errors in the Estimation of the Roughness Length. To estimate the magnitude of the error associated with uncertainties regarding the actual value of the roughness length, the response of buildings 1, 2, and 3 was calculated for coastal, open, suburban, center of town, and center of large city exposures. The zero plane displacement was in all cases assumed to be zero. The calculations showed that the sensitivity of the results to even large errors in the estimation of the roughness lengths (50 to 100%, say) is tolerably small (5 to 10% or less).

Spectra in the Lower Frequency Range and Along-Wind Response. It was shown in Sect. 2.3.3 that no universal relation exists describing the shape of the spectral curve in the lower frequency range, and that this shape appears to vary strongly between sites and between atmosphere and laboratory. To estimate the effect of this variation, the response of buildings 1, 2, and 3 (see Table 9.1.7) was calculated for open terrain and town exposures, using an expression for the lower frequency portion of the spectrum of the longitudinal velocity fluctuations that depends upon a parameter f_m, as in Eqs. 2.3.25. Ratios $[X_{max}]_{f_m}/[X_{max}]_{0.03}$ of the peak response calculated by assuming various values f_m to the peak response based on the value $f_m = 0.03$ are listed in Table 9.1.9.

The results of Table 9.1.9 show that the dependence of the peak response on the shape of the longitudinal spectrum in the low frequency range is relatively small, particularly for taller buildings.

It is also noted that as indicated by Eq. 5.3.41 the influence of the spectral curve shape in the lower frequency range upon the value of the accelerations is negligible.

TABLE 9.1.9.   Ratios $[X_{max}]_{f_m}/[X_{max}]_{0.03}$

Exposure	Building 1		Building 2		Building 3	
	Open	Town	Open	Town	Open	Town
$f_m = 0.01$	1.00	1.00	1.00	1.00	1.00	1.00
$f_m = 0.10$	0.98	0.97	0.97	0.94	0.95	0.91
$f_m = 0.19$	0.97	0.96	0.96	0.93	0.93	0.87

Across-Wind Correlation of the Pressures and Along-Wind Response. It was noted in Sects. 2.3 and 9.1.3 that uncertainties subsist with regard to the actual values in the atmosphere of the exponential decay coefficients C_y and C_z. It is therefore of interest to estimate the errors in the calculated along-wind

TABLE 9.1.10. Ratios of Response for Various Values C_z and C_y to Response Calculated Using $C_z = 10$ and $C_y = 16$

Case	C_z	C_y	Interval	Building 1 Open r_i	Building 1 Open $\tilde{r}_i$	Building 1 Town r_i	Building 1 Town $\tilde{r}_i$	Building 2 Open r_i	Building 2 Open $\tilde{r}_i$	Building 2 Town r_i	Building 2 Town $\tilde{r}_i$	Building 3 Open r_i	Building 3 Open $\tilde{r}_i$	Building 3 Town r_i	Building 3 Town $\tilde{r}_i$
1	10	16	$0 < n < \infty$	1	1	1	1	1	1	1	1	1	1	1	1
2	6.3	10	$0 < n < 0.9n_1$	1.01	1	1.01	1	1.01	1	1.01	1	1.01	1	1.01	1
	10	16	$0.9n_1 < n < \infty$												
3	4	6.4	$0 < n < 0.9n_1$	1.01	1	1.02	1	1.03	1	1.02	1	1.02	1	1.02	1
	10	16	$0.9n_1 < n < \infty$												
4	6.3	10	$0 < n < \infty$	1.05	1.2	1.06	1.3	1.10	1.4	1.05	1.4	1.03	1.7	1.04	1.8
5	4	6.4	$0 < n < 0.9n_1$	1.06	1.2	1.07	1.3	1.13	1.4	1.08	1.4	1.06	1.4	1.05	1.8
	6.3	10	$0.9n_1 < n < \infty$												
6	4	6.4	$0 < n < \infty$	1.12	1.5	1.11	1.5	1.17	1.3	1.18	1.8	1.08	2.1	1.10	2.6

response that correspond to possible errors in the values of these parameters. The along-wind response of buildings 1, 2, and 3 in open and town exposures was therefore calculated separately for $C_z = 10$, $C_y = 16$ (case 1 of Table 9.1.10), for $C_z = 4$, $C_y = 6.4$ (case 6), and for four intermediate cases in which C_z, C_y were assumed either constant throughout the frequency range (case 4) or to have lower values at low frequencies and higher values near and beyond the fundamental frequence n_1 (cases 2, 3, and 5). The ratios $r_i = [X_{max}]_i / [X_{max}]_1$ and $\tilde{r}_i = [\ddot{x}_{max}]_i / [\ddot{x}_{max}]_1$ in which cases 1 and i are denoted by the indices 1 and i, respectively ($i = 1, 2, 3, 4, 5, 6$), are listed in Table 9.1.10.

It can be seen from Table 9.1.10 that changes in the values of C_y and C_z in the lower frequency range have little effect on the response (cases 1, 2, and 3). If for frequencies near and beyond the fundamental frequency the values of these parameters are $C_z = 6.3$, $C_y = 10$ (cases 4 and 5), the total response is approximately 5% to 10% higher than if $C_z = 10$, $C_y = 16$ (cases 1, 2, and 3); however, the accelerations increase in the case of the taller buildings by 20% to 40%. If $C_z = 4$, $C_y = 6.4$—a situation that may be encountered in moderate winds such as occur during full-scale measurements of tall building response—then the total response is about 10% to 20% higher than in the case $C_z = 10$, $C_y = 16$, while the accelerations of the taller buildings are higher by 30% to 80%. The significant dependence of the exponential decay coefficients upon wind speed reflected in Figs. 2.3.5 and 2.3.6, and the sensitivity of the along-wind accelerations to variations in the values of these coefficients suggest that caution is in order in the interpretation of full-scale building acceleration measurements and the extrapolation of results based on such measurements to design situations.

9.2   ACROSS-WIND RESPONSE

Tall buildings are bluff (as opposed to streamlined) bodies that cause the flow to undergo separation, rather than follow the body contour. Depending upon conditions discussed for certain classical cases in Chapter 4, the wake flow thus created behind the building exhibits various degrees of periodicity, ranging from virtually periodic with a single frequency to fully turbulent. In each of these cases, at any given instant, the wake flow is asymmetrical (see for example, Fig. 4.4.3). The across-wind response is due principally to this asymmetry, although the lateral turbulent fluctuations in the oncoming flow may also contribute to the across-wind forces.

Expressions based on first principles for estimating the across-wind response of tall buildings do not currently exist. However, empirical information obtained from wind tunnel measurements is available concerning the across-wind response of tall buildings not subjected to interference effects, and expressions based on such information have been proposed in the literature. Different expressions are applicable according to whether or not the rms value of the across-wind oscillations at the tip of the building, σ_y, exceeds a critical value σ_{ycr}. If $\sigma_y > \sigma_{ycr}$, lock-in effects become significant, and the across-wind loads and oscillations increase as the wind speeds increase. Structures should be designed so that lock-in effects do not occur during their anticipated life.

For square tall buildings, experiments reported in [9-20, p. 81] and [9-24] suggest that it is conservative to assume:

$$\sigma_{ycr}/b \simeq 0.015 \text{ (open terrain, } z_0 \simeq 0.07 \text{ m)}$$

$$\sigma_{ycr}/b \simeq 0.025 \text{ (suburban terrain, } z_0 \simeq 1 \text{ m)}$$

$$\sigma_{ycr}/b \simeq 0.045 \text{ (city center, } z_0 \simeq 2.5 \text{ m)}$$

where $b =$ horizontal across-wind dimension of building. It is emphasized that these ratios are largely tentative.

Structures for which $\sigma_y < \sigma_{ycr}$. Several expressions for estimating σ_y are available in the literature. In all these expressions, the wind is assumed to blow from the most unfavorable directions (in the case of a square building, normal to a building face). Vickery [9-25] proposed the expression:

$$\frac{g_y \sigma_y(h)}{\sqrt{A}} = C \left[\frac{U(h)}{n_1 \sqrt{A}} \right]^n \frac{1}{\zeta_1^{1/2}} \frac{\rho}{\rho_b} \tag{9.2.1}$$

where $\sigma_y(h) =$ rms of across-wind oscillations at top of structure, $g_y =$ peak factor expressing the ratio of the peak response to rms response ($g_y \simeq 4.0$), $h =$ height of building, $A =$ cross-sectional area of building, $U(h) =$ mean wind speed at the top of the structure, $n_1 =$ fundamental frequency of vibration, $\zeta_1 =$ damping ratio, $\rho =$ air density, $\rho_b =$ bulk mass of building per unit volume, n and $C =$ constants determined empirically from wind tunnel measurements ($n = 3.5$, $C = 0.0006 \pm 0.00025$). The rms of the accelerations at the top of the structure, $\sigma_{\ddot{y}}(h)$, can be estimated by using Eq. 9.2.1 and the relation:

$$\sigma_{\ddot{y}}(h) = (2\pi n_1)^2 \sigma_y(h) \tag{9.2.2}$$

Equation 9.2.1 is based upon measurements of the response of building models with a linear fundamental modal shape and with geometric shapes, slenderness ratios $\sqrt{A}/h$, densities, and damping ratios shown in Fig. 9.2.1. It is noted in [9-25] that the use of Eq. 9.2.1 should be restricted to buildings with characteristics that do not differ drastically from those shown in Fig. 9.2.1.

The Supplement No. 4 to the National Building Code of Canada [9-12] proposes an expression that may be written in the form:

$$\sigma_{\ddot{y}}(h) \simeq n_1^2 (bd)^{1/2} \frac{1}{\zeta_1^{1/2}} \frac{\rho}{\rho_b} 0.0006 \left[\frac{U(h)}{n_1 (bd)^{1/2}} \right]^{3.3} \tag{9.2.3}$$

where $b =$ across-wind and $d =$ along-wind dimension of the structure. It can be seen that Eq. 9.2.3 is similar to Eq. 9.2.2 (where σ_y is given by Eq. 9.2.1) except that the exponent $n = 3.5$ is replaced by $n = 3.3$, and the coefficient $C = 0.0006 \pm 0.00025$ is replaced by $C \simeq 0.0006$. Equation 9.2.3 is based on measurements on models similar to those described in connection with Eq. 9.2.1.

Expressions in which measured modal force spectra are used follow from

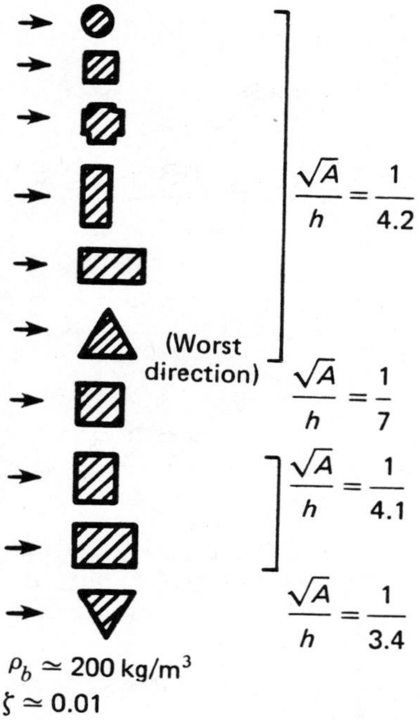

$$\frac{\sqrt{A}}{h}=\frac{1}{4.2}$$

(Worst direction)

$$\frac{\sqrt{A}}{h}=\frac{1}{7}$$

$$\frac{\sqrt{A}}{h}=\frac{1}{4.1}$$

$$\frac{\sqrt{A}}{h}=\frac{1}{3.4}$$

$\rho_b \simeq 200 \text{ kg/m}^3$

$\zeta \simeq 0.01$

FIGURE 9.2.1. Characteristics of models tested in the wind tunnel [9-25].

Eq. 5.3.32. If $S_F(n)$ = across-wind modal force, and the notation

$$\tilde{Y}^2\left[\frac{n_1 b}{U(h)}\right]=\frac{n_1 S_F[n_1 b/U(h)]}{[\frac{1}{2}\rho U^2(h)bh]^2} \tag{9.2.4}$$

is used, Eq. 5.3.32 becomes:

$$\sigma_y(h) \simeq \frac{\pi^{1/2}}{2\zeta_1^{1/2}}\frac{1}{(2\pi n_1)^2 M_1}\frac{1}{2}\rho U^2(h)bh\,\tilde{Y} \tag{9.2.5}$$

If it is assumed that the mass is uniformly distributed over the building height, that the building has a square shape in plan, and that the fundamental modal shape is linear, then

$$M_1 = \tfrac{1}{3}\rho_b b^2 h \tag{9.2.6}$$

(Eq. 5.2.6), and

$$\sigma_y(h) \simeq 0.0337\left[\frac{U(h)}{n_1 b}\right]^2 \frac{\rho}{\rho_b}\frac{1}{\zeta_1^{1/2}}\,b\tilde{Y} \tag{9.2.7}$$

Note that the quantities $\tilde{Y}$ implicit in Eqs. 9.2.1 and 9.2.3 are, respectively,

$$\tilde{Y} \simeq (4.45 \pm 1.80)10^{-3}\left[\frac{U(h)}{n_1 b}\right]^{1.5} \tag{9.2.8}$$

TABLE 9.2.1. Values $10^2\tilde{Y}$ [9-65]

Terrain	b/h	$n_1b/U(h)$							
		0.04	0.05	0.06	0.07	0.08	0.09	0.10	0.105
Urban[a]	1/9	4.7	5.7	7.2	9.7	12.0	17.0	23.0	24.0
	1/8.33	5.3[b] 5.3[d]	6.3[b] 6.5[d]	7.8[b] 8.1[d]	11.0[b] 11.0[d]	13.0[b] 13.0[d]	17.0[b] 14.0[d]	23.0[d] 12.0[d]	23.0[b] 10.0[d]
	1/6	8.4	9.7	11.0	13.0	15.0	19.0	21.0	21.0
	1/4	6.2[b] 7.9[c]	7.6[b] 9.9[c]	9.3[b]	11.0[b] 16.0[c]	12.0[b] 17.0[c]	17.0[b] 17.0[c]	15.0[b] 12.0[c]	15.0[b] 11.0[c]
	1/3	3.9	5.5	7.1	8.2	9.1	14.0	10.0	10.0
Suburban[a]	1/18	3.6	4.1	5.0	5.7	8.4	11.0	22.0	29.0
	1/9	5.1	6.0	7.5	10.0	13.0	19.0	28.0	29.0
	1/6	5.3	6.9	8.9	12.0	16.0	21.0	23.0	23.0
	1/4	5.0[b]	6.2[b]	8.0[b]	11.0[b]	15.0[b]	18.0[b]	19.0[b]	18.0[b]
	1/3	4.6	5.4	7.1	10.0	13.0	15.0	15.0	13.0
Open	1/4	6.2[c]	7.1[c]	8.7[c]	13.0[c]	17.0[c]	28.0[c]	29.0[c]	28.0[c]
All terrains[e]	1/3.4–1/7.1	56.0	39.0	30.0	24.0	20.0	16.0	14.0	13.0
All terrains[f]	1/3.4–1/7.1	29.0	22.0	17.0	14.0	12.0	10.0	8.9	8.3

$$n_1 b / U(h)$$

Terrain	b/h	0.115	0.125	0.14	0.16	0.18	0.20	0.25	0.30
Urban[a]	1/9	19.0	13.0	9.0	6.8	5.7	5.2	4.5	3.9
	1/8.33	19.0[b] 7.7[d]	14.0[b] 6.3[d]	10.0[b] 5.3[d]	6.9[b] 4.7[d]	6.3[b] 3.9[d]	5.7[b] 3.6[d]	4.7[b] 3.2[b]	3.8[b] 2.9[d]
	1/6	20.0	18.0	15.0	12.0	10.0	8.2	5.6	3.5
	1/4	15.0[b] 9.5[c]	13.0[b] 8.8[c]	12.0[b] 7.8[c]	9.6[b] 6.8[c]	13.0[b] 5.9[c]	6.8[b] 5.7[c]	4.9[b] 4.9[c]	3.4[b] 4.0[c]
	1/3	9.5	9.0	8.4	7.2	6.5	5.5	4.2	3.3
Suburban[a]	1/18	22.0	10.0	6.0	3.6	2.8	2.4	2.0	1.7
	1/9	13.0	7.8	6.0	4.5	3.5	2.8	2.6	2.3
	1/6	20.0	14.0	10.0	7.7	6.1	4.5	2.8	1.7
	1/4	16.0[b]	12.0[b]	9.5[b]	7.4[b]	5.9[b]	4.7[b]	3.7[b]	2.1[b]
	1/3	12.0	10.0	8.9	7.1	5.7	4.9	3.5	2.6
Open	1/4	11.0[c]	8.7[c]	7.5[c]	6.6[c]	5.9[c]	5.5[c]	4.8[c]	4.0[c]
All terrains[e]	1/3.4–1/7.1	11.0	10.0	8.5	7.0	5.8	5.0	3.6	2.7
All terrains[f]	1/3.4–1/7.1	7.1	6.6	5.9	4.9	4.1	3.6	2.8	2.1

[a]Results reported in [9-24] and [9-27], unless otherwise noted.
[b]Values obtained by interpolation between results reported in [9-27] for $b/h = 1/6$ and $b/h = 1/3$, and $b/h = 1/9$ and $b/h = 1/6$.
[c]Results reported in [9-28].
[d]Results reported in [9-26].
[e]Equation 9.2.8 (or Eq. 9.2.1 with $C = 0.0006$). Experimental values may differ from tabulated values by $\pm 40\%$. Application restricted to buildings with density of the order of 200 kg/m^3 and damping ratio of the order of 1%.
[f]Equation 9.2.9.

315

and

$$\tilde{Y} \simeq 4.45 \times 10^{-3} \left[\frac{U(h)}{n_1 b} \right]^{1.3} \tag{9.2.9}$$

Values of $\tilde{Y}$ based on measurements reported for square building models in [9-24], [9-26], [9-27], and [9-28] are listed in Table 9.2.1. Also included in Table 9.2.1 are values of $\tilde{Y}$ given by Eqs. 9.2.8 and 9.2.9.

Table 9.2.1 shows that, for any given $n_1 b/U(h)$, $\tilde{Y}$ is a function of terrain exposure and the slenderness ratio b/h. For example, the peak values of $\tilde{Y}$ for urban terrain appear to increase by a factor of approximately two if b/h decreases from 1/3 to 1/9. Also, according to data from [9-24], [9-27], and [9-28], the peak values of $\tilde{Y}$ increase as the terrain becomes smoother. This is the case

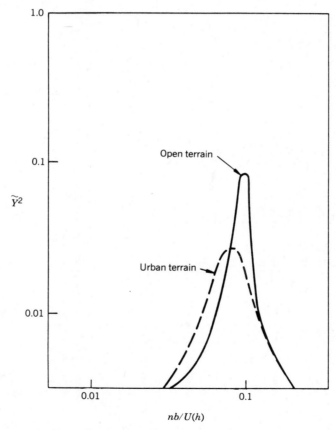

FIGURE 9.2.2. Shapes of $\tilde{Y}^2$ curve in open and urban terrains. After A. Kareem, "Across-Wind Response of Buildings," *Journal of the Structural Division*, ASCE, **108** (1982), 869–887.

because in rougher terrain the turbulence intensity is higher, which in turn causes the across-wind force to have a less peaked spectral density (Fig. 9.2.2), as well as a decreased coherence in the spanwise direction. Note that there are significant discrepancies among values of $\tilde{Y}$ obtained by various researchers. For example, for urban terrain, $n_1 b / U(h) = 0.105$, and $b/h = 1/8.33$, $\tilde{Y} = 0.10$ according to [9-26], versus $\tilde{Y} = 0.23$ on the basis of data from [9-27]; for urban terrain, $n_1 b / U(h) = 0.105$ and $b/h = 1/4$, $\tilde{Y} = 0.11$ according to [9-28], versus $\tilde{Y} = 0.15$ on the basis of data from [9-27]. Differences between values of $\tilde{Y}$ given by Eqs. 9.2.8 and 9.2.9 and those from [9-24], [9-26], [9-27], and [9-28] are also relatively large in several instances.

Numerical studies [9-29] show that, as in the case of along-wind response, the contribution to the total building deflections and accelerations of modes higher than the fundamental mode is negligible in practice, unless the ratios of natural frequencies in the higher modes to the fundamental frequency are close to unity, that is, substantially lower than those occurring in typical high rise buildings.

For a square building model with height to width ratios $h/b = 8.33$ located in urban terrain, it was found in [9-26] that the across-wind response decreases

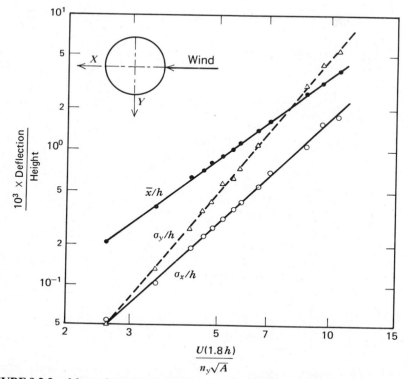

FIGURE 9.2.3.   Mean along-wind and root mean square of along-wind and across-wind deflections of a 64-story building model with a circular shape in plan [9-30].

from the maximum value that corresponds to wind normal to a building face, to about 50% of that value when the angle between the mean wind direction and the normal to a building face is about 15°. The peak across-wind response and the peak along-wind response induced by wind parallel to a diagonal of the cross section have approximately the same value, that is, they are approximately equal to 0.8 times the peak along-wind response induced by wind normal to a building face [9-26].

Numerical Example

The building considered in the numerical example of Sect. 9.1.2 is again assumed to be acted upon by wind corresponding to a fastest-mile speed at 10 m above ground in open terrain $U_f(10) = 78$ mph. The mean hourly wind speed at the top of the building is then $U(h) = 2.5u_* \ln(h/z_0)$, or $U(h) = 2.5 \times 2.98 \times \ln(200/1.00) = 39.4$ m/sec. The following results are obtained: $n_1 b/U(h) = 0.155$; $b/h = 1/5.7$; $\tilde{Y} \simeq 0.075$ (Table 9.2.1, suburban terrain); $\sigma_y = 0.23$ m (Eqs.

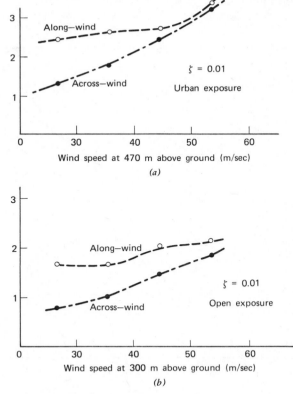

FIGURE 9.2.4. Ratios of peak along-wind and peak across-wind response to mean along-wind response for a 53-story building model with a square shape in plan in urban and open terrain [9-13].

9.2.7 or 9.2.5); $\sigma_{\ddot{y}} = 0.28$ m/sec^2 (Eq. 9.2.2). Assuming that the peak factors are $g_y \simeq 3.5$, $g_{\ddot{y}} \simeq 4.0$, it follows that the peak across-wind response and acceleration are $Y_{max} \simeq 0.805$ m, $\ddot{Y}_{max} \simeq 1.12$ m/sec^2. These values are larger than the corresponding values of the along-wind response calculated previously, that is $X_{max} \simeq 0.452$ m and $\ddot{X}_{max} \simeq 0.218$ m/sec^2.

Figure 9.2.3 shows the mean and rms along-wind response and the rms across-wind response of a 1/400 model of a 64-story building in urban terrain. The characteristics of the model were the following: $h = 0.658$ m, $\sqrt{A} = 0.154$ m (where A is the floor area), $n_x = 8.3$ Hz, $n_y = 8.49$ Hz, and $\zeta_x = \zeta_y \simeq 0.01$ (where n and ζ denote frequencies and damping ratios, respectively) [9-30]. Results of wind tunnel tests for the model of a 53-story building are shown in Fig. 9.2.4 for open and urban terrain [9-13].

9.3   TORSIONAL RESPONSE

Severe distortions due to the combined effects of across-wind loads and torsional moments occurred during the 1926 Florida hurricane in two Miami high rise structures, the 15-story Realty Building, and the 17-story Meyer-Kiser Building [9-31]. Both buildings had unusually narrow shapes in plan (the dimensions in plan of the Meyer-Kiser Building were about 14 m × 42 m). Their structural systems consisted of steel frames. The two transverse end frames of the Meyer-Kiser Building experienced horizontal deflections of about 0.60 m and −0.20 m, respectively.

Following these incidents engineers became concerned with wind-induced torsional effects, as shown by subsequent developments in the literature, including a 1939 ASCE report that dealt with such effects in some detail [9-32, 9-33]. Nevertheless, wind-induced torsion of tall buildings is not mentioned in the 1961 ASCE state-of-the-art report [9-34], or in any U.S. building code or standard. This deficiency may explain what appears to have been the absence of provisions against wind-induced torsion in the original design of the John Hancock Building in Boston, which by virtue of its shape is particularly sensitive to both across-wind and torsional effects.

Torsional effects are due to the fact that in any individual building the center of mass and/or the elastic center do not coincide with the instantaneous point of application of the aerodynamic loads. Ad hoc tests simulating these effects have been conducted for a number of years on individual building models. However, until recently, relatively little work has been performed toward the development of design information and analytical procedures for use by structural designers. A first attempt at studying analytically torsion induced on buildings by fluctuating wind loads was reported by Patrickson and Friedman [9-35]. More recently, Safak and Foutch have presented potentially useful methods for estimating the along-wind, across-wind, and torsional response of rectangular buildings [9-36, 9-37]. However, owing to the absence of sufficient information on aerodynamic loads, the methods are not presently usable for design purposes.

Wind tunnel and full-scale research studies of torsional response were first reported in [9-26] and [9-38]. Reference 9-26 includes information on wind-induced torsional moments in an isolated square building model having a height to width ratio $h/b = 8.33$ in flow that simulates urban conditions. According to the results of [9-26], torsional moments are largest when the mean wind velocity is normal to a building face. As the angle α between the mean wind velocity and the normal to the building face increases from $0°$ to $45°$, the torsional moments decrease from their maximum value corresponding to $\alpha = 0°$ to about 25% of that value for $\alpha = 45°$. Assuming that the mechanical properties of the model are similar to those of typical high rise structures, it was estimated in [9-26] that, for $\alpha = 0°$, the peak torsion-induced response of a corner column is approximately 65% of the peak along-wind response corresponding to $\alpha = 0°$. For $\alpha = 45°$, the peak torsion-induced response of a corner column is about 15% of the peak along-wind response corresponding to $\alpha = 0°$.

Systematic wind tunnel studies conducted at the University of Western Ontario were subsequently reported in [9-39] to [9-42].* These studies have led to the following empirical relation for estimating the peak base torque $T_{max}[U(h)]$ induced by winds with speed $U(h)$ at the top of the building:

$$T_{max}[U(h)] = \psi\{\bar{T}[U(h)] + g_T T_{rms}[U(h)]\} \tag{9.3.1}$$

where ψ is a reduction coefficient that is briefly discussed subsequently, $g_T \simeq 3.8$ is a torsional peak factor, and the linear and rms base torque, $\bar{T}[U(h)]$ and $T_{rms}[U(h)]$ are given by the expressions:

$$\bar{T}[U(h)] \simeq 0.038 \rho L^4 h n_T^2 U_r^2 \tag{9.3.2}$$

$$T_{rms}[U(h)] \simeq 0.00167 \frac{1}{\zeta_T^{1/2}} \rho L^4 h n_T^2 U_r^{2.68} \tag{9.3.3}$$

$$U_r = \frac{U(h)}{n_T L} \tag{9.3.4}$$

$$L = \frac{\int |r|\, ds}{A^{1/2}} \tag{9.3.5}$$

In Eqs. 9.3.2 to 9.3.5, ρ is the air density ($\rho \simeq 1.25$ kg/m^3), h is the height of the building, n_T and ζ_T are the natural frequency and the damping ratio in the fundamental torsional mode of vibration, ds is the elemental length of the building perimeter, $|r|$ is the torque arm of the element ds (i.e., the distance between the elastic center and the normal to the building boundary at the center of the element ds—see Fig. 9.3.1), and A is the cross-sectional area of the building. Equations 9.3.2 and 9.3.3 are based upon the experimental results shown in Figs. 9.3.2 and 9.3.3, in which the ordinates are the reduced mean and rms base torque, $T_r = \bar{T}/(\rho L^4 h n_T^2)$ and $\sigma_r = T_{rms}\zeta_T^{1/2}/(\rho L^4 h n_T^2)$, respectively. The torques $\bar{T}$ and T_{rms} are each induced by wind with reduced speed U_r and with

---

*The results of these studies were kindly provided to the authors by Dr. N. Isyumov.

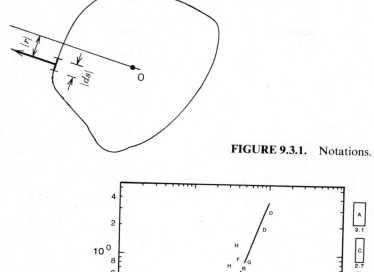

FIGURE 9.3.1. Notations.

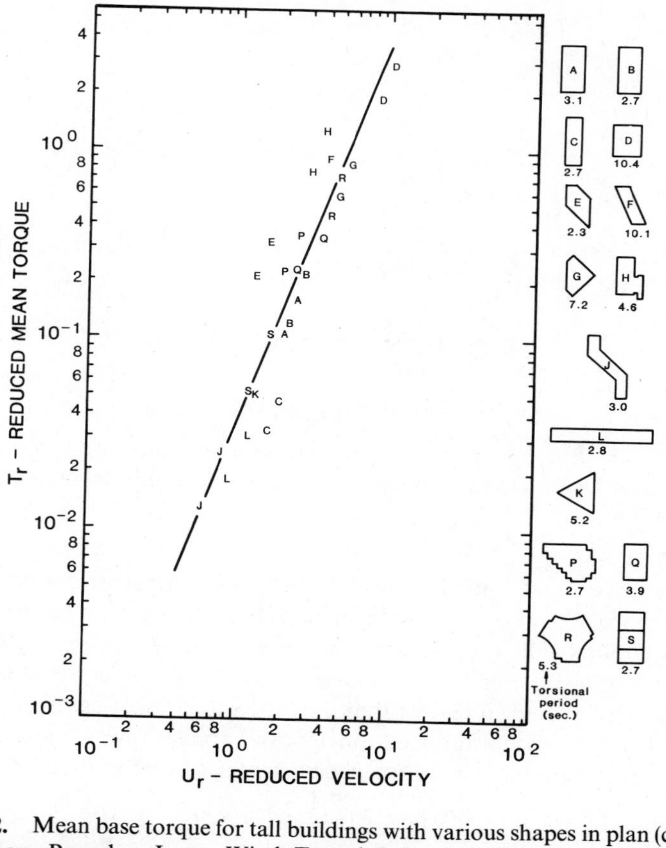

FIGURE 9.3.2. Mean base torque for tall buildings with various shapes in plan (courtesy Dr. N. Isyumov, Boundary-Layer Wind Tunnel Laboratory, University of Western Ontario).

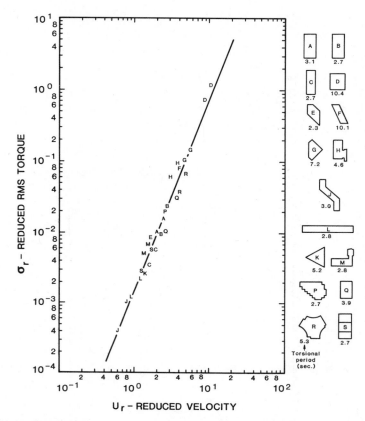

FIGURE 9.3.3. Root mean square of base torque for tall buildings with various shapes in plan (courtesy Dr. N. Isyumov, Boundary-Layer Wind Tunnel Laboratory, University of Western Ontario).

the respective most unfavorable direction. In general, the most unfavorable directions for $\bar{T}$ and T_{rms} do not coincide. In addition, in most cases neither of these directions will coincide with the direction of the extreme winds expected to occur at the site. For these reasons, the coefficient ψ in Eq. 9.3.1 is less than unity. It is estimated in [9-39] that $0.75 < \psi \leqslant 1$ in most cases.

The peak torsional-induced horizontal accelerations at the top of the building at a distance v from the elastic center can be written as

$$\ddot{\theta}v \simeq 2g_T T_{rms}v/(\rho_b bdhr_m^2) \qquad (9.3.6)$$

where $\ddot{\theta}$ is the peak angular acceleration and r_m is the radius of gyration. For a rectangular shape with uniform bulk mass per unit volume

$$r_m^2 = \frac{b^2 + d^2}{12} \qquad (9.3.7)$$

Eq. 9.3.6 was obtained in [9-39] assuming a linear fundamental modal shape and negligible contributions by higher torsional modes of vibration.

Numerical Example

For the building considered in the numerical examples of Sects. 9.1.2 and 9.2, $h = 200$ m, $b = d = 35$ m, $U(h) = 39.4$ m, $\rho_b = 200$ kg/m^3. It is assumed that the natural frequency and the damping ratio in the fundamental torsional mode of vibration are $n_T = 0.3$ Hz and $\zeta_T = 0.01$, respectively, and that the air density is $\rho = 1.25$ kg/m^3.

From Eq. 9.3.5, $L = 8(b/2)(b/4)/b = 35$ m. Then, $U_r = 39.4/(0.3 \times 35) = 3.75$ (Eq. 9.3.4), $\bar{T}[39.4] = 1.8 \times 10^7$ Nm (Eq. 9.3.2), $T_{rms}[39.4] = 1.95 \times 10^7$ Nm (Eq. 9.3.3), $T_{max} = 9.2 \times 10^7$ Nm (Eq. 9.3.1 in which it is assumed $\psi \simeq 1$, $g_T \simeq 3.8$). The peak torsion-induced horizontal acceleration at the top corner ($v = 35 \times \sqrt{2}/2 = 24.7$ m) is $\ddot{\theta}_{max} v = 0.37$ m/sec^2. Note that this exceeds the peak along-wind accelerations but is substantially less than the peak across-wind accelerations calculated in the previous numerical examples.

The peak combined effect of the along-wind, across-wind, and torsional loads can be obtained by summing up vectorially the individual peak effects of these loads and multiplying the result by a reduction factor (equal to 0.8, say) which accounts for the fact that, in general, individual peaks do not occur simultaneously. If the combined effect so calculated is less than an individual effect, it is the latter that should be considered in design.

More recently, based on aerodynamic data reported in [9-1], [9-66] presented a rigorous dynamic analysis of torsional moments which takes into account the effect of the distance between the elastic center and the center of mass of the structure.

9.4   TUNED MASS DAMPERS AND VISCOELASTIC DAMPING DEVICES

Two main types of device have been employed for the reduction of tall building vibrations in translation and/or torsion: tuned mass dampers, and viscoelastic dampers. Devices intended to reduce torsional vibrations must consist of at least two units located at sufficient distances from the elastic center of the structure.

9.4.1   Tuned Mass Dampers

The tuned mass damper (TMD) consists of a relatively small vibratory system (mass, spring, and dashpot) attached to a structure whose vibrations it is designed to mitigate. It was invented in 1909 by Frahm and has until recently been used primarily in mechanical engineering systems. In the last decade TMDs have increasingly been employed in wind-sensitive structures, including the Centrepoint Tower, Sidney, Australia [9-44], the CN Tower, Toronto [9-45], the John Hancock Tower, Boston (equipped with dual TMDs designed to control both torsional and lateral motions) [9-46, 9-47], and the Citicorp

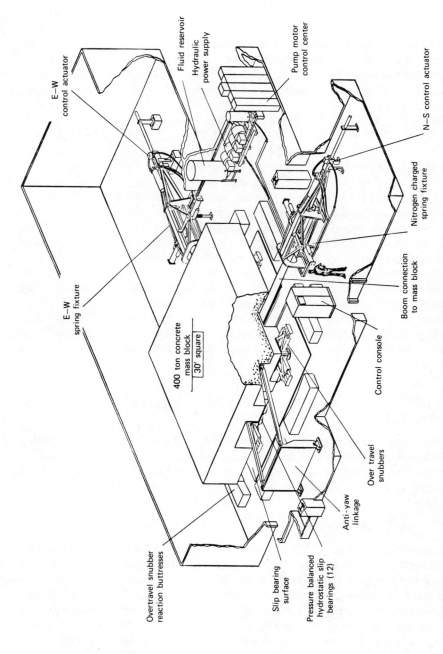

E–W
control actuator

Fluid reservoir

Hydraulic
power supply

Pump motor
control center

N–S control actuator

E–W
spring fixture

Nitrogen charged
spring fixture

400 ton concrete
mass block
30' square

Boom connection
to mass block

Control console

Overtravel snubber
reaction buttresses

Over travel
snubbers

Slip bearing
surface

Anti - yaw
linkage

Pressure balanced
hydrostatic slip
bearings (12)

FIGURE 9.4.1. MTS tuned mass damper system, Citicorp Center, New York City (courtesy MTS Systems Corp., Minneapolis, Minnesota).

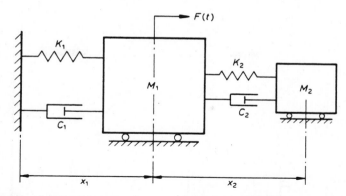

FIGURE 9.4.2.   Schematic representation of system equipped with a tuned mass damper.*

Center, New York City [9-48, 9-49, 9-50]. Generally, the purpose of the TMDs is to reduce building motions insofar as they affect occupant comfort, and the effect of the TMD is not taken into account in strength calculations [9-46, 9-47].

A schematic view of a TMD operating on the top floor of the Citicorp Center is shown in Fig. 9.4.1. The mass of the TMD consists in this case of a 400 ton concrete block bearing on a thin oil film. The TMD structural stiffness is provided by pneumatic springs which can be tuned to the actual frequency of the building as determined experimentally in the field. The TMD damping is provided by hydraulic shock absorbers. The system includes fail-safe devices to prevent excessive travel of the concrete block [9-49]. Additional information on TMD equipment and control systems is given in [9-46].

The theory of the tuned mass damper was developed by Den Hartog [9-51] for the system shown in Fig. 9.4.2, with $C_1 = 0$ and a harmonic load $F(t)$. On the basis of results given in [9-51], the theory was subsequently extended in [9-52] to include the case where $C_1 \neq 0$ and $F(t)$ is a random load with constant (white noise) spectral density. In Fig. 9.4.2, M_1, C_1, K_1, and M_2, C_2, K_2 are the mass, damping, and spring constant of the structure and of the TMD, respectively.

Effect of TMD Upon Deflections and Accelerations of Structure.   The effect of the TMD can be viewed as being equivalent to changing the damping ratio of the original system (not provided with a TMD) from the value $\zeta_1 = C_1/2\sqrt{K_1 M_1}$ to a larger value ζ_e. Thus, the deflections and accelerations of mass M_1 in the system of Fig. 9.4.2 can be obtained by calculating the deflections and accelerations of mass M_1 in the system with damping $C_e = 2\sqrt{K_1 M_1}\zeta_e$ shown in Fig. 9.4.3. Using results from [9-49] and [9-52] it can be shown that

$$\zeta_e = \frac{1}{2} \frac{\alpha_1(\alpha_2\alpha_3 - \alpha_1) - \alpha_0\alpha_3^2}{\alpha_0(\alpha_2\alpha_3 - \alpha_1) + \alpha_3(\beta_1^2 - 2\alpha_0) + \alpha_1} \qquad (9.4.1a)$$

*Figures 9.4.2 to 9.4.6 are reproduced from *Engineering Structures*, Vol. 4, No. 5, with permission of the publisher, Butterworth Scientific Ltd.

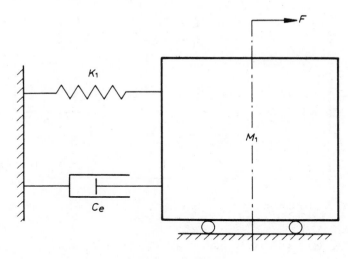

FIGURE 9.4.3. Notations.

where

$$\alpha_0 = f^2 \tag{9.4.1b}$$

$$\alpha_1 = 2f(\zeta_1 f + \zeta_2) \tag{9.4.1c}$$

$$\alpha_2 = 1 + f^2(1 + \mu) + 4f\zeta_1\zeta_2 \tag{9.4.1d}$$

$$\alpha_3 = 2\zeta_1 + 2\zeta_2 f(1 + \mu) \tag{9.4.1e}$$

$$\beta_1 = 2\zeta_2 f \tag{9.4.1f}$$

In Eqs. 9.4.1a through 9.1.4f:

$$\mu = M_2/M_1 \tag{9.4.1g}$$

$$f = \omega_2/\omega_1 \tag{9.4.1h}$$

$$\omega_i = \sqrt{K_i/M_i} \quad (i = 1, 2) \tag{9.4.1i}$$

$$\zeta_i = C_i/(2M_i\omega_i) \quad (i = 1, 2) \tag{9.4.1j}$$

For example, if $\mu = 0.01$, $f = 0.98$, $\zeta_1 = 0.01$, and $\zeta_2 = 0.0515$, then $\zeta_e = 0.03226$. The dependence of ζ_e upon ζ_2 is shown in Fig. 9.4.4 for $\zeta_1 = 0.01$, $f = 0.98$, and various values of μ [9-49]. Note that for each μ there exists an optimal (maximum) value of ζ_e (denoted by ζ_e^{opt}) which can be sought by representing Eq. 9.4.1a graphically. Alternatively, it is shown in [9-53] that, with negligible errors, the following approximate relations can be used for preliminary design purposes to obtain ζ_e^{opt} and the corresponding value of ζ_2 (denoted by ζ_2^{opt}):

$$\zeta_e^{\text{opt}} \simeq \frac{\sqrt{\mu}}{4} + 0.8\zeta_1 > \zeta_1 \tag{9.4.2}$$

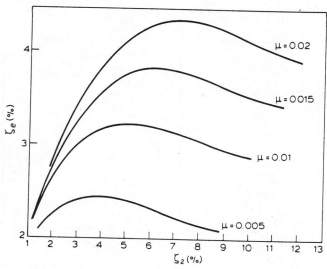

FIGURE 9.4.4.   Dependence of ζ_e upon ζ_2 and μ. After R. J. McNamara, "Tuned Mass Dampers in Buildings," *Journal of the Structural Division*, ASCE, **103** (1977), 1785–1798.

$$\zeta_2^{opt} \simeq \frac{\sqrt{\mu}}{2} \tag{9.4.3}$$

For example, if $\mu = 0.01$ and $\zeta_1 = 0.01$, then $\zeta_e^{opt} \simeq 0.033$ and $\zeta_2^{opt} \simeq 0.05$.

Displacements of TMD Mass.   In designing a TMD system, allowance must be made for the displacements (travel) of the TMD mass. These displacements are in practice relatively large. For example, in the case of Citicorp Center, TMD displacements induced by a storm with a 10-year return period were estimated from model tests to be on the order of 1.00 m.

Let the displacement of the TMD mass with respect to mass M_1 be denoted by x_2 (Fig. 9.4.2). The displacement due to resonant amplification effects only of mass M_1 in the original system (shown in Fig. 9.4.5) is denoted by $x_{1,0}$. (It is emphasized that $x_{1,0}$ does not include contributions due to mean or quasistatic loading.) Using results from [9-49] and [9-52], it can be shown that the ratio of the mean square values of x_2 and $x_{1,0}$, denoted by $\overline{x_{2norm}^2}$, is given by the relation

$$\overline{x_{2norm}^2} = \frac{\overline{x_2^2}}{\overline{x_{1,0}^2}} = \frac{2\zeta_1\alpha_1}{\alpha_1(\alpha_2\alpha_3 - \alpha_1) - \alpha_0\alpha_3^2} \tag{9.4.4}$$

For example, if $\mu = 0.01$, $f = 0.98$, $\zeta_1 = 0.01$, and $\zeta_2 = 0.0515$, then $\overline{x_{2norm}^2} = 13.7$. The dependence of $\overline{x_{2norm}^2}^{1/2}$ upon ζ_2 is shown in Fig. 9.4.6 for $\zeta_1 = 0.01$, $f = 0.98$, and various values of μ [9-49].

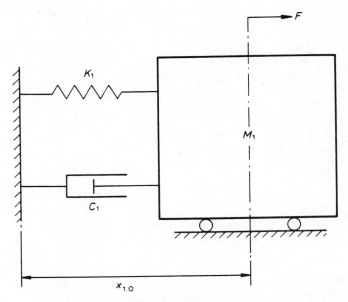

FIGURE 9.4.5. Notations.

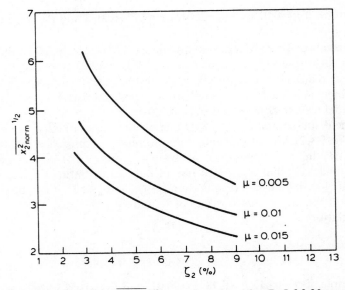

FIGURE 9.4.6. Dependence of $\overline{x^2_{2\mathrm{norm}}}^{1/2}$ upon ζ_2 and μ. After R. J. McNamara, "Tuned Mass Dampers in Buildings," *Journal of the Structural Division*, ASCE, **103** (1977), 1785–1798.

Design of TMDs for Actual Structures. Because buildings are multi-degree of freedom systems, the model shown in Fig. 9.4.2 is not a rigorous representation of a building with a TMD. The error inherent in the assumption that the building equipped with a TMD can be represented by the system of Fig. 9.4.2 (where M_1, K_1, and C_1 are equal, respectively, to the generalized mass, the stiffness, and the damping in the fundamental mode of the building not equipped with a TMD) was estimated for a particular structure in [9-53]. According to the approximate estimate of [9-53], the simplified model of Fig. 9.4.2 led in that particular case to an overestimation of the equivalent damping ratio of the structure by a factor of about 1.2.

It is noted that results reported in [9-54] on the dynamic response of light equipment attached to structures are applicable to the study of the errors inherent in the model of Fig. 9.4.2. These errors are generally negligible for structures with ratios of frequency in the second mode to frequency in the fundamental mode of the order of two or larger.

9.4.2 Viscoelastic Dampers

Viscoelastic dampers are passive devices that have the advantage of not requiring constant operational monitoring and of not depending on electric power.

The following factors need to be considered in the design of viscoelastic dampers*:

Overall damping required as a function of the specified mean recurrence interval of the wind loading (e.g., 10 years or 100 years)

Environment characteristics at damper locations (e.g., air temperature)

Space available for damper displacement, and requisite damper stiffness

Frequencies of vibration of building (translational and torsional)

The damper design includes the selection of the material properties (shear loss modulus, loss tangent, and their temperature dependence), and the size and number of dampers—see [9-67, 9-68] for details.

Buildings equipped with viscoelastic dampers include the World Trade Center, New York, and the Columbia Center Building, Seattle. Inspection of the viscoelastic dampers installed in the World Trade Center buildings appears to have indicated no perceptible aging effects after almost 15 years of service.

9.5 DESIGN OF CLADDING FOR WIND LOADS

The purpose of this section is to present a risk-consistent procedure for the design of glass cladding subjected to wind loads. The procedure is applicable to buildings with specified orientation, and requires the availability of sufficient

---

*Personal communication by Dr. P. Mahmoodi, 3M Company, St. Paul, Minnesota, 1985.

(1) wind speed data characterizing the extreme wind climate in the region of interest, and (2) aerodynamic pressure data obtained in the wind tunnel for various zones of the building facades.

The procedure presented here differs from conventional design practice in two respects. First, in conventional practice the design of each cladding panel is based on the requirement that the nominal wind load corresponding to a specified mean recurrence interval $\bar{N}$ (usually $\bar{N} = 50$ years) may not exceed a load capacity corresponding to a specified probability of failure P_f (usually $P_f = 0.008$). Second, in conventional design practice wind directionality is not taken into account. As shown in Sect. 8.1.2, this can lead to significant discrepancies between the nominal loads used in design and the actual loads. The safety level of the cladding can therefore be strongly nonuniform among the various zones of the building facades and among identical buildings having different orientations.* The purpose of the risk-consistent design procedure presented in this chapter is to eliminate or reduce such nonuniformities.

The conventional and the risk-consistent design procedures have a number of common steps. These are reviewed in Sect. 9.5.1, which also includes a description of the steps that distinguish the two procedures. Section 9.5.2 summarizes results of design applications that illustrate the economic and safety advantages inherent in the risk-consistent procedure.

9.5.1   Conventional and Risk-Consistent Procedures for Designing Cladding Glass

Procedures for conventional and risk-consistent design of cladding glass entail the following common steps:

1.   Obtaining information on the extreme wind climate.

2.   Converting basic wind speeds (i.e., fastest-mile wind speeds at 10 m above ground in open terrain) into wind speeds used for aerodynamic reference purposes (usually, mean hourly wind speeds at the top of the building).

3.   Obtaining from wind tunnel tests information on the time-dependent aerodynamic pressures acting at various points of the building facades.

4.   Converting the information on time-dependent aerodynamic pressures into equivalent wind loads with standardized time history, that is, loads whose effect upon the cladding panels is equivalent to that of the actual time-dependent loads.

5.   Estimating design wind loads using information on the wind climate and on the equivalent standardized wind loads.

6.   Obtaining information on the load capacity of the cladding panels.

---

*A measure of the cladding safety level for a zone (or building) is given by the ratio n_f/n_p between the expected number of panels that fail during the lifetime of the structure and the total number of panels for that zone (or building).

7. Adopting a design criterion relating the design wind loads to the load capacity of the panels.

8. Designing the cladding glass.

1. Extreme Wind Climate. The conventional design procedure uses information on extreme wind speeds regardless of direction. To apply the risk-consistent design procedure, the information needed to characterize the extreme wind climate in regions not subjected to hurricane winds consists of directional largest yearly fastest-mile wind speeds. Such information may be extracted from monthly Local Climatological Data summaries issued by the National Oceanic and Atmospheric Administration (see Sect. 3.4). These data are usually recorded over open terrain (airports) and should be reduced to a common elevation (usually 10 m above ground).

In hurricane-prone regions directional information on hurricane wind speeds can be obtained by Monte Carlo simulation (see Sect. 3.3.2) or from data stored in [8-9].

2. Conversion of Basic Speeds to Aerodynamic Reference Speeds. Given the basic wind speed $U_f(10, \theta)$ (i.e., the fastest-mile wind from direction θ at 10 m above ground in open terrain), the corresponding hourly mean speed, $U(h, \theta)$, at elevation h over the building site can be estimated by using Eqs. 9.1.6, 9.1.8, and 9.1.9 and the micrometerorological parameters of Tables 9.1.1, 9.1.2, and 9.1.3. For example if $U_f(10, \theta) = 78$ mph and the building has height $h = 200$ m and is located in a town with roughness length upwind of the building $z_0 \simeq 1.00$ m, $U(200, \theta) \simeq 39.4$ m/sec (see numerical examples in Sects. 9.1.2 and 9.2).

3. Aerodynamic Pressures on Building Facades. Information on aerodynamic pressures is obtained from wind tunnel tests. It is presented in terms of aerodynamic pressure coefficients defined as

$$C_p(M_j, \theta_k) = \frac{p(M_j, \theta_k)}{\frac{1}{2}\rho U^2(h, \theta_k)} \tag{9.5.1}$$

where $p(M_j, \theta_k)$ is the pressure at point M_j of the facade, induced by wind blowing from direction θ_k with a mean hourly speed at the top of the building, $U(h, \theta_k)$; ρ is the air density; and $C_p(M_j, \theta_k)$ is the pressure coefficient at point M_j corresponding to the wind direction θ_k. Pressure coefficients $C_p(M_j, \theta_k)$ are recorded as functions of time for various wind directions θ_k at various points of the building facades, including points near corners and eaves. Measurements are usually made for angles $\theta_k = k \times 15°$ ($k = 1, 2, \ldots, 24$) although occasionally the increments may be smaller than 15° to allow detection of directional maxima.

4. Equivalent 60-Sec Wind Loads. Wind pressures $p(M_j, \theta_k)$ and the corresponding pressure coefficients $C_p(M_j, \theta_k)$ are randomly fluctuating functions of time that depend upon the position M_j and the mean wind direction θ_k (e.g., see Fig. 4.7.2).

The load capacity of glass cladding panels depends upon the entire time history of the load. This dependence can in principle be taken into account by using basic fracture mechanics relations to describe the effect of fatigue caused by the fluctuating load, that is, the time-dependent growth in the size of flaws present on the surfaces and edges of the panels, and the consequent time-dependent decrease of the glass strength $S(t)$ [9-58]. The failure load is obtained from the condition that failure occurs when the tension stress $\sigma(t)$ [which is in general a nonlinear function of the load $p(t)$] is equal to the strength $S(t)$ [which is a function of the initial strength $S(0)$ and of the load $p(t)$] (Fig. 9.5.1). This approach entails Monte Carlo simulations of the initial strength from probability distributions obtained experimentally, as well as the calculation by numerical methods of the nonlinear relation between the loads $p(t)$ and the normal stresses $\sigma(M_j, \phi_l, t)$, for a sufficient number of points M_j and directions ϕ_l of the stresses. The approach is applied to panels subjected to: (1) loads with the time history $p(t)$, and (2) constant loads with a 60-sec duration, p_{60}, which are commonly used in North American design charts. Probability distributions of the load capacity of the panels are obtained for the loads $p(t)$ (indexed by their mean value $\overline{p(t)}$) and for the 60-sec load p_{60}. Let these distributions be denoted by $P_p[p(t)]$ and $P_{p_{60}}(p_{60})$, respectively. The 60-sec load p_{60}^{eq} equivalent to the load $p(t)$ is given by the relation

$$p_{60}^{eq} = P_{p_{60}}^{-1}\{P_p[\overline{p(t)}]\} \tag{9.5.2}$$

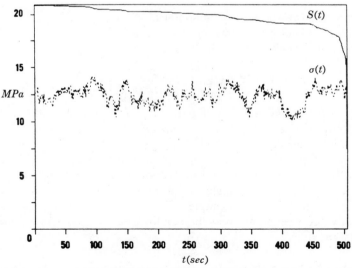

FIGURE 9.5.1.   Evolution of tension stress and strength with time at a point on the face of glass plate. Failure occurs at time $t \cong 506$ sec [9-58].

Thus, for any point M_j and wind direction θ_k, an equivalent aerodynamic coefficient can be defined as

$$C_{p_{60}}^{eq}(M_j, \theta_k) = \frac{p_{60}^{eq}(M_j, \theta_k)}{\frac{1}{2}\rho U^2(h, \theta_k)} \qquad (9.5.3)$$

The approach just described has so far been used only in exploratory investigations [9-58]. Currently, a simpler approach is used for design purposes, in which it is assumed that the actual fluctuating load causing failure is equivalent to a constant load with a small duration, t_{pk}, and a magnitude equal to the peak fluctuating load averaged over the time t_{pk}, p_{pk}. It is commonly assumed that $t_{pk} \simeq 1$ sec.*

The 1-sec constant load p_{pk} must in turn be converted into an equivalent 60-sec load p_{60}^{eq}. It can be shown from basic fracture mechanics relations that the stresses σ_{60} and σ_{pk} induced by the 60-sec load p_{60}^{eq} and the 1-sec load p_{pk}, respectively, are equivalent from the point of view of their effect on glass if

$$\sigma_{60}^n \times 60 = \sigma_{pk}^n \times 1 \qquad (9.5.4)$$

where n is the exponent in the phenomenological relation describing subcritical crack growth. For soda lime glass, it may be assumed for practical purposes that $n \simeq 16$ [9-59]. From Eq. 9.5.4 and the simplifying assumption that the load-stress relationship is linear it then follows that

$$p_{60}^{eq} \simeq 0.78 p_{pk} \qquad (9.5.5)$$

(e.g., see [9-60]). Thus, in this simplified approach, the equivalent 60-sec aerodynamic coefficient for any point M_j and wind direction θ_j has the expression

$$C_{p_{60}}^{eq}(M_j, \theta_k) \simeq \frac{0.78 p_{pk}(M_j, \theta_k)}{\frac{1}{2}\rho U^2(h, \theta_k)} \qquad (9.5.6)$$

5. Estimation of Design Wind Loads. From Eq. 9.5.3 if follows that the equivalent 60-sec loads p_{60}^{eq} are given by the relation

$$p_{60}^{eq}(M_j, \theta_k) = \frac{1}{2}\rho C_{p_{60}}^{eq}(M_j, \theta_k) U^2(h, \theta_k) \qquad (9.5.7)$$

which has the same form as Eq. 8.1.8. The design wind loads can therefore be estimated as shown in Sect. 8.1.2, in which $p_{60}^{eq}(M_j, \theta_k)$ and $C_{p_{60}}^{eq}(M_j, \theta_k)$ should be substituted for $Q(\theta)$ and $C(\theta)C_p(\theta)$, respectively.

The estimation of design wind loads differs according to whether the conventional or the risk-consistent design procedure is used. In *conventional design*

---

*The peak value p_{pk} depends upon the record length (or storm duration) T. Commonly it is assumed $T \simeq 20$ min to 1 hr (full-scale), to which there corresponds a laboratory record length $T_m = T(D_m/D)/(U_m/U)$, where D_m/D and U_m/U are the model geometric and velocity scale, respectively. For structural reliability calculations it is desirable to estimate the mean and standard deviation of the peak pressure p_{pk}, since, for any given value $\overline{p(t)}$, p_{pk} varies from record to record. This can be done from several records with length T_m, or by using techniques based on random processes theory (see Eq. A2.44, [4-75], [9-63], and especially [9-64], which contains useful practical results).

practice equivalent 60-sec loads with a 50-year mean recurrence interval, $p_{60,50}^{eq}(M_j)$ are estimated without considering the effects of wind directionality. For this reason the *actual* mean recurrence interval, $\bar{N}_{act}$, of the design load $p_{60,50}^{eq}$ varies from panel to panel. In the case of panels for which the direction of the most severe extreme winds coincides with the direction of the largest aerodynamic coefficient, $\bar{N}_{act}$ is indeed 50 years. However, for most other panels $\bar{N}_{act}$ exceeds 50 years, in some cases by one or even two orders of magnitude (see Sect. 8.1.2).

A second consequence of not accounting for wind directionality is that any two buildings that are identical in all respects but have different orientations will experience different numbers of panel failures during their lifetime. Indeed, since conventional practice does not account for wind directionality, it will yield exactly the same cladding design for the two buildings even though, owing to their different orientations with respect to the direction of the most severe extreme winds, the two buildings will exhibit different degrees of sensitivity to wind effects.

For the *risk-consistent design procedure* it is necessary to estimate the mean and the coefficient of variation of the equivalent 60-sec largest lifetime load. This is done as shown in Sect. 8.2.2 (Eqs. 8.2.7 to 8.2.14), in which $\bar{p}_{60_n}^{eq}$ and $V_{p_{60_n}^{eq}}$ should be substituted for $\bar{Q}_n$ and V_{Q_n}, respectively.

6. Load Capacity of Cladding Panels.   Information on the load capacity of cladding panels can be obtained from manufacturers' charts [9-55, 9-56]. These charts include estimates of the standard deviation and of the 0.8 percentage point of the load capacity of panels with different sizes for annealed, heat-strengthened, and tempered glass.* The charts of [9-55] and [9-56] exhibit mutual inconsistencies, and apparent internal inconsistencies have been noted in [9-55] (see [9-57, 9-58, 9-59], which report research aimed at improving these charts).

Owing to fatigue effects, the load capacity of glass panels depends upon the time history of the applied load [9-57, 9-58, 9-59]. The load capacities given in [9-55] and [9-56] have a standardized time history; that is, they are expressed in terms of constant loads with a 60-sec duration, denoted by p_{60}.

7. Design Criteria.   The *conventional design procedure* uses the following design criterion:

$$p_{60}(0.008) \geqslant p_{60,50}^{eq}(M_j) \tag{9.5.8}$$

where $p_{60,50}^{eq}(M_j)$ is the equivalent 60-sec wind load estimated without considering wind directionality effects (see discussion in item 5 above), and $p_{60}(0.008)$ is the 60-sec load capacity of the panel corresponding to a cumulative failure probability of 8 panels out of 1000.

---

*The 0.8 percentage point of the load capacity is the load to which there corresponds a probability of failure of 8 panels out of 1000 (see Sect. A1.5). Information on loads corresponding to other probabilities of failure is available in [9-56].

The *risk-consistent design procedure* is based on the requirement that the probability of failure of each panel during the lifetime of the building be less than a specified value P_f. It is now shown that this requirement leads to a design criterion expressed in terms of equivalent 60-sec loads and of 60-sec load capacities.

Consider the safety index β defined by Eq. A3.29. It is possible to write:

$$\beta = \frac{\ln(\bar{p}_{60}/\bar{p}_{60n}^{eq})}{(V_{p_{60n}^{eq}}^2 + V_{p_{60}}^2)^{1/2}} \qquad (9.5.9)$$

where $\bar{p}_{60n}^{eq}$ and $V_{p_{60n}^{eq}}$ are the mean and the coefficient of variation of the largest equivalent 60-sec wind load during the lifetime of the building, the subscript n represents the lifetime of the structure in years, and $\bar{p}_{60}$ and $V_{p_{60}}$ are the mean and coefficient of variation of the load capacity. From Eq. 9.5.9 it follows that $\bar{p}_{60}$ should satisfy the relation

$$\bar{p}_{60} \geqslant \bar{p}_{60n}^{eq}(M_j^*) \exp[\beta(V_{p_{60n}^{eq}}^2(M_j) + V_{p_{60}}^2)^{1/2}] \qquad (9.5.10)$$

where β is the value of the safety index corresponding to the failure probability P_f (see Eq. A3.37). Equation 9.5.9 is the design criterion used for risk-consistent design.

The question of the selection of the safety index β or, equivalently, of the failure probability P_f, is discussed under item 8 below.

8. Design.   For ease of construction it is necessary to divide the building facades into zones, each characterized by a single type of glass panel. Thus, for any given architectural pattern defined by the location, and by the height and width of the panels, the design of the cladding consists of: (1) dividing the building facades into such zones, and (2) selecting the type of glass (i.e., whether annealed, heat-strengthened, or tempered) and the panel thickness for each zone.

An example of division into zones of equal glass thickness, suggested in [9-19], is shown in Fig. 9.5.2. This division makes it possible to provide stronger panels at and near the edges and eaves, where aerodynamic pressures are usually largest. However, other possibilities exist. For example, the thickness of the glass panels may change as a function of elevation, as in the case of the John Hancock Building in Boston. In other cases the same glass thickness is used over an entire building face, or even over the entire building. For locations where wind-borne missiles, including roof gravel, may be expected to hit the cladding, special zones are suggested in [9-62].

We denote a zone in which the glass type and thickness is uniform by D_i $(i = 2, \ldots, n_D)$. If the conventional design method is used, Eq. 9.5.8 must be satisfied at all points M_j within D_i, that is

$$[p_{60}(0.008)]_i \geqslant \max_{D_i} [p_{60,50}^{eq}(M_j)] \qquad (9.5.11)$$

It is possible to estimate the expected number of failures inherent in the design based on Eq. 9.5.11 as follows. Each zone D_i is divided into subzones

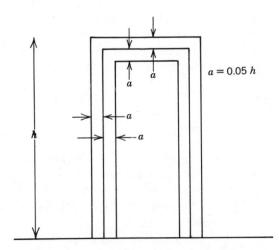

$a = 0.05\,h$

FIGURE 9.5.2.   Division of a high-rise building face into zones of equal glass thickness.

A_{ij} ($j = 1, 2, n_{A_i}$) over which it may be assumed that the wind loads do not vary significantly.* Using Eqs. 9.5.9 and A3.37, it is possible to calculate, for any given orientation of the building α_l, the safety index β_{ij}^l and the lifetime probability of failure P_{fij}^l of the panels within A_{ij}.

Let the number of panels within A_{ij} be denoted by n_{pij}. For the building designed by the conventional method (Eq. 9.5.11), the expected number of panel failures in subzone A_{ij} during the lifetime of the structure with orientation α_l is†

$$\bar{n}_{fij}^l = n_{pij} P_{fij}^l \tag{9.5.12}$$

The expected total number of panel failures per lifetime for the zone D_i and for the entire building are, respectively

$$\bar{n}_{fi}^l = \sum_j \bar{n}_{fij}^l \tag{9.5.13}$$

---

*In practice, A_{ij} are the tributary areas of the pressure taps on the wind tunnel model of the building (or, if a tributary area extends beyond the confines of D_i, the portion of that tributary area contained in the zone D_i).

†The failure condition for each panel of a zone is defined by the event $p_{60} - p_{60}^{cap} < 0$. Note that these events are not in all cases statistically independent, since the loads induced on various panels, and in some cases the load capacities of various panels, may be correlated. However, Eq. 9.5.12 holds regardless of whether the failure events are independent or not. This can be shown by considering the simple example of n_p coins. Let failure denote the occurrence of "heads." The expectation of the number of failures that would occur if the n_p coins were tossed once is $\bar{n}_f = 1/2 n_p$. This is true regardless of whether the failure events are independent (as in the case of coins having each an independent motion) or perfectly correlated (as in the case of a set of n_p coins, fixed onto a weightless board with all the "heads" on the same side, so that failure of one coin would entail failure of all the n_p coins). Note that while the expectations of n_f would be the same in the two cases, the standard deviations would not.

$$\bar{n}^l_f = \sum_i \bar{n}^l_{fi} \tag{9.5.14}$$

Experience appears to indicate that the cladding in any subzone A_{ij} designed in accordance with the conventional method (Eq. 9.5.8) is acceptable from a safety point of view if the aerodynamic and climatological data upon which the design was based are adequate. It might then be argued that the probability P_f corresponding to the safety index β used in Eq. 9.5.10 may have the value

$$P_f = \max_{i,j,l} \{P^l_{fij}\} \tag{9.5.15}$$

where $\max_{i,j,l}\{P^l_{fij}\}$ is the largest of the values P^l_{fij}.

However, such a choice of P_f for use in risk-consistent design might be imprudent. The authors believe that it is reasonable to adopt as a design objective an expected number of panel failures per lifetime for the entire building

$$\bar{n}_f = \max_l \bar{n}^l_f \tag{9.5.16}$$

Indeed, the conventional design procedure, ignoring as it does wind directionality effects, can be viewed as providing sufficient safety levels for all buildings, regardless of their orientation. This can be interpreted as meaning: (1) that the expected number of failures $\bar{n}^l_f$ inherent in the conventional design procedure is acceptable even for buildings with the most unfavorable orientation α_l, and (2) that if the conventional procedure is used, buildings with more favorable orientations are overdesigned.

If Eq. 9.5.16 is adopted as a design objective, the probability of failure of any one panel during the lifetime of the building is $P_f = \bar{n}_f / n_p$, where n_p is the total number of panels of the building. The safety index β is then calculated using Eq. A3.37, and the cladding for each zone D_i is designed by the risk-consistent procedure in accordance with Eq. 9.5.10, that is

$$[\bar{p}_{60}]_i \geqslant \max_{D_i} \{\bar{p}^{eq}_{60n}(M_k) \exp[\beta(V^2_{p^{eq}_{60n}}(M_k) + V^2_{p_{60}})^{1/2}] \tag{9.5.17}$$

A computer program for the design of cladding by Eq. 9.5.17 in conjunction with Eqs. 9.5.16 and A3.37 is referenced in [9-61]. Illustrative results obtained by using that program are presented in Sect. 9.5.2.

9.5.2 Economic and Safety Advantages of Risk-Consistent Design Procedure

To illustrate the potential advantages of the risk-consistent design procedure, results of computations taken from [9-61] are presented for a 200 m tall building represented in plan in Fig. 9.5.3. It was assumed that the building is located in terrain with uniform roughness in all directions ($z_0 = 1.00$ m), and that there are no neighboring structures influencing the building aerodynamics. Aerodynamic pressure coefficients obtained in the wind tunnel were extracted from [9-60]. The wind climate was assumed to be defined by the data of Table 8.1.2, for which summary statistics are given in Fig. 3.4.1. The facades were divided into zones of uniform glass thickness in accordance with Fig. 9.5.2. It was assumed

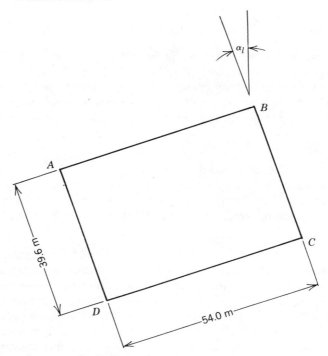

FIGURE 9.5.3. Dimensions of building in plan.

that the cladding consisted of annealed glass panels with dimensions 1.8 m × 1.8 m. The information on the load capacity of the panels was taken from [9-56]. Approximate typical prices per unit area of panels with various thicknesses were obtained from glass distributors. These were used as a basis for performing estimates of the nominal cost of cladding glass inherent in any given design.

From an inspection of Fig. 3.4.1 it is apparent that the wind effects are not equally severe for the parallel faces AD and BC (or AB and DC) of the building shown in Fig. 9.5.3. Nevertheless, as noted earlier, the conventional design method would result in this case in identical designs for those faces. It is also clear that the severity of the wind effects on the various faces depends upon the orientation α_l of the building. Again, this is not reflected in the conventional method, which results in identical cladding designs regardless of the building orientation α_l.

The cladding of the building shown in Fig. 9.5.3 was first designed in accordance with the conventional method. The nominal cost of the cladding so designed was estimated to be $361,000 for the entire building. Using the procedure described in the preceding section, the expected number of panel failures per lifetime inherent in the conventional design was estimated for various building orientations α_l. The results of the estimates are shown in col. 2 of Table 9.5.1.

Also shown in Table 9.5.1 (cols. 4 and 5) are nominal costs of cladding designed by the risk-consistent procedure on the basis of the following design

TABLE 9.5.1. Nominal Costs of Cladding for Various Designs.

| Building Orientation (1) | Conventional Practice | | Risk-Consistent Procedure | |
	Number of panel failures, $\bar{n}^l_f$ (2)	Nominal cost ($) (3)	Nominal cost, design based on value $\bar{n}^l_f$ from Col. 2 ($) (4)	Nominal cost, design based on value $\bar{n}^l_f = 12.0$ ($) (5)
0°	2.5	361,000	330,000	280,000
45°	11.9	361,000	345,000	345,000
90°	1.8	361,000	345,000	280,000
135°	12.1	361,000	345,000	345,000

objectives: the expected total number of failures per lifetime is equal (1) to the value $\bar{n}^l_f$ of col. 2 (see col. 4), and (2) to the value $\bar{n}^l_f = 12.0$, which corresponds approximately to the most unfavorable orientation of the building (see col. 5).

Consider the designs based on the first of these two objectives. It is seen that in this case the economies achieved are of the order of 5 to 10%. However, the fact that the conventional design is acceptable to building inspection authorities, regardless of the building orientation, implies that in the view of these authorities such a design is sufficiently safe even in those cases where the building orientation is unfavorable; as noted in Sect. 9.5.1, this observation leads to the adoption as a design objective of an expected number of failures per lifetime approximately equal to the largest of the estimated values $\bar{n}^l_f$ ($l = 1, 2, \ldots, 8$). A comparison between cols. 3 and 5 of Table 9.5.1 shows that, for buildings with favorable orientations, the use of the risk-consistent design procedure can then result in significant savings (in the case examined here, almost 25%).

As stated earlier, the results of Table 9.5.1 were obtained for building facades divided into zones in accordance with Fig. 9.5.2. As indicated in [9-61], similar conclusions hold for designs in which the glass thickness is constant over an entire building face.

REFERENCES

9-1 T. A. Reinhold et al., *Mean and Fluctuating Forces and Torques on a Tall Building Model of Square Cross-Section as a Single Model, in the Wake of a Similar Model, and in the Wake of a Rectangular Model*, Report VP1-E-79-11, Department of Engineering Science and Mechanics, Virginia Polytechnic Institute, Blacksburg, VA, March 1979.

9-2 J. W. Saunders and W. H. Melbourne, "Buffeting Effects of Upstream Buildings," *Proceedings Fifth International Conference on Wind Engineering*, Ft. Collins, Colorado, July 1979, Vol. I, Pergamon Press, 1980.

9-3 J. Blessman and J. D. Pierce, "Interaction Effects in Neighboring Tall Buildings," *Proceedings Fifth International Conference on Wind Engineering*, Ft. Collins, Colorado, July 1979, Vol. II, Pergamon Press, 1980.

9-4   H. Ruscheweyh, "Dynamic Response of High-Rise Buildings under Wind Action with Interference Effects from Surrounding Buildings of Similar Size," *Proceedings Fifth International Conference on Wind Engineering*, Ft. Collins, Colorado, July 1979, Vol. II, Pergamon Press, 1980.

9-5   J. A. Peterka and J. E. Cermak, "Adverse Wind Loading Induced by Adjacent Buildings," *J. Struct. Div.*, ASCE, **102**, No. ST3 (March, 1976), 533–548.

9-6   H. W. Liepmann, "On the Application of Statistical Concepts to the Buffeting Problem," *J. Aeronaut. Sci.*, **19**, 12 (Dec. 1952), 793–800, 822.

9-7   A. G. Davenport, "The Application of Statistical Concepts to the Wind Loading of Structures," *Proc. Inst. Civ. Eng.*, **19** (1961), 449–472.

9-8   A. G. Davenport, "Gust Loading Factors," *J. Struct. Div.*, ASCE, **93**, No. ST3, Proc. Paper 5255 (June 1967), 11–34.

9-9   J. Vellozzi and E. Cohen, "Gust Response Factors," *J. Struct. Div.*, ASCE, **94**, No. ST6, Proc Paper 5980 (June 1968), 1295–1313.

9-10   J. W. Reed, *Wind-Induced Motion and Human Discomfort in Tall Buildings*, Structures Publication No. 310, R71-42, Department of Civil Engineering, MIT, Cambridge, Mass., 1971.

9-11   E. Simiu, "Gust Factors and Along-Wind Pressure Correlations," *J. Struct. Div.*, ASCE, **99**, No. ST4, Proc. Paper 9686 (April 1973), 773–783.

9-12   *Canadian Structural Design Manual*, Supplement No. 4 to the National Building Code of Canada, Associate Committee on the National Building Code and National Research Council of Canada, Ottawa, 1975.

9-13   B. J. Vickery, "On the Reliability of Gust Loading Factors," in *Proceedings of the Technical Meeting Concerning Wind Loads on Buildings and Structures*, Building Science Series 30, National Bureau of Standards, Washington, D.C., 1970.

9-14   E. Simiu and D. W. Lozier, "The Buffeting of Structures by Strong Winds—Windload Program," NTIS Accession No. PB 294757/AS, Computer Program for Estimating Along-Wind Response, National Technical Information Service, Springfield, VA, 1979.

9-15   C. Soize, "Dynamique stochastique des structures élancées soumises aux charges du vent," *Revue Française de Mécanique*, **60** (1976), 57–65.

9-16   G. Solari, *DAWROS: A Computer Program for Calculating Along-Wind Response of Structures*, Pubblicazione dell'Istituto di Scienza delle Costruzioni, Serie IV, No. 1, University of Genova, Genova, Italy, 1981.

9-17   E. Simiu, "Revised Procedure for Estimating Along-Wind Response," *J. Struct. Div.*, ASCE, **106**, No. ST1 (Jan. 1980), 1–10.

9-18   G. Solari, "Along-Wind Response Estimation: Closed Form Solution," *J. Struct. Div.*, ASCE, **108**, No. ST1 (Jan. 1982), 225–244.

9-19   *American National Standard Building Code Requirement for Minimum Design Loads in Buildings and Other Structures*, A58.1, American Standards Institute, New York, 1982.

9-20   P. A. Rosati, *An Experimental Study of the Response of a Square Prism to Wind Load*, BLWT II-68, Faculty of Graduate Studies, University of Western Ontario, London, Ont., Canada, 1968.

9-21   W. C. Hurty and M. R. Rubinstein, *Dynamics of Structures*, Prentice-Hall, Englewood Cliffs, N.J., 1964.

9-22   J. A. Blume, N. M. Newmark, and L. H. Corning, *Design of Multistory Reinforced Concrete Buildings for Earthquake Motions*, Portland Cement Association, Chicago, Ill., 1961.

9-23   G. T. Taoka, M. Hogan, F. Khan, and R. H. Scanlan, "Ambient Response of Some Tall Structures," *J. Struct. Div.*, ASCE, **101**, No. ST1, Proc. Paper 11051 (Jan. 1975), 49–55.

9-24   C. S. Kwok and W. H. Melbourne, "Wind-Induced Lock-In Excitation of Tall Structures," *J. Struct. Div.*, ASCE, **107**, No. ST1 (Jan. 1981), 57–72.

9-25   B. J. Vickery, "Notes on Wind Forces on Tall Buildings," Annex to *Australian Standard 1170, Part 2-1973, SAA Loading Code Part 2—Wind Forces*, Standards Association of Australia, Sydney, 1973.

9-26   T. A. Reinhold and P. R. Sparks, "The Influence of Wind Direction on the Response of a

Square-Section Tall Building," *Proceedings Fifth International Conference on Wind Engineering*, Fort Collins, CO, July 1979, Pergamon Press, 1980.

9-27 J. W. Saunders, *Wind Excitation of Tall Buildings with Particular References to the Cross-Wind Motion of Tall Buildings of Constant Rectangular Cross-Section*, Doctoral Thesis, Dept. of Mechanical Engineering, Monash University, Victoria, Australia, 1975.

9-28 A. Kareem, "Across-Wind Response of Buildings," *J. Struct. Div.*, ASCE, **108** (April 1982), 869–887.

9-29 A. Kareem, "Wind-Excited Response of Buildings in Higher Modes," *J. Struct. Div.*, ASCE, **107**, No. ST4 (April 1981), 701–706.

9-30 A. G. Davenport, "The Response of Six Building Shapes to Turbulent Winds," *Phil. Trans. Roy. Soc. London*, **A269** (1971), 385–394.

9-31 F. E. Schmitt, "The Florida Hurricane and Some of Its Effects," *Eng. News Record*, **97**, 16 (Oct. 14, 1926), 624–627.

9-32 A. Smith, "Basis of Design for Hurricane Exposure," Report of Committee 308, *Proceedings*, ACI, **27** (1931), 903.

9-33 "Torsional Effects of Wind in Buildings," in *Wind Bracing in Steel Buildings*, Sixth Progress Report of Subcommittee No. 31, Committee on Steel of the Structural Division, *Reports*, June 1939, pp. 988–996.

9-34 "Wind Forces on Structures," *Trans. ASCE*, **126**, Part II (1961), 1124–1198.

9-35 C. P. Patrickson and P. Friedmann, *A Study of the Coupled Lateral and Torsional Response of Tall Buildings to Wind Loadings*, School of Engineering and Applied Science, University of California, Los Angeles, UCLA-ENG-76126, Dec. 1976.

9-36 D. A. Foutch and E. Safak, "Torsional Vibration of Wind Excited Symmetrical Structures," *J. Wind Eng. Ind. Aerodyn*, **7** (1981), 191–201.

9-37 D. A. Foutch and E. Safak, "Torsional Vibration of Along-Wind Excited Structures," *J. Eng. Mech. Div.*, ASCE, **107** (1981), 323–337.

9-38 G. C. Hart, R. M. DiJulio, and M. Lew, "Torsional Response of High Rise Buildings," *J. Struct. Div.*, ASCE, **101** (1975), 397–416.

9-39 G. L. Greig, *Toward an Estimate of Wind-Induced Dynamic Torque on Tall Buildings*, Master's Thesis, Dept. of Engineering, Univ. of Western Ontario, London, Ont., Sept. 1980.

9-40 D. Surry and G. Lythe, "Mean Torsional Loads on Tall Buildings," *Proceedings Fourth U.S. National Conference on Wind Engineering Research*, B. Hartz (Ed.), Dept. of Civil Engineering, Univ. of Washington, Seattle, WA, July 1981.

9-41 N. Isyumov, "The Aeroelastic Modeling of Tall Buildings," *Proceedings of International Workshop on Wind Tunnel Modeling for Civil Engineering Application*, Cambridge University Press, 1982.

9-42 N. Isyumov and M. Poole, "Wind-Induced Torque on Square and Rectangular Building Shapes," *Proceedings Sixth International Conference on Wind Engineering*, Elsevier, Amsterdam, 1984.

9-43 J. N. Yang and Y. K. Lin, "Along-Wind Motion of Multistory Building," *J. Eng. Mech. Div.*, ASCE, **107** (1981), 295–307.

9-44 "Tower Cables Handle Wind, Water Tank Damps It," *Eng. News Record*, Dec. 9, 1971, p. 23.

9-45 "Lead Hula-Hoops Stabilize Antenna," *Eng. News Record*, **197**, 4 (July 22, 1976), 10.

9-46 K. B. Wiesner, "Tuned Mass Dampers to Reduce Building Wind Motion," Preprint 3510, ASCE Convention and Exposition, Boston, April 2–6, 1979.

9-47 "Hancock Tower Now to Get Dampers," *Eng. News Record*, October 30, 1975, p. 11.

9-48 N. Isyumov, J. Holmes, and A. G. Davenport, "A Study of Wind Effects for the First National City Corporation Project—New York, U.S.A.," *University of Western Ontario Research Report BLWT-551-75*, London, Ont., Canada, 1975.

9-49 R. J. McNamara, "Tuned Mass Dampers for Buildings," *J. Struct. Div.*, ASCE, **103** (Sept. 1977), 1785–1798.

9-50 "Tuned Mass Dampers Sway Skyscrapers in Wind," *Eng. News Record*, Aug. 18, 1977, pp. 28–29.

9-51 J. P. Den Hartog, *Mechanical Vibrations*, McGraw-Hill, New York, 1956.

9-52   S. H. Crandall and W. D. Mark, *Random Vibrations in Mechanical Systems*, Academic Press, New York, 1963.

9-53   R. W. Luft, "Optimal Tuned Mass Dampers for Buildings," *J. Struct. Div.*, ASCE, **105**, (Dec. 1979), 2766–2772.

9-54   A. Der Kiureghian, J. L. Sackman, and B. Nour-Omid, *Dynamic Response of Light Equipment in Structures*, Report UCB/EERC-81/05, Earthquake Engineering Research Center, College of Engineering, University of California, Berkeley, CA, April 1981.

9-55   *PPG Glass Thickness Recommendations to Meet Architects Specified 1-Minute Wind Load*, PPG Industries, Pittsburg, 1981.

9-56   *LOF Technical Information—Strength of Glass Under Wind Loads*, Publication ATS-109, Libbey-Owens-Ford Company, Toledo, Ohio, 1980.

9-57   W. L. Beason, and J. R. Morgan, "Glass Failure Prediction Model," *J. Struct. Eng.*, **110** (Feb. 1984), 197–212.

9-58   D. A. Reed and E. Simiu, "Wind Loading and the Strength of Glass," *J. Struct. Eng.*, **110** (April 1984), 715–729.

9-59   E. Simiu and D. A. Reed, "Ring-on-Ring Tests and the Modeling of Cladding Glass Strength by the Weibull Distribution," *Proceedings IUTAM Symposium on Probabilistic Methods in the Mechanics of Solids and Structures*, Stockholm, June 19–21, 1984, N. C. Lind and S. Eggwertz (Eds.), Springer-Verlag, New York, 1985.

9-60   J. A. Peterka and J. E. Cermak, *Wind Tunnel Study of Atlanta Office Building*, Dept. of Civil Engineering, Colorado State University, Fort Collins, CO, Nov. 1978.

9-61   E. Simiu and A. Filotti, "Window Glass Facades as Structural Systems: An Improved Reliability-Based Design Procedure," *Proceedings International Conference on Structural Safety and Reliability*, May 27–29, 1985, I. Konishi and M. Shinozuka (Eds.), Kobe, Japan.

9-62   J. E. Minor, W. L. Beason, and P. L. Harris, "Designing for Windborne Missiles in Urban Areas," *J. Struct. Div.*, ASCE, **104** (1978), 1749–1760.

9-63   N. J. Cook and J. R. Mayne, "A Refined Working Approach to the Assessment of Wind Loads for Equivalent Static Design," *J. Wind Eng. Ind. Aerodyn.*, **6** (1980), 125–137.

9-64   J. A. Peterka, "Selection of Local Peak Pressure Coefficients for Wind Tunnel Studies of Buildings," *J. Wind Eng. Ind. Aerodyn.*, **13** (1983), 477–488.

9-65   E. Simiu, "Modern Developments in Wind Engineering, Part 4," *Eng. Struct.* **5** (1983), 273–281.

9-66   A. Tallin and B. Ellingwood, "Analysis of Torsional Moments on Tall Buildings," *J. Wind Eng. Ind. Aerodyn.* **18** (April 1985), 191–195.

9-67   P. Mahmoodi, "Design and Analysis of Viscoelastic Vibration Dampers for Structures", in *Proceedings, World Innovation Week Conference INOVA-73*, Eyrolles, Paris, 1974.

9-68   P. Mahmoodi, "Structural Dampers," *J. Struct. Div.*, ASCE, **95** (Aug. 1969), 1661–1672.

10

Slender Towers and Stacks with Circular Cross Section

Slender towers and stacks are designed to withstand the effects of both along-wind and across-wind loads. The along-wind response can be estimated by using the computer program of [9-14]. Simplified methods may be used if approximate estimates of the peak along-wind response in the fundamental mode are sought. Since the gust response factor G depends only weakly upon the fundamental modal shape (see Eq. 9.1.10), it can be calculated by using Table 9.1.5. The peak along-wind response is then obtained from the relation $X_{1\max}(z) = G\bar{x}_1(z)$, where $\bar{x}_1(z)$ is calculated using Eq. 5.3.4 (with $i=1$).

These procedures must be used with appropriate values for the average windward and leeward drag coefficients C_w and C_l. For slender towers and stacks with a circular shape in plan, it may be assumed in all cases that $C_l = 0$, so that the total drag coefficient $C_D = C_w$. Information on the magnitude of C_D and its dependence upon Reynolds number, surface roughness, and aspect ratio is provided in Sect. 10.2.2.

Several procedures for estimating across-wind response are currently available. Among these, the procedure developed by Rumman [10-1] has been widely applied to the design of reinforced concrete chimneys. The basis of this procedure is largely intuitive. Nevertheless, it appears that the results of its application have been satisfactory in practice.

It is generally agreed that the loading and response models inherent in Rumman's procedure are not entirely consistent with advances made over the last two decades in the fields of micrometeorology, aerodynamics, and aeroelasticity. According to [10-2], this could in certain situations lead to the underestimation of the across-wind response, particularly in the second mode of vibration. Procedures in which more advanced approaches are utilized were developed in [10-3], and by Vickery, Basu, and Clark in [10-2] and [10-4] to [10-9].

The ESDU procedure [10-3] is based on a modified version of the model consisting of Eqs. 6.1.4 and 6.1.5. It considers two response regions, one in which the forces associated with the motion are ignored and another in which the effect of these forces is taken into account. The response of the structure is estimated for each of these two regions, and the structure is designed for the higher of the two responses. It is suggested in [10-4] that a drawback of the ESDU procedure is the lack of a natural transition between the two response regimes, which introduces an element of arbitrariness in the application of the procedure.

The procedures developed by Vickery and coworkers imply in effect the following approach. A nominal response is calculated which corresponds to the assumption that aeroelastic effects do not occur. The actual response is then obtained through multiplication of the nominal response by an aeroelastic correction factor which varies continuously over the entire range of possible aeroelastic effects. The derivation of that factor is explained in some detail in Sect. 6.1.2. Because of uncertainties inherent in them, these procedures should be used with caution.

Structures that are light in weight and have low structural damping (e.g., certain steel stacks) could experience unacceptably severe aeroelastic effects unless provided with aerodynamic or mechanical devices for the alleviation of across-wind motions. Some of these devices have proven to be quite effective and are routinely incorporated in the design of steel stacks.

This chapter describes Rumman's procedure (Sect. 10.1) and the procedures developed by Vickery and coworkers (Sect. 10.2). These procedures are applicable to isolated structures.* Also presented in this chapter is information on aerodynamic and aeroelastic devices for the alleviation of the across-wind response (Sect. 10.3).

10.1   RUMMAN'S PROCEDURE

In this procedure it is assumed that towers or stacks with a circular cross section are subjected to a sinusoidal force per unit length with amplitude

$$F_0(z) = \tfrac{1}{2} C_L \rho U^2(z_{e_i}) D(z) \tag{10.1.1}$$

where ρ is air density ($\rho \simeq 1.25$ kg/m^3), $U(z_{e_i})$ is the critical wind speed at elevation z_{e_i}, C_L is the lift coefficient, and $D(z)$ is the diameter of structure at elevation z. The force $F_0(z)$ is assumed to be perfectly correlated spanwise. Reference 10-1 suggests $z_{e_i} \doteq h$, where h is the height of structure. Other references suggest

---

*If several stacks are grouped in a row, buffeting forces associated with vortex shedding can cause the response of stacks located downwind of the first structure in the row to be as high as four times the response of an identical but isolated stack. Limited wind tunnel data on the response of grouped stacks are available in [10-6] and [10-24]. For full scale information, see [10-25]. See also p. 234.

$z_{e_i} = 2/3h$ to $5/6h$ [10-10], or $z_{e_i} = 2/3h$ [10-11]. The wind speeds $U(z_{e_i})$ produce at elevation z_{e_i} vortex shedding with frequencies equal to the natural frequencies of the structure, so that

$$U(z_{e_i}) = \frac{1}{\mathscr{S}} n_i D(z_{e_i}) \qquad (i = 1, 2, \ldots) \tag{10.1.2}$$

The Reynolds number at elevation z_{e_i} is calculated as follows:

$$\mathscr{R}e = 67,000 U(z_{e_i}) D(z_{e_i}) \tag{10.1.3}$$

($U(z_{e_i})$ in m/sec and $D(z_{e_i})$ in m). For $\mathscr{R}e > 3 \times 10^6$ or so it is usually assumed $\mathscr{S} = 0.20$ to 0.25. From Eqs. 5.2.8, 5.2.10, 5.2.14, and 5.2.16 it follows that the peak deflection for the structure excited in the i-th mode may be written as

$$Y_i(z) = \frac{\rho}{16\pi^2} \frac{C_L}{\zeta_i} D^2(z_{e_i}) \int_0^h \frac{D(z)y_i(z)\,dz}{M_i \mathscr{S}^2} y_i(z) \tag{10.1.4}$$

where $y_i(z)$ is the i-th normal mode of vibration, ζ_i is the damping in i-th mode,

$$M_i = \int_0^h m(z) y_i^2(z)\,dz \tag{10.1.5}$$

is the generalized mass, and $m(z)$ is the mass of the structure per unit height. For reinforced concrete chimneys ratios $C_L/\zeta_i \simeq 13\text{-}16$ have in many instances been assumed for design purposes [10-1].

According to [10-11], it was determined from observations that tall reinforced concrete chimneys with constant or nearly constant diameter do not appear to experience unacceptably large motions if their Scruton number c_i, defined as

$$c_i = \frac{2M_i}{\displaystyle\int_0^h y_i^2(z)\,dz} \frac{\zeta_i}{\rho D^2(z_{e_i})} \tag{10.1.6}$$

is larger than four.

The peak moment at any elevation z is dominated by contributions due to inertial forces. Therefore, denoting the peak acceleration at elevation z by $\ddot{Y}_i(z)$, the peak moment associated with the i-th mode of vibration is

$$\mathscr{M}_i(z) \simeq \int_z^h m(z_1) \ddot{Y}_i(z_1)(z_1 - z)\,dz_1 \tag{10.1.7}$$

or

$$\mathscr{M}_i(z) \simeq (2\pi n_i)^2 \int_z^h m(z_1) Y_i(z_1)(z_1 - z)\,dz_1 \tag{10.1.8}$$

The shear force at elevation z_1 may similarly be written as

$$S_i(z_1) \simeq (2\pi n_i)^2 \int_z^h m(z_1) Y_i(z_1)\,dz_1 \tag{10.1.9}$$

Numerical Example

Consider a chimney with constant circular cross section for which $D = 17.63$ m, $h = 193.6$ m, and $n_1 = 0.364$ Hz [10-9]. It is assumed $C_L/\zeta_1 = 15$, $y_1(z/h) = (z/h)^{1.67}$, $m(z) = 58{,}000$ kg/m for $z < h/2$, $m(z) = 41{,}000$ kg/m for $z > h/2$, and $\mathscr{S} = 0.22$. From Eq. 10.1.5, $M_1 = 1.87 \times 10^6$ kg. The critical velocity in the first mode of vibration is $U(z_{e_i}) = 29.15$ m/sec (Eq. 10.1.2). The corresponding Reynolds number is $\mathscr{R}e = 3.4 \times 10^7$ (Eq. 10.1.3). The peak response at elevation z is $Y_1(z) = 0.52(z/h)^{1.67}$ m (Eq. 10.1.4), and the peak moment at the base is $\mathscr{M}(0) \simeq 1.17 \times 10^6$ kNm (Eq. 10.1.8). If it is assumed $\zeta_1 = 0.02$, then the Scruton number $c_1 = 4.3$ (Eq. 10.1.6).

10.2 PROCEDURES DEVELOPED BY VICKERY AND COWORKERS

It was mentioned earlier that these procedures may be viewed as, in effect, estimating the across-wind response in two phases. First, a nominal response is calculated by assuming that the structure is acted upon by the across-wind aerodynamic loads it would experience if it were at rest. The nominal response therefore does not reflect any aeroelastic effects, since the latter involve loads associated with the motion of the structure. The actual response is obtained through multiplication of the nominal response by a correction factor that accounts for the aeroelastic effects. The approach used to estimate the nominal response and the aerodynamic correction factor is described in Sect. 10.2.1. Information on the requisite aerodynamic and aeroelastic parameters is provided in Sect. 10.2.2. Approximate expressions for the across-wind response are given in Sect. 10.2.3.

It is emphasized that, although the procedures presented in this section are conceptually advanced, they yield results that may be uncertain to within at least 30%. This is the case in part because the structural damping is in most cases poorly known. In addition, much of the available information concerning the aerodynamic and aeroelastic parameters (see Sect. 10.2.2) is only tentative. Attempts to obtain such information from wind tunnel tests (as, for example, in [10-12]), are generally unsuccessful, owing to severe scale effects. For this reason it has been pointed out that wind tunnel simulations of the across-wind behavior of slender structures with circular cross section under wind loads cannot be used for design purposes unless carefully interpreted in the light of aeroelastic theory and of data obtained from full-scale tests [10-8].

10.2.1 Basic Approach to Estimation of the Across-Wind Response

Let $\sigma_{yi}^{\text{nom}}(z)$ denote the rms value of the nominal across-wind response at elevation z in the i-th mode of vibration. The rms value of the actual across-wind response at elevation z in the i-th mode is denoted by $\sigma_{yi}(z)$. The following relation holds:

$$\sigma_{yi}(z) = \left(\frac{\zeta_i}{\zeta_i + \zeta_{ai}}\right)^{1/2} \sigma_{yi}^{\text{nom}}(z) \tag{10.2.1}$$

where ζ_i is the structural damping ratio, ζ_{ai} is the aerodynamic damping ratio, and $[\zeta_i/(\zeta_i + \zeta_{ai})]^{1/2}$ is the aerodynamic correction factor in the i-th mode.

Estimation of Nominal Across-Wind Response. The nominal across-wind response is obtained by subjecting the structure to the across-wind aerodynamic loads it would experience if it were at rest. No aeroelastic effects are taken into account, and the only damping that affects the motion is the structural damping.

In a turbulent flow the structure at rest would experience a superposition of two across-wind loads. The first of these two loads, due to vortex shedding in the wake of the structure at rest, is denoted by $L_1(z, t)$. The second load, due to the lateral turbulence in the oncoming flow, is denoted by $L_2(z, t)$. The load $L_1(z, t)$ can be written in the form

$$L_1(z, t) = \tfrac{1}{2}\rho C_L(z, t)D(z)U^2(z) \tag{10.2.2}$$

so that its spectral density is

$$S_{L_1}(z, n) = [\tfrac{1}{2}\rho D(z)U^2(z)]^2 S_{C_L}(z, n) \tag{10.2.3}$$

According to [10-2], measurements indicate that the spectral density $S_{S_L}(z, n)$ can be represented by the bell-shaped function

$$\frac{nS_{C_L}(z, n)}{C_L^2} = \frac{1}{\sqrt{\pi}Bn_s} \exp\left\{-\left[\left(1 - \frac{n}{n_s}\right)\middle/ B\right]^2\right\} \tag{10.2.4}$$

where n denotes the frequency, n_s is the vortex shedding frequency given by the relation

$$n_s = \frac{\mathscr{S}U(z)}{D(z)} \tag{10.2.5}$$

$\mathscr{S}$ is the Strouhal number, and B is an empirical parameter that determines the spread (bandwidth) of the spectral curve. This model is compatible with results of full-scale measurements (Fig. 10.2.1).

The cross-spectral density of the load $L_1(z, t)$ can be expressed as [10-4]:

$$S_{L_1}(z_1, z_2, n) = S_{L_1}^{1/2}(z_1, n)S_{L_1}^{1/2}(z_2, n)R_0(z_1, z_2, n) \tag{10.2.6}$$

$$R_0(z_1, z_2, n) = \cos(2ar)\exp(-ar^2) \tag{10.2.7a}$$

$$r = 2|z_1 - z_2|/[D(z_1) + D(z_2)] \tag{10.2.7b}$$

The parameter a in Eq. 10.2.7a is a measure of the decay of the cross-spectral function $S_{L_1}(z_1, z_2, n)$ with the distance $|z_1 - z_2|$. Associated with the parameter a is a correlation length L which is a measure of the spanwise length beyond which the force fluctuations are no longer correlated.

The lift force $L_2(t)$ is the projection on the across-wind direction of the drag force induced by the resultant of the mean velocity $U(z)$ and of the lateral turbulent velocity $v(z, t)$. In large-scale turbulence this force has an angle of attack with respect to the along-wind direction equal to v/U, and its projection

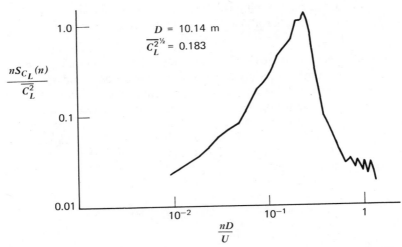

FIGURE 10.2.1. Power spectral density of lift force coefficient C_L measured on Hamburg television tower. From H. Ruscheweyh, "Wind Loadings on the Television Tower, Hamburg, Germany," *J. Ind. Aerodyn.*, **1** (1976), 315–333.

on that direction is

$$L_2(t) = \tfrac{1}{2}\rho C_D U^2(z)\,\frac{v(z,\,t)}{U(z)} \qquad (10.2.8)$$

Spectral and cross-spectral information on the lateral velocity fluctuations $v(t)$ is given in Sect. 2.3.5. Information on the aerodynamic parameters B, $\mathscr{S}$, $\overline{C_L^2}^{1/2}$, a, and C_D is given in Sect. 10.2.2.

The mean square value of the nominal response induced by each of the loads $L_1(t)$ and $L_2(t)$ can be estimated as in Sects. 5.2.7 or 5.3.2. The mean square value of the total nominal response is equal to the sum of the mean square values of the responses due to the loads $L_1(t)$ and $L_2(t)$. However, because these loads are uncorrelated, the peak value of the total nominal response is less than the sum of the individual peak responses due to $L_1(t)$ and $L_2(t)$.*

*It is of interest to estimate the extent to which the effect of the load $L_2(t)$ is significant from a practical point of view. Using the information of Sect. 2.3.5, it can be verified that the lateral velocity fluctuations differ from the longitudinal velocity fluctuations as follows: (a) the ordinates of the spectral density at high frequencies are larger by 33% for the lateral than for the longitudinal fluctuations; (b) the area under the spectral curve is lower by 50% for the lateral than for the longitudinal fluctuations; and (c) the exponential decay coefficients are lower by about 33% for the lateral than for the longitudinal fluctuations. Calculations then show that the peak nominal across-wind response due to $L_2(t)$ is of the order of 50% of the peak fluctuating part of the along-wind response or, roughly, about 25% of the peak total along-wind response.

It follows therefore that if the ratio between the along-wind response and the nominal across-wind response estimated without accounting for $L_2(t)$ is small, then taking $L_2(t)$ into consideration will have a negligible effect on the magnitude of the nominal response, particularly in view of the fact mentioned earlier that $L_1(t)$ and $L_2(t)$ are uncorrelated and that their peak values are therefore

TABLE 10.2.1. Suggested Structural Damping Ratios

Type of Structure	Structural Damping Ratio
Unlined steel stacks and similar structures	0.002–0.010
Lined steel stacks	0.004–0.016
Reinforced concrete chimneys and towers	0.004–0.020

Estimation of Aerodynamic Correction Factor $[\zeta_i/(\zeta_i + \zeta_{ai})]^{1/2}$. One of the difficulties that arises in the estimation of the aerodynamic correction factor is that relatively little reliable information is available on the structural damping ratios ζ_i. Ranges of values ζ_i suggested in [10-7] are listed in Table 10.2.1. The approach to the estimation of the aerodynamic damping ratio ζ_{ai} is described in some detail in Sect. 6.1.2. Information on the aeroelastic parameter K_{a0} needed to estimate ζ_{ai} (see Eqs. 6.1.36 to 6.1.38) is summarized in Sect. 10.2.2.

10.2.2 Aerodynamic and Aeroelastic Parameters

The purpose of this section is to provide information on the drag coefficient C_D, the Strouhal number $\mathscr{S}$, the rms lift coefficient $\overline{C_L^2}^{1/2}$, the bandwidth parameter B, the parameters describing the spanwise correlation of the across-wind load, and the aeroelastic parameter K_{a0} used to calculate the aerodynamic damping ratio ζ_{ai}.

The aerodynamic and aeroelastic parameters depend upon the Reynolds number

$$\mathscr{R}e = 67{,}000 U(z)D(z) \tag{10.2.9}$$

where $U(z)$ is the wind speed at elevation z in m/sec and $D(z)$ is the outside diameter in meters; upon the turbulence in the oncoming flow; upon the aspect ratio $h/D(h)$, where h is the height of the structure and $D(h)$ is the diameter at the tip; and upon the relative surface roughness k/D of the structure, where k is the height of the roughness elements. For steel stacks and reinforced concrete chimneys and towers $10^{-3} \gtrsim k/d \gtrsim 10^{-5}$ [10-7]. It is assumed herein that k/D varies only within this range.

---

not additive. On the other hand, if the ratio of along-wind to nominal across-wind response is high, the design will be governed by the along-wind response regardless of whether $L_2(t)$ is accounted for or not. Finally, if the ratio under consideration is close to unity, accounting for $L_2(t)$ would increase the peak nominal across-wind response by about 25% if $L_1(t)$ and $L_2(t)$ were correlated; however, because this is not the case, the increase will only be of the order of 10 to 15%. For these reasons, to a first approximation, the force induced by lateral turbulence fluctuations may be neglected, unless the estimated peak along-wind and across-wind response have approximately the same value, in which case the across-wind response should be augmented by roughly 10%.

Drag Coefficient C_D.   The dependence of C_D upon Reynolds number and surface roughness is represented in Fig. 4.5.5c for cylinders with aspect ratios $h/D(h) \gtrsim 20$. For structures with aspect ratios $10 < h/D(h) < 20$ it may be assumed that up to the elevation $h - D(h)$ the drag coefficient has the value

$$C_D = C_D^s \left[1 - 0.015 \left(20 - \frac{h}{D(h)} \right) \right] \tag{10.2.10}$$

where C_D^s is the value of the drag coefficient taken from Fig. 4.5.5c. From elevation $h - D(h)$ to the top of the structure the drag coefficient may be assumed to have the value $C_D = 1.4 C_D^s$ for all structures regardless of aspect ratio (see [10-13]). The main effect of turbulence in the oncoming flow is to decrease the Reynolds number corresponding to the onset of the critical region defined in Fig. 4.5.2.

Strouhal Number.   The following values of the Strouhal number are suggested for design purposes:

$$\mathscr{S} = 0.20 \qquad \mathscr{R}e \gtrsim 2 \times 10^5 \tag{10.2.11a}$$

$$0.2 \gtrsim \mathscr{S} \gtrsim 0.45 \qquad 2 \times 10^5 \gtrsim \mathscr{R}e \gtrsim 2 \times 10^6 \tag{10.2.11b}$$

$$\mathscr{S} \simeq c\{0.23 - 0.007[\log_{10}(k/d) + 5]\} \qquad \mathscr{R}e \gtrsim 2 \times 10^6 \tag{10.2.11c}$$

For $2 \times 10^5 \gtrsim \mathscr{R}e \gtrsim 2 \times 10^6$ the vortex shedding is random and the Strouhal number given by Eq. 10.2.11b corresponds to the predominant frequencies of the flow in the wake. In Eq. 10.2.11c the coefficient c depends upon aspect ratio as follows:

$$c = \begin{cases} 1.00 & h/D(h) \gtrsim 30 \tag{10.2.12a} \\ 0.736 + 0.012 \left[\dfrac{h}{D(h)} - 8.0 \right] & 8 < h/D(h) < 30 \tag{10.2.12b} \end{cases}$$

where h is the height of the structure and $D(h)$ is the diameter at the tip (see [10-13]).

RMS of Lift Coefficient $\overline{C_L^2}^{1/2}$.   The following values of the rms lift coefficient are suggested for design purposes (see [10-13]):

$$\overline{C_L^2}^{1/2} \simeq \begin{cases} 0.45 & \mathscr{R}e \gtrsim 2 \times 10^5 \tag{10.2.13a} \\ 0.14 & 2 \times 10^5 < \mathscr{R}e \gtrsim 2 \times 10^6 \tag{10.2.13b} \\ d \left\{ 0.15 + 0.035 \left[5 + \log_{10} \left(\dfrac{k}{D} \right) \right]^2 \right\} & \mathscr{R}e \geqslant 2 \times 10^6 \tag{10.2.13c} \end{cases}$$

In Eq. 10.2.13c the coefficient d has the expression

$$d \simeq \begin{cases} 1.00 & h/D(h) \gtrsim 12 \tag{10.2.14a} \\ 0.8 + 0.05 \left[\dfrac{h}{D(h)} - 8.0 \right] & 8 < \dfrac{h}{D(h)} < 12 \tag{10.2.14b} \end{cases}$$

The lift coefficient also appears to depend significantly upon turbulence intensity. However, little information on this dependence is available to date.

Bandwidth Parameter B.   Reference 10-4 suggests that

$$B^2 \simeq 0.08^2 + 2\overline{u^2}/U^2 \tag{10.2.15}$$

where $\overline{u^2}$ is the mean square value of longitudinal turbulence fluctuations and U is the mean wind speed. According to [10-9], for practical purposes it may be assumed $B \simeq 0.18$ for all flows.

Spanwise Correlation Parameters.   For Reynolds numbers $\mathcal{R}e > 2 \times 10^5$ it may be assumed that in Eq. 10.2.7 the coefficient $a = 1/3$, and that to this value there corresponds a correlation length $L \simeq D$ [10-4]. For $\mathcal{R}e < 2 \times 10^5$, $L \simeq 2.5$ [10-14]. Thus, using the notation $\mathcal{L} = L/D$,

$$\mathcal{L} = \begin{cases} 2.5 & \mathcal{R}e < 2 \times 10^5 & (10.2.16a) \\[2mm] 1.0 & \mathcal{R}e \geqslant 2 \times 10^5 & (10.2.16b) \end{cases}$$

Aeroelastic Parameter K_{a0}.   On the basis of tentative information from [10-4] and [10-13], the following approximate expressions may be used to obtain K_{a0}:

$$K_{a0}\frac{(U)}{U_{\mathrm{cr}}} \simeq \begin{cases} 0 & U/U_{\mathrm{cr}} < 0.85 & (10.2.17a) \\[2mm] a_t\left(3.5\dfrac{U}{U_{\mathrm{cr}}} - 2.95\right) & 0.85 \leqslant U/U_{\mathrm{cr}} < 1.0 & (10.2.17b) \\[2mm] 0.55a_t & 1.0 \leqslant U/U_{\mathrm{cr}} < 1.1 & (10.2.17c) \\[2mm] a_t\left(2.75 - 2\dfrac{U}{U_{\mathrm{cr}}}\right) & 1.1 \leqslant U/U_{\mathrm{cr}} < 1.3 & (10.2.17d) \\[2mm] a_t\left(0.46 - 0.25\dfrac{U}{U_{\mathrm{cr}}}\right) & 1.3 \leqslant U/U_{\mathrm{cr}} < 1.84 & (10.2.17e) \\[2mm] 0 & 1.84 \leqslant U/U_{\mathrm{cr}} & (10.2.17f) \end{cases}$$

where

$$a_t = a_1 a_2 a_3 a_4 \tag{10.2.18}$$

$$a_1 = \begin{cases} 1.0 & \mathcal{R}e < 10^4 & (10.2.19a) \\ 1.8 & 10^4 \leqslant \mathcal{R}e < 10^5 & (10.2.19b) \\ 1.0 & 10^5 \leqslant \mathcal{R}e & (10.2.19c) \end{cases}$$

$$a_2 = \begin{cases} 2.0 & U(10 \text{ m}) \gtrsim 12 \text{ m/sec} & (10.2.20a) \\ 1.0 & U(10 \text{ m}) \lesssim 12 \text{ m/sec} & (10.2.20b) \end{cases}$$

$$a_3 = 0.9 + 0.2\left[\log_{10}\left(\frac{k}{D}\right) + 5\right] \tag{10.2.21}$$

$$a_4 = \begin{cases} 1.0 & h/D(h) > 12.5 & (10.2.22a) \\[2ex] 1.0 - 0.04\left(12.5 - \dfrac{h}{D(h)}\right) & h/D(h) < 12.5 & (10.2.22b) \end{cases}$$

Equations 10.2.20 reflect the fact that if the wind speed at 10 m above ground is relatively low the atmospheric turbulence may be weak. This can lead to a considerable enhancement of the aeroelastic effects (see Fig. 6.1.10).

10.2.3   Approximate Expressions for the Across-Wind Response

The across-wind response in the i-th mode of vibration may be estimated as follows:

$$\sigma_{yi}(z) = \overline{\xi_i^2}^{1/2} y_i(z) \tag{10.2.23}$$

$$Y_i(z) = g_{yi}\sigma_{yi}(z) \tag{10.2.24}$$

$$g_{yi} = [2\ln(3600n_i)]^{1/2} + \frac{0.577}{[2\ln(3600n_i)]^{1/2}} \tag{10.2.25}$$

$$\overline{\xi_i^2}^{1/2} = \overline{\xi_{\text{nom},i}^2}^{1/2}\left[\frac{\zeta_i}{(\zeta_i + \zeta_{ai})}\right]^{1/2} \tag{10.2.26}$$

$$S_i(z) \simeq (2\pi n_i)^2 \int_z^h m(z_1)Y_i(z_1)\,dz_1 \tag{10.2.27}$$

$$\mathcal{M}_i(z) \simeq (2\pi n_i)^2 \int_z^h m(z_1)Y_i(z_1)(z_1 - z)\,dz_1 \tag{10.2.28}$$

where $\sigma_{yi}(z)$ is the rms of the deflection at elevation z in the i-th mode of vibration, $\overline{\xi_i^2}^{1/2}$ is the rms of the corresponding generalized coordinate, $y_i(z)$ in the i-th modal shape, $Y_i(z)$ is the peak deflection in the i-th mode of vibration, g_{yi} is the peak factor, n_i is the natural frequency in the i-th mode in Hz, $\overline{\xi_{\text{nom},i}^2}^{1/2}$ is the rms nominal generalized coordinate in the i-th mode (which corresponds to the response estimated by assuming that no aeroelastic effects occur and that the motion is affected only by structural damping), $[\zeta_i/(\zeta_i + \zeta_{ai})]^{1/2}$ is the aeroelastic correction factor, ζ_i is the structural damping in the i-th mode, ζ_{ai} is the aerodynamic damping in the i-th mode, $S_i(z)$ and $\mathcal{M}_i(z)$ are, respectively, the shear force and the bending moment at elevation z due to the across-wind response in the i-th mode, and $m(z)$ is the mass of the structure per unit length.

To estimate the across-wind response, expressions are needed for the rms of the nominal generalized coordinate in the i-th mode, $\overline{\xi_{\text{nom},i}^2}^{1/2}$, and the aerodynamic damping in the i-th mode, ζ_{ai}. These expressions are given below separately for: (1) structures with constant cross section, and (2) tapered structures. In both cases the expressions are valid only for relatively small ratios $\sigma_{yi}(h)/D(h)$, for example 3% or less (to which there would correspond negligible values of the second term within the bracket of Eq. 6.1.36). It is noted that, in practice, the design of a structure will be acceptable only if the ratios $\sigma_{yi}(h)/D(h)$ inherent in that design are indeed small.

Structures with Constant Cross Section.   The following approximate expressions based on the approach described in Sect. 10.2.1 were proposed in [10-9]:

$$\overline{\xi^2_{\text{nom},i}}^{1/2} \simeq \frac{0.035 \overline{C_L^2}^{1/2} \mathscr{L}^{1/2}}{\zeta_i^{1/2} \mathscr{S}^2} \frac{\rho D^3}{M_i} \left[D \int_0^h y_i^2(z)\, dz \right]^{1/2} \qquad (10.2.29)$$

$$\zeta_{ai} \simeq -\frac{\rho D^2}{M_i} K_{a0}(1) \int_0^h y_i^2(z)\, dz \qquad (10.2.30)$$

where ρ is the air density ($\rho \sim 1.25$ kg/m^3), M_i is the generalized mass in the i-th mode (Eq. 10.1.5), and D is the outside diameter. The critical wind speed corresponding to the i-th mode of vibration has the expression:

$$U_{\text{cr},i} = \frac{n_i D}{\mathscr{S}} \qquad (10.2.31)$$

Information on the structural damping ratios ζ_i is given in Table 10.2.1. Information on the parameters, $\mathscr{S}$, $\overline{C_L^2}^{1/2}$, $\mathscr{L}$, and K_{a0} is given in Sect. 10.2.2. Note that in Eq. 10.2.20a the speed $U(10$ m$)$ corresponding to the i-th mode is

$$U(10\text{ m}) = \frac{\ln(10/z_0)}{\ln[(5/6)h/z_0]} U_{\text{cr},i} \qquad (10.2.32)$$

where h is the height of the structure in meters, and z_0 is the roughness length in meters for the terrain that determines the wind profile over the upper half of the chimney (see Table 2.2.1 and Section 2.4.1).

Numerical Example

Consider the chimney described in the numerical example of Sect. 10.1 ($h = 193.6$, $D = 17.63$ m, $n_1 = 0.364$ Hz, $y_1(z/h) = (z/h)^{1.67}$, $m(z) = 58,000$ kg/m for $z < h/2$, $m(z) = 41,000$ kg/m for $z > h/2$, $M_1 = 1.87 \times 10^6$ kg). It is assumed $\zeta_1 = 0.02$, $k/D = 10^{-5}$, and $z_0 = 0.05$ m. We seek the response in the first mode.

Assuming tentatively $\mathscr{S} = 0.22$, the critical wind speed at elevation $5h/6 = 161.3$ m is $U_{\text{cr},1} = 0.364 \times 17.63/0.22 = 29.16$ m/sec (Eq. 10.2.31), to which there corresponds a Reynolds number $\mathscr{R}e = 3.4 \times 10^7 > 2 \times 10^6$ (Eq. 10.2.9). The aspect ratio is $h/D \simeq 11$. It follows that

$\mathscr{S} \simeq 0.178$	(Eqs. 10.2.11c, 10.2.12b)
$\mathscr{L} = 1.0$	(Eq. 10.2.16b)
$\overline{C_L^2}^{1/2} \simeq 0.143$	(Eqs. 10.2.13c and 10.2.14b)
$\displaystyle\int_0^h y_1^2(z)\, dz = 44.7$ m	
$\overline{\xi^2_{\text{nom},1}}^{1/2} = 0.115$ m	(Eq. 10.2.29)
$U(10) > 12$ m/sec	(Eq. 10.2.31 and 10.2.32)
$K_{a0}(1) \simeq 0.465$	(Eq. 10.2.17c, 10.2.18, 10.2.19c, 10.2.20b, 10.2.21, and 10.2.22b)

$$\zeta_{a1} = -0.0043 \qquad \text{(Eq. 10.2.30)}$$

$$\overline{\xi_1^2}^{1/2} = 0.130 \text{ m} \qquad \text{(Eq. 10.2.26)}$$

$$g_{y1} = 3.94 \qquad \text{(Eq. 10.2.25)}$$

$$\sigma_{y1}(z) = 0.130(z/193.6)^{1.67} \text{ m} \qquad \text{(Eq. 10.2.23)}$$

$$Y_1(z) = 0.51(z/193.6)^{1.67} \text{ m} \qquad \text{(Eq. 10.2.24)}$$

$$\mathcal{M}_1(0) = 1150 \times 10^6 N \qquad \text{(Eq. 10.2.28)}$$

Note that the results of the calculations depend strongly upon, in particular, the assumed value of the structural damping ratio ζ_i. Had the value $\zeta_1 = 0.01$ been appropriate, the results obtained would have been larger than those obtained in this example by a factor of $[(0.02 - 0.0043)/(0.01 - 0.0043)]^{1/2} \simeq 1.66$.

Tapered Structures. The following approximate expressions based on the approach described in Sect. 10.2.1 were proposed in [10-9]:

$$\overline{\xi_{\text{nom},i}^2(z_{e_i})}^{1/2} \simeq \frac{0.016 \overline{C_L^2}^{1/2} \mathscr{L}^{1/2} \rho D^4(z_{e_i}) y_i(z_{e_i})}{\zeta_i^{1/2} \mathscr{S}^2 M_i \beta^{1/2}(z_{e_i})} \qquad (10.2.33)$$

$$\beta(z_{e_i}) \simeq \frac{0.1 D(z_{e_i})}{z_{e_i}} - \left. \frac{dD(z)}{dz} \right|_{z = z_{e_i}} \qquad (10.2.34)$$

$$\zeta_{ai}(z_{e_i}) = -\frac{\rho D_0^2}{M_i} \int_0^h K_{a0} \left(\frac{U(z; z_{e_i})}{U_{\text{cr}}(z_{e_i})} \right) \left[\frac{D(z)}{D_0} \right]^2 y_i^2(z) \, dz \qquad (10.2.35)$$

where the notations of Eq. 10.2.29 are used, $D_0 =$ outside diameter at base, z_{e_i} is the elevation corresponding to the critical velocity

$$U_{\text{cr}}(z_{e_i}) = \frac{n_i D(z_{ei})}{\mathscr{S}}, \qquad (10.2.36)$$

$$\frac{U(z; z_{e_i})}{U_{\text{cr}}(z_{e_i})} = \frac{\ln(z/z_0)}{\ln(z_{e_i}/z_0)} \qquad (10.2.37)$$

and z_0 is the roughness length for the terrain that determines the wind profile over the upper half of the chimney (see Table 2.2.1 and Sect. 2.4.1).

Since, as in Eq. 10.2.26

$$\overline{\xi_i^2(z_{e_i})}^{1/2} = \overline{\xi_{\text{nom},i}^2(z_{e_i})}^{1/2} \left(\frac{\zeta_i}{\zeta_i + \zeta_{ai}(z_{e_i})} \right)^{1/2}$$

it follows that the maximum response in the i-th mode corresponds to the maximum value taken on by the function

$$F_i(z_{e_i}) = D^4(z_{e_i}) y_i(z_{e_i})/\{\beta(z_{e_i})[\zeta_i + \zeta_{ai}(z_{e_i})]\}^{1/2} \qquad (10.2.38)$$

To determine that value it is in practice necessary to calculate $F(z_{e_i})$ [and, in particular, $\zeta_{ai}(z_{e_i})$] for a sufficiently larger number of elevations $0 < z_{e_i} < h$.

As pointed out in [10-8], if the structure is very lightly tapered (i.e., if

$dD(z)/dz|_{z=z_{e_i}}$ and, therefore, $\beta(z_{e_i})$ is small—see Eq. 10.2.34), then the approximations on which Eq. 10.2.33 is based are no longer valid and Eq. 10.2.33 ceases to be applicable. In that case, the chimney is assumed to behave as if it had a constant outside diameter D equal to the average diameter of its top third [10-9], and Eqs. 10.2.29 to 10.2.31 are applied with the same values of the parameters $\mathscr{S}$, $\overline{C_L^2}^{1/2}$, and $\mathscr{L}$ as those used in Eq. 10.2.33. In practice it is therefore necessary to calculate both the value of the response yielded by Eqs. 10.2.33 and 10.2.35 and the value yielded by Eqs. 10.2.29 and 10.2.31. It follows from [10-2] that the response to be assumed for structural design purposes is the *smaller* of these two values.

Numerical Example

Consider a chimney with height $h=365.8$ m, outside diameter at the base $D_0=37.8$ m, outside diameter at the tip $D(h)=12.6$ m, constant taper (i.e., $-dD(z)/dz=[D_0-D(h)]/h$), fundamental frequency $n_1=0.252$ Hz [10-9]. It is assumed that the fundamental modal shape $y_1(z/h)=(z/h)^2$, the mass per unit length $m(z)=180{,}000(1-0.9z/h)$ kg/m, the structural damping ratio in the first mode $\zeta_1=0.01$, the relative surface roughness of the structure $k/D\simeq 10^{-5}$, and the roughness of the terrain $z_0=0.008$ m. We seek the response of the chimney in the first mode of vibration.

Assuming tentatively $\mathscr{S}\simeq 0.2$, the critical speed $U_{cr}>0.252\times 12.6/0.2=15.9$ m/sec (Eq. 10.2.36), to which there corresponds $\mathscr{R}e>67{,}000\times 15.9\times 12.6=1.3\times 10^7$ (Eq. 10.2.9). The aspect ratio is $h/D(h)\simeq 29.0$. It follows that

$$\mathscr{S}\simeq 0.23 \qquad \text{(Eq. 10.2.11c)}$$
$$\overline{C_L^2}^{1/2}=0.15 \qquad \text{(Eqs. 10.2.13c and 10.2.14a)}$$
$$\mathscr{L}=1.0 \qquad \text{(Eq. 10.2.16b)}$$
$$M_1=3.3\times 10^6 \text{ kg} \qquad \text{(Eq. 10.1.5)}$$
$$a_1=1.0 \qquad \text{(Eq. 10.2.19c)}$$
$$a_3=0.9 \qquad \text{(Eq. 10.2.21)}$$
$$a_4=1.0 \qquad \text{(Eq. 10.2.22a)}$$

The coefficient a_2 (Eqs. 10.2.20) depends upon the wind speed $U(10; z_{e_i})$.

As mentioned earlier, the function $F_i(z_{e_i})$ must be calculated for a sufficiently large number of elevations z_{e_i} to obtain the value that maximizes the response. We show here calculations for $z_{e_1}=365.8$ m and $z_{e_1}=182.9$ m.

For $z_{e_1}=365.8$ m, $U_{cr}(365.8)=13.70$ m/sec (Eq. 10.2.36) and $U(10; 365.8)=9.1$ m/sec <12 m/sec (Eq. 10.2.37). Therefore, $a_2=2.0$ (Eq. 10.2.20a). It can be verified that $\zeta_{a1}(365.8)\simeq -0.0065$ (Eqs. 10.2.17 and 10.2.35), and $F_1(365.8)\simeq 1.6\times 10^6$ m^4 (Eqs. 10.2.38 and 10.2.34).

For $z_{e_1}=182.9$ m, $U_{cr}=27.61$ m/sec (Eq. 10.2.36), $U(10; 182.9)=19.16$ m/sec >12 m/sec (Eq. 10.2.37), $a_2=1.0$ (Eq. 10.2.20b), $\zeta_{a1}(182.9)=-0.0042$

(Eqs. 10.2.17 and 10.2.35), $F_1(182.9) \simeq 4.6 \times 10^6 \text{ m}^4$ (Eqs. 10.2.38 and 10.2.34). It can be verified that the largest value of $F_1(z_{e_1})$ and, therefore, the highest response in the first mode occurs for $z_{e_1} \simeq 182.9 \text{ m}$. It follows that

$$\overline{\xi_{\text{nom},1}^2(182.9)}^{1/2} = 0.06 \text{ m} \qquad \text{(Eqs. 10.2.33 and 10.2.34)}$$

$$\overline{\xi_1^2(182.9)}^{1/2} = 0.079 \text{ m} \qquad \text{(Eq. 10.2.26)}$$

$$\sigma_{y1}(z) = 0.079(z/365.8)^2 \text{ m} \qquad \text{(Eq. 10.2.23)}$$

$$Y_1(z) = 0.304(z/365.8)^2 \text{ m} \qquad \text{(Eqs. 10.2.24 and 10.2.25)}$$

$$\mathscr{M}_1(0) = 1290 \times 10^6 \text{ Nm} \qquad \text{(Eq. 10.2.28)}$$

The response will now be estimated by using Eqs. 10.2.29–10.2.31. The average outside diameter of the top third of the chimney is $D = 16.8 \text{ m}$. It follows that

$$\overline{\xi_{\text{nom},1}^2}^{1/2} \simeq \frac{0.035 \times 0.15 \times 1.0}{0.01^{1/2} \times 0.23^2} \frac{1.25 \times 16.8^3}{3.3 \times 10^6} \left(16.8 \frac{365.8}{5}\right)^{1/2}$$

$$= 0.0625 \text{ m} \qquad \text{(Eq. 10.2.29)}$$

$$U_{\text{cr}} = 18.4 \text{ m/sec} \qquad \text{(Eq. 10.2.31)}$$

$$U(10) > 12 \text{ m/sec} \qquad \text{(Eq. 10.2.32)}$$

$$\zeta_{a1} = -\frac{1.25 \times 16.8^2}{3.3 \times 10^6}(0.9 \times 0.55)\frac{365.8}{5}$$

$$\simeq -0.0039 \qquad \text{(Eqs. 10.2.30, 10.2.17}c\text{, 10.2.18,}$$
$$\text{10.2.19}c\text{, 10.2.20}b\text{, 10.2.21, and}$$
$$\text{10.2.22}a\text{)}$$

$$\sigma_{y1}(z) = 0.079(z/365.8)^2 \text{ m} \qquad \text{(Eqs. 10.2.23 and 10.2.26)}$$

$$Y_1(z) = 0.304(z/365.8)^2 \text{ m} \qquad \text{(Eqs. 10.2.24 and 10.2.25)}$$

$$\mathscr{M}_1(0) = 1280 \times 10^6 \text{ Nm} \qquad \text{(Eq. 10.2.28)}$$

It is seen that in this case the response yielded by Eqs. 10.2.29–10.2.31 is approximately the same as that obtained by Eqs. 10.2.33–10.2.35.

10.3   ALLEVIATION OF VORTEX-INDUCED OSCILLATIONS

Aerodynamic Devices. A common method of alleviating vortex-induced oscillations is the provision of "spoiler" devices that destroy or reduce the coherence of the shed vortices [10-26, 10-27].

Of the various types of such devices, one of the most effective is the helical strake system first described in [10-15]. The system consists of three thin rectangular strakes with a pitch of one revolution in 5 diameters and a strake (radial) height of 0.10 diameter (to 0.13 diameter for very light or lightly damped structures) applied over the top 33% to 40% of the stack height. The effectiveness

FIGURE 10.3.1.   Steel chimney with helical strakes. From G. Hirsch and H. Ruscheweyh, "Full-Scale Measurements on Steel Chimney Stacks," *J. Ind. Aerodyn.*, **1** (1976), 341–347.

of the system is not impaired by a gap of $0.005D$ between the strake and the cylinder surface [10-16]. Reference 10-17 reports the remarkable results obtained by using this system (with 5 mm thick strakes, 0.6 m strake height, and 30 m pitch) in the case of a 145 m tall and 6 m diameter steel stack (Fig. 10.3.1).

For Reynolds numbers $\mathscr{R}e < 2 \times 10^5$ or so, in flow with about 15% turbulence intensity, helical strakes were found to reduce the peak of the across-wind resonant oscillations by a factor of about two, as opposed to a factor of about 100 in the case of smooth flow [10-23]. Strakes may also be expected to be less effective at Reynolds numbers $2 \times 10^5 < \mathscr{R}e < 2 \times 10^6$ or so than outside this range, since for such Reynolds numbers the vortex shedding is random (see Fig. 4.4.4). It appears that the performance of strakes can be unsatisfactory in the case of stacks grouped in a row [10-28, 10-29]. It is noted that the strakes increase drag, as shown in Fig. 10.3.2 [10-18].

Shrouds can also be effective in reducing the coherence of shed vortices. A schematic view of a shroud fitted to a stack is shown in Fig. 10.3.3. Results of wind tunnel experiments reported in [10-16] showed that oscillations were substantially reduced with only the top 25% of the model height shrouded. The most effective shrouds were found to be those with a gap width $w \simeq 0.12D$ and an open-area ratio between 20% and 36% (with length of square $s = 0.052D$ to $0.070D$).

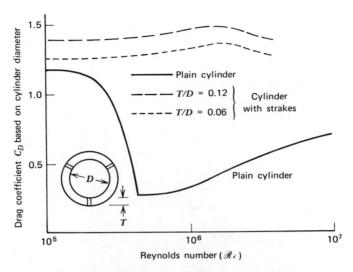

FIGURE 10.3.2.   Effect of strakes on drag coefficient. From L. R. Wooton and C. Scruton, "Aerodynamic Stability," in *The Modern Design of Wind-Sensitive Structures*, Construction Industry Research and Information Association, London, U.K., 1971, pp. 65–81. By permission of the Director of the National Physical Laboratory, U.K., and the Director of the Construction Industry Research and Information Association, U.K.

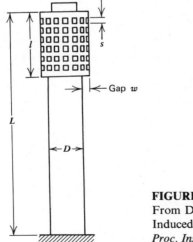

FIGURE 10.3.3.   View of shroud fitted to a stack [10-16]. From D. E. Walshe and L. R. Wooton, "Preventing Wind-Induced Oscillations of Structures of Circular Section," *Proc. Inst. Civ. Eng.*, **47** (1970), 1–24.

Mechanical Devices.   Such devices include hydraulic dampers and tuned mass dampers (TMDs).

The use of hydraulic dampers to reduce vortex-induced oscillations is discussed in [10-19]. An example of such an application is given in [10-17], which mentions the use of three hydraulic automotive shock absorbers installed at

120° angles in a plane view between a 47 m high stack and a separate structure at the 18 m level.

The tuned mass damper (TMD) consists of a secondary vibratory system attached to the structure and located near its top (see Sect. 9.4). If excited by harmonic (or quasiharmonic) oscillations of the structure, the TMD will vibrate in opposition to these motions and thereby reduce the amplitude of the structural response. The basic theory of the TMD is discussed in [10-20]. One of the first tuned mass dampers used in a large structure was designed for the Centerpoint Tower in Sydney, Australia. The mass for the damper was in this case provided by the water tank of the tower [10-21]. Further applications of TMDs to reduce tower oscillations are discussed in [10-22].

REFERENCES

10-1   W. S. Rumman, "Basic Structural Design of Concrete Chimneys," *J. Power Div.*, ASCE, **96** (June 1970), 309–318.

10-2   B. J. Vickery and A. W. Clark, "Lift or Across-Wind Response of Tapered Stacks," *J. Struct. Div.*, ASCE, **98**, No. ST1 (Jan. 1972), 1–20.

10-3   ESDU, *Across-Wind Vibrations of Structures of Circular Cross-Section in Wind or Gas Flows*, Item 78006, Engineering Science Data Unit, London, 1978.

10-4   B. J. Vickery and R. I. Basu, "Across-Wind Vibrations of Structures of Circular Cross-Section, Part I, Development of a Two-Dimensional Model for Two-Dimensional Conditions," *J. Wind Eng. Ind. Aerodyn.*, **12** (1983), 49–73.

10-5   R. I. Basu and B. J. Vickery, "Across-Wind Vibrations of Structures of Circular Cross-Section, Part 2, Development of a Mathematical Model for Full Scale Application," *J. Wind Eng. Ind. Aerodyn.*, **12** (1983), 75–97.

10-6   B. J. Vickery, "Across-Wind Buffeting in a Group of Four In-Line Model Chimneys," *J. Wind Eng. Ind. Aerodyn.*, **8** (1981), 177–193.

10-7   R. I. Basu and B. J. Vickery, "A Comparison of Model and Full-Scale Behavior in Wind of Towers and Chimneys," *Proceedings Wind Tunnel Modeling for Civil Engineering Applications*, Gaithersburg, MD, April 1982, Cambridge University Press, New York, 1982.

10-8   B. J. Vickery, "The Aeroelastic Modeling of Chimneys and Towers," *Proceedings Wind Tunnel Modeling for Civil Engineering Applications*, Gaithersburg, MD, April 1982, Cambridge University Press, New York, 1982.

10-9   B. J. Vickery and R. I. Basu, "Simplified Approaches to the Evaluation of the Across-Wind Response of Chimneys," *Proceedings 6th International Conference on Wind Engineering*, March 1983, Gold Coast, Australia, in *J. Wind Eng. Ind. Aerodyn.*, **14** (1983), 153–166.

10-10   L. C. Maugh and W. S. Rumman, "Dynamic Design of Reinforced Concrete Chimneys," *Journal Am. Concrete Inst.*, Sept. 1967.

10-11   G. M. Pinfold, *Reinforced Concrete Chimneys and Towers*, Viewpoint Publications, Scholium International, Inc., Flushing, NY, 1975.

10-12   K. C. S. Kwok and W. H. Melbourne, "Wind-Induced Lock-In Excitation of Tall Structures," *J. Struct. Div.*, ASCE, **107** (1981), 57–72.

10-13   R. I. Basu, *Across-Wind Response of Slender Structures of Circular Cross Section to Atmospheric Turbulence*, Vol. 1, Research Report BLWT-2-1983, University of Western Ontario, Faculty of Engineering Science, London, Ont., Canada, 1983.

10-14   A. G. Davenport and M. Novak, "Vibration of Structures Induced by Wind," Chapter 29-II in *Shock and Vibration Handbook*, 2nd ed., C. M. Harris and C. E. Crede (Eds.), McGraw-Hill, New York, 1976.

10-15   C. Scruton, *Note on a Device for the Suppression of the Vortex-Excited Oscillations of Flexible Structures of Circular or Near Circular Section, with Special Reference to Its Application to*

Tall Stacks, NPL Aero Report No. 1012, National Physical Laboratory, Teddington, U.K., 1963.

10-16 D. E. Walsh and L. R. Wooton, "Preventing Wind-Induced Oscillations of Structures of Circular Section," *Proc. Inst. Civ. Eng.*, **47** (1970), 1–24.

10-17 G. Hirsch and H. Ruscheweyh, "Full-Scale Measurements on Steel Chimney Stacks," *J. Ind. Aerodyn.*, **1**, 4 (Aug. 1976), 341–347.

10-18 L. R. Wooton and C. Scruton, "Aerodynamic Stability," in *Modern Design of Wind-Sensitive Structures*, Construction Research and Information Association, London, 1970.

10-19 A. Brunner, "Amortisseur d'oscillations hydraulique pour cheminées," *Journées de l'Hydraulique*, **8**, Part III, Lille, France (1964).

10-20 J. P. Den Hartog, *Mechanical Vibrations*, 4th ed., McGraw-Hill, New York, 1956.

10-21 "Tower's Cables Handle Wind, Water Tank Damps It," *Eng. News Record*, **187**, 24 (Dec. 1971), 23.

10-22 R. H. Scanlan and R. L. Wardlaw, "Reduction of Flow-Induced Vibrations," in *Isolation of Mechanical Vibration Impact and Noise*, AMD, Vol. 1, Section 2, ASME, New York, 1973, 35–63.

10-23 I. S. Gartshore, J. Khanna, and S. Laccinole, "The Effectiveness of Vortex Spoilers on a Circular Cylinder in Smooth and Turbulent Flow," in *Wind Engineering*, Proceedings of the Fifth International Conference, Fort Collins, Colorado, July 1979, J. E. Cermak (Ed.), Pergamon Press, Oxford and New York, 1980.

10-24 W. Hanenkamp and W. Hammer, "Transverse Vibration Behavior of Cylinders in Line", *J. Wind Eng. Ind. Aerodyn.*, **7** (1981), 37–53.

10-25 H. Ruscheweyh, "Problems With In-Line Stacks: Experience With Full-Scale Objects", *Eng. Struct.*, **6** (1984), 340–343.

10-26 M. Zdravkovich, "Review and Classification of Various Aerodynamic and Hydrodynamic Means for Suppressing Vortex Shedding", *J. Wind Eng. Ind. Aerodyn.*, **7** (1981), 145–189.

10-27 M. Zdravkovich, "Reduction of Effectiveness of Means for Suppressing Wind-Induced Oscillation", *Eng. Struct.*, **6** (1984), 344–349.

10-28 H. Ruscheweyh, "Straked In-Line Steel Stacks With Low Mass Damping", *J. Wind Eng. Ind. Aerodyn.*, **8** (1981), 203–210.

10-29 H. Ruscheweyh, "Dynamische Windwirkung an Bauwerken", Bauverlag, Wiesbaden, 1982.

11

Hyperbolic Cooling Towers

Much research into the wind loading of hyperbolic cooling towers has been conducted following the wind-induced collapse in 1965 of three out of a group of eight cooling towers at the Ferrybridge Power Station in England [11-1]. Principal areas of investigation have been: (a) the spatial distribution and the variation with time of the wind loading on the tower surface, and (b) the response of the tower to wind loads, including the dynamic effects induced by fluctuating wind loads. The purpose of this chapter* is to summarize and reference principal results currently available in these two areas. These results are presented in Sects. 11.1 and 11.2 for towers that are not significantly affected aerodynamically by the presence of neighboring structures. Information on groups of cooling towers is presented in Sect. 11.3.

11.1 DESCRIPTION OF WIND LOADING

Wind-induced pressures acting on a tower are determined by the characteristics of the oncoming flow, the tower geometry, and the features of the tower surface. In addition, the pressures depend upon the Reynolds number of the flow, which is in most cases of the order of 10^7 to 10^8 for the prototype, and by about two orders of magnitude smaller in the wind tunnel. On account of this dependence it has been necessary to complement wind tunnel test by full scale measurements.

As usual, it is convenient to describe the pressures in terms of the sum of a mean and a fluctuating part.

11.1.1 Mean Pressures

The mean pressure at a point defined by the height above ground z and the

---

*The authors would like to acknowledge the valuable contributions to this chapter by Professor D. A. Reed.

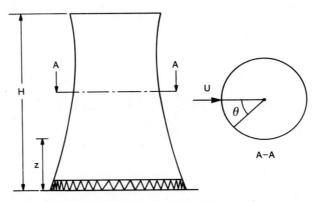

FIGURE 11.1.1.   Hyperbolic cooling tower—notations.

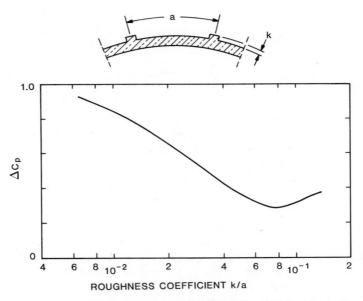

FIGURE 11.1.2.   Approximate pressure difference ΔC_p as a function of roughness coefficient k/a for towers with 36 to 144 ribs. After H. J. Niemann, "Wind Effects on Cooling-Tower Shells," *J. Struct. Div.*, ASCE, **106** (1980), 643–661.

angular coordinate θ (Fig. 11.1.1) can be expressed as:

$$p(z, \theta) \simeq \tfrac{1}{2}\rho[C_p(z, \theta)U^2(z) + C_{pi}U^2(H)] \tag{11.1.1}$$

where ρ is the air density ($\rho \simeq 1.25\,\text{kg/m}^3$), $U(z)$ is the mean wind speed at elevation z in the undisturbed oncoming flow, $C_p(z, \theta)$ is the mean external pressure coefficient, H is the height of the tower, and C_{pi} is the internal pressure coefficient. Based on results of full-scale measurements, [11-2] suggests $C_{pi} \simeq$

0.4.* The following tentative relations, based on wind tunnel and full-scale measurements, have been proposed for the external pressure coefficient $C_p(z, \theta)$ [11-3]:

$$C_p(z, \theta) \simeq 1 \qquad\qquad \theta = 0^\circ \qquad (11.1.2a)$$

$$C_p(z, \theta) \simeq 1 - B \sin^C\left(90\,\frac{\theta}{\theta_1}\right) \quad 0 < \theta \leqslant \theta_b \qquad (11.1.2b)$$

$$C_p(z, \theta) = C_p(z, \theta_b) \qquad\qquad \theta > \theta_b \qquad (11.1.2c)$$

$$B \simeq 1 + 0.4\left(\frac{H}{z}\right)^{2\alpha} + \Delta C_p \qquad (11.1.2d)$$

$$C \simeq -\frac{\ln B}{\ln\left[\sin\left(90\,\dfrac{\theta_o}{\theta_1}\right)\right]} \qquad (11.1.2e)$$

where H = height of tower, ΔC_p is a function of the ratio k/a of the rib height, k, to the distance between ribs, a, represented in Fig. 11.1.2, α is an exponent characterizing the mean wind profile (Table 2.2.2), and the angle θ is expressed in degrees. The angles θ_o, θ_1, and θ_b are represented in the schematic pressure distribution diagram of Fig. 11.1.3. Values for these angles are given in Fig. 11.1.4a (based on full scale measurements on the Schmehausen tower) and in Figs. 11.1.4b and 11.1.4c (based on wind tunnel measurements) [11-20].

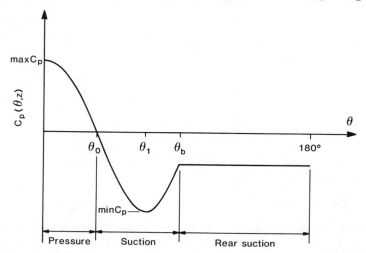

FIGURE 11.1.3. Distribution of pressure coefficient C_p. After H. Pröpper and J. Welsch, "Wind Pressures on Cooling Tower Shells," in *Wind Engineering*, Proceedings of the Fifth International Conference, Fort Collins, CO, July 1979, J. E. Cermak (Ed.), Pergamon Press, 1980.

*Wind tunnel measurements quoted in [11-2] suggest that somewhat higher internal pressures occur in unvented towers, that is, $C_{pi} \simeq 0.5$ or even 0.6.

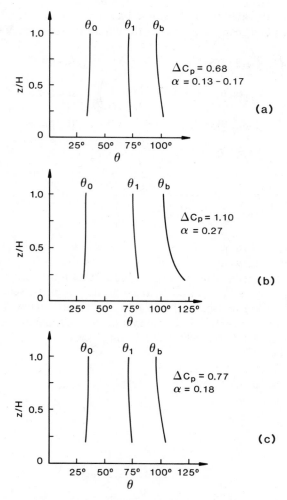

FIGURE 11.1.4. Angles θ_0, θ_1, and θ_b (After [11-20]).

Numerical Example

Assume that, as in the case of the Martin's Creek tower, $H = 127$ m, $k/a \simeq 0.02$, and $\alpha \simeq 0.17$. We seek the values of $C_p(z, \theta)$ for $z = 95.4$ m and $\theta = 35°$, $70°$, and $97°$.

From Fig. 11.1.2, $\Delta C_p \simeq 0.65$. For $z = 95.4$ m, Fig. 11.1.4a yields $\theta_o \simeq 35°$, $\theta_1 \simeq 70°$, and $\theta_b \simeq 97°$. It follows that $B = 2.1$, $C = 2.14$ (Eqs. 11.1.2d and 11.1.2e). From Eq. 11.1.2b, $C_p(95.4$ m, $35°) \simeq 0$, $C_p(95.4$ m, $70°) \simeq -1.1$, and $C_p(95.4$ m, $97°) \simeq -0.38$.

Values of the external pressure coefficient C_p at the tower throat obtained from full-scale measurement by a number of investigators are shown in Fig.

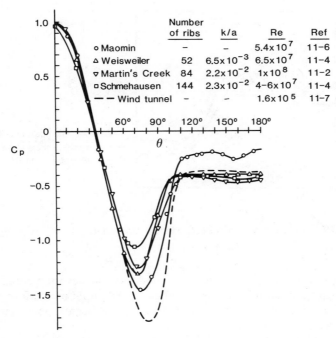

FIGURE 11.1.5. Mean pressure coefficient around the throat section of hyperbolic cooling tower for four full-scale data sets and one wind tunnel set. After Tien-fun Sun and Liang-mao Zhou, "Wind Pressure Distribution Around a Ribless Hyperbolic Cooling Tower," *J. Wind Eng. Ind. Aerodyn.*, **14** (1983), 181–192.

11.1.5. Note that the values obtained for the tower of [11-6] differ appreciably from the other sets of values. This is due to the absence of ribs on the external surface of that tower. Note also the agreement to within about 15% between the values obtained in the numerical example and the values measured on the Martin's Creek tower at the throat elevation $z = 95.4$ m [11-2]. Figure 11.1.5 also shows an example of differences between values of C_p obtained from a set of wind tunnel tests on the one hand and full scale measurements on the other.

11.1.2   Fluctuating Pressures

Stresses induced by fluctuating pressures are usually comparable in value to stresses induced by the mean loads. The purpose of this section is to present descriptions of fluctuating pressures for use in the estimation of tower response. Additional information on fluctuating pressures is presented in [11-5] and [11-22].

RMS of Fluctuating Wind Pressures. The rms of the fluctuating wind pressures, $\sigma_p(z, \theta)$, may be written as:

$$\sigma_p(z, \theta) = \tfrac{1}{2}\rho C_p'(z, \theta)U^2(z) \qquad (11.1.3)$$

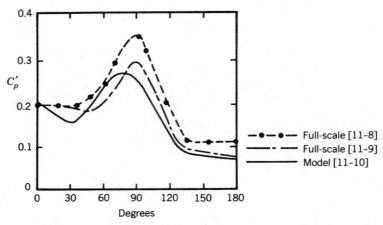

FIGURE 11.1.6. Fluctuating pressure coefficient around the throat of a hyperbolic cooling tower.

where ρ is the air density, $U(z)$ is the mean wind speed at elevation z, and $C'_p(z, \theta)$ is an empirical fluctuating pressure coefficient. Attempts to relate $\sigma_p(z, \theta)$ to the turbulence intensity of the oncoming flow have been reported in [11-3] and [11-8]. According to [11-3]

$$C'_p(z, 0) \simeq 1.8 \frac{\sigma_u}{U(z)} \tag{11.1.4}$$

where σ_u is the rms of the longitudinal velocity fluctuations. The variation of $C'_p(z, \theta)$ with θ at the elevation of the throat is shown for three sets of measurements in Fig. 11.1.6 [11-21]. According to [11-3] and [11-11], this variation depends upon the ratio k/D, where k is the height of the ribs and D is the diameter of the tower at the throat. Note that the coefficients C'_p are lower for the rougher than for the smoother towers in the region $60° < \theta < 120°$ (Fig. 11.1.7).

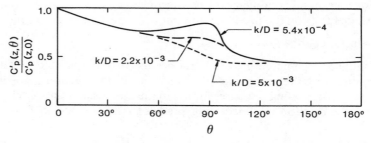

FIGURE 11.1.7. Ratios $C'_p(z, \theta)/C'_p(z, 0)$ for towers with various roughness parameters k/D at elevation $z = 0.7 H$. From H. Pröpper and J. Welsch, "Wind Pressures on Cooling Tower Shells," in *Wind Engineering*, Proceedings of the Fifth International Conference, Fort Collins, CO, July 1979, J. E. Cermak (Ed.), Pergamon Press, 1980.

Spectra of Fluctuating Pressures.   The following expression for the spectra of fluctuating pressures was proposed in [11-3]:

$$\frac{nS_p(z, \theta, n)}{\sigma_p^2(z, \theta)} = \frac{\beta_p(\theta)X_p(\theta)}{[1 + \gamma_p(\theta)X_p^2(\theta)]^{d(\theta)}}$$ (11.1.5)

where

$$d(\theta) = \frac{1 - a_1(\theta)}{2}$$ (11.1.6)

$$\gamma_p(\theta) = \left[\frac{\beta_p(\theta)}{0.115}\right]^{1/d(\theta)}$$ (11.1.7)

$$X_p(\theta) = \left[b_0^{1/\alpha}(\theta)\left(\frac{D}{L_u^x}\right)^{2/3}\right]^{1/a_1(\theta)} \frac{nD}{U(z)}$$ (11.1.8)

where n is the frequency, the parameters $a_1(\theta)$, $b_0(\theta)$, and $\beta_p(\theta)$ are given in Fig. 11.1.8, α is the power law exponent (Table 2.2.2), D is the diameter at the throat, and L_u^x is the integral scale of turbulence (Sect. 2.3.2).

Cross-Spectra of Fluctuating Pressures.   According to [11-12], quadrature spectra are negligible, that is, the cross-spectra are for practical purposes equal

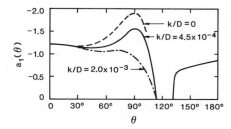

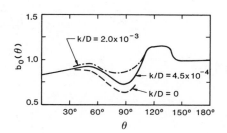

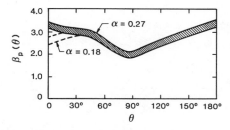

FIGURE 11.1.8.   Parameters $a_1(\theta)$, $b_0(\theta)$, and $\beta_p(\theta)$. From H. Pröpper and J. Welsch, "Wind Pressures on Cooling Tower Shells," in *Wind Engineering*, Proceedings of the Fifth International Conference, Fort Collins, CO, July 1979, J. E. Cermak (Ed.), Pergamon Press, 1980.

to the co-spectra. The following relations were proposed in [11-12] for the cross-spectra of the pressure fluctuations:

1. Windward Region ($\theta \gtrsim 100°$, $\theta' \gtrsim 100°$)

$$S_p(z, \theta, z', \theta', n) = R_v(z, z', n) R_f(\theta, \theta', n) S_p^{1/2}(z, \theta, n) S_p^{1/2}(z', \theta', n) \quad (11.1.9)$$

2. Leeward Region ($\theta \gtrsim 100°$, $\theta' \gtrsim 100°$)

$$S_p(z, \theta, z', \theta', n) = R_v(z, z', n) R_r(\theta, \theta', n) S_p^{1/2}(z, \theta, n) S_p^{1/2}(z', \theta', n) \quad (11.1.10)$$

According to [11-12], cross-spectra of pressures on the windward region on the one hand and pressures on the leeward region on the other are negligible. This is a simplifying assumption which is not entirely consistent with results reported in [11-3].

In Eqs. 11.1.9 and 11.1.10

$$R_v(z, z', n) = \exp(-\beta_1 \tilde{f}_1) \quad\quad\quad (11.1.11)$$

$$R_r(\theta, \theta', n) = \exp(-\beta_2 \tilde{f}_2) \quad\quad\quad (11.1.12)$$

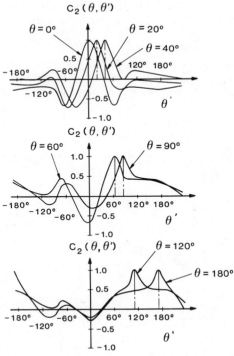

Figure 11.1.9.   Correlation coefficients $C_2(\theta, \theta')$. From H. Pröpper and J. Welsch, "Wind Pressures on Cooling Tower Shells," in *Wind Engineering*, Proceedings of the Fifth International Conference, Fort Collins, CO, July 1979, J. E. Cermak (Ed.), Pergamon Press, 1980.

$$R_f(\theta, \theta', n) = C_2(\theta, \theta')R(|\theta - \theta'|, n) \tag{11.1.13}$$

$$R(|\theta - \theta'|, n) = \exp(-\beta_3 \tilde{f}_2^2) \tag{11.1.14}$$

$$\tilde{f}_1 = \frac{n|z - z'|}{U(\delta)} \tag{11.1.15}$$

$$\tilde{f}_2 = \frac{\pi n D \dfrac{|\theta - \theta'|}{360°}}{U(\delta)} \tag{11.1.16}$$

where $\beta_1 \simeq 7$, $\beta_2 \simeq 11$, $\beta_3 \simeq 25$ [11-12], $U(\delta)$ is the mean wind speed at the gradient height δ listed in Table 2.2.2, and $C_2(\theta, \theta')$, as obtained in [11-3], is given in Fig. 11.1.9.

11.2   ESTIMATION OF TOWER RESPONSE

Several approaches to the estimation of tower response have been proposed. For towers that exhibit no significant resonant amplification effects, [11-7] employs expressions for the meridional and circumferential correlations of the fluctuating pressures to obtain the variances of the meridional, circumferential, and normal displacements of the tower shell.

This approach was superseded by [11-12], which employs a spectral approach in which models of spectra and cross-spectra of pressure fluctuations (see Sect. 11.1) are used to obtain the spectral densities of the response by methods fundamentally similar to those of Sects. 5.2.7 and 5.3. The spectral approach is applicable to towers for which resonant amplification effects are significant, as well as to towers which—as is most commonly the case—are sufficiently stiff that resonant amplification effects are negligible. In both cases the calculations can be carried out by using a computer program similar to that listed in [9-14], but modified to account for differences in geometry and in the modeling of pressures, as well as for the fact that a typical response of the tower, rather than having the form of Eq. 5.2.1, is written as

$$y(s, \theta, t) = \sum_m \sum_i [q_{m,i}(t) \cos m\theta + q'_{m,i}(t) \sin m\theta] y_{m,i}(s) \tag{11.1.17}$$

where s is the distance along the meridian, θ is the angular coordinate, t is the time, $q_{m,i}$ and $q'_{m,i}$ are the time-dependent symmetric and antisymmetric generalized coordinates for mode m, i, respectively, and $y_{m,i}(s)$ is the vertical modal shape. An attempt to use a spectral approach to estimate the response was also reported in [11-13].

In [11-14] and [11-15] finite element methods of analysis are used in conjunction with step by step integrations in the time domain. One advantage of such an approach is that it can accommodate nonlinearities and changes of the physical properties of the structure during the loading process. Time histories of fluctuating pressures used in this approach can consist of measured data, as

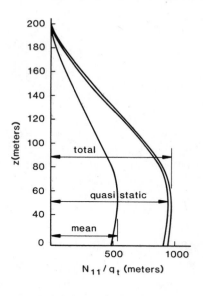

FIGURE 11.2.1. Ratios of meridional stress, N_{11}, at stagnation point to dynamic pressure, q_t, at elevation of tower throat. After H.-J. Niemann, "Wind Effects on Cooling-Tower Shells," *J. Struct. Div.*, ASCE, **106** (1980), 643–661.

in [11-14]* and [11-15], or can be simulated by Monte Carlo methods from spectral and cross-spectral information. More recently, ARIMA (Auto Regressive Integrated Moving Average) methods have been used for representing fluctuating loads in the time domain [11-16]. Time-domain solutions, though potentially useful for research purposes, are costly and generally impractical for routine design.

Spectral methods, as developed in [11-12], were applied in [11-4] to study the response of typical reinforced concrete towers with ratio $H/D = 2.0$ ($D =$ diameter at throat). The results obtained indicated that the resonant amplification effects contributed less than 5% to the total response. A typical diagram of the ratio N_{11}/q_t at the stagnation point is shown in Fig. 11.2.1 for $U(z_{\text{throat}}) = 45.4$ m/sec (N_{11} is the meridional stress, $q_t = (1/2)\rho U^2(z_{\text{throat}})$, ρ is the air density, and $U(z_{\text{throat}})$ is the mean wind speed at the elevation of the tower throat). It is seen that in this case the peak total response differs insignificantly from the peak quasistatic response (obtained by neglecting resonant amplification effects). The latter is approximately twice as large as the mean response.

It is shown in [11-4] that for the type of towers studied therein, the design may be based on an equivalent static pressure

$$p(z, \theta) = C_p(z, \theta) q_D(z) \tag{11.2.1}$$

where, in open country,

$$q_D(z) \simeq 5 \left(\frac{z}{\delta} \right)^{0.23} [\tfrac{1}{2} \rho U^2(10)] \phi \tag{11.2.2}$$

*Because measured data were available only for the throat section, it was assumed in [11-14] that the vertical distribution of the loads is uniform.

where $\delta \simeq 280$ m, ρ is the air density ($\rho \simeq 1.25$ kg/m^2), $U(10)$ is the hourly mean wind speed at 10 m above ground, and ϕ is a factor accounting for resonant amplification effects ($1 \gtrsim \phi \gtrsim 1.1$).

An equivalent static pressure approach is also included in [11-17], in which the expression for the equivalent pressures is consistent with the format used for dynamic pressures in the American National Standard A58.1-1972 [11-18]. Reference 11-17 recommends the use in this expression of aerodynamic coefficients obtained from wind tunnel or full-scale tests, and of a gust loading factor to be determined by a dynamic analysis. It is specified that the gust loading factor, which is applied to a load corresponding to the fastest-mile wind speed, shall not be less than unity. This corresponds to a gust loading factor corresponding to an hourly mean speed of about $(1.24)^2 = 1.54$, where 1.24 is the conventional ratio between fastest-mile wind speed and hourly wind speed assumed in [11-18].

The use of a single gust loading factor implies the assumption that the stress amplification due to wind gustiness may be considered for practical purposes to be the same at all points of the tower and for all types of stress. As shown in [11-23], this assumption is not necessarily correct in all cases.

11.3 GROUPS OF HYPERBOLIC COOLING TOWERS

Wind-induced stresses in the tower shells can be considerably more severe in the case of groups of towers than for isolated structures. This was borne out by the behavior during the November 1, 1965 storm* of the eight towers of the Ferrybridge C Generating Station (Fig. 11.3.1), three of which collapsed, while five survived. The inquiry of [11-1] indicated that failure was due to large tensions in the windward face of the towers. On the basis of wind tunnel tests and of information on the design of the towers, it was estimated in [11-19] that the mean hourly wind speeds at 10 m above ground, $U(10)$, at which failure of the towers could be expected to occur had the values shown in Table 11.3.1. The wind speeds $U(10)$ during the storm were reported to rise from about 18 m/sec to about 20 m/sec. The reported sequence of tower failures was found to be consistent with the results of Table 11.3.1 [11-19].

It is noted in [11-19] that higher mean and fluctuating loads often occur when the wind blows through a gap between upstream towers. Details on distributions of mean fluctuating pressures on the surface of towers placed in

TABLE 11.3.1. Estimated Wind Speeds Corresponding to Tower Failures (m/sec) [11-19]

Tower	1A	1B	2A	2B	3A	3B	4A	4B
$U(10)$	19.1	19.1	19.7	23.4	21.5	23.8	21.6	21.3

*The approximate mean wind direction is shown in Fig. 11.3.1.

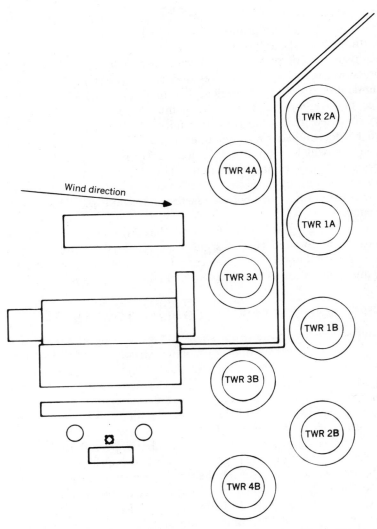

FIGURE 11.3.1. Tower locations at Ferrybridge C Generating Station. From J. Armitt, "Wind Loading on Cooling Towers," *J. Struct. Div.*, ASCE, **106** (1980), 623–641.

the wake of other buildings or in groups are given for specific configurations in [11-19] on the basis of both wind tunnel and full-scale measurements, and in [11-8] on the basis of full-scale tests.

Stress amplifications due to interference effects can also occur in the case of pairs of cooling towers (Fig. 11.3.2). Laboratory data on such amplifications are shown in Fig. 11.3.3 for various wind directions and distances between the towers in a pair (max n_{11} is the maximum hoop tension; max n_{22} is the maximum meridional tension; min n_{22} is the maximum meridional compression;

FIGURE 11.3.2.   Cooling towers, Limerick Generating Station, Limerick, Pennsylvania. Courtesy of Philadelphia Electric Company, Limerick Generating Station.

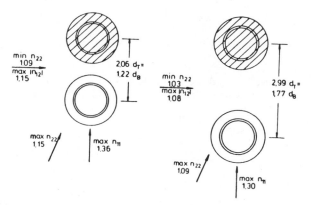

FIGURE 11.3.3.   Ratios of stresses amplified by interference effects to corresponding stresses on isolated tower (d_T is the diameter at throat; d_B is the diameter at base). After H.-J. Niemann, "Reliability of Current Design Methods for Wind-Induced Stresses" in *Natural Draft Cooling Towers*, P. L. Gould, W. B. Krätzig, I. Mungan, and U. Wittek, (Eds.), Springer-Verlag, Berlin, 1984.

max $|n_{12}|$ is the maximum shear force). It is seen that in some instances the amplifications are considerable (over 30%). It is noted in [11-23] that cooling towers can also be adversely affected by the presence of adjacent buildings within a power plant.

REFERENCES

11-1   *Report of the Committee of Inquiry into the Collapse of Cooling Towers at Ferrybridge on Monday, 1 November 1965*, Central Electricity Generating Broad, H.M. Stationary Office, London, England, 1966.

11-2   N. J. Sollenberger, R. H. Scanlan, and D. P. Billington, "Wind Loading and Response of Cooling Towers," *J. Struct. Div.*, ASCE, **106** (1980), 601–621.

11-3   H. Pröpper and J. Welsch, "Wind Pressures on Cooling Tower Shells," in *Wind Engineering*, Proceedings of the Fifth International Conference, Fort Collins, CO, July 1979, J. E. Cermak (Ed.), Pergamon Press, 1980.

11-4   H. J. Niemann, "Wind Effects on Cooling-Tower Shells," *J. Struct. Div.*, ASCE, **106** (1980), 643–661.

11-5   J. F. Sageau, *In Situ Measurement of the Mean and Fluctuating Pressure Fields Around a 122 Meters Smooth, Isolated Cooling Tower*, Electricité de France, Direction des Etudes et Recherches, 6 quai Watier, Chatou, France, Sept. 1979.

11-6   T. F. Sun and L. M. Zhou, "Wind Pressure Distributions on a Ribless Hyperbolic Cooling Tower," *Proceedings 6th International Conference on Wind Engineering*, Gold Coast, Australia, in *J. Wind Eng. Ind. Aerodyn.*, **14** (1983), 181–192.

11-7   S. H. Abu-Sitta and M. G. Hashish, "Dynamic Wind Stresses in Hyperbolic Cooling Towers," *J. Struct. Div.*, ASCE, **99** (Sept. 1973), 1823–1935.

11-8   J. F. Sageau, *Caractérisation des champs de pression moyens et fluctuants à la surface des grands aérorefrigérants*, Electriçité de France, Direction des Etudes et Recherches, 6 quai Watier, Chatou, France, July 1979.

11-9   H. Ruscheweyh, "Wind Loadings on Hyperbolic Natural Draught Cooling Towers," *J. Ind. Aerodyn.*, **1** (1976), 335–340.

11-10   A. G. Davenport and N. Isyumov, *The Dynamic and Static Action of Wind on Hyperbolic Cooling Towers*, Vol. 1, Research Report No. BLWTI-66, Univ. of Western Ontario, London, Ont., Canada, 1966.

11-11   M. Pirner, "Wind Pressure Fluctuations on a Cooling Tower," *J. Wind Eng. Ind. Aerodyn.*, **10** (1982), 343–360.

11-12   M. G. Hashish and S. H. Abu-Sitta, "Response of Hyperbolic Cooling Towers to Turbulent Wind," *J. Struct. Div.*, ASCE, **100** (1974), 1037–1051.

11-13   M. P. Singh and A. K. Gupta, "Gust Factors for Hyperbolic Cooling Towers," *J. Struct. Div.*, ASCE, **102** (1978), 371–386.

11-14   P. K. Basu and P. L. Gould, "Cooling Towers Using Measured Wind Data," *J. Struct. Div.*, ASCE, **106** (1980), 579–600.

11-15   R. L. Steinmetz, D. P. Billington, and J. F. Abel, "Hyperbolic Cooling Tower Dynamic Response to Wind," *J. Struct. Div.*, ASCE, **104** (1978), 35–53.

11-16   D. A. Reed and R. H. Scanlan, "Cooling Tower Wind Loading," in *Proceedings of the 4th U.S. National Conference on Wind Engineering Research*, Department of Civil Engineering, University of Washington, Seattle, Washington, July July 26–29, 1981, Vol. 1, pp. 254–261.

11-17   *Reinforced Concrete Cooling Tower Shells—Practice and Commentary*, ACI 334, IR-77, American Concrete Institute, Detroit, Michigan, 1977.

11-18   *American National Standard Building Code Requirements for Minimum Design Loads in Buildings and Other Structures*, A58.1, American National Standards Institute, New York, 1972.

11-19   J. Armitt, "Wind Loading on Cooling Towers," *J. Struct. Div.*, ASCE, **106** (1980), 623–641.

11-20   J. Welsch, *Der Einfluss des Windprofils auf die statischen Windbeanspruchungen von rotations-hyperbolischen Kühlturmschalen*, Lehrstuhl I, Institut für konstruktiven Ingenieurbau, Ruhr-Universität Bochum, February 1978.

11-21   D. A. Reed and E. Simiu, "Wind Loads on Cooling Towers," Draft State of the Art Report on Wind Effects on Structures, Committee on Wind Effects, American Society of Civil Engineers, 1984.

11-22   Y. Kawarabata, S. Nakae, and M. Harada, "Some Aspects of the Wind Design of Cooling Towers", *J. Wind Eng. Ind. Aerodyn.* **14** (1983), 167–180.

11-23   H.-J. Niemann, "Reliability of Current Design Methods for Wind-Induced Stresses," in *Natural Drought Cooling Towers*, Proceedings of the 2nd International Symposium, Ruhr-Bochum, Germany, P. L. Gould, W. B. Krätzig, I. Mungan, and U. Wittek, Springer-Verlag, Berlin, 1984.

12

Trussed Frameworks and Plate Girders

Trussed frameworks subjected to wind loads have routinely been used in structural engineering applications for more than a century. Nevertheless, the state of knowledge concerning the effects of wind on this type of structure is still imperfect, and provisions concerning such effects included in various standards, codes, and design guides are in some cases mutually inconsistent and in disagreement with experimental data [12-1].

For any given wind speed, the principal factors that determine the wind load acting on a trussed framework are:

- The aspect ratio λ, that is, the ratio of the length of the framework to its width. If end plates or abutments are provided, the flow around the framework is essentially two-dimensional, so that for aerodynamic purposes the length of a framework may be considered to be infinite.
- The solidity ratio ϕ, that is, the ratio of the effective to the gross area of the framework.* For any solidity ratio ϕ the wind load is for practical purposes independent of the truss configuration, that is, of whether a diagonal truss, a K-truss, and so forth, is involved.
- The shielding of portions of the framework by other portions located upwind. The degree to which shielding occurs depends on the configuration of the spatial framework. If the framework consists of parallel trusses (or girders), the shielding depends on the number and spacing of the trusses (or girders).
- The shape of the members, that is, whether the members are rounded or have sharp edges. Forces on rounded members depend on Reynolds number $\mathcal{R}e$

---

*The effective area of a plane truss is the area of the shadow projected by its members on a plane parallel to the truss, the projection being normal to that plane. The gross area of a plane truss is the area contained within the outside contour of that truss. The effective area and the gross area of a spatial framework are defined, respectively, as the effective area and the gross area of its upwind face.

and on the roughness of the member surface (see Fig. 4.5.5). For trusses with sharp-edged members the effect of the Reynolds number and of the shape and surface roughness of the member is, in practice, negligible.

- The turbulence in the oncoming flow. As noted in Sect. 4.5, the effect of turbulence on the drag force acting on frameworks with sharp-edged members is relatively small in most cases of practical interest [12-2, 12-5, 12-14]. A similar conclusion appears to be valid for frameworks composed of members with circular cross section in flows with subcritical Reynolds members. For this reason, and owing to scaling difficulties, in most cases wind tunnel tests for trussed frameworks are to this day conducted in smooth flow [12-1, 12-5, 12-6].

- The orientation of the framework with respect to the mean wind direction.

The purpose of this chapter is to present a review of information on the aerodynamic behavior of trussed frameworks and plate girders including single trusses and girders, systems consisting of two or more parallel trusses or girders, and square and triangular towers. In several instances test results from several sources are presented with a view to allowing an assessment of the errors that may be expected in typical wind tunnel measurements.

Throughout this chapter the aerodynamic coefficients are referred to, and should be used in conjunction with, the effective area of the framework, A_f. Note that wind forces on ancillary parts (e.g., ladders, antenna dishes) must be taken into account in design in addition to the wind forces on the trussed frameworks themselves [12-7, 12-17].

For recent studies on wind effects on cranes, see [12-15] and [12-16].

12.1 SINGLE TRUSSES AND GIRDERS

Figure 12.1.1 summarizes measurements of the drag coefficient $C_D^{(1)}$ for a single truss with infinite aspect ratio normal to the wind. The data of Fig. 12.1.1 were obtained in the 1930s in Göttingen for trusses with sharp-edged members [12-2, 12-3],* and in the late 1970s at the National Maritime Institute, U.K. (NMI), both for trusses with sharp-edged members and trusses with members of circular cross section.† It is seen that differences between the Göttingen and the NMI results for frameworks with sharp-edged members do not exceed 15% or so. For single trusses normal to the wind and composed of sharp-edged members, ratios $C_D^{(1)}(\lambda)/C_D^{(1)}(\lambda=\infty)$ of the drag coefficients corresponding to an aspect ratio λ on the one hand and to an infinite aspect ratio on the other are shown in Fig. 12.1.2 [12-3].

Drag coefficients $C_D^{(1)}$ reported in [12-5] for trusses normal to the wind, composed of sharp-edged members, and having aspect ratios $1/6 < \lambda < 6$, are

---

*References 12-2, 12-3, and 12-4 are available in English as Building Research Establishment Library Translation No. IT2202, Building Research Station, Garston, Watford, U.K.
†The NMI measurements for trusses with members of circular cross section referred to in this chapter were carried out at Reynolds numbers $10^4 \gtrsim \mathcal{R}e \gtrsim 7 \times 10^4$ [12-6].

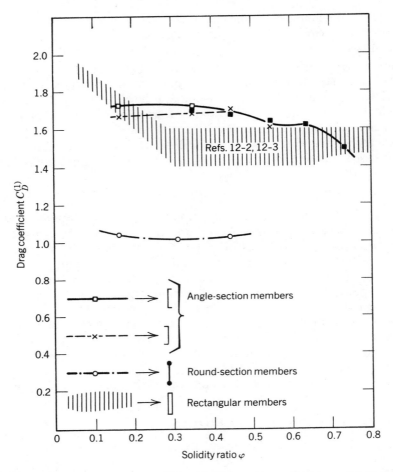

FIGURE 12.1.1. Drag coefficient $C_D^{(1)}$ for single truss, $\lambda = \infty$, wind normal to truss. From R. E. Whitbread, "The Influence of Shielding on the Wind Forces Experienced by Arrays of Lattice Frames," in *Wind Engineering, Proceedings of the Fifth International Conference*, Fort Collins, CO, July 1979, J. E. Cermak (Ed.), Vol. 1, Pergamon Press, New York, 1980, pp. 405–420.

listed in Table 12.1.1. Also listed in Table 12.1.1 are values $C_D^{(1)}(\lambda = \infty)$ obtained from the drag coefficients of [12-5] through multiplication by the appropriate correction factor taken from Fig. 12.1.2. It can be seen that differences between the values $C_D^{(1)}(\lambda = \infty)$ based on [12-5] and the corresponding Flachsbart and NMI values of Fig. 12.1.1 do not exceed 20%.

Figure 12.1.3 [12-7] summarizes results of tests on trusses with members of circular cross section ($\lambda = \infty$) conducted in the subsonic wind tunnel at Porz-Wahn, West Germany [12-8] and in the compressed air tunnel of the National

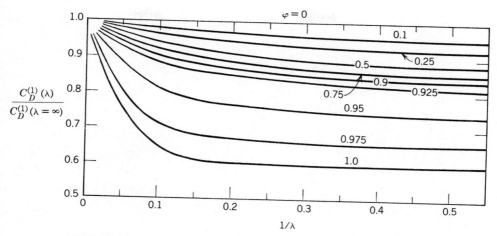

FIGURE 12.1.2.   Ratios $C_D^{(1)}(\lambda)/C_D^{(1)}(\lambda=\infty)$, wind normal to truss [12-3].

TABLE 12.1.1.   Drag Coefficients for Simple Trusses

		0.14	0.29	0.47	0.77	1.0
(1)	$C_D^{(1)}\left(\dfrac{1}{6}<\lambda<6\right)^a$	$1.40\pm5\%$	$1.54\pm5\%$	$1.27\pm5\%$	$1.18\pm5\%$	$1.28\pm5\%$
(2)	$C_D^{(1)}(\lambda=\infty)$	~1.45	~1.65	~1.45	~1.35	~2.10

*a*Reference 12-5.

Physical Laboratory, U.K. [12-10].* Note that for Reynolds numbers $\mathscr{R}e<10^5$ the drag coefficients in Fig. 12.1.3 differ by about 5% or less from the corresponding results of Fig. 12.1.1.

A framework whose solidity ratio is $\phi=1$ is a solid plate (or a girder). The drag coefficient corresponding to wind normal to the plate can be obtained from Figs. 12.1.1 and 12.1.2. Additional information on the aerodynamic behavior of rectangular plates is given in Sects. 4.5 and 4.6.

It was shown in Sect. 4.6 that the aerodynamic force normal to a rectangular plate with aspect ratio $\lambda\simeq5$ to 10 is larger when the yaw angle† is $\alpha\cong40°$ than

---

*Figures 12.1.3 and 12.4.5 to 12.4.8 are reproduced with permission of CIDECT (Comité International pour le Développement et l'Etude de la Construction Tubulaire) from *Wind Forces on Unclad Tubular Structures*, H. B. Walker (Ed.), Constrado Publication 1/75, Constructional Steel Research and Development Organization, Croydon, U.K., 1975. They are based in part on research work carried out by CIDECT and reported in [12-8] and [12-9].

†The yaw angle is the horizontal angle between the mean wind direction and the normal to the trusses.

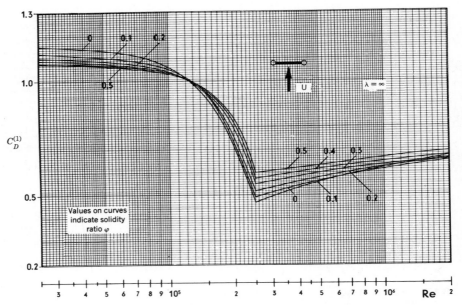

FIGURE 12.1.3. Drag coefficient $C_D^{(1)}$ for single truss with members of circular cross section , $\lambda = \infty$, wind normal to truss [12-7] (courtesy Comité International pour le Développement et l'Etude de la Construction Tubulaire, and Constructional Steel Research and Development Organisation).

if the wind is normal to the plate (Fig. 4.6.2). However, for trusses with solidity ratios $\phi < 0.4$ or so the maximum drag occurs when the wind is normal to the truss [12-2].

12.2   PAIRS OF TRUSSES AND OF PLATE GIRDERS

We consider a pair of identical, parallel trusses, and denote the drag coefficient corresponding to the total aerodynamic force normal to the trusses by $C_D^{(2)}(\alpha)$, where α is the yaw angle. For brevity, the notation $C_D^{(2)}(0) = C_D^{(2)}$ is used. The cases where the wind is normal to the truss ($\alpha = 0$) and where $\alpha \neq 0$ are considered in Sects. 12.2.1 and 12.2.2, respectively.

12.2.1   Trusses Normal to the Wind

Two parallel trusses normal to the wind affect each other aerodynamically, so that the drag on the upwind and on the downwind truss will have drag coefficients $\Psi_I C_D^{(1)}$ and $\Psi_{II} C_D^{(1)}$, respectively, where $C_D^{(1)}$ is the drag coefficient for a single truss normal to the wind and, in general, $\Psi_1 \neq \Psi_2 \neq 1$. It follows that

$$C_D^{(2)} = C_D^{(1)}(\Psi_I + \Psi_{II}) \tag{12.2.1}$$

Figure 12.2.1 shows values of Ψ_I and Ψ_{II}, reported in [12-4] for three types of truss, all with sharp-edged members and infinite aspect ratio, as functions of the solidity ratio ϕ, and of the ratio between the truss spacing in the along-wind direction, e, and the truss width, d. Values of Ψ_I and Ψ_{II}, also reported in [12-4], for four types truss of truss with sharp-edged members and aspect ratio $\lambda = 9.5$

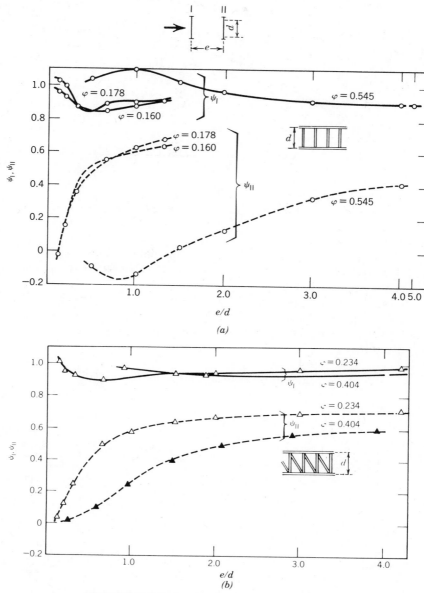

FIGURE 12.2.1a,b.   Factors Ψ_I and Ψ_{II} [12-4].

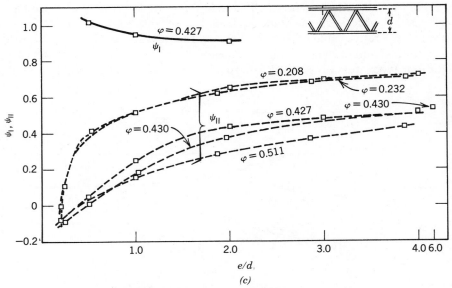

FIGURE 12.2.1c.   Factors Ψ_I and Ψ_{II} [12-4].

are shown in Fig. 12.2.2. On the basis of the data of Figs. 12.2.1 and 12.2.2, [12-4] suggested the use for design purposes of the conservative values $C_D^{(2)}/C_D^{(1)}$ given, for $e/d > 1.0$, in Fig. 12.2.3.

Recent measurements conducted at the National Maritime Institute, U.K., (NMI) on trusses with infinite aspect ratio are summarized in Fig. 12.2.4. Reference 12-6 suggests the following approximate expressions based on the results of Fig. 12.2.4:

$$\frac{C_D^{(2)}}{C_D^{(1)}} = 2 - \phi^{0.45}(e/d)^{\phi - 0.45} \quad \text{for } 0 < \phi < 0.5 \tag{12.2.2}$$

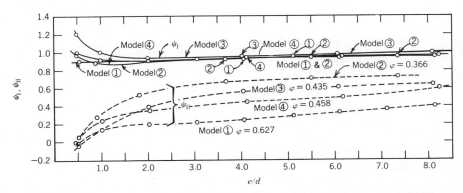

FIGURE 12.2.2.   Factors Ψ_I and Ψ_{II} for four sets of two parallel trusses with sharp-edged members, $\lambda = 9.5$, wind normal to trusses [12-4].

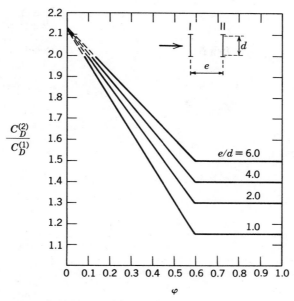

FIGURE 12.2.3. Approximate ratios $C_D^{(2)}/C_D^{(1)}$ proposed for design purposes by Flachsbart [12-4].

for trusses with sharp-edged members, and

$$\frac{C_D^{(2)}}{C_D^{(1)}} = 2 - \phi_e^{0.45}(e/d)^{\phi_e - 0.45} \tag{12.2.3}$$

for trusses composed of members with circular cross section. The nominal solidity ratio ϕ_e in Eq. 12.2.3 is related to the actual solidity ratio as shown in Fig. 12.2.5.

Figure 12.2.6 shows ratios $C_D^{(2)}/C_D^{(1)}$ for trusses with sharp-edged members and aspect ratio $\lambda = 8$ [12-1].

Examples:

1. Consider a truss with sharp-edged members, solidity ratio $\phi = 0.18$, spacing ratio $e/d = 1.0$, and aspect ratio $\lambda = \infty$. From Fig. 12.1.1, $C_D^{(1)} \simeq 1.70$ according to both Flachsbart and the NMI tests. From both Flachsbart's and the NMI tests, $C_D^{(2)}/C_D^{(1)} = \Psi_I + \Psi_{II} \cong 1.55$ (Figs. 12.2.1a and 12.2.4a), so that $C_D^{(2)} \simeq 1.70 \times 1.55 \simeq 2.65$. Note that according to Fig. 12.2.3—proposed by Flachsbart as a deliberately conservative design chart—$C_D^{(2)}/C_D^{(1)} \simeq 1.83$, which exceeds the value based on Figs. 12.2.1a and 12.2.4a by about 20%.

2. Consider a truss with sharp-edged members, solidity ratio $\phi = 0.46$, spacing ratio $e/d = 1.0$, and aspect ratio $\lambda \cong 9.0$. Approximate values of drag coefficients $C_D^{(1)}$, ratios $C_D^{(2)}/C_D^{(1)} = \Psi_I + \Psi_{II}$, and corresponding calculated drag coefficients $C_D^{(2)}$, based on the Göttingen [12-4], NMI [12-6] and Western Ontario [12-5] information, are listed in Table 12.2.1. It is seen that while the difference between the values $C_D^{(1)}$ based on [12-4] and [12-5] is about 12%,

TABLE 12.2.1. Drag Coefficients Based on the Göttingen, NMI, and Western Ontario Studies

	Flachsbart	NMI	Western Ontario
References	12-4	12-6	12-5
$C_D^{(1)}$	$1.5 \times 0.95 \cong 1.43^a$	$1.7 \times 0.95 \cong 1.62^c$	1.27^e
$C_D^{(2)}/C_D^{(1)}$	$0.23 + 0.92 = 1.15^b$	1.29^d	1.30^f
$C_D^{(2)}$	$1.15 \times 1.43 = 1.64$	2.08	1.65

[a]Figs. 12.1.1 and 12.1.2.
[b]Fig. 12.2.2.
[c]Figs. 12.1.1 and 12.1.2.
[d]Eq. 12.2.3a or Fig. 12.2.4a.
[e]Table 12.1.1.
[f]Fig. 12.2.6.

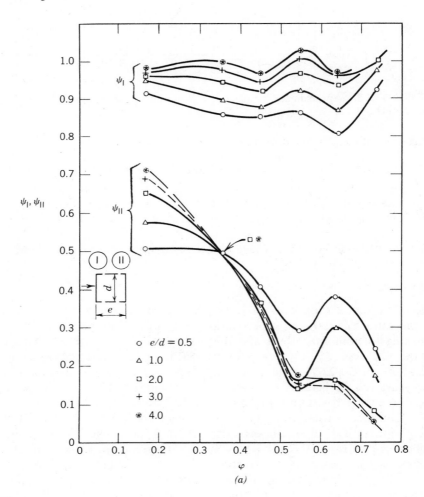

(a)

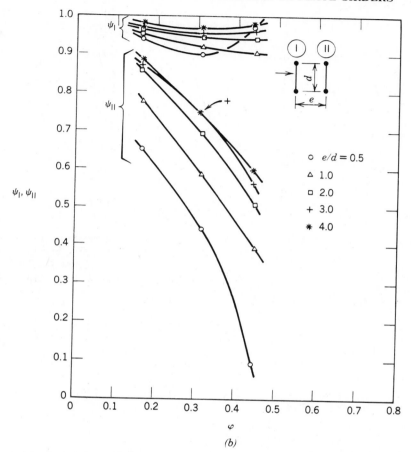

FIGURE 12.2.4. Factors Ψ_I and Ψ_{II} for two parallel trusses with (*a*) sharp-edged members and (*b*) members of circular cross section, $\lambda = \infty$, wind normal to trusses. From R. E. Whitbread, "The Influence of Shielding on the Wind Forces Experienced by Arrays of Lattice Frames," in *Wind Engineering, Proceedings of the Fifth International Conference,* Fort Collins, CO, July 1979, J. E. Cermak (Ed.), Vol. 1, Pergamon Press, New York, 1980, pp. 405–420.

the corresponding values $C_D^{(2)}$ are virtually identical in this case. Note also that the difference between the values $C_D^{(2)}$ based on [12-6] on the one hand and on [12-4] or [12-5] on the other is about 25%.

12.2.2 Trusses Skewed with Respect to Wind Direction

We now consider the case in which the yaw angle is $\alpha \neq 0$. For certain values of α the effectiveness of the shielding decreases, and the drag coefficient $C_D^{(2)}(\alpha)$ characterizing the total force normal to the trusses is larger than the value $C_D^{(2)}$. (Recall that, by definition, $C_D^{(2)}(0) = C_D^{(2)}$.)

Ratios $\max \{C_D^{(2)}(\alpha)\}/C_D^{(1)}$ reported in [12-5] for trusses with sharp-edged

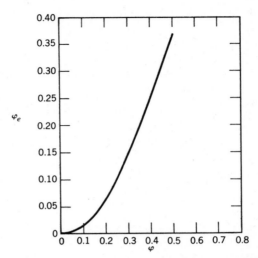

FIGURE 12.2.5. Equivalent solidity ratio ϕ_e for trusses with members of circular cross-section and solidity ratio ϕ. From R. E. Whitbread, "The Influence of Shielding on the Wind Forces Experienced by Arrays of Lattice Frames," in *Wind Engineering*, *Proceedings of the Fifth International Conference*, Fort Collins, CO, July 1979, J. E. Cermak (Ed.), Vol. 1, Pergamon Press, New York, 1980, pp. 405–420.

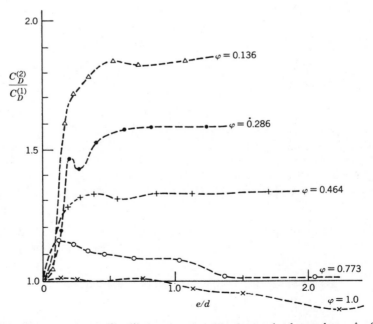

FIGURE 12.2.6. Ratios $C_D^{(2)}/C_D^{(1)}$ for trusses with sharp-edged members, $\lambda = 8$, wind normal to trusses. From P. N. Georgiou and B. J. Vickery, "Wind Loads on Building Frames," in *Wind Engineering*, *Proceedings of the Fifth International Conference*, Fort Collins, CO, July 1979, J. E. Cermak (Ed.), Vol. 1, Pergamon Press, New York, 1980, pp. 421–433.

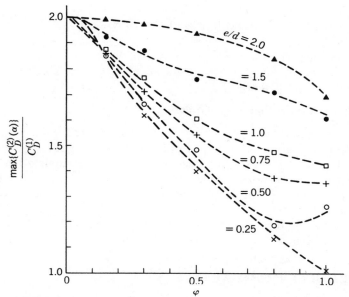

FIGURE 12.2.7. Ratios max $\{C_D^{(2)}(\alpha)\}/C_D^{(1)}$ for trusses with sharp-edged members, $\lambda = 8$. From P. N. Georgiou and B. J. Vickery, "Wind Loads on Building Frames," *Wind Engineering, Proceedings of the Fifth International Conference*, Fort Collins, CO, July 1979, J. E. Cermak (Ed.), Vol. 1, Pergamon Press, New York, 1980, pp. 421–433.

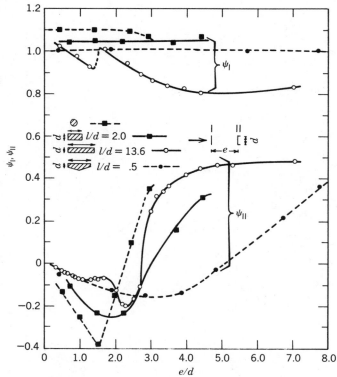

FIGURE 12.2.8. Factors Ψ_I and Ψ_{II} for two parallel solid plates (girders) [12-4, 12-11].

members and aspect ratio $\lambda = 8$ are shown in Fig. 12.2.7. For example, for $e/d = 1.0$, $\phi = 0.286$, and $\lambda = 8$, the ratio max $\{C_D^{(2)}(\alpha)\}/C_D^{(1)} \cong 1.77$, versus $C_D^{(2)}/C_D^{(1)} = 1.59$ (Fig. 12.2.6).

12.2.3   Pairs of Solid Plates and Girders

Figure 12.2.8 shows the dependence of the factors Ψ_I and Ψ_{II} (see Eq. 12.2.1) upon the spacing ratio e/d for a solid disk and for three girders normal to the wind [12-4, 12-11]. For certain values of the horizontal angle α between the wind direction and the normal to the plates the ratio $C_D^{(2)}(\alpha)/C_D^{(2)}$ may be larger than unity. For example, for a plate with aspect ratio $\lambda = 4$ and spacing ratio $e/d = 0.5$, if $40° \gtrsim \alpha \gtrsim 65°$, then $C_D^{(2)}(\alpha)/C_D^{(2)} \cong 1.20$ [12-1].

Data concerning the effect of bridge decks on the aerodynamic forces acting on pairs of plate girders are available in [12-12].

12.3   MULTIPLE-FRAME ARRAYS

The first attempts to measure aerodynamic forces on multiple frame arrays were reported in [12-1] and [12-6].

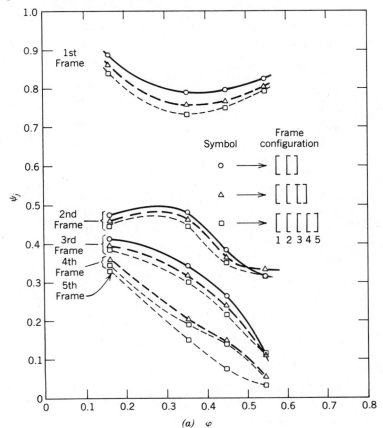

(a) φ

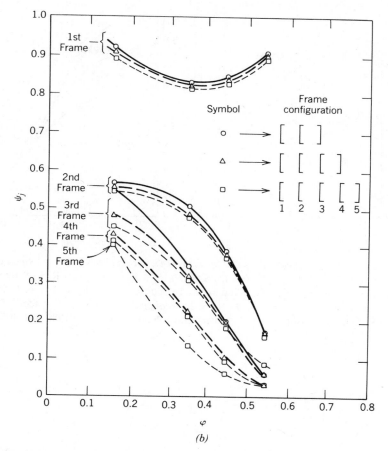

FIGURE 12.3.1.   Factors $\Psi_j (j=1, 2, \ldots, n)$ for arrays of n parallel trusses ($n=3$, 4, and 5) with sharp-edged members, $\lambda = \infty$, wind normal to trusses. (*a*) Spacing ratio $e/d=0.5$. (*b*) Spacing ratio $e/d=1.0$. From R. E. Whitbread, "The Influence of Shielding on the Wind Forces Experienced by Arrays of Lattice Frames," *Wind Engineering, Proceedings of the Fifth International Conference*, Fort Collins, CO, July 1979, J. E. Cermak (Ed.), Vol. 1, Pergamon Press, New York, 1980, pp. 405–420.

For frames normal to the wind, the drag coefficients for the first, second, . . . , n-th frame may be written as $\Psi_1 C_D^{(1)}$, $\Psi_2 C_D^{(1)}$, . . . , $\Psi_n C_D^{(1)}$, where $C_D^{(1)}$ is the drag coefficient for a single frame normal to the wind. The drag coefficient for the array of frames normal to the wind is then

$$C_D^{(n)} = C_D^{(1)}(\Psi_1 + \Psi_2 + \cdots + \Psi_n) \qquad (12.3.1)$$

Factors $\Psi_j (j=1, 2, \ldots, n)$ for arrays of three, four, and five parallel trusses with sharp-edged members and infinite aspect ratio are given in Figs. 12.3.1*a* and 12.3.1*b* for spacing ratios $e/d=0.5$ and $e/d=1$, respectively [12-6]. Drag coefficients $C_D^{(n)}$ for the same arrays are shown in Figs. 12.3.2*a* and 12.3.2*b*

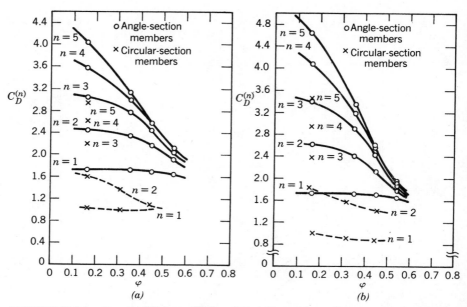

FIGURE 12.3.2. Drag coefficients $C_D^{(n)}$ for arrays of n parallel trusses, $\lambda = \infty$, wind normal to trusses. (a) Spacing ratio $e/d = 0.5$. (b) Spacing ratio $e/d = 1.0$. From R. E. Whitbread, "The Influence of Shielding on the Wind Forces Experienced by Arrays of Lattice Frames," *Wind Engineering, Proceedings of the Fifth International Conference*, Fort Collins, CO, July 1979, J. E. Cermak (Ed.), Vol. 1, Pergamon Press, New York, 1980, pp. 405–420.

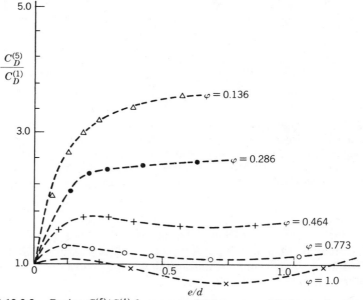

FIGURE 12.3.3. Ratios $C_D^{(5)}/C_D^{(1)}$ for arrays of five trusses with sharp-edged members, $\lambda = 8$, wind normal to trusses. From P. N. Georgiou and B. J. Vickery, "Wind Loads on Building Frames," *Wind Engineering, Proceedings of the Fifth International Conference*, Fort Collins, CO, July 1979, J. E. Cermak (Ed.), Vol. 1, Pergamon Press, New York, 1980, pp. 421–433.

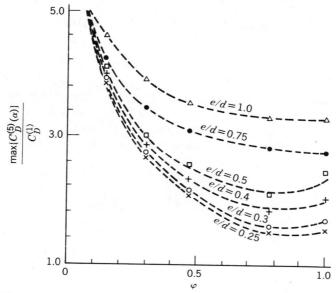

FIGURE 12.3.4. Ratios max $\{C_D^{(5)}(\alpha)/C_D^{(1)}\}$ for arrays of five trusses with sharp-edged members, $\lambda = 8$. From P. N. Georgiou and B. J. Vickery, "Wind Loads on Building Frames," *Wind Engineering, Proceedings of the Fifth International Conference,* Fort Collins, CO, July 1979, J. E. Cermak (Ed.), Vol. 1, Pergamon Press, New York, 1980, pp. 421–433.

[12-6]. Also shown in Figs. 12.3.2 are measurements of $C_D^{(n)}$ for trusses with infinite aspect ratio and members with circular cross section [12-6].

Ratios $C_D^{(5)}/C_D^{(1)}$ measured in [12-1] for trusses with sharp-edged members and aspect ratio $\lambda = 8$ are shown in Fig. 12.3.3. As pointed out in Sect. 12.2, the drag force normal to the trusses does not reach a maximum when the trusses are normal to the wind, but for some yaw angle $\alpha \neq 0$. Ratios max $\{C_D^{(5)}(\alpha)\}/C_D^{(1)}$ measured for the trusses just described are shown in Fig. 12.3.4 [12-1].

12.4 SQUARE AND TRIANGULAR TOWERS

As pointed out earlier, the aerodynamic coefficients given in this chapter are in all cases referred to, and should be used in conjunction with, the effective area of the framework, A_f. For square and triangular towers, A_f is the effective area of one of the identical faces of the tower. The influence of wind gustiness on the tower loading and response can be determined by using the methods for estimating along-wind response discussed in Chapter 9.*

*The width of the structure used as an input in these methods should be equal to the actual width of the framework. This assures that the lateral coherence of the load fluctuations is taken into account. On the other hand, the depth (along-wind dimension) of the framework should be assumed to be equal to zero in order not to overestimate the favorable effect of the along-wind cross-correlations of the fluctuating loads (see Sect. 4.7.4). Finally, the area of the framework per unit height at any given elevation, used to estimate the mean and the fluctuating drag forces, should be equal to the effective area per unit height at that elevation.

For information on guyed tower response and design, see [4-10], [4-17], and [12-17] to [12-26].

12.4.1   Aerodynamic Data for Square and Triangular Towers

The results of wind force measurements on square towers can be expressed in terms of the aerodynamic coefficients $C_N(\alpha)$ and $C_T(\alpha)$ associated, respectively, with the wind force components N and T $(N \geqslant T)$ normal to the faces of the tower (Fig. 12.4.1), and of the aerodynamic coefficient $C_F(\alpha)$ associated with the total wind force F acting at a yaw angle $\alpha = \tan^{-1}(T/N)$. Note that $C_F(\alpha) = [C_N^2(\alpha) + C_T^2(\alpha)]^{1/2}$ since, as indicated earlier, all aerodynamic coefficients are referenced to the effective area of one face of the framework, A_f.

For a triangular tower (which has in practice and is therefore assumed here to have equal sides in plan), the results of the measurements can be expressed in terms of the aerodynamic coefficients $C_F(\alpha)$ (Fig. 12.4.2). The aerodynamic coefficients $C_F(0°)$ and $C_F(60°)$ correspond, respectively, to wind forces acting in a direction normal to a side and along the direction of a median (Figs. 12.4.2a and 12.4.2b).

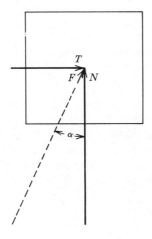

FIGURE 12.4.1.   Notations.

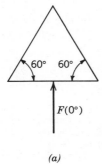

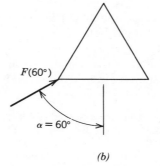

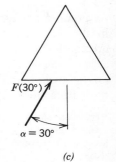

(a)   (b)   (c)

FIGURE 12.4.2.   Notations.

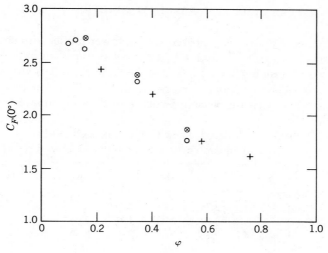

o Angle members—smooth flow.
⊗ Angle members—turbulent flow.
+ Square shaped members—smooth flow.

FIGURE 12.4.3. Drag coefficients $C_F(0°)$ for square tower with sharp-edged members measured at National Maritime Institute, U.K. From A. R. Flint and B. W. Smith, "The Development of the British Draft Code of Practice for the Loading of Lattice Towers," *Wind Engineering, Proceedings of the Fifth International Conference*, Fort Collins, CO, July 1979, J. E. Cermak (Ed.), Vol. 2, Pergamon Press, New York, 1980, pp. 1293–1304.

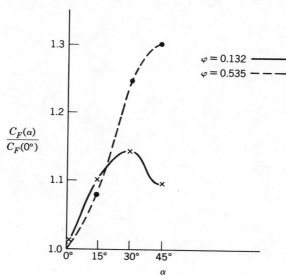

FIGURE 12.4.4. Ratios $C_F(\alpha)/C_F(0°)$ for square tower with sharp-edged members measured at National Maritime Institute, U.K. From A. R. Flint and B. W. Smith, "The Development of the British Draft Code of Practice for the Loading of Lattice Towers," *Wind Engineering, Proceedings of the Fifth International Conference*, Fort Collins, CO, July 1979, J. E. Cermak (Ed.), Vol. 2, Pergamon Press, New York, 1980, pp. 1293–1304.

Measurements of loads on a tapered square tower model with sharp-edged members, aspect ratio $\lambda \cong \infty$, and solidity ratio averaged over the height of the tower $\phi \cong 0.19$ (ranging from $\phi = 0.13$ at the base to $\phi = 0.47$ at the tip) were reported in the 1930s by Katzmayr and Saitz [12-13]. Until recently these measurements have been the principal source of data on square towers. The

TABLE 12.4.1. Aerodynamic Coefficients: $C_N(\alpha)$, $C_T(\alpha)$, and $C_F(\alpha)$ for a Square Tower with $\phi \cong 0.19$ and $\lambda \cong \infty$ [12-13]

α	0°	9°	18°	27°	36°	45°
$C_N(\alpha)$	2.54	2.75	2.97	3.01	2.84	2.60
$C_T(\alpha)$	—	0.19	0.70	1.36	2.05	2.49
$C_F(\alpha)$	2.54	2.76	3.05	3.30	3.50	3.60

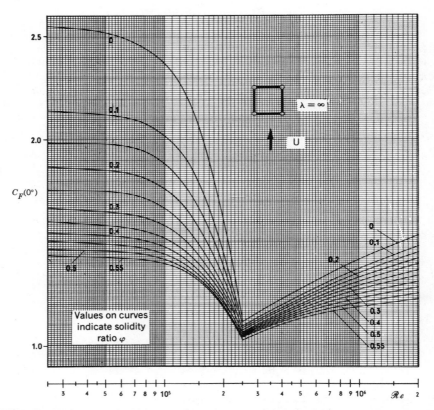

FIGURE 12.4.5. Drag coefficients $C_F(0°)$ for square tower with members of circular cross section [12-7] (courtesy Comité International pour le Développement et l'Etude de la Construction Tubulaire, and Constructional Steel Research and Development Organisation).

coefficients $C_N(\alpha)$, $C_T(\alpha)$, and $C_F(\alpha)$ obtained in [12-13] are listed for various angles α in Table 12.4.1 [12-13].

For $\alpha = 45°$, the values of $C_N(\alpha)$ and $C_T(\alpha)$ should be equal; as pointed out in [12-13], the 4% difference between these values in Table 12.4.1 is due to measurement errors. Note that the value $C_N(0°) = 2.54$ is close to the values inferred from [12-5] and [12-6], which are, respectively, $C_N(0°) = C_D^{(2)} \simeq 1.5 \times 1.73 = 2.60$ (as obtained by linear interpolation for $\phi = 0.19$ and $e/d = 1.0$ from Table 12.1.1 and Fig. 12.2.6), and $C_N(0°) = C_D^{(2)} \simeq 1.7(0.93 + 0.58) = 2.57$ (Eq. 12.2.1, and Figs. 12.1.1 and 12.2.4a). Note also that while the largest tension (compression) in the tower columns is caused by winds acting in the direction $\alpha = 45°$, the largest stresses in the bracing members occur for $\alpha = 27°$.

Measurements of forces on square towers with sharp-edged members ($\lambda = \infty$) were more recently conducted at the National Maritime Institute, U.K. (NMI)

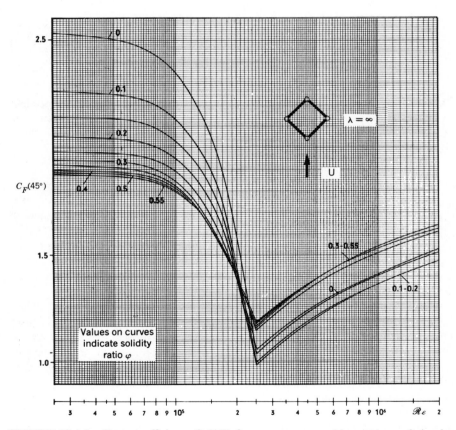

FIGURE 12.4.6. Drag coefficients $C_F(45°)$ for square tower with members of circular cross section [12-7] (courtesy Comité International pour le Développement et l'Etude de la Construction Tubulaire, and Constructional Steel Research and Development Organisation).

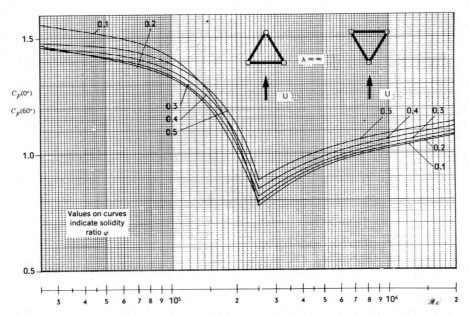

FIGURE 12.4.7.   Drag coefficients $C_F(0°)$ and $C_F(60°)$ for triangular tower with members of circular cross section [12-7] (courtesy Comité International pour le Développement et l'Etude de la Construction Tubulaire, and Constructional Steel Research and Development Organisation).

[12-14]. Coefficients $C_F(0°)$ and ratios $C_F(\alpha)/C_F(0°)$ based on these measurements are shown in Figs. 12.4.3 and 12.4.4, respectively. Note, for example, that for $\phi \cong 0.19$, $C_F(0°) \cong 2.60$ (Fig. 12.4.3), vs. $C_F(0°) = 2.54$, as obtained in [12-13] (Table 12.4.1). The agreement is less good for the ratio $C_F(45°)/C_F(0°)$, which is about 1.12 according to Fig. 12.4.4, and about 1.40 according to the data of Table 12.4.1. As shown subsequently in this section, data on square towers composed of members with circular cross section suggest that the NMI results are more reliable than those of [12-13].

Square Towers Composed of Members with Circular Cross Section.   Figures 12.4.5 and 12.4.6 [12-7] represent, respectively, proposed aerodynamic coefficients $C_F(0°)$ and $C_F(45°)$ as functions of Reynolds number $\mathscr{R}e$ for towers with aspect ratio $\lambda = \infty$, based on recent wind tunnel test results reported in [12-8] and [12-9]. The values $C_F(45°)$ of Fig. 12.4.6 may be regarded as conservative envelopes that account for the loadings in the most unfavorable directions. Results of tests conducted at NMI in both smooth and turbulent flow at Reynolds numbers $\mathscr{R}e \cong 2 \times 10^4$ for solidity ratios $\phi \cong 0.17$, $\phi \cong 0.23$, and $\phi \cong 0.31(\lambda = \infty)$ match the curves of Fig. 12.4.5 and 12.4.6 to within about 5% or less [12-14].

Note that for $0 < \phi < 0.3$ the ratio $C_F(45°)/C_F(0°)$ is considerably closer to

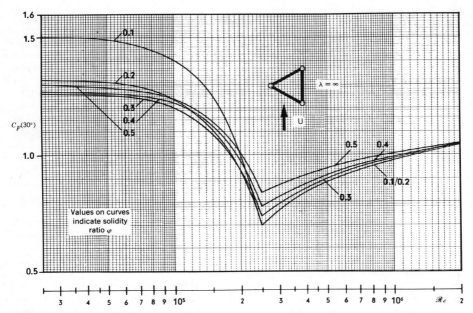

FIGURE 12.4.8. Drag coefficients $C_F(30°)$ for triangular tower with members of circular cross section [12-7] (courtesy Comité International pour le Développement et l'Etude de la Construction Tubulaire, and Constructional Steel Research and Development Organisation).

1.1 than to the value 1.4 inherent in Table 12.4.1. This would tend to confirm the broad validity of the NMI results on square towers with sharp-edged members discussed earlier in this section.

Triangular Towers Composed of Members with Circular Cross Section.   Figures 12.4.7 and 12.4.8 [12-7] represent proposed aerodynamic coefficients $C_F(0°) \simeq C_F(60°)$ and $C_F(30°)$ as functions of Reynolds number $\mathscr{R}e$ for towers with aspect ratio $\lambda = \infty$, based on measurements reported in [12-8], [12-9), and [12-10].

REFERENCES

12-1   P. N. Georgiou and B. J. Vickery, "Wind Loads on Building Frames," *Wind Engineering, Proceedings of the Fifth International Conference*, Fort Collins, CO, July 1979, J. E. Cermak (Ed.), Vol. 1, Pergamon Press, New York, 1980, pp. 421–433.

12-2   O. Flachsbart, "Modellversuche über die Belastung von Gitterfachwerken durch Windkräfte. 1. Teil: Einzelne ebene Gitterträger," *Der Stahlbau*, **7** (April 1934), 65–69.

12-3   O. Flachsbart, "Modellversuche über die Belastung von Gitterfachwerken durch Windkräfte. 1. Teil: Einzelne ebene Gitterträger," *Der Stahlbau*, **7** (May 1934) 73–79.

12-4   O. Flachsbart, and H. Winter, "Modellversuche über die Belastung von Gitterfachwerken durch Windkräfte. 2. Teil: Räumliche Gitterfachwerke," *Der Stahlbau*, **8** (April 1935), 57–63.

12-5 P. N. Georgiou, B. J. Vickery, and R. Church, "Wind Loading on Open Framed Structures," *Proceedings Third Canadian Workshop on Wind Engineering*, Vancouver, April 1981.

12-6 R. E. Whitbread, "The Influence of Shielding on the Wind Forces Experienced by Arrays of Lattice Frames," *Wind Engineering, Proceedings of the Fifth International Conference*, Fort Collins, CO, July 1979, J. E. Cermak (Ed.), Vol. 1, Pergamon Press, New York, 1980, pp. 405–420.

12-7 *Wind Forces on Unclad Tubular Structures*, H. B. Walker (Ed.), Constrado Publication 1/75, Constructional Steel Research and Development Organization, Croydon, U.K., 1975.

12-8 G. Schulz, *The Drag of Lattice Structures Constructed from Cylindrical Members (Tubes) and its Calculation*, CIDECT Report No. 69/21, Düsseldorf, West Germany, 1969 (in German).

12-9 G. Schulz, International Comparison of Standards on the Wind Loading of Structures, CIDECT Report No. 69/29, Düsseldorf, West Germany, 1969 (in German).

12-10 R. W. F. Gould and W. G. Raymer, *Measurements Over a Wide Range of Reynolds Numbers of the Wind Forces on Models of Lattice Frameworks*, National Physical Laboratory Sc. Rep. No. 5-72, Teddington, U.K., May 1972.

12-11 G. Eiffel, *La Résistance de l'Air et l'Aviation*, H. Dunod & E. Pinat, Paris, 1911.

12-12 J. M. Biggs, S. Namyet, and J. Adachi, "Wind Loads on Girder Bridges," *Transactions*, ASCE, **121** (1956), 101–113.

12-13 D. Katzmayr and H. Seitz, "Winddruck auf Fachwerktürme von quadratischem Querschnitt," *Der Bauingenieur*, **21/22** (1934), 218–221.

12-14 A. R. Flint and B. W. Smith, "The Development of the British Draft Code of Practice for the Loading of Lattice Towers," *Wind Engineering, Proceedings of the Fifth International Conference*, Fort Collins, CO, July 1979, J. E. Cermak (Ed.), Vol. 2, Pergamon Press, New York, 1980, pp. 1293–1304.

12-15 J. F. Eden, A. Iny, and A. J. Butler, "Cranes in Storm Winds," *Eng. Struct.*, **3** (1981), 175–180.

12-16 J. F. Eden, A. J. Butler, and J. Patient, "Wind Tunnel Tests on Model Crane Structures," *Eng. Struct.*, **5** (1983), 289–298.

12-17 G. A. Savitskii, *Calculations for Antenna Installations*, Technical Translation TT 79-52040, published for the National Science Foundation by Amerind Publishing Co., New Delhi, 1982, available from National Technical Information Service, Springfield, VA 22161.

12-18 V. Koloušek, M. Pirner, O. Fischer, and J. Náprstek, *Wind Effects on Civil Engineering Structures*, Elsevier, Amsterdam, 1984.

12-19 R. J. McCaffrey and A. J. Hartmann, "Dynamics of Guyed Towers," *J. Struct. Div.*, ASCE, **98** (1972), 1309–1323.

12-20 J. W. Vellozzi, "Tall Guyed Tower Response to Wind Loading," *Proceedings Fourth International Conference on Wind Effects on Buildings and Structures*, Heathrow, September 1975, Cambridge University Press, New York, 1976.

12-21 R. A. Williamson, "Stability Study of Guyed Tower Under Ice Loads," *J. Struct. Div.*, ASCE, **99** (1973), 2391–2408.

12-22 J. E. Goldberg and J. T. Gaunt, "Stability of Guyed Towers," *J. Struct. Div.*, ASCE, **99** (1973), 741–756.

12-23 R. A. Williamson and M. N. Margolin, "Shear Effects in Design of Guyed Towers," *J. Struct. Div.*, ASCE, **92** (1966), 213–260.

12-24 F. Rosenthal and R. A. Skop, "Method for the Analysis of Guyed Towers," *J. Struct. Div.*, ASCE, **108** (1982), 543–558.

12-25 D. M. Brown and J. W. Melin, *Guyed Tower Program Listings and User's Manual*, Technical Report sponsored by United States Coast Guard, U.S. Department of Transportation (Contract DOT-CG-52604-A), J. W. Mellin and Assoc., Urbana, Ill., 1975.

12-26 H. A. Bucholdt and P. Spinelli, *Static and Dynamic Analysis of Guyed Masts*, Progress Report UFIST/04/1983, University of Florence, Department of Civil Engineering, Florence, Italy, 1983.

13

Suspended-Span Bridges, Tension Structures, and Power Lines

Structures that consist of or depend for their integrity on cables or membranes may exhibit an increased susceptibility to wind effects. The purpose of this chapter is to present information and references concerning such structures, including suspension and cable-stayed bridges, cable roofs, fabric structures (air-supported or otherwise subjected to tension), and power lines.

13.1 SUSPENDED-SPAN BRIDGES

Suspended-span (i.e., suspension and cable-stayed) bridges must be designed to withstand the drag forces induced by the mean wind. In addition, such bridges are susceptible to aeroelastic effects, which include torsional divergence (or lateral buckling), vortex-induced oscillation, flutter, galloping, and buffeting in the presence of self-excited forces. The study of these effects is possible only on the basis of information provided by wind tunnel tests. Various types of such tests are briefly described in Sect. 13.1.1. Procedures for analyzing the susceptibility of suspended-span bridge decks to aeroelastic effects and pertinent design considerations are presented in Sects. 13.1.2 through 13.1.6. A brief survey of investigations into the behavior of suspended-span bridges under the action of wind is included in Sect. 13.1.7.

It is noted that the action of wind must be taken into account not only for the completed bridge, but for the bridge in the construction stage as well. In general, the same methods of testing and analysis apply in the two cases. To decrease the vulnerability of the partially-completed bridge to wind, temporary ties and damping devices are used. Also, to minimize the risk of strong wind loading during construction, the latter usually takes place in seasons with low probabilities of occurrence of severe storms.

Aeroelastic phenomena may affect, in addition to the deck, bridge towers, hangers, and cables. Problems related to the design of these, or similar, elements are dealt with in Sect. 13.1.8.

13.1.1   Types of Suspended-Span Bridge Wind Tunnel Tests

The following three types of wind tunnel tests are currently being used to obtain information on the aerodynamic behavior of suspended-span bridges:

1.   *Tests on models of the full bridge.*   In addition to being geometrically similar to the full bridge, such models must satisfy similarity requirements pertaining to mass distribution, reduced frequency, mechanical damping, and shapes of vibration modes (see Chapter 7). The construction of full-bridge models is thus elaborate and their cost is therefore relatively high. The usual scale of such models is of the order of 1/300, although scales of 1/100 have been used in a few cases [13-1] to [13-6]. A view of a full-bridge model in a wind tunnel is shown in Fig. 13.1.1.

2.   *Tests on "taut strip" models* [13-7]. In these models two wires stretched across the wind tunnel serve as the basic inner structure, which is then externally clad to resemble a given bridge geometrically. The tensioned wires permit the duplication to model frequency scale of the fundamental bending and torsional frequencies of the bridge. Such models then respond to the laboratory wind flow in a manner similar to the center span of a suspension bridge.

FIGURE 13.1.1.   Model of the Halifax Narrows Bridge (courtesy Boundary-Layer Wind Tunnel Laboratory, University of Western Ontario).

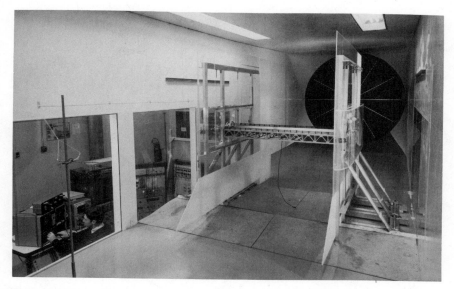

FIGURE 13.1.2. Section model of the Halifax Narrows Bridge (courtesy Boundary-Layer Wind Tunnel Laboratory, University of Western Ontario).

3. *Tests on section models.* Section models consist of representative span-wise sections of the deck constructed to scale, spring-supported at the ends to allow both vertical and torsional motion, and, usually, enclosed between end plates to reduce aerodynamic end effects (Fig. 13.1.2). Section models are relatively inexpensive. They can be constructed to scales of the order of 1/50 to 1/25 so that the discrepancies between full-scale and model Reynolds number* are smaller than in the case of full-bridge tests. Section models are quite useful for making initial assessments, based on simple tests, of the extent to which a bridge deck shape is aeroelastically stable. Finally, section models have the important advantage of allowing the measurement of the fundamental aerodynamic characteristics of the bridge deck on the basis of which comprehensive analytical studies can then be carried out. These characteristics include:

(a) The steady-state drag, lift, and moment coefficients, defined as follows:

$$C_D = \frac{\bar{D}}{\frac{1}{2}\rho U^2 B} \tag{13.1.1}$$

$$C_L = \frac{\bar{L}}{\frac{1}{2}\rho U^2 B} \tag{13.1.2}$$

$$C_M = \frac{\bar{M}}{\frac{1}{2}\rho U^2 B^2} \tag{13.1.3}$$

---

*For a discussion of Reynolds number similarity requirements, see Chapters 4 and 7.

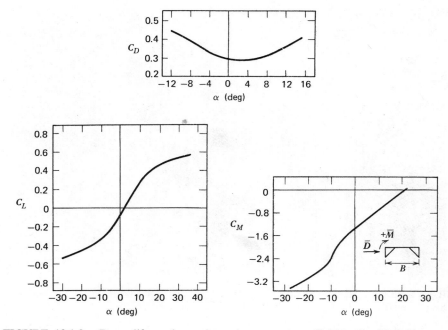

FIGURE 13.1.3.   Drag, lift, and aerodynamic moment coefficients for New Tacoma Narrows Bridge [13-1].

where $\bar{D}$, $\bar{L}$, and $\bar{M}$ are the mean drag, lift, and moment per unit span, respectively, ρ is the air density, B is the deck width, and U is the mean wind speed in the oncoming flow at the deck elevation. These coefficients are usually plotted as functions of the angle α between the horizontal plane and the plane of the bridge deck. Coefficients C_D, C_L, and C_M are shown in Fig. 13.1.3 for the open-truss bridge deck of the new Tacoma Narrows Bridge [13-1] and in Fig. 13.1.4 for a proposed streamlined box section of the New Burrard Inlet Crossing [13-8].

   (b) The motional aerodynamic coefficients H_1^*, H_2^*, H_3^*, A_1^*, A_2^*, A_3^*. These coefficients characterize the self-excited forces acting on the oscillating bridge and are discussed in some detail in Sect. 6.5.2. Examples of coefficients H_i^*, A_i^* ($i = 1, 2, 3$) for various types of bridge decks are given in Sect. 13.1.4. Questions pertaining to the laboratory determination of H_i^*, A_i^* are reviewed in [13-9].

   (c) The Strouhal number $\mathscr{S}$ (see Sect. 4.4).

13.1.2   Torsional Divergence or Lateral Buckling

Lateral buckling of a bridge deck may be viewed as that condition wherein, given a slight deck twist, the drag load and the self-excited aerodynamic moment will precipitate a torsional divergence instability. The torsional divergence phenomenon has been analyzed in Sect. 6.4 in the case of a two-dimensional structure. In this section, the analysis of Sect. 6.4 is extended to the case of a full bridge.

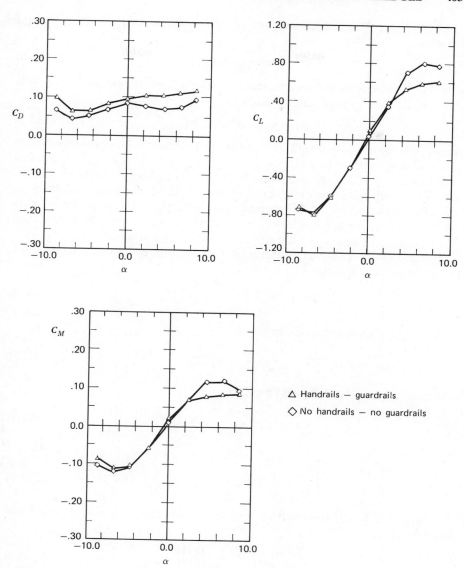

FIGURE 13.1.4. Drag, lift, and aerodynamic moment coefficients for proposed deck of New Burrard Inlet Crossing [13-8]. Courtesy of the National Aeronautical Establishment, National Research Council of Canada.

The data needed for the analysis are the experimentally measured moment coefficient $C_M(\alpha)$ and the torsional flexibility matrix C_T of the deck. Let x_i and x_j $(i, j = 1, 2, \ldots, N)$ denote values of the coordinate x along the span. The elements of the matrix C_T are denoted by c_{ij} and represent the angle of twist α_i at $x = x_i$ induced by a unit torsional moment acting at $x = x_j$.

Let $\{\alpha\}$ represent the column matrix of the angles of twist α_i. In matrix notation

$$\{\alpha\} = C_T\{M\} \tag{13.1.4}$$

where $\{M\}$ represents the column matrix of the torsional moments M_j applied at $x = x_j$. These moments can be written as

$$M_j = \tfrac{1}{2}\rho U^2 B^2 \Delta L_j C_M(\alpha_j) \tag{13.1.5}$$

where ΔL_j is the span length associated with point x_j. The problem is now susceptible of solution by iteration on Eqs. 13.1.4 and 13.1.5. First it is assumed $\alpha_j = 0$ for all j and M_j are calculated from Eq. 13.1.5. Inserting these results in Eq. 13.1.4 yields a column of values α_i; reinserting these into Eq. 13.1.5 develops new moments, and so on. The process will converge for any chosen velocity less than the *critical divergence velocity* that will, conceptually, be approached in an asymptotic manner by the iterative method suggested.

The process is simplified, however, in the case where $C_M(\alpha)$ can be approximated by a linear function

$$C_M(\alpha) \equiv \frac{dC_M}{d\alpha}\alpha + C_{M0} \tag{13.1.6}$$

where $C_{M0} = C_M(0)$. Using the notation

$$\frac{1}{p} = \tfrac{1}{2}\rho U^2 B^2 \Delta L_i \tag{13.1.7}$$

and assuming $\Delta L_i = \Delta L$ for all i yields

$$\{\alpha\} = C_T \frac{1}{p}\left\{\frac{dC_M}{da}\alpha + C_{M0}\right\} \tag{13.1.8}$$

or

$$\left[pI - \frac{dC_M}{d\alpha}C_T\right]\{\alpha\} = C_T\{C_{M0}\} \tag{13.1.9}$$

Equation 13.1.9 will have infinite (torsionally divergent) solutions when the determinant

$$\left|pI - \frac{dC_M}{d\alpha}C_T\right| = 0 \tag{13.1.10}$$

Equation 13.1.10 yields a set of characteristic values p of which the largest $p = p_c$ corresponds to the lowest velocity $U = U_c$ for torsional divergence:

$$U_c = \left[\frac{2}{p_c\rho B^2 \Delta L}\right]^{1/2} \tag{13.1.11}$$

In general it is found that only torsionally weak bridges incur the actual danger of torsional divergence/lateral buckling at wind speeds attainable in practice. It should also be noted that, for many bridge decks, the moment induced by the horizontal wind is *negative* (i.e., it twists the bridge deck so as to create a negative angle of attack, the wind then approaching the upper side of the deck). Such decks are not highly susceptible to torsional divergence at wind speeds in the usual range; however, if the slope of the curve $dC_M/d\alpha$ vs. α is positive, a theoretical torsional divergence is still possible.

13.1.3 Locked-In Vortex-Induced Response

Open truss sections generally "shred" the oncoming flow to such an extent that large, concerted vortices cannot occur and vortex-induced oscillations of the deck are weak. However, in the case of bluff deck sections of the box— or open box—type, instances of severe vortex-induced response are known to have occurred.

One such instance is cited in [13-10]. To reduce the oscillations, fairings were added to the section as shown in Fig. 13.1.5, which includes results of wind tunnel measurements. It is noted that in this case the water surface is close to the underside of the projected prototype and could thus be expected to affect significantly the flow around the deck. For this reason the water surface was also modeled in the laboratory.

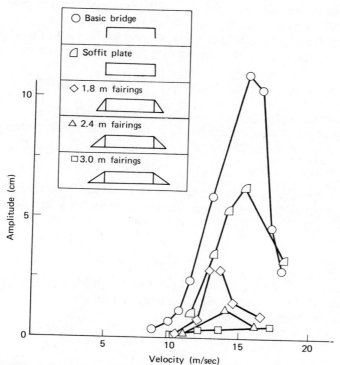

FIGURE 13.1.5. Vertical amplitudes of vortex-induced deflections for various bridge deck sections of the proposed Long Creek's Bridge [13-10]. Courtesy of the National Aeronautical Establishment, National Research Council of Canada.

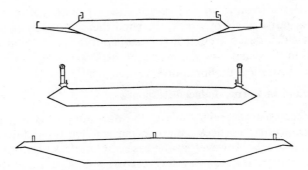

FIGURE 13.1.6.   Streamlined bridge deck forms.

Additional examples of streamlined bridge deck forms are shown in Fig. 13.1.6.

Analytical Procedure for Estimating the Vertical Vortex-Induced Response.
Under the action of the mean flow and of the shed vortices, the model section
will be subjected to a self-excited and to a vortex-induced lift. With notations
used in Sect. 6.5 and assuming that the vertical and torsional modes are uncoupled aerodynamically, the equation of motion of the section will be

$$m[\ddot{h}+2\zeta_h\omega_h\dot{h}+\omega_h^2 h]=\tfrac{1}{2}\rho U^2 B\left[KH_0^*(K)\frac{\dot{h}}{U}+C_{LV}\sin\omega t\right] \quad (13.1.12)$$

where ω is the vortex-shedding circular frequency and H_0^* and C_{LV} are coefficients to be determined. If the model is given some initial vertical deformation, its response will have the form

$$h\simeq(h_0\pm h_t e^{-\gamma\omega t})\sin(\omega t+\phi) \quad (13.1.13)$$

where h_0 is the steady-state amplitude, ϕ is a phase angle, and γ and h_t are
constants identifiable from the experimental observations. It can then be shown
easily that

$$C_{LV}=\frac{4\gamma m\omega_h^2 h_0}{\rho U_s^2 B} \quad (13.1.14)$$

where $U_s=n_h A/\mathcal{S}$, $n_h=\omega_h/2\pi$, A is the net area of bridge deck projected on a
vertical plane normal to the mean wind (per unit span), $\mathcal{S}$ is the Strouhal
number for the bridge deck, and that

$$H_0^*=\frac{4m}{\rho B^2}\left[\zeta_h\frac{\omega_h}{\omega}-\gamma\right] \quad (13.1.15)$$

It is noted that at lock-in $\omega\simeq\omega_h$.

The dimensionless quantities C_{LV} and H_0^* are applied to the prototype
bridge in the following manner. If ζ_p is the assumed mechanical damping

ratio of the prototype, the total (aerodynamic plus mechanical) damping in the prototype case can be written as

$$\tilde{\zeta}_p = \zeta_p - \frac{\rho B^2}{4m} H_0^* \tag{13.1.16}$$

The prototype being assumed to respond in an early bending mode $h_1(x)$ according to the relation

$$h(x, t) = h_1(x) q_1(t) \tag{13.1.17}$$

$q_1(t)$ is governed by the following equation:

$$M_1[\ddot{q}_1 + 2\tilde{\zeta}_p \omega_1 \dot{q}_1 + \omega_1^2 q_1] = \tfrac{1}{2}\rho U_s^2 B C_{LV} \left[\int_0^L h_1(x)\, dx \right] \sin(\omega_1 t + \phi) \tag{13.1.18}$$

In Eq. 13.1.18, ω_1 is the circular frequency of the chosen mode and M_1 is the generalized mass of that mode:

$$M_1 = \int_0^L h_1^2(x) m(x)\, dx \tag{13.1.19}$$

where $m(x)$ is the mass per unit span and L is the span of the prototype bridge. The maximum amplitude at vortex-induced resonance is then given by

$$[h(x)]_{max} = \frac{\rho U_s^2 B C_{LV} \int_0^L h_1(x)\, dx}{4 M_1 \omega_1^2 \tilde{\zeta}_p}\, h_1(x) \tag{13.1.20}$$

For example, if $h_1(x)$ is a half sine wave over the span of a bridge with a uniformly distributed mass, then the deflection at the span center is

$$\left[h\left(\frac{L}{2}\right) \right]_{max} = \frac{\rho U_s^2 B C_{LV}}{m \pi \omega_1^2 \tilde{\zeta}_p} \tag{13.1.21}$$

The accuracy of the above procedure is acceptable only if the difference between the mechanical damping ratios of the model and of the prototype is small. If this difference is large, the procedure may become inapplicable on account of strong nonlinear effects.

An alternate, nonlinear model (see Sect. 6.1.1) may also be employed. If the description of section activity as given by Eq. 13.1.12 is modified to the following (Van der Pol) form:

$$m[\ddot{h} + 2\zeta_h \omega_h h + \omega_n^2 h] = \rho U^2 B K H^* \left(1 - \varepsilon \frac{h^2}{B^2} \right) \frac{h}{U} \tag{13.1.12a}$$

then H^* and ε become the aerodynamic parameters. These are presumed to be evaluated from section model tests as described in Chapter 6. It can then easily be demonstrated that the maximum amplitude in the mode $h_1(x)$ is approximated by

$$[h_1(x)]_{max} = \eta_0 h_1(x) \tag{13.1.20a}$$

where

$$\eta_0 = \left[\frac{4\rho B^2 H^* I_1 - 8M_1\zeta}{\rho B^2 H^* \varepsilon I_2} \right]^{1/2} \tag{13.1.20b}$$

with

$$I_1 = \int_0^L h_1^2(x)\, dx$$

$$I_2 = \int_0^L \frac{h_1^4(x)\, dx}{B^2}$$

and M_1 as defined in Eq. 13.1.19. In case $h_1(x)$ is a half sine wave over the span, then $I_1 = L/2$ and $I_2 = 3L/8B^2$.

13.1.4   Flutter

Prediction of Flutter Velocity for a Full-Span Bridge.   The flutter phenomenon was studied in some detail in Sect. 6.5 under the assumption that two-dimensional geometrical conditions hold. In the case of a full-span bridge, the deformations of the deck are functions of position along the span so that this assumption is no longer valid. A generalization of the results of Sect. 6.5 to the case of the full-span bridge is presented herein.

Since flutter, in its most critical condition (i.e., corresponding to the lowest wind velocity) involves the lowest frequency modes of vibration of the bridge, it generally suffices to assume single modes $h(x)$ and $\alpha(x)$ as participants:

$$h(x, t) = h(x)p(t) \tag{13.1.22}$$

$$\alpha(x, t) = \alpha(x)q(t) \tag{13.1.22a}$$

It then follows from Eqs. 6.5.2 and 6.5.4 that the flutter equations of a symmetric bridge deck take the form

$$M_1[\ddot{p} + 2\zeta_h\omega_h\dot{p} + \omega_h^2 p]$$

$$= \tfrac{1}{2}\rho U^2(2B)\left[KC_{11}H_1^* \frac{\dot{p}}{U} + KC_{12}H_2^* \frac{B\dot{q}}{U} + K^2 C_{12}H_3^* q \right] \tag{13.1.23a}$$

$$I_1[\ddot{q} + 2\zeta_\alpha\omega_\alpha\dot{q} + \omega_\alpha^2 q]$$

$$= \tfrac{1}{2}\rho U^2(2B^2)\left[KC_{12}A_1^* \frac{\dot{p}}{U} + KC_{22}A_2^* \frac{B\dot{q}}{U} + K^2 C_{22}A_3^* q \right] \tag{13.1.23b}$$

where notations from Sect. 6.5 are used, and where

$$M_1 = \int_0^L m(x)h^2(x)\, dx \tag{13.1.24a}$$

$$I_1 = \int_0^L I(x)\alpha^2(x)\, dx \tag{13.1.24b}$$

$$C_{11} = \int_0^L h^2(x)\,dx \tag{13.1.24c}$$

$$C_{12} = \int_0^L h(x)\alpha(x)\,dx \tag{13.1.24d}$$

$$C_{22} = \int_0^L \alpha^2(x)\,dx \tag{13.1.24e}$$

In Eqs. 13.1.24, L is the bridge span, $m(x)$ is the mass of deck per unit length, and $I(x)$ is the moment of inertia of deck per unit length.

It is noted that the motion of the full bridge obeys equations similar in form to those of the section model. Therefore, after substitution of the generalized aerodynamic coefficients

$$C_{11}H_1^*;\ C_{12}H_2^*;\ C_{12}H_3^*;\ C_{12}A_1^*;\ C_{22}A_2^*;\ C_{22}A_3^*;$$

for the aerodynamic coefficients used in the two-dimensional case of Sect. 6.5.3, that is,

$$H_1^*;\ H_2^*;\ H_3^*;\ A_1^*;\ A_2^*;\ A_3^*;$$

respectively, the solution of the flutter equations is identical to that outlined in Sect. 6.5.3.

As indicated in Sect. 6.5.3, for certain types of bridges the values of the aerodynamic coefficients are such that the vertical and torsional motions are uncoupled and single-degree of freedom flutter in torsion occurs. For such bridges, the critical velocity (i.e., the velocity at which flutter occurs) is determined from the condition that the total (mechanical plus aerodynamic) damping is zero in the mode concerned, that is,

$$A_2^* \simeq \frac{2I_1\zeta_\alpha}{\rho B^4 C_{22}} \tag{13.1.25}$$

Let the value of the reduced velocity for which Eq. 13.1.25 holds be denoted by K_c. From the definition of the reduced velocity, it follows that the critical velocity is

$$U_c = \frac{B\omega_\alpha}{K_c}$$

Representative plots of the coefficients H_i^* and A_i^* for various bridge deck types are shown in Figs. 13.1.7 and 13.1.8. These coefficients are nondimensional reflections of the flutter proclivities of the bridge deck due to the *geometry* of the latter and are independent of structural characteristics, such as frequency. It is noted that, for any given positive value of A_2^*, the lowest value of the reduced velocity in Fig. 13.1.7 is that corresponding to the deck section 1 (original Tacoma Narrows bridge). Susceptibility to flutter is typical of solid girder or "H-section" forms, which are therefore no longer used in suspended-span bridge design.

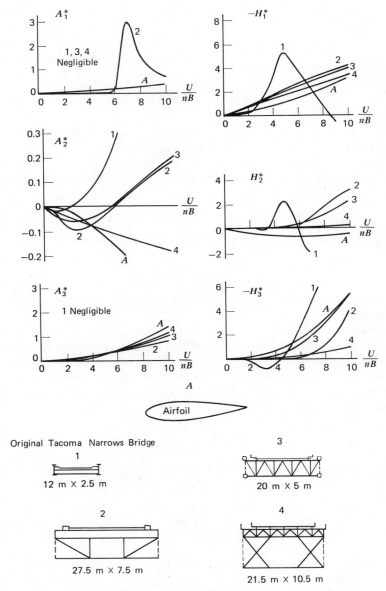

FIGURE 13.1.7. Coefficients H_i^* and A_i^* for various sections.

The question arises of whether experimental values of H_i^* and A_i^* obtained under laminar flow conditions are appropriate for the analysis of full-scale bridges, which are subjected to turbulent atmospheric flows. While no definitive answers to this question are available to date [13-11], some information has recently been made available in [13-12]. Therein H_i^* and A_i^* were obtained in

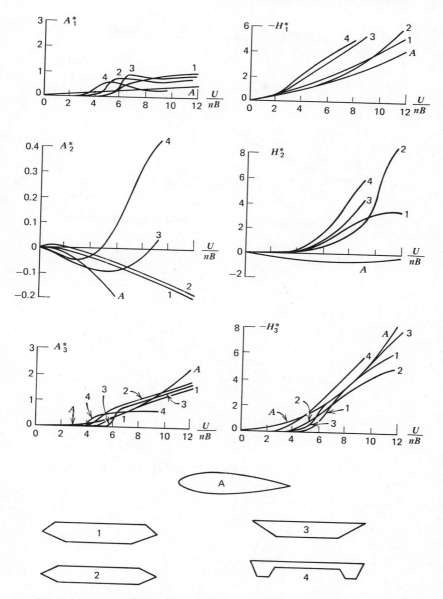

FIGURE 13.1.8. Coefficients H_i^* and A_i^* for various box girder sections.

turbulent flows with 11% turbulence intensity and with longitudinal and lateral integral scales of the order of the deck width and half that width, respectively. Results obtained in [13-12] are shown in Figs. 13.1.9 to 13.1.13. It is noted that in the case of the coefficient A_2^*, which determines the flutter velocity for bridges with aerodynamically uncoupled vertical and torsional motions, the disparity

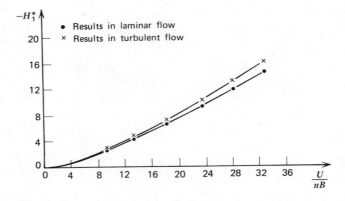

FIGURE 13.1.9.   Coefficients H_1^* in laminar and turbulent flow.

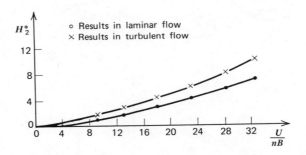

Note: Coefficients H_3^* are, for practical
purposes, identical in laminar and
turbulent flow

FIGURE 13.1.10.   Coefficients H_2^* in laminar and turbulent flow.

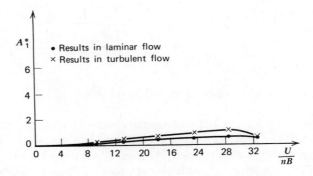

FIGURE 13.1.11.   Coefficients A_1^* in laminar and turbulent flow.

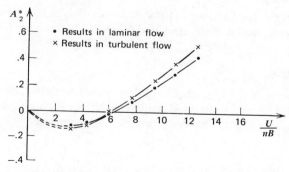

FIGURE 8.4.12.   Coefficients A_2^* in laminar and turbulent flow.

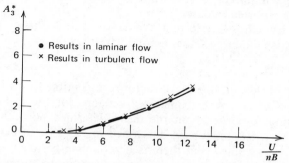

FIGURE 13.1.13.   Coefficients A_3^* in laminar and turbulent flow.

between results obtained under turbulent and laminar conditions for reduced velocities $U/nB < 12$ is of the order of 15% or less.

Dependence of Aeroelastic Stability Upon Bridge Characteristics.   The aeroelastic stability of a bridge is controlled by three factors:

1. *Geometry of the bridge deck.*   Unstable shapes include solid girder or "H-section" types of deck form; open-truss deck sections with closed, unslotted or unvented roadways; and certain very bluff cross sections. On the other hand, stability is enhanced by streamlined forms and by open-truss sections that contain vents or grills through the roadway surface.

2. *Frequencies of vibration of the bridge.*   High torsional frequencies tend to enhance stability. Examples of torsionally stiff shapes are closed torsion-box sections, or deep trusses closed by roadway and wind bracing to constitute a latticed tube. On the other hand "H-sections" are torsionally weak. Stability is also enhanced if the torsion-to-bending frequency is high.

3. *Mechanical damping of the bridge.*   Aeroelastic stability is clearly enhanced if the mechanical damping ratios of the bridge are high.

We also mention the possibility of enhancing the aeroelastic stability of a bridge by vibration reduction devices. Such a device, consisting of tuned mass dampers (TMDs—see Sect. 9.4.1) provided with disk brakes and not requiring any power source, was recently designed for the 1939 Bronx Whitestone Bridge [13-85].

13.1.5 Galloping

The susceptibility of a bridge deck to galloping can be determined by inspecting the plots of the lift and drag coefficients C_L and C_D versus α (see, for example, Figs. 13.1.3 and 13.1.4). The condition for incipient galloping instability is (see Sect. 6.2):

$$\frac{dC_L}{d\alpha} + C_D < 0 \tag{13.1.26}$$

Cases of large-amplitude across-wind galloping of suspended-span bridges have not been reported to date.

It follows from Eq. 13.1.26 that avoidance of deck shapes with regions of strongly negative lift curve slopes is conducive to stability.

13.1.6 Buffeting

The buffeting of slender, line-like structures susceptible to aeroelastic effects is discussed in Sect. 6.6. Approximate expressions for the mean square value of the buffeting response about the mean deformed position, based on the relations of Sect. 6.6, have been developed in [13-13]. In deriving these expressions it was assumed that there is no coupling between vertical and torsional modes. In addition, approximations similar to those of [6-85] were used, among which the assumption that the only significant contributions to the response are those due to eddies with frequencies equal, or nearly equal, to the natural frequencies of the bridge. Because of their relative simplicity and direct usefulness for design applications, the expressions of [13-13] are reproduced below.

Consider a bridge with the following characteristics: L is the bridge span, B is the deck width, $\alpha_r(x)$ and $h_r(x)$ are the rth vibration modes in torsion and bending, x is the coordinate along the span, $n_{\alpha r}$ and n_{hr} are the rth natural frequencies of vibration in torsion and bending, $\zeta_{\alpha r}$ and ζ_{hr} are the mechanical damping in rth torsion and bending modes, $m(x)$ is the mass of bridge deck per unit length, $I(x)$ is the mass moment of inertia per unit length about effective structural rotation axis (elastic axis), r is the distance from center of mass to elastic axis of bridge deck, A is the net area per unit span of bridge deck projected on a vertical plane normal to the mean wind, and z is the height of bridge deck above ground (water) level. Let $\alpha(x)$ denote the fluctuation of the twist angle about the mean position α_0 given by Eq. 6.6.22 and let $h(x)$ denote the vertical deflection. The mean square values of $\alpha(x)$ and $h(x)/B$ can be written approximately as

$$\sigma_\alpha^2(x) \cong \sum_{r=1}^{N} \frac{\pi \alpha_r^2(x)}{4\tilde{\zeta}_{\alpha r} \tilde{K}_{\alpha r}^3} \left(\frac{\rho B^4 L}{I_r} \right)^2 \left(\int_0^L \frac{\alpha_r^2(x)\, dx}{L} \right) \times \frac{2(c_\alpha - 1)}{c_\alpha^2} \frac{S_{MU}(\tilde{n}_{\alpha r})}{U^2} \tag{13.1.27}$$

$$\sigma_{h/B}^2(x) \simeq \sum_{r=1}^{N} \frac{\pi h_r^2(x)}{4\tilde{\zeta}_{hr} K_{hr}^3} \left(\frac{\rho B^2 L}{M_r}\right)^2 \left(\int_0^L \frac{h_r^2(x)\,dx}{L}\right) \times \frac{2(c_h-1)}{c_h^2} \frac{S_{LU}(\tilde{n}_{hr})}{U^2} \quad (13.1.28)$$

where N is the number of vibration modes in torsion or bending, U is the mean wind speed, and ρ is the air density ($\simeq 1.25 \text{ kg/m}^3$). In Eq. 13.1.27

$$\tilde{K}_{ar} \cong K_{ar}\left[1 - \frac{I_{ar}}{I_r} A_3^* \left(\frac{U}{n_{ar}B}\right)\right]^{1/2} \quad (13.1.27a)$$

$$K_{ar} = \frac{B(2\pi n_{ar})}{U} \quad (13.1.27b)$$

$$I_{ar} = \int_0^L \rho B^4 \alpha_r^2(x)\,dx \quad (13.1.27c)$$

(aerodynamic moment of inertia in rth mode)

$$I_r = \int_0^L I(x)\alpha_r^2(x)\,dx \quad (13.1.27d)$$

(generalized moment of inertia in rth mode)

$$\tilde{\zeta}_{ar} = \frac{1}{2\tilde{K}_{ar}}\left[2\zeta_{ar}K_{ar} - \frac{I_{ar}}{I_r}\tilde{K}_{ar}A_2^*\left(\frac{U}{\tilde{n}_{ar}B}\right)\right] \quad (13.1.27e)$$

$$\tilde{n}_{ar} = \frac{U\tilde{K}_{ar}}{2\pi B} \quad (13.1.27f)$$

$$c_\alpha = \frac{15\tilde{n}_{ar}L}{U} \quad (13.1.27g)$$

$$S_{MU}(\tilde{n}_{ar}) \cong C_{ME}^2 S_u(\tilde{n}_{ar}) + \tfrac{1}{4}C_{M_0}'^2 S_w(\tilde{n}_{ar}) \quad (13.1.27h)$$

$$C_{ME} = C_M(\alpha_0) + \frac{Ar}{B^2}C_D(\alpha_0) \quad (13.1.27i)$$

$$S_u(\tilde{n}_{ar}) = \frac{u_*^2}{\tilde{n}_{ar}}\frac{200\tilde{f}_{ar}}{(1+50\tilde{f}_{ar})^{5/3}} \quad (13.1.27j)$$

(spectrum of longitudinal velocity fluctuations, see Eq. 2.3.21)

$$S_w(\tilde{n}_{ar}) = \frac{u_*^2}{\tilde{n}_{ar}}\frac{3.36\tilde{f}_{ar}}{1+10\tilde{f}_{ar}^{5/3}} \quad (13.1.27k)$$

(spectrum of vertical velocity fluctuations, see Eq. 2.3.33)

$$\tilde{f}_{ar} = \frac{\tilde{n}_{ar}z}{U} \quad (13.1.27l)$$

$$u_* = \frac{U(z_{\text{ref}})}{2.5\ln(z_{\text{ref}}/z_0)} \quad (13.1.27m)$$

The quantities A_2^*, A_3^*, $C_M(\alpha_0)$, $C_D(\alpha_0)$, $C_{M_0}' = dC_M/d\alpha|_{\alpha=\alpha_0}$ (Eqs. 13.1.27a, 27e, 27h, and 27i) are aerodynamic properties of the bridge deck sections that must be determined by experiment. In Eq. 13.1.27m, z_{ref} is any reference height at which a mean wind velocity $U(z_{ref})$ is specified (for example, the bridge height above ground or water); z_0 is the roughness length characterizing the terrain roughness (see Table 2.2.1).

In Eq. 13.1.28,

$$K_{hr} = \frac{B(2\pi n_{hr})}{U} \tag{13.1.28a}$$

$$\tilde{\zeta}_{hr} \cong \zeta_{hr} - \frac{M_{ar}}{2M_r} H_1^* \left(\frac{U}{n_{hr}B} \right) \tag{13.1.28b}$$

$$M_{ar} = \int_0^L \rho B^2 h_r^2(x) \, dx \tag{13.1.28c}$$

(aerodynamic mass in rth mode)

$$M_r = \int_0^L m(x) h_r^2(x) \, dx \tag{13.1.28d}$$

(generalized mass in rth mode)

$$c_h = \frac{15 n_{hr} L}{U} \tag{13.1.28e}$$

$$S_{LU}(n_{hr}) \cong C_L^2(\alpha_0) S_u(n_{hr}) + \tfrac{1}{4} C_{LE}'^2 S_w(n_{hr}) \tag{13.1.28f}$$

$$C_{LE}' = C_{L_0}' + \frac{A}{B} C_D(\alpha_0) \tag{13.1.28g}$$

$$S_u(n_{hr}) = \frac{u_*^2}{n_{hr}} \frac{200 f_{hr}}{(1 + 50 f_{hr})^{5/3}} \tag{13.1.28h}$$

$$f_{hr} = \frac{n_{hr} z}{U} \tag{13.1.28i}$$

$$S_w(n_{hr}) = \frac{u_*^2}{n_{hr}} \frac{3.36 f_{hr}}{1 + 10 f_{hr}^{5/3}} \tag{13.1.28j}$$

The quantities H_1^*, $C_L(\alpha_0)$, $C_D(\alpha_0)$, and $C_{L_0}' = dC_L/d\alpha|_{\alpha = \alpha_0}$ (Eqs. 13.1.28b, 13.1.28f, and 13.1.28g) are experimentally determined aerodynamic properties of the bridge. For the quantity u_* in Eq. 13.1.28h, see Eq. 13.1.27m.

The principal features of the bridge that bear upon buffeting response are reflected in the parameters of Eqs. 13.1.27 and 13.1.28. It is noted that wide bridge decks generally exhibit higher buffeting response. Bridges with deck sections having high drag, or for which the slopes of the lift and moment curves are steep, also tend to be more susceptible to buffeting. As far as the mechanical properties of the bridge are concerned, it is clear that increased mechanical damping and increased stiffness result in a reduction of buffeting response.

13.1.7   Survey of Investigations into Suspended-Span Behavior

Following the Tacoma Narrows disaster in 1940, investigations into suspension bridge aerodynamics were carried out by Farquharson, Vincent, von Kármán, and Dunn [13-1] and, independently, by Steinman [13-14, 13-15]. The approach based on identifying the aeroelastic behavior of bridge sections by experimentally determined motional aerodynamic coefficients of freely oscillating models was later developed and applied by Princeton investigators [13-16] to [13-18]. Additional studies based on this approach have been carried out by the Federal Highway Administration [13-13, 13-19]. Bridge section model studies in turbulent flow as well as studies of wake effects have recently been conducted at Virginia Polytechnic Institute [13-20, 13-21, 13-22].

In Great Britain model and full-scale investigations by Scruton et al. at the National Physical Laboratory have led to the design, among others, of the Severn bridge [13-2, 13-3, 13-23 to 13-27]. A vigorous research effort pursued since the 1960s in Japan has included work on the following: techniques for obtaining motional aerodynamic coefficients, the application of such coefficients in analytical studies, and nonlinear aerodynamic effects [13-4, 13-28 to 13-35]. In Canada, studies of full-scale, "taut-strip," and section models have been carried out at the University of Western Ontario and at the National Research Council [13-5, 13-7, 13-8, 13-36 to 13-39]. Significant investigations conducted in Norway, Germany, France, and Belgium have been reported in [13-40 to 13-47].

13.1.8   Structural Members

Bridge members of circular, square, I- or H-section may be susceptible to wind-induced vibrations, particularly under the action of shed vortices.

If susceptibility to vortex-induced vibration is a problem, one of three types of solution can, in general, be used. First, the stiffness of the member can be increased so that the critical wind velocity exceeds the velocities that might be expected to occur during the life of the structure. To calculate the critical velocity U_{cr}, the following relation is used:

$$U_{cr} = \frac{n_1 D}{\mathscr{S}}$$

(13.1.29)

where n_1 is the fundamental frequency of vibration in the across-wind direction, D is the across-wind dimension, and $\mathscr{S}$ is the Strouhal number of the member. Because the dimension D of an individual member is small compared to the integral scale of the atmospheric turbulence, it may be assumed that the member behaves aerodynamically as if the flow were smooth so that the Strouhal number can be taken from Table 4.4.1.

Secondly, devices may be used that spoil the coherence of the shed vortices. Helical strakes and shrouds of the same design and with the same proportions as indicated in Sect. 10.3 may be employed on circular members. Figure 13.1.14 shows a spoiler device consisting of staggered fins that was successfully used

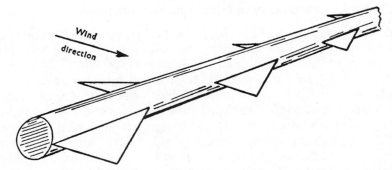

FIGURE 13.1.14.   Staggered fins on a pipeline suspension bridge [13-48, 10-23].

to suppress the oscillations of a pipeline suspension bridge [10-23, 13-48]. It is noted that this device would not be effective if the member were exposed to winds blowing from any direction (as would be the case if the member were vertical) rather than from just the direction parallel to the plane of the fins. Figure 13.1.15 shows perforations in the web of an I-section member that are also useful in reducing vortex-induced response.

Finally, in certain cases tuned mass damper (TMD) devices may be employed. The principle of these devices was discussed in Sect. 9.4.1. An example of a TMD used to control the oscillations of bridge I-beam members is described in detail in [13-49]. The device consists of a cantilevered rubber-shank pendulum

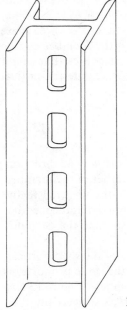

FIGURE 13.1.15.   Perforated web of I-section member.

weighted at the lower end. The weight employed may be of the order of 0.75% or more of the weight of the structural member.

To reduce vortex-induced oscillations of individual cables such as those in cable-stayed bridges or the deck hangers of suspension bridges, cable-to-cable ties, friction or hydraulic dampers, or TMD devices may be employed. In cases in which the oscillations cannot be prevented, fatigue-free cable terminations may have to be used to avert damage at the supports.

13.2   TENSION STRUCTURES AND POWER LINES

13.2.1   Cable Roofs

Vibrations of cable-supported roofs are caused principally by buffeting forces due to incident and structure-induced turbulence. It is likely that flutter (self-excited oscillation) of cable roofs is rare since most roof structures do not permit enough deflection to induce significant changes in the aerodynamic forces. The magnitude of the buffeting forces can be investigated in wind tunnel testing of aeroelastic or rigid models [13-50]. In the latter case, loading functions to be used in dynamic studies can be developed from the recorded time-dependent pressures.

Unwanted vibrations will not occur if the cable roof is sufficiently stiff. Stiffness is achieved by the provision of sufficient weight, for example, in the form of precast concrete roof panels; by pretensioning of cables; and/or by the provision of a stiffening system of tensioned cables with curvature opposite to that of the main, load-bearing system. In double-curvature roofs, the load-bearing and stiffening cables form a network—in most cases, orthogonal. Unless carefully designed, such roofs may exhibit serious vibration problems that have, in the past, necessitated the provision of additional ties and the lubrication of cable intersections to reduce noise caused by cable-to-cable friction. In single curvature roofs, stiffening cables may be provided at some distance underneath the load-bearing system, as in the case of the well-known Utica, N.Y. auditorium. The two layers of cables and the vertical members joining them form elements with considerable stiffness that prevent the occurrence of any significant wind-induced oscillations.

Recent studies on wind effects on cable roofs are reported in [13-51] to [13-53].

13.2.2   Air-Supported and Tensioned Fabric Structures

Long-span fabric structures, especially of the air-supported type, are a relatively new architectural and engineering development [13-54] to [13-58]. In many instances their design has been based on rudimentary representations of the wind loading [13-58]. Attempts to develop more realistic and elaborate wind loading criteria or wind tunnel modeling procedures are reported in [13-59] to [13-63].

13.2.3   Power Lines

The main vibration problems of long-span cables and power lines are associated with vortex-shedding, full-span galloping, and sub-span wake-induced galloping. These problems are briefly discussed below. For additional information and studies on wind effects on power lines, see [13-64] to [13-71], and [3-49].

Vortex-Induced Oscillations. Vortex-induced or Aeolian oscillations in long-span cables are generally caused by winds with speeds of the order of 2 to 10 m/sec. The oscillations generate packets of narrow-band random waves arriving at the cable supports. Since the cable is not perfectly flexible, the waves cause oscillatory bending stresses near the supports that result in fatigue damage unless protective measures are taken [13-72, 13-73, 13-74]. In the case of stranded wires fatigue damage can be produced by shear-induced friction, which affects mainly the inner wires.

Approaches used to prevent fatigue failure include the provision of special cushioned supports that alleviate the bending stresses and applications of the tuned mass damper (TMD) concept such as the classical Stockbridge damper [13-75, 13-76, 13-77]. The Stockbridge damper (Fig. 13.2.1) consists of a reactive counter-vibrating mass with a fairly wide band of frequency possibilities. The effect of the mass is to suppress to a large extent the last half-wave (nearest the support) generated by the cable oscillation. Like all TMD devices (see Sect. 9.4.1), the Stockbridge damper is not an energy dissipation device to any appreciable extent; it is, instead, an anti-resonant spring-mass device. Stockbridge dampers or similar devices can be readily purchased for a wide range of specific applications.

Full-Span Galloping.   Full-span power line galloping occurs most characteristically when ice forms on conductors and creates a new surface contour that is prone to galloping [13-78] to [13-80].

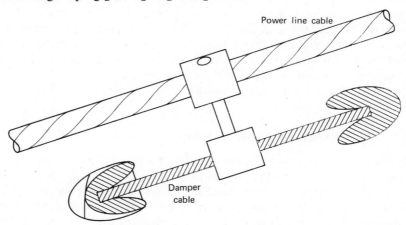

Power line cable

Damper
cable

FIGURE 13.2.1.   Stockbridge damper.

Means of alleviating the galloping of power lines have included melting of ice by carrying higher currents temporarily and thus heating the cables; installation of galloping warning sensors located at support towers in regions where cable icing is known to take place; anti-galloping aerodynamic devices designed to spoil the local flow; and tuned mass dampers at the center of cable spans.

Subspan Galloping. Subspan galloping, or wake-induced lateral galloping (see Sect. 6.3), has occurred repeatedly in grouped or bundled conductors [13-81] to [13-84]. Countermeasures have included (a) detuning the various cables in a bundle from each other by means of special spacers and (b) increasing damping by providing energy-dissipating spacers or by lowering the cable tensions, a measure that results in an increase of the inherent self-damping of the cable. None of these solutions has been fully effective. In particular, some highly complex and costly spacers with articulated and spring loaded arms have been found to be unsatisfactory.

Conceptually simpler—although again costly—solutions have included: a large increase in the number of spacers used between supports so that sub-span lengths are cut down and the corresponding frequencies are raised; and a continuous twist of the conductor bundle from support to support, which breaks the spanwise coherence of the vortices shed in the wake of the windward conductor. This latter solution has been judged, so far, to be impractical for application in the field.

REFERENCES

13-1   F. B. Farquharson (Ed.), *Aerodynamic Stability of Suspension Bridges*, Parts I–V, Bulletin No. 116, University of Washington Engineering Experiment Station, Seattle, 1949–1954.

13-2   C. Scruton, "Severn Bridge Wind Tunnel Tests," *Surveyor*, **107**, No. 2959 (Oct. 1948), 555.

13-3   C. Scruton, "Experimental Investigation of Aerodynamic Stability of Suspension Bridges with Special Reference to Proposed Severn Bridge," *Proc. Inst. Civ. Eng.*, **1**, Part 1, No. 2 (Mar. 1952), 189–222.

13-4   A. Hirai, I. Okauchi, and T. Miyata, "On the Behavior of Suspension Bridges Under Wind Action," in *Proceedings of the International Symposium on Suspension Bridges*, Laboratorio Nacional de Engenharia Civil, Lisbon, 1966, pp. 249–256.

13-5   A. G. Davenport, N. Isyumov, D. J. Fader, and C. F. P. Bowen, *A Study of Wind Action on a Suspension Bridge During Erection and Completion*, Report No. BLWT-3-69 with Appendix BLWT-4-70, Faculty of Engineering Science, University of Western Ontario, London, Canada, May 1969 and March 1970.

13-6   W. H. Melbourne, *West Gate Bridge Wind Tunnel Tests*, Internal Report, Department of Mechanical Engineering, Monash University, Clayton, Victoria, Australia.

13-7   A. G. Davenport, "The Use of Taut Strip Models in the Prediction of the Response of Long-Span Bridges to Turbulent Wind Flow-Induced Structural Vibrations," in *Proceedings of the IUTAM–IAHR Symposium on Flow-Induced Structural Vibrations*, Karlsruhe, West Germany, 1972, Springer-Verlag, Berlin, 1974, pp. 373–382.

13-8   R. L. Wardlaw, *Static Force Measurements of Six Deck Sections for the Proposed New Burrard Inlet Crossing*, Report No. LTR-LA-53, NAE, National Research Council, Ottawa, Canada, 1970.

13-9   R. H. Scanlan, *Recent Methods in the Application of Test Results to the Wind Design of Long, Suspended-Span Bridges*, Report No. FHWA-RD-75-115, Federal Highway Administration, Office of Research and Development, Washington, D.C., 1975.

13-10   R. L. Wardlaw and L. L. Goettler, *A Wind Tunnel Study of Modifications to Improve the Aerodynamic Stability of the Long's Creek Bridge*, Report No. LTR-LA-8, NAE, National Research Council, Ottawa, Canada, 1968.

13-11   Y. K. Lin, "Stochastic Analysis of Bridge Motion in Large-Scale Turbulent Winds," in *Wind Engineering*, Proceedings of the Fifth International Conference, J. E. Cermak (Ed.), Pergamon Press, New York, 1980.

13-12   W. H. Lin, "Forced and Self-Excited Responses of a Bluff Structure in a Turbulent Wind," Doctoral Dissertation, Department of Civil Engineering, Princeton University, 1977.

13-13   R. H. Scanlan and R. H. Gade, "Motion of Suspended Bridge Spans Under Gusty Wind," *J. Struct. Div.*, ASCE, **103** (1977), 1867–1883.

13-14   D. B. Steinman, "Aerodynamic Theory of Bridge Oscillations," *Proc. ASCE*, **75**, 8 (Oct. 1949), 1147–1184.

13-15   D. B. Steinman, "Aerodynamic Theory of Bridge Oscillations," *Proc. ASCE*, **76**, 1 (Jan. 1950), 154–158.

13-16   R. H. Scanlan and A. Sabzevari, "Experimental Aerodynamic Coefficients in the Analytical Study of Suspension Bridge Flutter," *J. Mech. Eng. Sci.*, Institution of Mechanical Engineers, London, **11**, 3 (June 1969), 234–242.

13-17   R. H. Scanlan and J. J. Tomko, "Airfoil and Bridge Deck Flutter Derivatives," *J. Eng. Mech. Div.*, ASCE, **97**, No. EM6 (Dec. 1971), 1717–1737.

13-18   R. H. Scanlan, J. G. Béliveau, and K. S. Budlong, "Indicial Aerodynamic Functions for Bridge Decks," *J. Eng. Mech. Div.*, ASCE, **100**, No. EM4 (Aug. 1974), 657–672.

13-19   R. H. Gade, H. R. Bosch, and W. Podolny, Jr., "Recent Aerodynamic Studies of Long-Span Bridges," *J. Struct. Div.*, ASCE, **102**, No. ST7 (July 1976), 1299–1315.

13-20   T. A. Reinhold, H. W. Tieleman, and F. J. Maher, *The Torsional Response of a Suspension Bridge Stiffening Truss Model to Turbulence and to an Upstream Obstacle*, Report No. VPI-E-74-28, Department of Engineering Science and Mechanics, Virginia Polytechnic Institute and State University, Blacksburg, Va., 1974.

13-21   T. A. Reinhold, H. W. Tieleman, and F. J. Maher, *Decay of the Wake of a Box Girder*, Report No. VPI-E-75-18, Department of Engineering Science and Mechanics, Virginia Polytechnic Institute and State University, Blacksburg, Va., 1975.

13-22   T. A. Reinhold, H. W. Tieleman, and F. J. Maher, *Wake Study of a Suspension Bridge Stiffening Truss*, Report No. VPI-E-75-17, Department of Engineering Science and Mechanics, Virginia Polytechnic Institute and State University, Blacksburg, Va., 1975.

13-23   G. Roberts, "The Severn Bridge. A New Principle of Design," *Proceedings of the International Symposium on Suspension Bridges*, Laboratorio Nacional de Engenharia Civil, Lisbon, 1966, pp. 629–639.

13-24   C. Scruton, *Experimental Investigation of the Aerodynamic Stability of Suspension Bridges*, Report No. 165, National Physical Laboratory, Teddington, U.K., 1948.

13-25   C. Scruton, *Experiments on the Aerodynamic Stability of Suspension Bridges: Results of Tests of a Full Model in Horizontal Winds*, Report No. 185, National Physical Laboratory, Teddington, U.K., 1950.

13-26   D. E. Walshe, *The Aerodynamic Investigation for a Box-Type Roadway Deck for a Suspension Bridge Proposed for the Humber Crossing*, Special Report No. 012, National Physical Laboratory, Teddington, U.K., 1968.

13-27   D. E. Walshe, *The Aerodynamic Investigation for the Suspended Structure of the Proposed Bosporus Bridge*, Special Report No. 020, National Physical Laboratory, Teddington, U.K., 1969.

13-28   A. Hirai and T. Okubo, "On the Design Criteria Against Wind Effects for Proposed Honshu-Shikoku Bridges," in *Proceedings of the International Symposium on Suspension Bridges*, Laboratorio Nacional de Engenharia Civil, Lisbon, 1966, pp. 265–272.

13-29 M. Ito, "On the Wind-Resistant Design of Truss-Stiffened Suspension Bridges," in *Proc. Second USA–Japan Research Seminar on Wind Effects on Structures*, Kyoto, 1974, University of Tokyo Press, 1976, pp. 285–296.

13-30 N. Shiraishi, "On the Aerodynamic Responses of Truss-Stiffened Bridge Sections in Fluctuating Wind Flows," in *Proceedings of the IUTAM-IAHR Symposium on Flow-Induced Structural Vibrations*, Karlsruhe, West Germany, 1972, Springer-Verlag, Berlin, 1974, pp. 401–405.

13-31 I. Konishi, N. Shiraishi, and M. Matsumoto, "Aerodynamic Response Characteristics of Bridge Structures," in *Proceedings of the Fourth International Conference on Wind Effects on Buildings and Structures*, London, 1975, Cambridge Univ. Press, Cambridge, U.K., 1976, pp. 199–208.

13-32 T. Okubo, N. Narita, and K. Yokoyama, "Some Approaches for Improving Wind Stability for Cable-Stayed Girder Bridges," in *Proceedings of the Fourth International Conference on Wind Effects on Buildings and Structures*, London, 1975, Cambridge Univ. Press, Cambridge, U.K., 1976, pp. 241–249.

13-33 Y. Nakamura and T. Yoshimura, "Binary Flutter of Suspension Bridge Deck Sections," *J. Eng. Mech. Div.*, ASCE, **102**, No. EM4 (Aug. 1976), 685–700.

13-34 T. Miyata, Y. Kubo, and M. Ito, "Analysis of Aeroelastic Oscillations of Long-Span Structures by Nonlinear Multi-Dimensional Procedures," in *Proceedings of the Fourth International Conference on Wind Effects on Buildings and Structures*, London, 1975, Cambridge Univ. Press, Cambridge, U.K., 1976, pp. 215–225.

13-35 M. Ito and Y. Nakamura, "Aerodynamic Stability of Structures in Wind." *IABSE Surveys* S-20/82, International Association for Bridge and Structural Engineering, ETH- Hönggerberg, Zürich, 1982.

13-36 R. L. Wardlaw, *A Preliminary Wind Tunnel Study of the Aerodynamic Stability of Four Bridge Sections for the Proposed New Burrard Inlet Crossing*, Report No. LTR-LA-31, NAE, National Research Council, Ottawa, Canada, 1969.

13-37 R. L. Wardlaw, "Some Approaches for Improving the Aerodynamic Stability of Bridge Road Decks," in *Proceedings of the Third International Conference on Wind Effects on Buildings and Structures*, Tokyo, 1971, Saikon, Tokyo, 1972, pp. 931–940.

13-38 R. L. Wardlaw, *A Wind Tunnel Study of the Aerodynamic Stability of the Proposed Pasco-Kennewick Intercity Bridge*, Report No. LTR-LA-163, NAE National Research Council, Ottawa, Canada, 1974.

13-39 H. P. A. H. Irwin, *Wind Tunnel and Analytical Investigation of the Response of the Lions' Gate Bridge to a Turbulent Wind*, Report No. LTR-LA-210, NAE, National Research Council, Ottawa, Canada, 1978.

13-40 C. Ostenfeld, G. Haas, and A. G. Frandsen, "Motorway Bridge Across Lillebaelt. Model Tests for the Superstructure of the Suspension Bridge," in *Proceedings of the International Symposium on Suspension Bridges*, Laboratorio Nacional de Engenharia Civil, Lisbon, 1966, pp. 587–608.

13-41 K. Kloeppel and G. Weber, "Teilmodellversuche zur Beurteilung des aerodynamischen Verhaltens von Bruecken," *Der Stahlbau*, **32**, 3 (1963), 65–79.

13-42 K. Kloeppel and F. Thiele, "Modellversuche im Windkanal zur Bemessung von Bruecken gegen die Gefahr winderregter Schwingungen," *Der Stahlbau*, **32**, 12 (1967), 353–365.

13-43 F. Leonhardt, "Zur Entwicklung aerodynamisch stabiler Haengebruecken," *Bautech.*, **45**, 10–11 (1968), 1–21.

13-44 F. Leonhardt, "Latest Developments of Cable-Stayed Bridges for Long Spans," *Bygningsstakiske Medd.*, **45**, 4 (1974), 89–143.

13-45 H. Loiseau and E. Szechenyi, *Etude du comportement aéroélastique du tablier d'un pont à haubans*, T.P. 1975–75, Office National d' Études et de Recherches Aérospatiales, Chatillon, France.

13-46 J. Roche, "Les Méthodes d'étude aérodynamique des ponts à haubans," in *Proceedings of the Conference on Cable-Stayed Bridges*, Paris, 1974, pp. 75–86.

13-47   D. Olivari and F. Thiry, "Wind Tunnel Tests of the Aeroelastic Stability of the Heer–Agimont Bridge," Tech. Note 113, von Kármán Institute for Fluid Dynamics, Rhode St. Genèse, Belgium, 1975.

13-48   R. C. Baird, "Wind-Induced Vibration of a Pipe-Line Suspension Bridge and its Cure," *Trans. ASME*, **77** (Aug. 1955), 797–804.

13-49   H. P. A. H. Irwin, K. R. Cooper, and R. L. Wardlaw, *Application of Vibration Absorbers to Control Wind-Induced Vibration of I-Beam Truss Members on the Commodore Barry Bridge*, Laboratory Technical Report No. LTR-LA-194, NAE, National Research Council, Ottawa, Canada, 1976.

13-50   S. Kawamura, J. E. Cermak, and E. Kimoto, *Aerodynamic Behavior of Hanging Roofs*, Technical Report No. CEP70-71SK-JEC-EK76, Department of Civil Engineering, Colorado State University, Structures Division, ASCE, 1971.

13-51   T. Matsumoto, "An Investigation on the Response of Pretensioned One-Way-Type Suspension Roofs to Wind Action," *J. Wind Eng. Ind. Aerodyn.*, **13** (1983), 383–394.

13-52   E. Kimoto and S. Kawamura, "Aerodynamic Behavior of One-Way Type Hanging Roofs," *J. Wind Eng. Ind. Aerodyn.*, **13** (1983), 395–406.

13-53   I. Elashkar and M. Novak, "Wind Tunnel Studies of Cable Roofs," *J. Wind Eng. Ind. Aerodyn.*, **13** (1983), 407–420.

13-54   T. Herzog, *Pneumatic Structures*, Oxford University Press, New York, 1976.

13-55   *Practical Applications for Air-Supported Structures*, International Conference held at Las Vegas, Oct. 1974, Canvas Products Association International, St. Paul, Minnesota, 1974.

13-56   "Technics: Fabric Structures," *Progressive Architecture*, **51** (1980), 110–120.

13-57   A. Morrison, "The Fabric Roof," *Civ. Eng.*, **50** (1980), 60–65.

13-58   *Air-Supported Structures*, State of the Art Report, American Society of Civil Engineers, New York, 1979.

13-59   M. Horcic, *Windbelastung und Berechnung des Spannungs- und Verformungszustandes im zylindrischen Teil von Traglufthallen mit besonderer Berücksichtigung des Konstruktionsmaterials* (H. von Gunten, Referent, H. H. Thomann, Korreferent), Eidgenössische Technische Hochschule, Zürich, 1974.

13-60   B. V. Tryggvason and N. Isyumov, *A Study of the Wind-Induced Response of the Air Supported Roof for the Dalhousie University Sports Complex*, BLWT-SS7-1977, Boundary-Layer Wind Tunnel Laboratory, University of Western Ontario, London, Ont., Canada, 1977.

13-61   B. V. Tryggvason, "Aeroelastic Modeling of Pneumatic and Tensioned Structures," in *Wind Engineering*, Proceedings of the Fifth International Conference, Fort Collins, CO, July 1979, J. E. Cermak (Ed.), Pergamon Press, New York, 1980.

13-62   H. P. A. H. Irwin and R. L. Wardlaw, "A Wind Tunnel Investigation of a Retractable Fabric Roof for the Montreal Olympic Stadium," in *Wind Engineering*, Proceedings of the Fifth International Conference, Fort Collins CO, July 1979, J. E. Cermak (Ed.), Pergamon Press, New York, 1980.

13-63   R. J. Kind, "Aeroelastic Modeling of Membrane Structures," in *Wind Tunnel Modeling for Civil Engineering Applications*, T. A. Reinhold (Ed.), Cambridge Univ. Press, Cambridge, U.K., 1982.

13-64   J. C. R. Hunt and D. J. W. Richards, "Overhead-Line Oscillations and the Effect of Aerodynamic Dampers," *Proc. Inst. Elec. Eng.*, **116** (1969), 1869–1874.

13-65   J. C. R. Hunt and M. D. Rowbottom, "Meteorological Conditions Associated with the Full-Span Galloping Oscillations of Overhead Transmission Lines," *Proc. Inst. Elec. Eng.*, **120** (1973), 874–876.

13-66   V. J. Brzozowski and R. J. Hawks, "Wake-Induced Full-Span Instability of Bundle Conductor Transmission Lines," *AIAA J.*, **14** (1976), 179–184.

13-67   A. N. Hoover and R. J. Hawks, "Role of Turbulence in Wake-Induced Galloping of Transmission Lines," *AIAA J.*, **15** (1977), 66–70.

13-68   A. S. Richardson, Jr., *Dynamic Load Model Study for Overhead Transmission Lines*, HCP/T-

2063/2, U.S. Department of Energy, Division of Electric Energy Systems, Washington, D.C., 1977.

13-69 A. H. Peyrot and A. M. Goulois, "Analysis of Flexible Transmission Lines," *J. Struct. Div.*, ASCE, **104** (1978), 763–769.

13-70 "Device Reins in Galloping Power Lines," *Eng. News-Record*, Nov. 30, 1978, p. 17.

13-71 A. G. Davenport, "Gust Response Factors for Transmission Line Loading," in *Wind Engineering*, Proceedings of the Fifth International Conference, Fort Collins, CO, 1979, Pergamon Press, York, 1980.

13-72 J. S. Carroll, "Laboratory Studies of Conductor Vibration," *Trans. AIEE*, **55**, 5 (May 1936), 543–547.

13-73 F. B. Farquharson and R. E. McHugh, Jr., "Wind Tunnel Investigation of Conductor Vibration with Use of Rigid Models," *Trans. AIEE*, **75**, Part 3 (Oct. 1956), 871–878.

13-74 J. S. Tompkins, L. L. Merrill, and B. L. Jones, "Quantitative Relationships in Conductor Vibration Damping," *Trans. AIEE*, **75** (Oct. 1956), 879–896.

13-75 G. H. Stockbridge, "Overcoming Vibration in Transmission Cables," *Electr. World*, **86**, 26 (Dec. 1925), 1304–1305.

13-76 R. G. Sturm, "Vibration of Cables and Dampers—I, II," *Electr. Eng.*, **55** (1936), 455–465, 673–688.

13-77 R. A. Komenda and R. L. Swart, "Interpretation of Field Vibration Data," *Trans. IEEE*, **PAS-87**, 4 (April 1968), 1066–1073.

13-78 A. S. Richardson, J. R. Martuccelli, and W. S. Price, "Research Study on Galloping of Electric Power Transmission Lines," in *Proceedings of the Symposium on Wind Effects on Buildings and Structures*, Vol. 2, National Physical Laboratory, Teddington, U.K., 1963, pp. 611–686.

13-79 A. S. Richardson, "Design and Performance of an Aeroelastic Anti-Galloping Device," IEEE Summer Power Meeting, Chicago, Conference Paper No. 68, CP-670-PWR, 1968.

13-80 M. D. Rowbottom and R. R. Aldham-Hughes, *Sub-Span Oscillation: A Review of Existing Knowledge*, Report No. 22-71(SC)-02, Central Electricity Research Laboratories, Leatherhead, Surrey, U.K., 1971.

13-81 O. Nigol and G. J. Clarke, "Conductor Galloping and Control Based on Torsional Mechanism," IEEE, Power Engineering Meeting, New York, Conference Paper No. C74 016-2, 1974.

13-82 K. R. Cooper and R. L. Wardlaw, "Aeroelastic Instabilities in Wakes," in *Proceedings of the Third International Conference on Wind Effects on Buildings and Structures*, Tokyo, 1971, Saikon, Tokyo, 1972, pp. 647–655.

13-83 R. L. Wardlaw, K. R. Cooper, and R. H. Scanlan, "Observations on the Problem of Subspan Oscillation of Bundled Power Conductors," *DME/NAE Q. Bull.* 1973 (1), National Research Council, Ottawa, Canada.

13-84 R. H. Scanlan, *A Wind Tunnel Investigation into the Aerodynamic Stability of Bundled Power Line Conductors for Hydro-Québec*, Part VI, Report No. LTR-LA-121 NAE, National Research Council, Ottawa, Canada, 1972.

13-85 "Harmonizing with the Wind," *Eng. News Record*, Oct. 25, 1984, pp. 12–13.

14

Offshore Structures

Wind loads affect offshore structures during construction, towing, and in service. They are a significant structural design factor, especially in the case of large compliant platforms, such as guyed towers and tension leg platforms.

Wind can also affect the flight of helicopters near offshore platform landing decks [14-1, 14-2, 14-3], as potentially dangerous conditions may be created by flow separation (see Sect. 4.3) at the edges of the platform. Let the horizontal distance between the upstream edge of the platform and the upstream edge of the helideck be denoted by d, and let the depth of the upstream surface producing the separated flow be denoted by t. On the basis of wind tunnel tests it has been suggested that the elevation, h, of the helideck with respect to the upstream platform edge should vary from at least $h \simeq 0.2t$ if $d \simeq 0$ to at least $h \simeq 0.5t$ if $d \simeq t$ [14-2].

This chapter includes information on wind loads on offshore structures of various types (Sect. 14.1), and on the treatment of dynamic wind effects in the case of compliant structures (Sect. 14.2).

14.1 WIND LOADING ON OFFSHORE STRUCTURES

Methods for calculating wind loads on offshore platforms are recommended in [14-4] to [14-8]. However, laboratory and full-scale measurements indicate that these methods may, in some instances, have serious limitations, particularly insofar as they do not account for the presence of lift forces, and account insufficiently—or not at all—for shielding and mutual interference effects. For example, according to wind tunnel test results obtained for a jack-up (self-elevating) platform [14-9], the methods of [14-4] and [14-5] overestimate wind loads on jack-up units by at least 35%. Estimates based on full-scale data for an anchored semisubmersible platform [14-10] suggest that the method of [14-5] overpredicts wind loads by as much as 100%. It has therefore become common practice to obtain design information on wind forces on platforms from laboratory tests. Most tests provide data on mean, as opposed to gusting loads. In using such data the effect of gustiness should be accounted for by analytical means (see for example, Sect. 14.2). Possible Reynolds number effects should also be assessed with care.

This section briefly reviews a number of wind tunnel tests conducted for semisubmersible units and for a large guyed tower platform. Wind tunnel test information on jack-up units, on jacket structures in the towing mode, and on two types of concrete platform is available in [14-9], [14-11] to [14-14], and [14-35].

14.1.1   Wind Loading on Semisubmersible Units

A schematic view of the model of a semisubmersible unit used for tests reported in [14-15] is shown in Fig. 14.1.1.*

The side force and heeling moment coefficients are defined as

$$CY = \frac{Y}{\frac{1}{2}\rho U^2(50)A_s} \tag{14.1.1}$$

$$CK = \frac{K}{\frac{1}{2}\rho U^2(50)A_s H_s} \tag{14.1.2}$$

where Y is the side force, K is the heeling moment, ρ is the air density, $U(50)$ is the mean wind speed at 50 m above sea level, A_s is the projected side area, and H_s is the elevation of the center of gravity of A_s. Coefficients CY and CK are obtained separately for the overwater and for the underwater part of the unit. The overwater coefficients reflect the action of wind and should be obtained in a flow simulating the atmospheric boundary layer. The underwater coefficients account for hydrodynamic effects and should therefore be measured in uniform flow.

Figures 14.1.2 and 14.1.3 show values of CY and CK measured in [14-15] for the case of an upright draft† $T_{M0°} = 10.85$ m (corresponding, for the unit being modeled, to a displacement‡ of 17,729 tons). As noted in [14-15], the purpose of the tests for the underwater part is to determine the elevation of the center of reaction (i.e., the point of application of the resultant of the underwater forces) for the free-floating unit. For an anchored or dynamically positioned unit the center of reaction is determined by the anchoring forces or by the thrusters [14-15].

Figure 14.1.4 shows estimated values of the heeling forces induced by 100-

---

*Figure 14.1.1 through 14.1.6 are excerpted from E. Bjerregaard and S. Velschou, "Wind Overturning Effect on a Semisubmersible," Paper OTC 3063, *Proceedings*, Offshore Technology Conference, Houston, TX, May 1978. Copyright 1978 Offshore Technology Conference.

†The upright draft $T_{M0°}$ is the depth of immersion of the unit in the even heel condition (i.e., for an angle of heel $\phi = 0°$).

‡The displacement is the volume of water displaced by the immersed part of the unit.

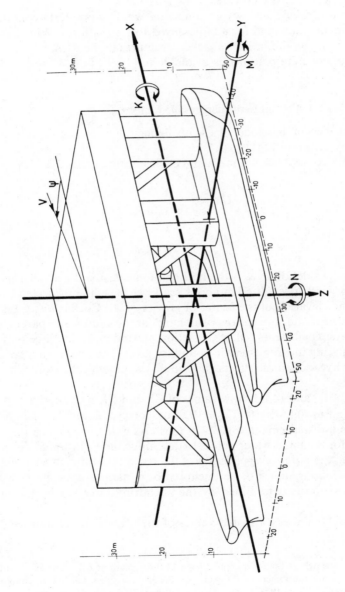

FIGURE 14.1.1. Schematic view of a semisubmersible unit model [14-15].

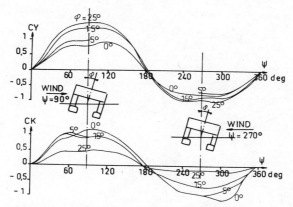

FIGURE 14.1.2. Values CY and CK as functions of wind direction ψ at different angles of heel ϕ for the overwater part [14-15].

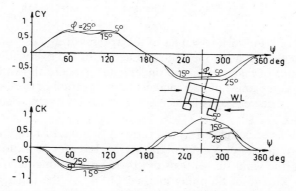

FIGURE 14.1.3. Values CY and CK as functions of direction ψ at different angles of heel ϕ for the underwater part [14-15].

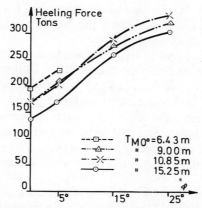

FIGURE 14.1.4. Wind heeling forces corresponding to 100-knot beam winds [14-15].

429

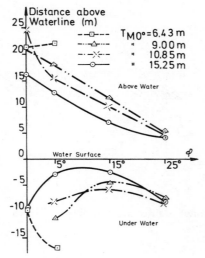

FIGURE 14.1.5. Elevation of center of action of wind forces and corresponding center of reaction on the underwater part [14-15].

knot beam winds* for various values of the upright draft $T_{M0°}$ and of the angle of heel ϕ. The elevations of the center of action of the overwater (wind) force and of the center of reaction on the underwater part are shown in Fig. 14.1.5. It is seen that as the angle of heel increases the elevation of the center of action of the wind force decreases. This decrease is due to lift forces arising at nonzero angles of heel ϕ.

The heeling lever is defined as the ratio of the overturning moment to the displacement of the vessel. Values of the heeling lever for 100-knot beam winds, obtained from the wind tunnel tests of [14-15] on the one hand and by using the American Bureau of Shipping method [14-4] on the other, are shown in Fig. 14.1.6. (The displacements listed in [14-15] for the 6.43 m, 9.00 m, and 15.25 m drafts are 12,740 tons, 16,963 tons, and 19,495 tons, respectively.) It is seen that for large angles of heel the differences between the two sets of values are considerable. This is largely due to the failure of [14-4] to account for the effects of lift. It is noted in [14-16] that the largest overturning moments are commonly induced by quartering winds.

In the tests of [14-15] and [14-16] the water surface was modeled by the rigid horizontal surface of the wind tunnel floor. Following the method described in [14-17], tests reported in [14-18] were also conducted by placing the model in a tank filled with viscoelastic material up to the level of the wind tunnel floor. This facilitates the testing of models of partially submerged units. Reference 4-18 also includes results of tests conducted in the presence of rigid obstructions aimed at representing water waves. The results revealed that waves could increase the overturning moments substantially. This suggests

---

*Winds blowing along the axis Y (Fig. 14.1.1). Wind blowing along the axis X are referred to as head (or bow) winds. Winds whose directions bisect the angles between axes X and Y are referred to as quartering winds.

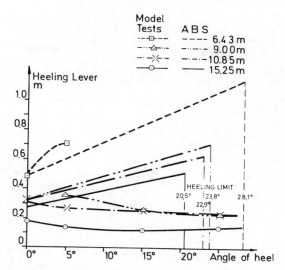

FIGURE 14.1.6. Wind heeling levers obtained from wind tunnel tests and from American Bureau of Shipping (ABS) method [14-15].

the need for improving the simulation of the sea surface in laboratory tests.

Additional wind tunnel tests of semisubmersible units are reported in [14-14] and [14-19] to [14-22].

14.1.2   Wind Loads on a Guyed Tower Platform

Reference 14-23 presents results of wind tunnel measurements on a 1:120 scale model of the overwater part of a structure similar to Exxon's Lena guyed tower platform. (A schematic view of the platform, installed in over 300 m of water in the Gulf of Mexico, is shown in Fig. 14.1.7 [14-24]—see also Fig. 14.1.8.) The mean wind profile created in the laboratory matched closely both the power law:

$$U(z) = U(10) \left(\frac{z}{10}\right)^{1/12}$$
(14.1.3)

and the expression for sustained winds (i.e., winds averaged over at least one minute) recommended by the United States Geological Survey [14-7] for use within the Gulf of Mexico:

$$U(z) = U(10) \left(\frac{z - z_d}{10 - z_d}\right)^{0.1128}$$
(14.1.4)

where $z_d = 2.2$ m and z is the elevation above the still water level in meters. The air/water boundary was modeled by the rigid horizontal surface of the wind tunnel floor. Force and moment coefficients were defined by relations of the type

$$C_F = \frac{F}{\frac{1}{2}\rho U^2 (16) A_R}$$
(14.1.5)

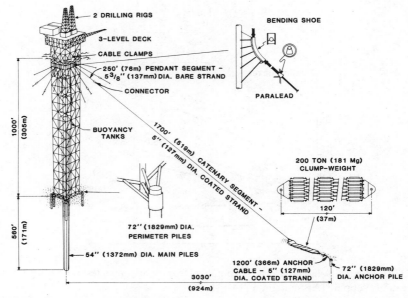

FIGURE 14.1.7.   Schematic view of Lena guyed tower platform. From M. S. Glasscock and L. D. Finn, "Design of a Guyed Tower for 1000 ft of Water in the Gulf of Mexico," *J. Struct. Eng.*, **110** (1984), 1083–1098.

$$C_M = \frac{M}{\frac{1}{2}\rho U^2 (16) A_R L_R} \tag{14.1.6}$$

where F and M are the mean force and the moment of interest, ρ is the air density, $U(16)$ is the mean wind speed at 16 ft above the water surface, and the reference area A_R and length L_R were chosen as 1 ft^2 and 1 ft, respectively. The force and moments obtained in [14-23] are represented in Fig. 14.1.9, which also shows the notations for the respective aerodynamic coefficients. The moments characterized by the coefficients CMD and CMT were taken with respect to a distance of 6.2 in (62 ft full-scale) below the still water level. The measured values of the aerodynamic coefficients are represented in Fig. 14.1.10 for several platform configurations. The configuration for the base case was the same as in Fig. 14.1.8, except that the deck structure was not enclosed. Additional results in [14-23] show that the effect of enclosing the deck is negligible for practical purposes, as is the effect of the well conductors. Removing the flare boom results in torsional moment reductions, but has negligible effects otherwise. It is shown in [14-23] that drag forces and drag moments based on wind tunnel measurements are smaller by about 30% and 20%, respectively, than the calculated values based on [14-7].

   To check the extent to which the results depend upon the laboratory facility being used, the same structure was subsequently tested independently in a

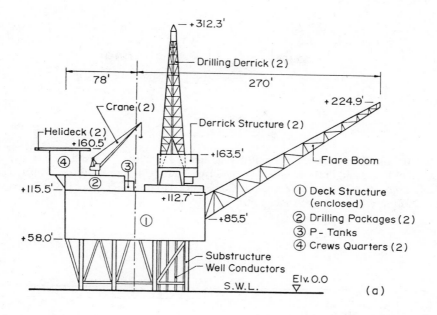

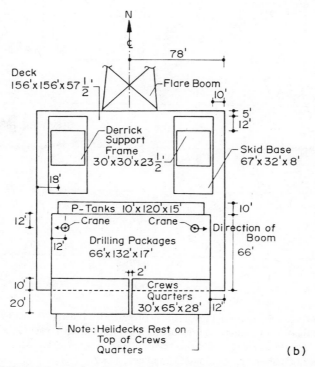

FIGURE 14.1.8. Guyed tower platform: (*a*) side elevation; (*b*) plan [14-25].

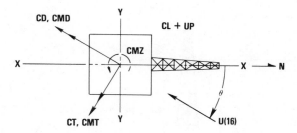

FIGURE 14.1.9. Notations. From P. J. Pike and B. J. Vickery, "A Wind Tunnel Investigation of Loads and Pressure on a Typical Guyed Tower Offshore Platform," Paper OTC 4288, *Proceedings*, Offshore Technology Conference, Houston, TX, May 1982. Copyright 1982 Offshore Technology Conference.

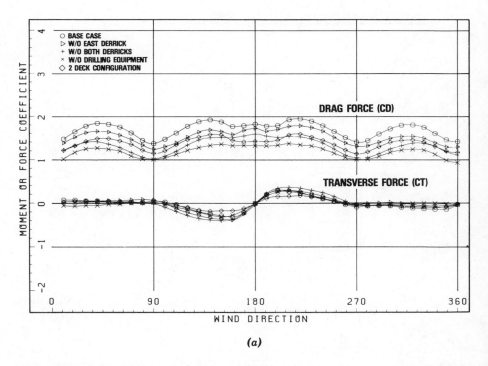

(a)

FIGURE 14.1.10. Wind tunnel test results. From P. J. Pike and B. J. Vickery, "A Wind Tunnel Investigation of Loads and Pressure on a Typical Guyed Tower Offshore Platform," Paper OTC 4288, *Proceedings*, Offshore Technology Conference, Houston, TX, May 1982. Copyright 1982 Offshore Technology Conference.

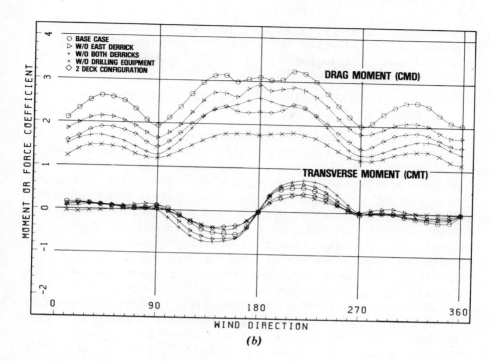

(b)

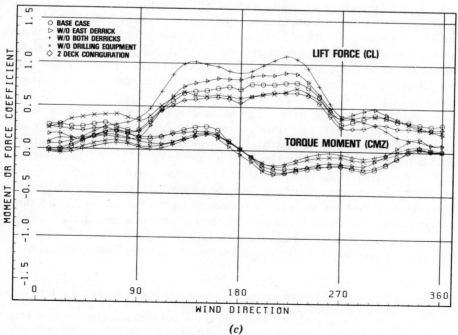

(c)

different wind tunnel [14-25]. In most cases of significance from a design viewpoint the results obtained in [14-25] were larger than those of [14-23] by amounts that did not exceed 20 to 30%.*

14.2   DYNAMIC WIND EFFECTS ON COMPLIANT OFFSHORE STRUCTURES

Compliant offshore platforms are designed to experience significant motions under load. An important advantage of compliance is that the forces of inertia associated with these motions contribute to counteracting the external loads.

In the case of large offshore structures installed in deep water, compliance has the additional advantage of making it possible to design platforms with very low natural frequencies in the surge, sway, and yaw degrees of freedom† (e.g., 1/30 Hz to 1/150 Hz, depending upon type of platform and water depth). Wave motions have narrow spectra centered about relatively high frequencies (from 1/15 Hz, say, for extreme events to about 1 Hz for service conditions). Thus, aside from possible second-order effects, compliant platforms generally do not exhibit any dynamic amplification of the wave-induced response.

Unlike wave motions, wind speed fluctuations in the atmospheric boundary layer are characterized by broad-band spectra (see Fig. 2.3.4). For this reason it has been stated in the literature that wind-induced dynamic amplification effects on compliant structures are significant [14-23, 14-26]. A more guarded assessment of the effects of wind gustiness was presented in [14-27] as part of an evaluation of the response to environmental loads of the North Sea Hutton tension leg platform (Fig. 14.2.1, see also [14-28]). According to [14-27]: "Wind gusts are typically broad-banded and may contain energy which could excite surge motions at the natural period. These would be controlled by surge damping. Theoretical and experimental research is required to clarify the importance of this matter."

Investigations into the behavior of tension leg platforms under wind loads reported in [14-29] and [14-30] were based on the assumption that the response to wind is described by a system with proportional damping, the damping ratio being of the order of 5%. However, it was shown in [14-31] that for structures comparable to the Hutton platform the effective hydrodynamic damping is considerably stronger, and that the wind induced dynamic amplification for low frequency motions are for this reason negligible. Section 14.2.1 describes the approach used in [14-31] to estimate the response of a tension leg platform to wind in the presence of current and waves, and a simple method for estimating the order of magnitude of the damping inherent in the hydrodynamic loads.

---

*Differences between results of aerodynamic measurements conducted independently in different facilities are also noted in Sect. 4.6.

†Linear motions in the longitudinal, transverse, and vertical direction are referred to as surge, sway, and heave, respectively. Angular motions in a transverse, longitudinal, and horizontal plane are referred to as roll, pitch, and yaw, respectively.

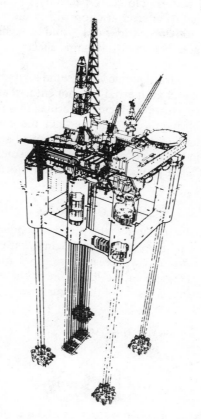

FIGURE 14.2.1.   Schematic view of the Hutton tension leg platform. From N. Ellis, J. H. Tetlow, F. Anderson, and A. L. Woodhead, "Hutton TLP Vessel—Structural Configuration and Design Features," Paper OTC 4427, *Proceedings*, Offshore Technology Conference, Houston, TX, May 1982. Copyright 1982 Offshore Technology Conference.

14.2.1   Turbulent Wind Effects on Tension Leg Platform Surge

Under the assumption that the external loads are parallel to one of the sides of the platform shown in Fig. 14.2.1, the equation of surge motion can be written as

$$M\ddot{x} = F_x(t) \tag{14.2.1a}$$

where

$$F_x(t) \simeq F_u(t) + F_h(t) + R(t) \tag{14.2.1b}$$

In equation 14.2.1b, $F_u(t)$, $F_h(t)$, and $R(t)$ denote the wind force, the hydrodynamic force, and the restoring force, respectively. Not included in Eq. 14.2.1b is the damping force due to internal friction within the structure, which corresponds to a damping ratio of the order of 1% and is negligible compared to the damping forces associated with hydrodynamic effects.

Wind Loads.   Like the hydrodynamic loads, wind loads consist of a component due to flow separation, and an inertial component associated with the relative fluid-body accelerations. However, it can be verified that the inertial component is about two orders of magnitude smaller than the component due to flow separation, and can therefore be neglected in practical applications.

To estimate the wind-induced drag force it is assumed that the elemental drag force per unit of area projected on a plane P normal to the mean wind speed can be written as:

$$p(y, z, t) = \tfrac{1}{2}\rho_a C_p(y, z)[u(y, z, t) - \dot{x}(t)]^2 \tag{14.2.2}$$

where ρ_a is the air density, $C_p(y, z)$ is the pressure coefficient at elevation z and horizontal coordinate y in the plane P, t is the time, x is the surge displacement, the dot denotes differentiation with respect to time, and $u(y, z, t)$ is the wind speed upwind of the structure in the direction of the mean wind. The speed $u(y, z, t)$, can be expressed as a sum of the mean speed $U(z)$ and the fluctuating speed $u'(y, z, t)$, that is

$$u(y, z, t) = U(z) + u'(y, z, t) \tag{14.2.3}$$

The total wind-induced drag force is

$$F_u(t) = \int_{A_a} p(y, z, t) \, dy \, dz \tag{14.2.4}$$

where A_a is the projection of above-water part of the platform on a plane normal to the mean wind speed.

The mean speeds can be modeled by the logarithmic law (Sect. 2.2.3). The spectra of the longitudinal velocity fluctuations can be modeled by Eqs. 2.3.25.* The cross-spectra of the longitudinal velocity fluctuations are modeled by Eq. 2.3.30. The effect of longitudinal separation should also in principle be taken into account. However, it follows from information presented in [2-89] that this effect is negligible as far as fluctuating aerodynamic loads on offshore structures are concerned.

It can be verified that the mean square values of u' and $\dot{x}$ and the mean value of the product $u'\dot{x}$ are small compared to the square of U. It then follows from Eqs. 14.2.2, 14.2.3, and 14.2.4 that the mean drag load can be written as

$$\overline{F_u(t)} \simeq \tfrac{1}{2}\rho_a C_a A_a U^2(z_a) \tag{14.2.5}$$

*Note that for the frequency $n=0$ Eqs. 2.3.25 yield a spectral ordinate $S(0)$ proportional to the integral turbulence scale L_u^x, in accordance with fundamental principles (see Eq. 2.3.19). On the other hand, Eq. 2.3.23 (quoted, for example, in [14-23]) yields $S(0)=0$, and therefore underestimates the spectral ordinates in the range of typical natural frequencies for compliant structures (Fig. 2.3.4).

where the overall aerodynamic drag coefficient is

$$C_a = \frac{1}{A_a U^2(z_a)} \int_{A_a} C_p(y, z) U^2(z) \, dy \, dz \qquad (14.2.6)$$

and z_a is the elevation of the aerodynamic center of the above-water part of the platform. From Eqs. 14.2.2 to 14.2.5 it follows that the fluctuating part of the wind drag load that would act on the platform at rest (i.e., with $\dot{x} = 0$) is

$$F'_{u,r}(t) = \rho_a \int_{A_a} C_p(y, z) U(z) u'(z, t) \, dy \, dz \qquad (14.2.7)$$

where the subscript r refers to the fact that the platform is at rest. The Fourier transform of the autocovariance function of $F'_{u,r}(t)$ yields

$$S_{F_{u,r}}(n) = \rho_a^2 \int_{A_a} \int_{A_a} C_p(y_1, z_1) C_p(y_2, z_2) U(z_1) U(z_2)$$
$$\times S^C_{u_1 u_2}(y_1, y_2, z_1, z_2) \, dy_1 \, dy_2 \, dz_1 \, dz_2 \qquad (14.2.8)$$

The spectrum $S_{F_{u,r}}(n)$ can be estimated numerically by assuming $C_p(y_i, z_i) \simeq C_a \, (i = 1, 2)$. An equivalent wind fluctuation spectrum can then be defined as

$$S_{u,eq}(n) = \frac{S_{F_{u,r}}(n)}{[\rho_a C_a A_a U(z_a)]^2} \qquad (14.2.9)$$

From $S_{u,eq}(n)$ it is possible to generate by Monte Carlo simulation realizations of the process

$$u'_{eq}(t) = \sum_j u'_{eq\,j} \cos(2\pi n_j t + \phi_j) \qquad (14.2.10)$$

In Eq. 14.2.10 the phase angle ϕ_j is generated by random sampling from a uniform distribution in the interval $0 < \phi_j < 2\pi$.

Let the spectrum of the force $F'_{equ,r}(t)$, defined as

$$F'_{equ,r}(t) = \rho_a C_a A_a U(z_a) u'_{eq}(t) \qquad (14.2.11)$$

be denoted by $S_{F_{equ,r}}(n)$. Clearly,

$$S_{F_{equ,r}}(n) \simeq S_{F_{u,r}}(n) \qquad (14.2.12)$$

Thus, $u'_{eq}(t)$ can be viewed as an equivalent wind speed fluctuation which is perfectly coherent over the area A_a and whose effect upon the structure at rest is the same as that of the actual fluctuating wind field.

The total wind load acting on the platform can thus be expressed as

$$F_u(t) = \tfrac{1}{2} \rho_a C_a A_a [U(z_a) + u'_{eq}(t) - \dot{x}(t)]^2 \qquad (14.2.13)$$

Numerical calculations have shown that if the difference between the elevation of the helideck (or the top of the crew quarters) and the underside of the lower deck in a typical drilling and production platform is less than about two-thirds to three-quarters of the width of the main deck, the term $C_z^2 (z_1 - z_2)^2$ of Eq. 2.3.30 can be neglected when evaluating the integral in Eq. 14.2.8. This is a consequence of the fact that C_z is smaller than C_y by a factor of about 1.5. The approximation inherent in neglecting $C_z^2 (z_1 - z_2)^2$ is slightly conservative from a structural engineering point of view (though insignificantly so). Noting, then, that for any arbitrary function, Φ,

$$\int_0^1 \int_0^1 \Phi(|Y_1 - Y_2|)\, dY_1\, dY_2 = \int_0^1 \Phi(t)(1 - t)\, dt \qquad (14.2.14)$$

(Fig. 14.2.2), and assuming $C_p(y_i, z_i) \simeq C_a$, $U(z_i) \simeq U(z_a)$, and $S_{u_i}(n) \simeq S_u(z_a, n)$, $(i = 1, 2)$, it follows after some algebra from Eqs. 14.2.8, 2.3.30, and 14.2.9, that

$$S_{u,eq}(n) \simeq S_u(z_a, n)J(n) \qquad (14.2.15)$$

where $S_u(z_a, n)$ is the spectrum of longitudinal velocity fluctuations at elevation z_a, and $J(n)$ is a reduction factor accounting for the imperfect coherence among the fluctuating wind pressures at different points of the platform, given by the expression

$$J(n) = -2/E\{-\exp(-E) + (1 - 1/E)[\exp(-E) - 1]\} \qquad (14.2.16)$$

$$E = C_y b\, \frac{n}{U(z_a)} \qquad (14.2.17)$$

In Eq. 14.2.17, b is the width of main deck. Equation 14.2.15 can be used in lieu of Eqs. 14.2.8 and 14.2.9 for the Monte Carlo simulation of the equivalent velocity fluctuations $u'_{eq}(t)$ (see Eq. 14.2.10) needed in the expression of the total wind load acting on the platform, $F_u(t)$.

Hydrodynamic Loads. The total hydrodynamic load F_h can be written in the form

$$F_h = F_v + F_e - A\ddot{x} - B\dot{x} \qquad (14.2.18)$$

where F_v is the total hydrodynamic viscous force, F_e is the total wave-induced exciting force, A is the surge added mass, and B is the surge wave-radiation damping coefficient. It was assumed for convenience in [14-31] that the wave motion is monochromatic, hence the absence of second-order drift forces in

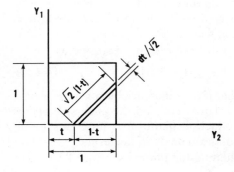

FIGURE 14.2.2. Integration domain and transformation of variables.

Eq. 14.2.18 [14-33]. It was assumed in addition that $B=0$, since the radiation damping at low frequencies is negligible [14-32].

The total wave-induced exciting force and the surge added mass can be estimated numerically on the basis of potential theory. Alternatively, these two terms may be assumed to be given by the inertia component of the Morison equation, that is

$$A \simeq \rho_w \sum_i \sum_j \Psi_{ij}(C_{m_{ij}}-1) \qquad (14.2.19)$$

$$F_e \simeq \rho_w \sum_i \sum_j \Psi_{ij} C_{m_{ij}} \left\{ \frac{\partial v_{ij}}{\partial t} + [\bar{v}_i + v_{ij} - \dot{x}] \frac{\partial v_{ij}}{\partial X} \right\} \qquad (14.2.20)$$

[14-34, p. 31], where ρ_w is the water density, Ψ_{ij} is the elemental volume of the submerged structure, $C_{m_{ij}}$ is the surge inertia coefficient corresponding to Ψ_{ij}, X is the horizontal distance from some arbitrary origin to the center of Ψ_{ij} along the direction parallel to surge motion, $\bar{v}_i$ and v_{ij} are the current velocity and horizontal particle velocity due to wave motion, respectively, at center of Ψ_{ij}. Equations 14.2.19 and 14.2.20 may be employed if for the component being considered the ratio of diameter to wave length, $D/L \leqslant 0.2$ [14-34, p. 283]. Since for $T_w \simeq 15$ sec, $L = gT_w^2/2\pi \simeq 350$ m [14-34, p. 283], where T_w is the wave period and g is the acceleration of gravity, it follows that for members of typical tension leg platform structures, for which $D < 20$ m or so, the use of Eqs. 14.2.19 and 14.2.20 is acceptable if three-dimensional flow effects are not taken into account. The wave motion can be described by deep water linear theory, so that

$$v_{ij} = \frac{\pi H}{T_w} e^{-k_w z i} \cos\left(k_w X_j - \frac{2\pi t}{T_w}\right) \qquad (14.2.21)$$

where H is the wave height and k_w is the wave number given by

$$k_w = \frac{1}{g}\left(\frac{2\pi}{T_w}\right)^2 \qquad (14.2.22)$$

[14-34, p. 157]. The total hydrodynamic viscous load may be described by the viscous component of Morison's equation

$$F_v = 0.5\rho_w \sum_i \sum_j C_{d_{ij}} A_{p_{ij}} |\bar{v}_i + v_{ij} - \dot{x}| [\bar{v}_i + v_{ij} - \dot{x}] \qquad (14.2.23)$$

where $A_{p_{ij}}$ is area of elemental volume Ψ_{ij} projected on a plane normal to the direction of the current, and $C_{d_{ij}}$ is the drag coefficient corresponding to $A_{p_{ij}}$.

If the relative motion of the body with respect to the fluid is harmonic, the drag and inertia coefficients in Morison's equation can be determined on the basis of experimental results as functions of local oscillatory Reynolds number, $\mathcal{R}e = 2\pi D^2/(\nu T_f)$, Keulegan-Carpenter number, $K = V T_f/D$, and relative body surface roughness, where D is the diameter of the body, ν is the kinematic viscosity, V and T_f are the amplitude and period of the relative fluid-body velocity. However, actual relative fluid-body motions are not harmonic. This introduces uncertainties in the determination of the drag and inertia coefficients even if experimental information for harmonic relative motions were available in terms of $\mathscr{R}e$ and K. Unfortunately, such information is not available for the small K numbers (of the order of 2) and large Reynolds numbers (of the order of 10^6) of interest in tension leg platform design. For this reason calculations should be carried out for various sets of values C_d, C_m, and investigations should be conducted into the sensitivity of the results to changes in these values.

Restoring Force.   The surge restoring force in a tension leg platform is supplied by the horizontal projection of the total tension force in the tethers. Most of this force is the result of pretensioning, which is achieved by ballasting the floating platform, tying it by means of the tethers to the foundations at the sea floor, then deballasting it. The tension forces in the tethers should exceed the compression forces induced by pitching and rolling moments due to extreme loads.

Under the assumption that the tethers are straight at all times, the restoring force can be written as

$$R(t) \simeq -(T+\Delta T)\frac{x}{l_n+\Delta l_n} \qquad (14.2.24)$$

where T is the initial pretensioning force, ΔT is the incremental tension due to surge motion, l_n is the nominal length of the tethers at $x=0$, Δl_n is the incremental length, and

$$\frac{T+\Delta T}{l_n+\Delta l_n} \simeq \frac{T}{l_n}+C_{NL}[1-\sqrt{1-(x/l)^2}] \qquad (14.2.25)$$

where C_{NL} is the downdraw coefficient, equal to the weight of water displaced as the draft is increased by a unit length [14-32] (Fig. 14.2.3).

In reality, hydrodynamic and inertia forces cause the tethers to deform transversely. The angle between the horizontal and the tangent to the tether

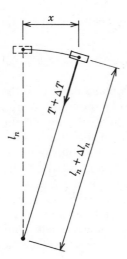

FIGURE 14.2.3.   Notations.

axis at the platform heel can therefore differ significantly from the values corresponding to the case of a straight tether. Nevertheless, owing to the relatively small role of the restoring force in the dynamics of typical tension leg platforms, the effect of such differences on the motion of the platforms appears to be negligible for practical purposes [14-36, 14-37, 14-38].

Surge Response. The surge response is obtained by solving eq. 14.2.1. This equation is nonlinear, the strongest contribution to the nonlinearity being due to the hydrodynamic viscous load F_v (Eqs. 14.2.1, 14.2.18, and 14.2.23). For this reason it is appropriate to solve Eq. 14.2.1 in the time domain.

The nominal natural period in surge is

$$T_n = 2\pi \left(\frac{M_{\text{eff}}}{k} \right)^{1/2} \tag{14.2.26}$$

where M_{eff} is the coefficient of the term in $\ddot{x}$ and k the coefficient of the term in x in Eq. 14.2.1. From Eqs. 14.2.1, 14.2.18, and 14.2.24, it follows that

$$T_n = 2\pi \left[\frac{(M+A)l_n}{T} \right]^{1/2} \tag{14.2.26a}$$

A calculated time history of the surge response is represented in Fig. 14.2.4 as a function of time for a platform with the geometrical configuration of Fig. 14.2.5, under the following assumptions: platform mass $M = 34.3 \times 10^6$ kg; total initial tension in legs $T = 1.56 \times 10^5$ kN; Morison equation coefficients

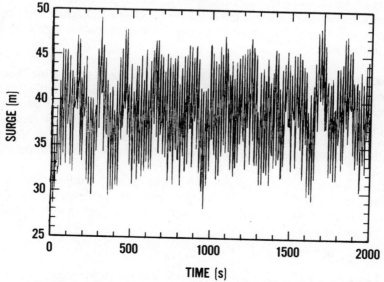

FIGURE 14.2.4. Calculated time history of surge response [14-31].

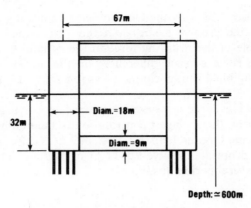

FIGURE 14.2.5. Geometry of platform.

$C_{m_{ij}} = 1.8$,* $C_{d_{ij}} = 0.6$; wave height and period $H = 25$ m and $T_w = 15$ sec, respectively; current speed varying from 1.4 m/sec at the mean water level to 0.15 m/sec at 550 m depth; aerodynamic parameter $C_a A_a = 4320$ m^2; elevation of aerodynamic center $z_a = 50$ m; atmospheric boundary-layer flow parameters $\kappa = 0.002$, $\beta = 6.0$, $L_u^x = 180$ m, $f_m = 0.07$, $f_s = 0.2$, $C_y = 16$ (see Eqs. 2.2.23, 2.3.2, 2.3.4, 2.3.25, and 14.2.17); and mean wind speed $U(z_a) = 45$ m/sec. It is shown in [14-31] that the contributions of the mean wind and of the wind fluctuations to the peak response of Fig. 14.2.4 are about 40% and 12%, respectively. It can be verified that this conclusion is equivalent to stating that wind-induced resonant amplification effects are negligible in the cases investigated in [14-31]. Sensitivity studies showed that the results were affected insignificantly by uncertainties with respect to the actual values of the atmospheric boundary-layer flow parameters.

Nominal Damping Ratio of Pseudo-Linear System Representing the Response to Wind. It was indicated previously that the estimation of the effect of turbulent wind on surge was attempted in the literature on the basis of the assumption that the equation of surge motion represents a linear system with a viscous damping term characterized by a nominal damping ratio, ζ. The effect of this term is postulated to be equivalent to the damping effect of the hydrodynamic viscous force.

Such an approach is acceptable if the order of magnitude of the nominal damping ratio is consistent with the hydrodynamic behavior of the system. Calculations are now presented that illustrate how such nominal damping ratios can be estimated.

---

*For the platform of Fig. 14.2.5, it follows from these assumptions and from Eqs. 14.2.19 and 14.2.26a that the nominal natural frequency is $T_n \simeq 100$ sec.

It is assumed that $u'_{eq}(t)$ (see Eq. 14.2.10) is given by the harmonic function

$$u'_{eq}(t) = u'_{eq1} \cos \frac{2\pi}{T_n} t \qquad (14.2.27)$$

where T_n is the natural period in surge, and the system under consideration is linear with mass $M + A$, natural period T_n, and damping ratio ζ. The amplitude of the contribution to the surge response of a harmonic force $\rho_a C_a A_a U(z_a) u'_{eq1} \cos 2\pi n t$ is denoted by x_{umax}, and is given by the relation

$$x_{umax} = \frac{\rho_a C_a A_a U(z_a) u'_{eq1}}{(M+A)\left(\dfrac{2\pi}{T_n}\right)^2 \{[1-(nT_n)^2]^2 + (2\zeta n T_n)^2\}^{1/2}} \qquad (14.2.28)$$

The nominal damping ratio ζ is estimated from the condition that x_{umax} be equal to the contribution to the surge response of the force $\rho_a C_a A_a U(z_a) u'_{eq1} \cos (2\pi/T_n)t$, as obtained by solving Eqs. 14.2.1. By substituting $1/T_n$ for n in Eq. 14.2.28 it follows that

$$\zeta = \frac{\frac{1}{2}\rho_a C_a A_a U(z_a) u'_{eq1}}{x_{umax}(M+A)\left(\dfrac{2\pi}{T_n}\right)^2} \qquad (14.2.29)$$

Calculations were carried out in [14-31] for the platform shown in Fig. 14.2.5 (with tether lengths $l_n \simeq 600$ m) and for a similar platform with tether lengths $l_n \simeq 150$ m, using the mechanical, hydrodynamic, aerodynamic, and atmospheric boundary-layer flow parameters listed previously. Equation 14.2.29 yielded $\zeta \simeq 0.5$ and $\zeta \simeq 0.2$ for the platforms with $l_n \simeq 600$ m and $l_n \simeq 150$ m, respectively.

The values $C_{d_{ij}} = 0.6$ and $C_{m_{ij}} = 1.8$ on which these results were based may not be realistic for members with large diameters, such as those depicted in Fig. 14.2.5 [14-34]. For this reason calculations were carried out in [14-31] for a number of possible sets of value $C_{d_{ij}}$, $C_{m_{ij}}$. The calculated nominal damping ratios ζ were in all cases found to be sufficiently large to prevent the occurrence of strong wind-induced dynamic amplification effects.

REFERENCES

14-1   M. E. Davies, L. R. Cole, and P. G. G. O'Neill, *The Nature of Air Flows Over Offshore Platforms*, NMI R14 (OT-R-7726), National Maritime Institute, Feltham, U.K., June 1977.

14-2   M. E. Davies, *Wind Tunnel Modelling of the Local Atmospheric Environment of Offshore Platforms*, NMI R58 (OT-R-7935), National Maritime Institute, Feltham, U.K., May 1979.

14-3   K. H. Littleburg, "Wind Tunnel Testing Techniques for Offshore Gas/Oil Production Platforms," Paper OTC 4125 *Proceedings*, Offshore Technology Conference, Houston, TX, 1981.

14-4   *Rules for Building and Classing Mobile Offshore Drilling Units*, American Bureau of Shipping, New York, 1980.

14-5   *Rules for the Construction and Classification of Mobile Offshore Units*, Det norske Veritas, Oslo, 1981.

14-6   *Rules for the Design, Construction, and Inspection of Offshore Structures*, Appendix B, Loads, Det norske Veritas, Oslo, 1977 (Reprint with Corrections 1979).

14-7   *Requirements for Verifying the Structural Integrity of OCS Platforms*, Appendices, United States Geological Survey, OCS Platform Verification Division, Reston, VA, 1979.

14-8   *API Recommended Practice for Planning, Designing, and Constructing Fixed Offshore Platforms*, API RP 2A, American Petroleum Institute, Washington, D.C., 1981.

14-9   D. J. Norton and C. V. Wolff, "Mobile Offshore Platform Wind Loads," Paper OTC 4123, *Proceedings*, Offshore Technology Conference, Houston, TX, 1981.

14-10 H. Boonstra, "Analysis of Full-Scale Wind Forces on a Semisubmersible Platform Using Operator's Data," *J. Petroleum Technol.* (1980), 771–776.

14-11 P. J. Ponsford, *Measurements of the Wind Forces and Measurements on an Oil Production Jacket Structure in Tow-Out Mode*, NMI R30 (OT-R-7801), National Maritime Institute, Feltham, U.K., January 1978.

14-12 C. F. Cowdrey, *Time-Averaged Aerodynamic Forces and Moments on a Model of a Three-Legged Concrete Production Platform*, NMI R35 (OT-R-7808), National Maritime Institute, Feltham, U.K., June 1982.

14-13 C. F. Cowdrey, *Time-Averaged Aerodynamic Forces and Moments on a Model of a Three-Legged Concrete Production Platform*, NMI R36 (OT-R-7808), National Maritime Institute, Feltham, U.K., June 1982.

14-14 B. L. Miller and M. E. Davies, *Wind Loading on Offshore Structures—A Summary of Wind Tunnel Studies*, NMI R136 (OT-R-8225), National Maritime Institute, Feltham, U.K., September 1982.

14-15 E. Bjerregaard and S. Velschou, "Wind Overturning Effect on a Semisubmersible," Paper OTC 3063, *Proceedings*, Offshore Technology Conference, Houston, TX, May 1978.

14-16 E. Bjerregaard and E. Sorensen, "Wind Overturning Effects Obtained from Wind Tunnel Tests with Various Submersible Models," OTC Paper 4124, *Proceedings*, Offshore Technology Conference, Houston, TX, May 1981.

14-17 J. H. Ribbe and J. C. Brusse, "Simulation of the Air/Water Interface for Wind Tunnel Testing of Floating Structures," *Proceedings Fourth U.S. National Conference, Wind Engineering Research*, B. J. Hartz (Ed.), Department of Civil Engineering, University of Washington, Seattle, July 1981.

14-18 J. M. Macha and D. F. Reid, *Semisubmersible Wind Loads and Wind Effects*, Paper no. 3, Annual Meeting, New York, Nov. 1984. The Society of Naval Architects and Marine Engineers, New York, 1984.

14-19 R. F. W. Gould, *The Prediction of the Wind Loading on a Wind Tunnel Model of an Offshore Drilling Unit Based on the SEDCO 700 Design*, NMI R18 (OT-R-7729), National Maritime Institute, Feltham, U.K., July 1979.

14-20 C. F. Cowdrey and R. F. W. Gould, *Time-Averaged Aerodynamic Forces and Moments on a National Model of a Semisubmersible Offshore Rig*, NMI R26 (OT-R-7748), National Maritime Institute, Feltham, U.K., September 1982.

14-21 P. J. Ponsford, *Wind-Tunnel Measurements of Aerodynamic Forces and Moments on a Model of a Semisubmersible Offshore Rig*, NMI R34 (OT-R-7807), National Maritime Institute, Feltham, U.K., June 1982.

14-22 A. W. Troesch, R. W. Van Gunst, and S. Lee, "Wind Loads on a 1:115 Model of a Semi-submersible," *Marine Technol.*, **20** (July 1983), 283–289.

14-23 P. J. Pike and B. J. Vickery, "A Wind Tunnel Investigation of Loads and Pressure on a Typical Guyed Tower Offshore Platform," Paper OTC 4288, *Proceedings*, Offshore Technology Conference, Houston, TX, May 1982.

14-24 M. S. Glasscock and L. D. Finn, "Design of a Guyed Tower for 1000 ft of Water in the Gulf of Mexico," *J. Struct. Eng.* **110** (1984), 1083–1098.

14-25 T. A. Morreale, P. Gergely, and M. Grigoriu, *Wind Tunnel Study of Wind Loading on a Compliant Offshore Platform*, NBS-GCR-84-465, National Bureau of Standards, Washington, D.C., December 1983.

14-26 J. R. Smith and R. S. Taylor, "The Development of Articulated Buoyant Column Systems as an Aid to Economic Offshore Production," *Proceedings*, European Offshore Petroleum Conference Exhibition, London, England, October 1980, pp. 545–557.

14-27 J. A. Mercier, S. J. Leverette, and A. L. Bliault, "Evaluation of Hutton TLP Response to Environmental Loads," Paper OTC 4429, *Proceedings*, Offshore Technology Conference, Houston, TX, May 1982.

14-28 N. Ellis, J. H. Tetlow, F. Anderson, and A. L. Woodhead, "Hutton TLP Vessel—Structural Configuration and Design Features," Paper OTC 4427, *Proceedings*, Offshore Technology Conference, Houston, TX, May 1982.

14-29 A. Kareem and C. Dalton, "Dynamic Effects of Wind on Tension Leg Platforms," Paper OTC 4229, *Proceedings*, Offshore Technology Conference, Houston, TX, May 1982.

14-30 B. J. Vickery, "Wind Loads on Compliant Offshore Structures," *Proceedings*, Ocean Structural Dynamic Symposium, Oregon State University, Department of Civil Engineering, Corvallis, Oregon, September 1982, pp. 632–648.

14-31 E. Simiu and S. D. Leigh, "Turbulent Wind and Tension Leg Platform Surge," *J. Struct. Eng.*, **110** (1984), 785–802.

14-32 N. Salvesen et al., "Computations of Nonlinear Surge Motions of Tension Leg Platforms," Paper OTC 4394, *Proceedings*, Offshore Technology Conference, Houston, TX, May 1982.

14-33 J. A. Pinkster and G. Van Oortmerssen, "Computation of First- and Second-Order Forces On Oscillating Bodies in Regular Waves," *Proceedings*, Second International Conference on Ship Hydrodynamics, Univ. of California, Berkeley, 1977.

14-34 T. Sarpkaya and M. Isaacson, *Mechanics of Wave Forces on Offshore Structures*, Van Nostrand Reinhold Co., New York, 1981.

14-35 A. G. Davenport and E. C. Hambly, "Turbulent Wind Loading and Dynamic Response of Jackup Platform," Paper OTC 4824, *Proceedings*, Offshore Technology Conference, Houston, TX, May 1984.

14-36 E. R. Jefferys and M. H. Patel, "On the Dynamics of Taut Mooring Systems," *Eng. Struct.*, **4** (1982), 37–43.

14-37 E. Simiu, A. Carasso, and C. E. Smith, "Tether Deformation and Tension Leg Platform Surge," *J. Struct. Eng.*, **110** (1984), 1419–1422.

14-38 E. Simiu and A. Carasso, "Interdependence Between Dynamic Surge Motions of Platform and Tethers for a Deep Water TLP," *Proceedings*, Fourth International Conference on Behavior of Offshore Structures (BOSS), 1–5 July 1985, Delft, The Netherlands.

15

Wind-Induced Discomfort in and Around Buildings

It is required that structures subjected to wind loads be sufficiently strong to perform adequately from a structural safety viewpoint. Recent experience has shown that in the case of tall, flexible buildings, the designer must also take into account wind-related serviceability requirements. The latter may be formulated, in general terms, as follows: structures shall be so designed that their wind-induced motions will not cause unacceptable discomfort to the building occupants.

Wind-induced discomfort is also of concern in the altogether different context of the serviceability of outdoor areas within a built environment. Certain building and open space configurations may give rise to relatively intense local wind flows. It is the designer's task to ascertain in the planning stage the possible existence of zones in which such flows would cause unacceptable discomfort to users of the outdoor areas of concern. Appropriate design decisions must be made to eliminate such zones if they exist.

The notion of unacceptable discomfort, which is seen to play a central role in the statement of serviceability requirements, may be defined as follows. In any given design situation various degrees of wind-induced discomfort may be expected to occur with certain frequencies that depend upon the degree of discomfort, the features of the design and the wind climate at the location in question. The discomfort is unacceptable if any of these frequencies is judged to be too high. Statements specifying maximum acceptable mean frequencies of occurrence for various degrees of discomfort are known as comfort criteria.

In comfort criteria developed for use in design it is impractical to refer explicitly to degrees of discomfort as such. Rather, reference is made to a suitable parameter, various values of which are associated with various degrees of discomfort. In the case of wind-induced building motions, this parameter is the building acceleration. In criteria pertaining to the serviceability of pedestrian areas, the parameter employed is an appropriate measure of the wind speed near the ground at the location of concern. Clearly, to develop comfort criteria it is required that parameter values be established that correspond to various degrees of human discomfort. Furthermore, it is necessary that to various degrees of discomfort—or, equivalently, to the parameter values that correspond to them—there be assigned maximum acceptable probabilities of occurrence.

Verifying the compliance of a design with requirements set forth in a given set of comfort criteria involves two steps. First, an estimate must be obtained of the wind velocities under the action of which the parameter of concern will exceed the values specified by the comfort criteria (these values may be referred to as critical). Second, the frequencies of occurrence of these velocities must be estimated on the basis of appropriate wind climatological information. If the frequencies thus estimated are lower than maximum acceptable frequencies specified by the comfort criteria, then the design is regarded as adequate from a serviceability viewpoint.

The development of comfort criteria for the design of tall buildings and questions related to their practical use are discussed in Sect. 15.1. Comfort criteria for the design of pedestrian areas and related design information are dealt with in Sects. 15.2 through 15.4.

15.1 SERVICEABILITY OF TALL BUILDINGS UNDER THE ACTION OF WIND

15.1.1 Human Response to Wind-Induced Vibrations

Studies of human response to mechanical vibrations have been conducted within the last two decades mainly by the aerospace industry. Because the frequencies of vibration of interest in aerospace applications are relatively high (usually 1 Hz to 35 Hz), the usefulness of these studies to the structural engineer is generally limited. Nevertheless, results obtained for high frequencies have been extrapolated in [15-1] to frequencies lower than 1 Hz, with the following correspondence being proposed between various degrees of user discomfort and the accelerations causing them:

Degree of Discomfort	Acceleration (in percentages of the acceleration of gravity g)
Imperceptible	$<\frac{1}{2}\%g$
Perceptible	$\frac{1}{2}\%g-1\frac{1}{2}\%g$
Annoying	$1\frac{1}{2}\%g-5\%g$
Very Annoying	$5\%g-15\%g$
Intolerable	$>15\%g$

Results of experiments aimed at establishing perception thresholds for periodic motions of 0.067 Hz to 0.2 Hz have been reported in [15-2]. The experiments, carried out on 112 subjects in motion simulators representative of an office environment, were designed to take into account the influence upon perception thresholds of body orientation, body movement, body posture, and the extent to which the motion is anticipated by the subjects of the experiments. The perception thresholds as reported by 50% of the subjects were found to be approximately 1%g, 0.9%g, and 0.6%g for frequencies of vibration of 0.067 Hz, 0.1 Hz, and 0.2 Hz, respectively. It is noted that—within this frequency range—the perception thresholds decrease as the frequencies increase. Additional experimental results are used in [15-2] as a basis for a tentative relation between the horizontal acceleration of a floor and the percentage of the individuals on that floor for whom the acceleration will be perceptible.

Studies of human response to vibrations of a motion simulator have also been reported in [15-3] and [15-4] for frequencies in the range 0.1 Hz to 1 Hz. Average perception thresholds were found to vary from about 0.6%g for frequencies of 0.1 Hz to about 0.3%g for frequencies of 0.25 Hz. Motions were distinctly perceptible and the subjects were annoyed while working at their desks if the accelerations exceeded 1.2%g. Beyond accelerations of 4%g the perceptions were described as strong and the subjects experienced difficulties in walking. The motions were described as extremely annoying or intolerable beyond accelerations of the order of 5%g to 6%g. Similar results have been reported in [15-5].

A study presented in [15-6] and [15-7] is based on observations of human response to actual rather than simulated wind-induced accelerations. The investigation covered the behavior during a storm of two buildings and of their occupants. Estimates of the rms value of the top floor accelerations during the storms were based on response measurements for one of the buildings, and on wind speed measurements and wind tunnel testing for the second. These estimates represented averages (a) in time over the periods of highest storm intensity (20 min to 30 min)* and (b) in space over the entire area of the floor—the space averaging being performed to account for wind-induced torsional motions. (For the effect of torsional motions see also [15-32].) The rms values thus obtained were 0.2%g for the first, and 0.5%g for the second of the two buildings. Interviews with building occupants then revealed that about 35% of the persons on the higher floors in the first building experienced motion sickness symptoms during the storm. For the second building, the reported percentage was about 45%. It is noted in [15-7] that creaking noises that occur during the building motion may increase significantly the feeling of discomfort and should therefore be minimized by proper structural detailing.

Results of surveys conducted among occupants of tall buildings in Japan are reported in [15-8].

---

*Time averages were also effected over longer periods [15-6].

15.1.2   Comfort Criteria

Comfort criteria should in principle be based on an extensive knowledge of the degree to which building users are prepared to accept discomfort associated with wind-induced accelerations. However, at present such knowledge is scarce.

A simple comfort criterion has been proposed in [15-9], believed by its authors to be justified by the results of [15-2]. This criterion, which limits the average number of occurrences of $1\%g$ accelerations at the top occupied floor to at most 12 per year, has been applied to the design of the World Trade Center [15-9]. In [15-6] an attempt is presented to develop comfort criteria on the basis (a) of recorded objections by building users to the recurrence of wind-induced building vibrations and (b) of estimates by owners or developers of the possible economic repercussions of user dissatisfaction with the building performance. From interviews with building occupants who had experienced motions with an rms value of the top floor accelerations of about $0.5\%g$ it was estimated that about 2% of the people in the top one-third of a building would object to more than one occurrence of such motions in six years. Interviews with building owners and developers suggested, on the other hand, that rental or sales of office space would not be affected significantly if at most 2% of the occupants in the top one-third of the building found the sway objectionable. On the basis of those findings, it is suggested in [15-6] that the following design criterion appears to be reasonable: "The return periods, for storms causing an rms horizontal acceleration at the building top which exceeds $0.5\%g$, shall not be less than six years. The rms shall represent an average over the 20-min period of highest storm intensity and be spatially averaged over the building floor." This criterion is presented in [15-6] as tentative and in possible need of adjustment as additional information becomes available.

15.1.3   Relation Between Wind Velocities and Building Accelerations

A first step in verifying the compliance of a design with requirements set forth in comfort criteria consists in the estimation, for each possible direction, of the wind speeds that would induce the acceleration levels of interest. Wind tunnel test results may be used to obtain plots of speed versus direction for the wind velocities that induce critical building accelerations (that is, accelerations equal to those specified by the comfort criteria). An example of such a plot is shown in Fig. 15.1.1. (Note that the methods of Chapter 8 can be applied in this context.) Speeds corresponding to points outside the curve of Fig. 15.1.1 will induce accelerations such that—if a criterion of the type proposed in [15-6] is used—$\bar{\sigma} > \bar{\sigma}^*$, where $\bar{\sigma}$ is the spatially averaged rms value of the top floor accelerations and $\bar{\sigma}^*$ is the critical value of $\bar{\sigma}$ specified by the comfort criteria (e.g., in [15-6], $\bar{\sigma}^* = 0.5\%g$). For estimates of building accelerations, see also Chapter 9.

15.1.4   Frequencies of Occurrence of Winds Inducing Critical Accelerations

The second step in verifying the adequacy of a design from a serviceability viewpoint is to estimate the frequency of occurrence of accelerations $\bar{\sigma}$ higher

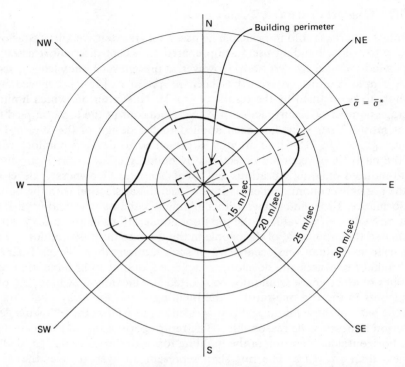

FIGURE 15.1.1.   Wind speeds inducing critical building accelerations.

than the critical value $\bar{\sigma}^*$ specified by the comfort criteria. As shown in [15-6], it is reasonable to define this frequency as the mean number per year $N_S(\bar{\sigma} > \bar{\sigma}^*)$ of storms causing accelerations $\bar{\sigma} > \bar{\sigma}^*$. It is acceptable, in practice, to approximate $N_S(\bar{\sigma} > \bar{\sigma}^*)$ by the number of days per year $N_D(\bar{\sigma} > \bar{\sigma}^*)$ during which the maximum wind speeds exceed the values corresponding to the curve of Fig. 15.1.1. It may be argued that, for office buildings, high speeds occurring at night should not be counted in estimating the mean frequency N_D. However, in view of the many uncertainties inherent in the design for building serviceability, such refinements do not appear to be warranted even though they might reduce N_D by a factor of the order of two.

The number of days per year $N_D(\bar{\sigma} > \bar{\sigma}^*)$ during which wind velocities exceed certain specified values (that is, the values defined by the curve of Fig. 15.1.1) can be obtained readily from Local Climatological Data (L.C.D.) sheets for the weather station closest to the location in question (see Sects. 3.1, 3.4, and 8.3.1). The L.C.D.s contain daily records of the fastest-mile speeds and of the corresponding wind directions. To use the information obtained from the L.C.Ds in conjunction with Fig. 15.1.1, proper adjustments must be made to account for anemometer elevation, roughness of terrain, and averaging of the wind speed with respect to time, as shown in Sect. 3.1.

The estimated mean yearly frequency $N_D(\bar{\sigma} > \bar{\sigma}^*)$ must be compared with the maximum acceptable annual frequency of occurrence of accelerations $\bar{\sigma} > \bar{\sigma}^*$ specified by the comfort criteria. Let this frequency be denoted by $N_A(\bar{\sigma} > \bar{\sigma}^*)$ (e.g., the value of $N_A(\bar{\sigma} > \bar{\sigma}^*)$ proposed in [15-6] is 1/6 per year). If $N_D < N_A$, the design is regarded as adequate from a serviceability viewpoint.

15.2 COMFORT CRITERIA FOR PEDESTRIAN AREAS WITHIN A BUILT ENVIRONMENT

The problem of wind-induced discomfort in pedestrian areas is not new (see Fig. 15.2.1 and p. 173). However, in recent years new types of building and open space configurations have evolved. These may exhibit under certain unfavorable conditions zones of intense surface winds causing unacceptable discomfort to users of pedestrian areas. Typically,* such configurations involve tall buildings rising well above the surrounding built environment and adjacent to open spaces such as plazas or malls. As indicated previously, to define the notion of unacceptable discomfort quantitatively it is required (1) that a correspondence be established between various degrees of pedestrian discomfort and the wind speeds causing them; and (2) that maximum acceptable frequencies of occurrence be specified for these wind speeds. The present section is devoted to a brief discussion of these two requirements.

FIGURE 15.2.1. The Gust. Lithograph by Marlet, collection of the Bibliothèque de la Ville de Paris (photo Roger-Viollet, Paris).

*But not exclusively; see [15-16].

15.2.1   Wind Speeds and Pedestrian Discomfort

Let V denote the mean wind speed measured at approximately 2 m above ground and averaged over 10 min to 1 hr. Observations of wind effects on people and calculations involving the rate of working against the wind suggest that the following degrees of discomfort are induced by various speeds V [15-10, 15-11]:

$$V = 5 \text{ m/sec} \qquad \text{onset of discomfort}$$
$$V = 10 \text{ m/sec} \qquad \text{definitely unpleasant}$$
$$V = 20 \text{ m/sec} \qquad \text{dangerous}$$

A more detailed description of effects of winds of various intensities (as defined by the classical Beaufort scale) is presented in Table 15.2.1 [15-10]. Tentative information on comfort of strolling pedestrians under various sun exposure, ambient temperature, clothing, and wind speed conditions is provided in [15-12].

Experiments reported in [15-13] and [15-14] suggest that pedestrian comfort is a function not only of the mean speed V, but of wind gustiness as well.

TABLE 15.2.1. Summary of Wind Effects [15-10]

Beaufort Number	Description of Wind	Speed (m/sec)	Description of Wind Effects
0	Calm	Less than 0.4	No noticeable wind
1	Light airs	0.4–1.5	No noticeable wind
2	Light breeze	1.6–3.3	Wind felt on face
3	Gentle breeze	3.4–5.4	Wind extends light flag Hair is disturbed Clothing flaps
4	Moderate breeze	5.5–7.9	Wind raises dust, dry soil, and loose paper Hair disarranged
5	Fresh breeze	8.0–10.7	Force of wind felt on body Drifting snow becomes airborne Limit of agreeable wind on land
6	Strong breeze	10.8–13.8	Umbrellas used with difficulty Hair blown straight Difficulty to walk steadily Wind noise on ears unpleasant Windborne snow above head height (blizzard)
7	Moderate gale	13.9–17.1	Inconvenience felt when walking
8	Fresh gale	17.2–20.7	Generally impedes progress Great difficulty with balance in gusts
9	Strong gale	20.8–24.4	People blown over by gusts

It is therefore reasonable, in principle, to study wind effects on people in terms of an effective wind speed V^e defined as follows:

$$V^e = V\left[1 + k\frac{\overline{v'^2}^{1/2}}{V}\right] \qquad (15.2.1)$$

where V is the mean speed, $\overline{v'^2}^{1/2}$ is the rms of longitudinal velocity fluctuations, and k is a constant reflecting the degree to which the effects of the fluctuations are significant. According to the results of [15-13] and [15-14], an appropriate value for this constant is $k \simeq 3.0$. However, other investigators use the value $k = 1.5$ [15-15] or $k = 1.0$ [15-16]. According to [15-14], wind tunnel experiments and observations of pedestrian performance suggest the following correspondence between speeds V^e (with $k \simeq 3.0$) and various degrees of discomfort.

$V^e = 6$ m/sec	onset of discomfort
$V^e = 9$ m/sec	performance affected
$V^e = 15$ m/sec	control of walking affected
$V^e = 20$ m/sec	dangerous

Subsequent observations of pedestrian performance in a large wind tunnel and at the base of a high-rise building, conducted in Japan on over 2000 pedestrians, have led to the development of the following proposed criteria:

$V_3 < 5$ m/sec	performance not affected
5 m/sec $< V_3 <$ 10 m/sec	performance affected
10 m/sec $< V_3 <$ 15 m/sec	performance seriously affected
15 m/sec $< V_3$	performance very seriously affected

where V_3 is the wind speed averaged over 3 sec [15-17]. As noted in [15-17], these criteria are equivalent to or marginally more severe than those of [15-14].

The ability of pedestrians to adjust to strong winds is affected adversely if the exposure to such winds is relatively sudden, as is the case in zones with flows that are highly nonuniform in space. It is therefore noted in [15-14] that if the mean speed varies by 70% over a distance of less than 2 m or so, the effects of wind on people are more severe than suggested above.

Measurements of wind drag on people are reported in [15-29].

15.2.2 Comfort Criteria

Comfort criteria were previously defined as statements specifying maximum acceptable frequencies of occurrence for various degrees of discomfort. The following simple criterion based on extensive experience with the study of ground level wind effects in built environments is suggested in [15-11]. Complaints about wind conditions are not likely to arise if, in pedestrian areas, winds with mean speeds $V > 5$ m/sec are estimated to occur less than 10% of the time. Complaints might arise if such speeds are estimated to occur between 10% and 20% of the time. Estimated frequencies higher than 20% correspond

TABLE 15.2.2. Comfort Criteria for Various Pedestrian Areas

Criterion	Area Description	Limiting Wind Speed	Frequency of Occurrence
1	Plazas and Parks	Occasional gusts to about 6 m/sec	10% of the time or about 1000 h/yr
2	Walkways and other areas subject to pedestrian access	Occasional gusts to about 12 m/sec	1 or 2 times per month or about 50 h/yr
3	All of above	Occasional gusts to about 20 m/sec	About 5 h/yr
4	All of above	Occasional gusts to about 25 m/sec	Less than 1 h/yr

broadly to situations where in existing shopping centers remedial action had to be taken to reduce wind speeds.

More detailed comfort criteria reflecting individual opinions on acceptable frequencies of occurrence of various wind speeds have been proposed in [15-15], [15-18], and [15-19]. An example of such criteria is given in Table 15.2.2 [15-18].

The first criterion in Table 15.2.2 is roughly equivalent to the criterion previously quoted of [15-11]. The limiting gust speed of 25 m/sec corresponds to winds that could knock a frail person to the ground [15-19]. Otherwise, as indicated in [15-18], the values of Table 15.2.2 are subjective and have been arrived at in the absence of reliable data.

As shown in Sect. 15.4, the calculated frequency of occurrence of wind speeds in pedestrian areas depends very strongly upon the estimation procedure being used. It is noted that the comfort criteria of [15-11]—and similar criteria suggested by other authors—are applicable only if the wind speed frequencies are estimated by the simplified procedure of Sect. 15.4. These criteria are no longer applicable if the detailed procedure of Sect. 15.4 is used.

In the absence of established criteria, decisions regarding the acceptability of comfort conditions in a pedestrian area are left, in practice, to the judgment of the site owners [15-20].

15.3   ZONES OF HIGH SURFACE WINDS WITHIN A BUILT ENVIRONMENT

15.3.1   Wind Flow Near Tall Buildings

As noted in [15-11], high wind speeds occurring at pedestrian level around tall buildings are in general associated with the following types of flow:

1. Vortex flows that develop near the ground, as shown in Fig. 15.3.1.

FIGURE 15.3.1.   Wind flow in front of a tall building (wind blowing from left to right.).*

FIGURE 15.3.2.   Wind flow near the windward face of a tall building (wind blowing from left to right).

*Figures 15.3.1 through 15.3.14, 15.3.24, and 15.3.25 contributed by permission of the Director, Building Research Establishment, U.K. Copyright, Controller of Her Britannic Majesty's Stationery Office.

2.   Descending air flows passing around windward corners, as shown in Fig. 15.3.2.

3.   Air flows through ground floor openings connecting the windward to the leeward side of a building (Fig. 15.3.2) or cross flows from the windward side of one building to the leeward side of a neighboring building.

The flow visualization in Figs. 15.3.1 and 15.3.2 was obtained by injecting smoke in the airstream. It is seen that the flow patterns in the immediate vicinity of the windward face are consistent with the pressure distributions shown on the windward face in Fig. 4.6.7*b* (i.e., the air flows from zones of high to zones of low pressures). Part of the air deflected downwards by the building forms a vortex (Fig. 15.3.1) and thus sweeps the ground in a reverse flow (area *A*, marked "vortex flow" in Fig. 15.3.3). Another part is accelerated around the building corners (Fig. 15.3.2) and forms jets that sweep the ground near the building sides (areas *B*, marked "corner streams" in Fig. 15.3.3). If an opening connecting the windward to the leeward side is present at or near the ground level, part of the descending air will be sucked from the zone of relatively high pressures on the windward side into the zone of relatively low pressures (suctions) on the leeward side (Fig. 15.3.2). A through-flow will thus sweep the area *C* shown in Fig. 15.3.3. Through-flows of this type have caused serious discomfort to users of the MIT Earth Sciences Building in Cambridge, Mass., a structure about 20 stories high [15-21]. Cross flows between pairs of buildings are caused by similar pressure differences, as shown in Fig. 15.3.4.

The pattern of the surface wind flow within a site depends in a complex way upon the relative location, the dimensions, the shapes, and certain of the architectural features (such as ground floor openings) of the buildings involved, upon the roughness and the topographical features of the terrain around the site, and upon the possible presence near the site of one or several tall buildings.

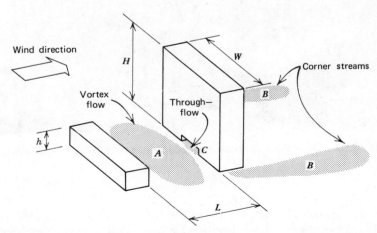

FIGURE 15.3.3.   Regions of high surface wind speeds around a tall building (after [15-11]).

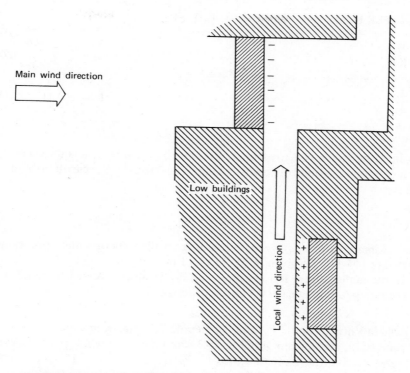

FIGURE 15.3.4.   Cross-flow between two tall buildings (after [15-11]).

To study the surface wind flow in any given built environment it is therefore necessary, in general, to conduct wind tunnel tests. Nevertheless, as indicated in [15-11], experience has shown that information based on aerodynamic studies of the basic reference case represented in Fig. 15.3.3 is useful for the prediction of surface winds in a wide range of practical situations. Such information is presented in [15-11] and will be summarized below. Its range of applicability includes built environments that retain a basic similarity with the configuration shown in Fig. 15.3.3 and in which the height of the buildings does not exceed 100 m or so. Detailed information on the wind environment around single buildings and around groups of buildings is presented in [15-30].

15.3.2   Wind Speeds at Pedestrian Level in a Basic Reference Case [15-11]

Surface winds around models of the tall building shown in Fig. 15.3.3 were measured in wind tunnel tests conducted at a 1/120 scale. The roughness conditions simulated in the tests were typical of a suburban environment, the mean wind profile being given, approximately, by a power law with exponent $\alpha = 0.28$. The surface winds depend upon the dimensions H, W, L, and h defined in Fig. 15.3.3 and are expressed in terms of ratios V/V_H, where V and V_H are mean speeds at pedestrian level and at elevation H, respectively. In certain

TABLE 15.3.1. Approximate Ratios V_H/V_0 [15-11]

H (m)	20	30	40	50	60	70	80	90	100
$\dfrac{V_H}{V_0}$	0.73	0.82	0.89	0.94	0.99	1.04	1.08	1.11	1.14

applications it is useful to estimate the ratio V/V_0, where V_0 is the mean speed at 10 m above ground in open terrain. The ratios V/V_0 can be obtained as follows:

$$\frac{V}{V_0} = \frac{V}{V_H} \frac{V_H}{V_0} \tag{15.3.1}$$

Approximate ratios V_H/V_0 corresponding to the experimental conditions reported in [15-11] are given in Table 15.3.1 for various heights H.

In the material that follows, the wind direction is assumed to be normal to the building face (angle $\theta = 0°$) unless otherwise stated.

Speeds in Vortex Flow. V_A and V_H denote the maximum mean wind speed at pedestrian level in zone A of Fig. 15.3.3 and the mean wind speed at elevation H, respectively. Approximate ratios V_A/V_H are given in Fig. 15.3.5 as functions of W/H for various ratios L/H and for the ranges of values H/h shown. The height h corresponded in all the model tests to typical heights of suburban buildings (7 m to 16 m). It is noted that as the building becomes more slender (as the ratio W/H becomes lower) the ratio V_A/V_H decreases.

Typical examples of the variation of V_A with individual variables are shown in Fig. 15.3.6. If the distance L between the low-rise and high-rise building is small, the vortex cannot penetrate effectively between the buildings and V_A is small (Fig. 15.3.6b). If L is very large or if h is very small, the vortex that forms upwind of the tall building will be poorly organized and weak; V_A will therefore be relatively low (Figs. 15.3.6b and 15.3.6d). If h approaches the value of H, the taller building will in effect be sheltered and the speed V_A will thus be low.

It is noted that the ratio V_A/V_H is of the order of 0.5 for a range of practical situations.

Speeds in Corner Streams.   Figure 15.3.7 shows the approximate dependence upon H/h of the ratio V_B/V_H, where V_B and V_H denote the maximum mean speed at pedestrian level in the zones swept by the corner streams and the mean speed at elevation H, respectively. A typical example of the variation of V_B with the variables H, L, W, and h is given in Fig. 15.3.8. The speed V_B is seen to depend weakly upon the angle θ between the mean wind direction and the normal to the building face. However, the orientation of the corner streams

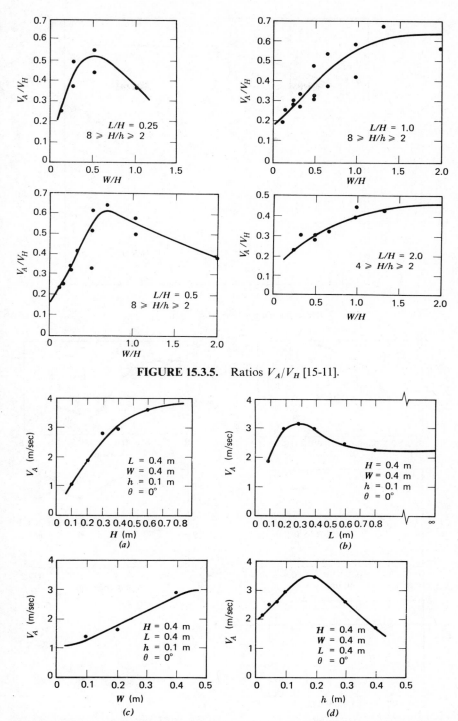

FIGURE 15.3.5. Ratios V_A/V_H [15-11].

FIGURE 15.3.6. Examples of the variation of V_A with individual parameters [15-11].

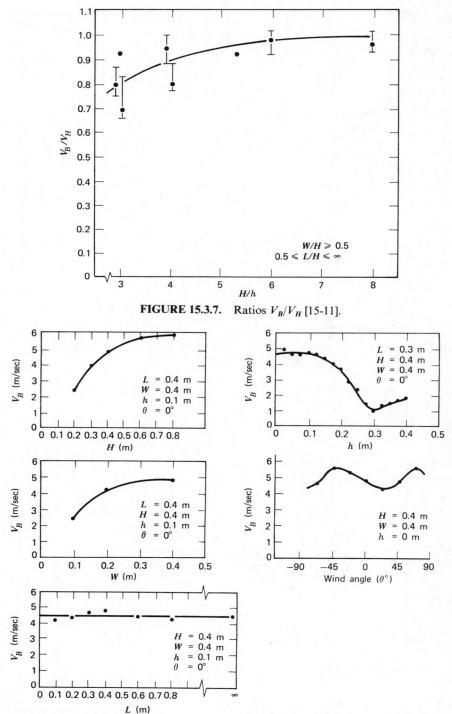

FIGURE 15.3.7. Ratios V_B/V_H [15-11].

FIGURE 15.3.8. Examples of the variation of V_B with individual parameters [15-11].

and, hence, the position of the point of maximum speed V_B may depend significantly upon the direction θ of the mean wind.

Information on the wind speed field around the corner of a wide building model ($H = 0.4$ m, $W = 0.4$ m, $L = 0.3$ m) is given in Fig. 15.3.9. The wind speed decreases rather slowly within a distance from the building corner equal, approximately, to H. The ratio $Y/(D/2)$, where Y is defined as in Fig. 15.3.9 and D is the building depth, provides an approximate measure of the position of the

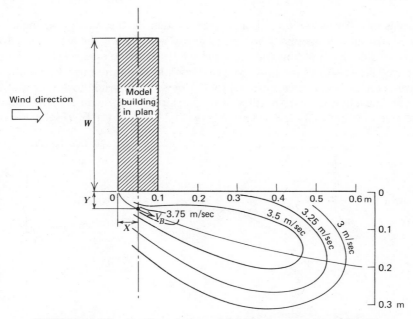

FIGURE 15.3.9.   Surface wind speed field in a corner stream [15-11].

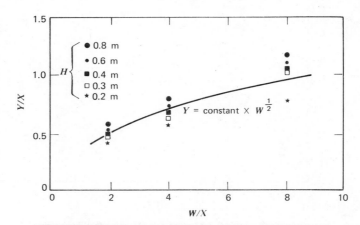

FIGURE 15.3.10.   Empirical curve Y/X vs. W/X [15-11].

corner stream. Measured values of this ratio for various values of H and of $W/(D/2)$ are shown in Fig. 15.3.10. It is seen that the points of Fig. 15.3.10 are fit reasonably well by a curve of the form $Y = \text{constant} \times W^{1/2}$. For example, if $W = 45$ m and $D = 15$ m, then $W/(D/2) = 6$, $Y/(D/2) \simeq 0.8$ (Fig. 15.3.10), and the maximum speed on the centerline of the building would occur at $Y \simeq 0.8 \times D/2 = 6$ m.

It is noted that the ratio V_B/V_H is of the order of 0.95 for a range of practical situations.

Speeds in a Through-Flow.   Let V_C and V_H denote the maximum mean wind speed through a ground floor passageway connecting the windward to the leeward side of a building and the mean wind speed at elevation H, respectively. Figure 15.3.11 shows the approximate dependence of the ratio V_C/V_H upon the parameter H/h as determined in [15-11] by semiempirical formulas and wind tunnel measurements. Examples of the variation of V_C with H, W, L, h, and θ are given in Fig. 15.3.12. It is seen in Fig. 15.3.12b that for $W/H < 0.5$ the ratios

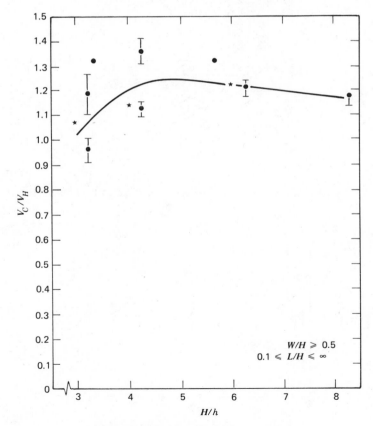

FIGURE 15.3.11.   Ratios V_C/V_H [15-11].

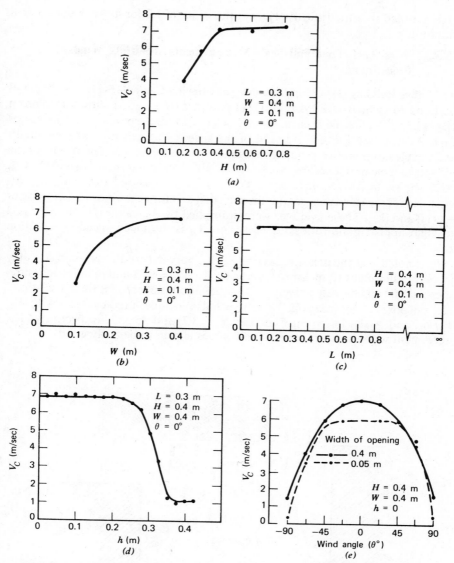

FIGURE 15.3.12.   Examples of the variation of V_C with individual parameters [15-11].

V_C/V_H are lower than in Fig. 15.3.11. Figure 15.3.12e shows for various values of θ the range of variation of V_C with opening width.

The graphs of Figs. 15.3.11 and 15.3.12 are based on measurements in and near passageways with sharp-edged entrances. If the edges of the entrance are rounded to form a bellmouth shape, the speeds V_C can be reduced with respect to those of Figs. 15.3.11 and 15.3.12 by as much as 25% or so [15-11].

It is noted that the ratio V_C/V_H is of the order of 1.2 for a range of practical situations.

15.3.3   Wind Tunnel and Full-Scale Measurements of Surface Winds: Case Studies*

a.   Office Building (H = 31 m) Spanning a Shopping Center [15-11].   A 31 m tall building for which $H/h = 4.4$, $W/H = 1.6$, and $L/H \simeq 0.85$ is shown in plan in Fig. 15.3.13. Full-scale measurements of ground level speeds $V_{(i)}$ at locations $i = 1, 2, \ldots, 9$ (see Fig. 15.3.13) and of the speeds V_{36} measured at location 10 at 36 m elevation above ground were made on ten occasions. The results of the ten sets of measurements are expressed in Table 15.3.2 in terms of ratios $V_{(i)}/V_{36}$. Also shown in Table 15.3.2 are averages of the measured ratios $V_{(i)}/V_{36}$ for west winds [measurement sets (a) through (h)] and for east winds [measurement sets (j) and (k)]. These averages were multiplied by the factor $(36/31)^{0.28} = 1.04$ to yield approximate ratios $V_{(i)}/V_H$, where V_H is the mean speed at elevation $H = 31$ m.

It is noted that the measured ratios $V_{(i)}/V_{36}$ vary in certain cases considerably from measurement to measurement [e.g., $V_{(5)}/V_{36} = 1.33$ and 0.56 for measurement sets (e) and (f), respectively]. No explanation is offered for these variations. For purposes of comparison, Table 15.3.2 also includes predicted ratios V_A/V_H, V_B/V_H, and V_C/V_H based on Figs. 15.3.5, 15.3.7, and 15.3.11, respectively. The agreement with the average measured values is seen to be fairly good.

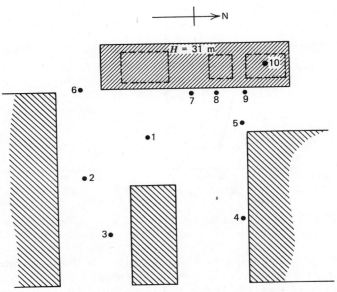

FIGURE 15.3.13.   Plan view, case study (a).

---

*The source of the material is indicated by reference numbers in each case. For additional case studies, see [15-31].

TABLE 15.3.2. Wind Speeds Near Office Building

Locations	West Wind $V_{(i)}/V_{36}$ (a)	(b)	(c)	(d)	(e)	(f)	(g)	(h)	Average	$V_{(i)}/V_H$	$V_{(i)}/V_H$ Predicted	East Wind $V_{(i)}/V_{36}$ (j)	(k)	Average	$V_{(i)}/V_H$	$V_{(i)}/V_H$ Predicted
1	0.36	0.54	0.61	0.37	0.70	0.35	0.34	0.42	0.46	0.48		0.31	0.56	0.43	0.45	
2	0.13	0.25	0.17	0.15	0.30	0.11	0.12	0.21	0.18	0.19		0.21	0.32	0.26	0.27	
3	0.16	0.36	0.30	0.16	0.27	0.11	0.11	0.26	0.22	0.23		0.31	0.40	0.35	0.36	0.59
4	0.15	0.22	0.22	0.16	0.25	0.10	0.12	0.23	0.18	0.19		0.26	0.35	0.30	0.31	Vortex
5	0.81	0.95	0.93	0.59	1.33	0.56	0.56	0.68	0.80	0.83		0.30	0.45	0.37	0.39	
6	0.54	0.93	0.80	0.58	0.88	0.53	0.49	0.79	0.69	0.72	0.92	0.95	0.99	0.97	1.01	0.92
7	0.48	0.79	0.85	0.59	1.05	0.41	0.42	1.00	0.70	0.73	Corner-stream	0.74	0.91	0.82	0.85	Corner-stream
8	0.74	1.01	0.86	0.67	1.05	0.46	0.40	0.39	0.70	0.73	1.06	0.79	1.32	1.05	1.09	1.06
9	0.98	1.05	1.13	0.88	1.77	0.61	0.76	1.02	1.02	1.06	Through-flow	0.90	1.16	1.03	1.07	Through-flow
Speed at Location 10 (m/sec)	9.8	9.7	9.5	8.6	7.3	11.6	12.6	5.7				8.4	6.3			

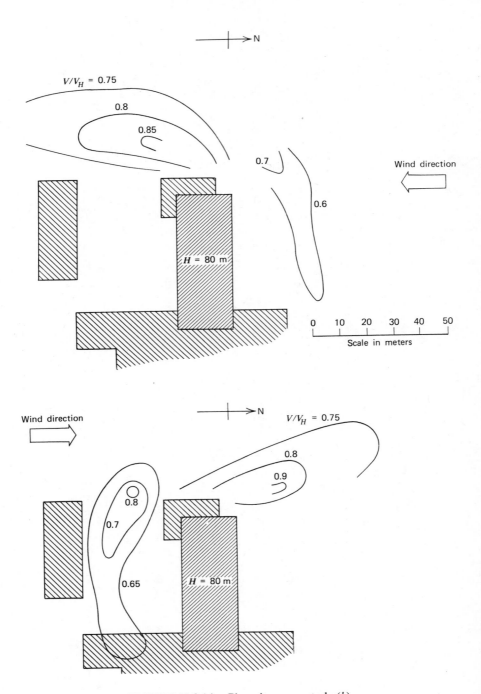

FIGURE 15.3.14. Plan view, case study (b).

b. Model of a Building in Utrecht, Netherlands [15-11].   A proposed 80 m tall building with width $W = 50$ m, depth $D = 22$ m, and for which $H/h = 8.0$, $W/H = 0.63$, and $L/H = 0.5$, is shown in plan in Fig. 15.3.14. Contours of ratios V/V_H, shown in Fig. 15.3.14 for south and for north winds, were obtained in [15-11] using wind tunnel data reported in [15-22]. Measured ratios V_A/V_H and V_B/V_H are about 0.65 (at the center line of the building) and 0.90, respectively. Predicted ratios V_A/V_H and V_B/V_H based on Figs. 15.3.5 and 15.3.7 are about 0.60 and 1.00, respectively. The agreement between predicted and measured values is seen to be reasonably good. It is noted, however, that the vortex flow is asymmetrical and contains regions in which the ratios V/V_H are as high as 0.8.

c. Models of Place Desjardins, Montreal [15-23].   Figure 15.3.15 shows a model (1/400 scale) of one among several designs considered for a development

FIGURE 15.3.15. Place Desjardins model (courtesy of the National Aeronautical Establishment, National Research Council of Canada).

in Place Desjardins, Montreal. The predominant wind direction, determined from measurements at the top of a tall building near the site, is shown in Fig. 15.3.16. Wind tunnel tests were conducted for that direction only. Surface flow patterns were observed by using thread tufts taped to the model surfaces, a wool tuft on the end of a hand-held rod, and a liquid mixture of kerosene-chalk (china clay) sprayed over the horizontal surfaces of the model. As the wind blows over the model, the mixture is swept away from high speed zones and accumulates in zones of stagnating flow. After the evaporation of the kerosene, the white accumulations of chalk indicate zones of low speeds while areas that are dark represent zones where surface winds are high. Wind speed measurements were made in these latter zones. The numbers given in Fig. 15.3.16

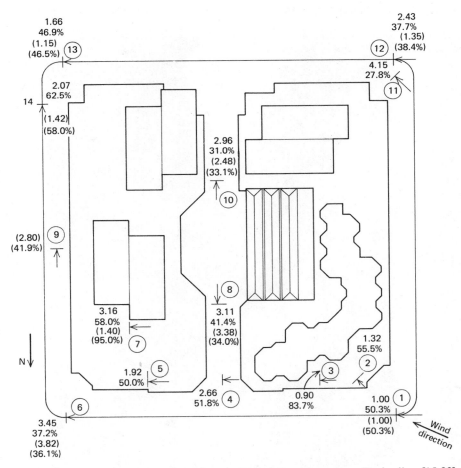

FIGURE 15.3.16. Wind speeds and turbulence intensities, place Desjardins [15-23] (courtesy of the National Aeronautical Establishment, National Research Council of Canada).

FIGURE 15.3.17. Commerce Court Model [15-24] (courtesy Boundary-Layer Wind Tunnel Laboratory, The University of Western Ontario).

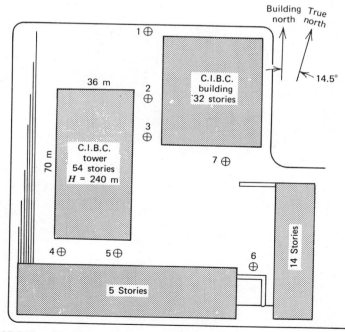

FIGURE 15.3.18. Plan view, Commerce Court. After N. Isyumov and A. G. Davenport, "Comparison of Full-Scale and Wind Tunnel Wind Speed Measurements in the Commerce Court Plaza," *J. Ind. Aerodyn.*, **1** (1975), 201–212.

represent ratios of mean wind speeds at the locations shown to the mean speed at 1.8 m above ground at the northwest corner of the development. The percentages of Fig. 15.3.16 represent turbulence intensities, and the arrows show the direction of the wind component that was measured by the probe. The quantities that are not between parentheses correspond to measurements made in the absence of a projected 50-story tower near the southwest corner of the development. To investigate the effect of the tower upon the surface winds, measurements were also made with the model of the tower in place. Results of these measurements are shown between parentheses in Fig. 15.3.16.

d. Commerce Court Plaza, Toronto [15-24]. A 1/400 scale model and a plan view of the Commerce Court project in Toronto are shown in Figs. 15.3.17 and 15.3.18, respectively. Surface flow patterns obtained by smoke visualization are shown for two wind directions in Figs. 15.3.19 and 15.3.20 [15-25]. Ratios V/V_H, where V and V_H are mean wind speeds at 2.7 m and 240 m above ground, were obtained from measurements in the wind tunnel and, after the completion of the structures, on the actual site. The results of the measurements are shown in Fig. 15.3.21 as functions of wind direction for locations 1 through 7 (see Fig. 15.3.18). The agreement between wind tunnel and full-scale values is seen

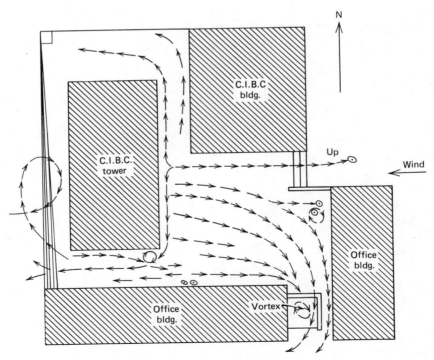

FIGURE 15.3.19. Surface wind flow pattern, Commerce Court (east wind) [15-25].

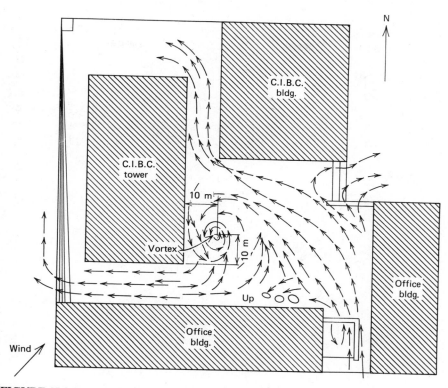

FIGURE 15.3.20.   Surface wind flow pattern, Commerce Court (southwest wind) [15-25].

to be generally acceptable, although differences of the order of 30%, 50% and even more can be noted in certain cases.

e.   Model of the DMA Tower, Paris [15-26].   Models of the 120 m tall DMA tower and of adjacent projected structures are photographed in Fig. 15.3.22 against the background of the actual site. Let V^e and V^e_H denote speeds defined as in Eq. 15.2.1 with $k=1$ and measured at 2 m and 120 m above ground, respectively. Ratios V^e/V^e_H obtained in wind tunnel tests for the southwest wind direction are shown in Fig. 15.3.23. It is noted that for this direction the highest winds occur between the two curved buildings located northwest of the tower (circled value $V^e/V^e_H = 1.08$ in Fig. 15.3.23) rather than in the immediate vicinity of the tower itself. The increase of the wind speeds by the channeling of the flow between buildings forming an angle in plan is sometimes referred to as a Venturi effect [15-16].

15.3.4   Improvement of Surface Wind Conditions

If at certain locations surface winds are judged to be too high and thus to cause unacceptable discomfort to pedestrians, ways must be sought to improve

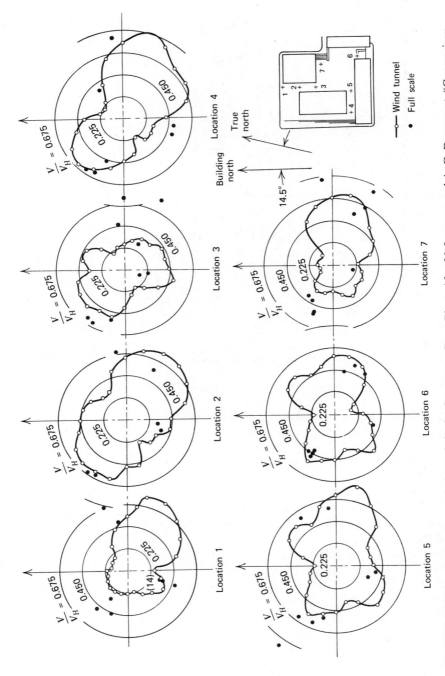

FIGURE 15.3.21. Surface mean wind speeds in the Commerce Court Plaza. After N. Isyumov and A. G. Davenport. "Comparison of Full-Scale and Wind Tunnel Wind Speed Measurements in the Commerce Court Plaza," *J. Ind. Aerodyn,* **1** (1975), 201–212.

FIGURE 15.3.22.   DMA Tower (courtesy Centre Scientifique et Technique du Bâtiment, Etablissement de Nantes).

environmental wind conditions or otherwise protect pedestrians from un-pleasant wind effects. In certain extreme cases it may be necessary to design buildings of lower height or of different configurations than were originally intended. If possible, open areas should be so designed as to prevent pedestrian traffic through high wind zones. Also, as suggested in [15-12], handrails should be provided in potentially dangerous areas. In certain extreme cases, it may be necessary to enclose windy areas frequently used for pedestrian traffic.

Local improvements of surface wind conditions can be achieved by pro-viding (a) roofs over pedestrian areas and/or (b) solid or porous screens at suitable locations. Studies of sheltering effects due to screens are reported in [15-27] and [15-28]. However, no general design rules exist to date on the basis of which sheltering effects could be predicted reliably within a built environment. Also, as noted in [15-12], solid screens merely deflect the wind from one location to another so that the consequences of their use must be investigated carefully.

A few case studies illustrating remedial measures aimed at reducing pedes-trian level wind speeds are presented below.*

---

*The source of the material is indicated by reference numbers in each case.

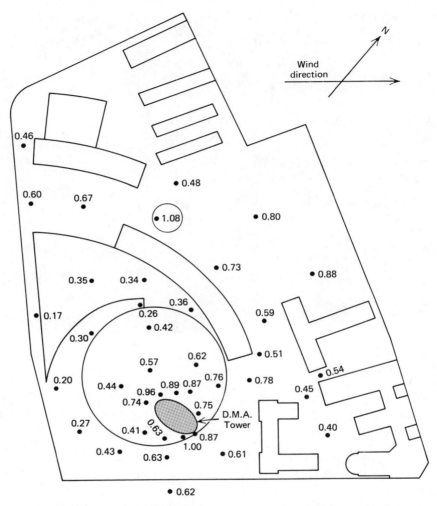

FIGURE 15.3.23.   Surface wind speeds near the DMA Tower [15-26].

a.   Shopping Center, Croydon, England [15-11].   Figure 15.3.24 is a view from the west of an office building, 75 m tall, 70 m wide, and 18 m deep adjoining a shopping center 75 m long. A passageway 12 m high and 3.7 m high connects the shopping center on the west side of the building to the street on the east side (Fig. 15.3.25). The shopping center was designed and built without the curved roof over the shopping mall that can be seen in Fig. 15.3.24. After the completion of the building complex it became apparent that remedial measures were necessary to reduce wind speeds in the passageway and in the shopping mall. The ground level wind flow was investigated in the wind tunnel, first for the complex as initially built (that is, with the mall not covered) and then with various arrangements of roofs over the mall and of screens within the

FIGURE 15.3.24.   Tall building and shopping center, Croydon [15-11].

passageway. Ratios V/V_H measured in the wind tunnel (V and V_H are the mean speeds at 1.8 m and 75 m above ground, respectively) are shown in Fig. 15.3.25 in three cases. For the complex as first built, the highest values of the ratio V/V_H were 0.68 in the vortex flow zone and 1.01 in the through-flow zone. The provision of a full roof over the mall but of no screens within the passageway reduced considerably pedestrian level speeds caused by west winds. However, with east winds, the flow was trapped under the roof and the wind speeds within the mall were, for this reason, high; as shown in Fig. 15.3.25, the speeds were also high at the east entrance of the passageway. A solid roof close to the tall building followed by a partial roof over the rest of the mall, and a screen obstructing 75% of the passageway area resulted in a significant reduction of surface winds, as shown in Fig. 15.3.25. It is noted that, to protect the mall from

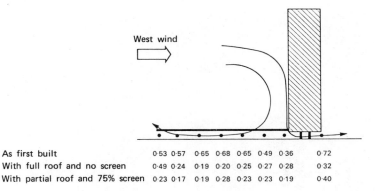

West wind

As first built	0·53 0·57	0·65 0·68 0·65 0·49 0·36	0·72
With full roof and no screen	0·49 0·24	0·19 0·20 0·25 0·27 0·28	0·32
With partial roof and 75% screen	0·23 0·17	0·19 0·28 0·23 0·23 0·19	0·40

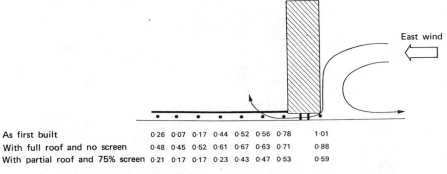

East wind

As first built	0·26 0·07	0·17 0·44 0·52 0·56 0·78	1·01
With full roof and no screen	0·48 0·45	0·52 0·61 0·67 0·63 0·71	0·88
With partial roof and 75% screen	0·21 0·17	0·17 0·23 0·43 0·47 0·53	0·59

FIGURE 15.3.25.   Model test results, Croydon [15-11].

strong vortex flows caused by west winds, the solid roof had to extend for at least 18 m from the building face.

The solution actually applied consisted of providing (a) a full roof over the entire mall (Fig. 15.3.24) and (b) screens with 75% blockage in the passageway. This solution proved effective in ensuring a comfortable wind environment.

b.   Models of Place Desjardins, Montreal [15-23].   It is seen in Fig. 15.3.16 that the ground level winds in the Place Desjardins mall (Fig. 15.3.15) are relatively high: with the 50-story tower southeast of the development not installed, $V_{(8)}/V_{(1)} = 3.11$ and $V_{(10)}/V_{(1)} = 2.96$; with the tower in place, $V_{(8)}/V_{(1)} = 3.38$ and $V_{(10)}/V_{(1)} = 2.48$. Wind tunnel measurements of pedestrian level wind speeds are also reported in [15-23] for the case in which the mall was covered. With the 50-story tower in place, the effect of covering the mall was to reduce the mean wind speeds by a factor of five at location 8 and by a factor of about 1.67 at location 10. However, with the tower not installed, while the mean speeds were reduced by a factor of almost three at location 8, the reduction at location 10 was insignificant.

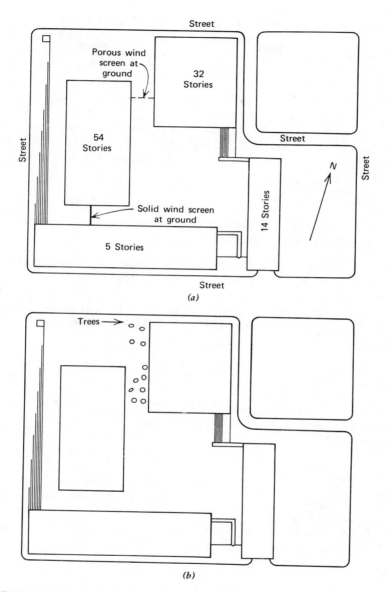

FIGURE 15.3.26. Remedial measures at Commerce Court: (*a*) screens; (*b*) trees. After N. Isyumov and A. G. Davenport, "The Ground Level Wind Environment in Built-up Areas," in *Proceedings of the Fourth International Conference on Wind Effects on Buildings and Structures*, London, 1975, Cambridge Univ. Press, Cambridge, U.K., 1976, pp. 403–422.

c.   Commerce Court Plaza, Toronto [15-15].   After the completion of the building complex shown in Fig. 15.3.18, conditions were found to be particularly annoying on windy days for pedestrians walking from the relatively protected zone north of the 32-story tower into the flow funneled through the passageway 2-3. Wind tunnel tests indicated that the provision of screens at the ground level as shown in Fig. 15.3.26a would result at locations 2, 5, and 6 in reductions of undesirable mean speeds of the order of 40%. However, while effective aerodynamically this solution was rejected for architectural reasons. Instead, potted evergreens about 3 m high were placed as shown in Fig. 15.3.26b. This reduced the mean winds by about 20% at location 2, 10% at location 5, and 33% at location 6.

15.4   FREQUENCIES OF OCCURRENCE OF UNPLEASANT WINDS WITHIN A BUILT ENVIRONMENT

Detailed Estimation Procedure.   Let $V_0(V, \theta)$ denote the wind speeds at 10 m above ground in open terrain that induce pedestrian level wind speeds V at a given location in a built environment, and let the angle θ define the direction of the velocity vector with speed V_0. The frequency of occurrence at the location concerned of wind speeds larger than V, denoted by f^V, can be written approximately as

$$f^V = \sum_{i=1}^{n} f_i^{V_0} \qquad (15.4.1)$$

in which $f_i^{V_0}$ are the frequencies of occurrence in open terrain of winds with speeds larger than $V_0(V, \theta_i)$ and with directions $\theta_i - \pi/n < \theta < \theta_i + \pi/n$, the angle θ_i being defined as

$$\theta_i = \frac{2\pi i}{n} \qquad (i = 1, 2, \ldots, n) \qquad (15.4.2)$$

In practical applications a 16-point compass is commonly used so that in Eqs. 15.4.1 and 15.4.2 $n = 16$.

To obtain f^V it is necessary, first, to estimate the values of $V_0(V, \theta_i)$. From wind climatological data it is then possible to estimate the frequencies $f_i^{V_0}$.

The speed $V_0(V, \theta_i)$ can be written as follows:

$$V_0(V, \theta_i) = \frac{1}{V/V_H(\theta_i)} \frac{V_0(\theta_i)}{V_H(\theta_i)} V \qquad (15.4.3)$$

The ratios $V_0(\theta_i)/V_H(\theta_i)$ characterize the site from a micrometeorological standpoint. For standard roughness conditions in open terrain, these ratios depend upon the elevation H and upon the roughness conditions upwind of the site, as shown in Sects. 2.2 and 3.1. The ratios $V/V_H(\theta_i)$ at a given location are an aerodynamic property of the wind environment and are estimated on the basis of wind tunnel tests, as seen in Sect. 15.3 (e.g., Fig. 15.3.21).

A useful basis for the estimation of frequencies $f_i^{V_0}$ is provided by weather station records of wind speeds and directions, observed at three-hour intervals and published in monthly Local Climatological Data sheets (see Sect. 3.1). Consider, for example, all the three-hour interval observations in a year (8 obs/day × 365 days = 2920 obs), and assume that 58 out of these observations represent NNW winds with speeds in excess of 6 m/sec. The frequency of occurrence of such winds can then be estimated as follows:*

$$f_1^6 = \frac{58}{2920} \simeq 2\% \tag{15.4.4}$$

It is desirable, in practice, to base frequency estimates on several years of data. This is the case for two reasons. First, one year of data might not reflect the wind climate in a representative way. Second, the observations taken at three-hour intervals are instantaneous values, which are sometimes lower, sometimes higher than the mean speeds. The estimation error associated with such differences is small if the sample size is large.

In certain applications it may be of interest to estimate frequencies for individual seasons, or for a grouping of seasons (e.g., spring, summer, and fall). In such cases, the only data used to estimate wind frequencies are those that cover the season (or seasons) of interest. It is also noted that winds occurring, say, from 11 p.m. to 5 a.m. are, in many cases, of little concern from the standpoint of pedestrian comfort. In estimating wind frequencies, midnight and 3 a.m. observations can then be eliminated from the data set.

Information on frequencies of wind speeds at a weather station $f_i^{V_0}$ may be presented either diagrammatically or in the form illustrated by Table 15.4.1.

An example is now presented of the calculation of frequencies f^V. The calculations are carried out for location 4 of Fig. 15.3.18 for which the plot V/V_H is given in Fig. 15.3.21. It is assumed that the ratio $V_0/V_H \equiv 1.5$ and that the wind climate is described by Table 15.4.1. The frequency f^V is sought for pedestrian level winds with speeds $V > 5$ m/sec. Equation 15.4.3 can then be written as

$$V_0(5, \theta_i) = \frac{7.5}{V/V_H(\theta_i)} \tag{15.4.5}$$

The calculations are given in Table 15.4.2.

Simplified Estimation Procedure. A simplified version of the procedure just presented is suggested in [15-11] for built environments similar in configuration to the basic reference case (Fig. 15.3.3) dealt with in Sect. 15.3. In this version, the aerodynamic information used, rather than being a function of wind direction (as, for example, in Fig. 15.3.21), is limited to the results given

---

*The superscript in the notation f_1^6 represents the speed $V_0 = 6$ m/sec while the subscript corresponds to the value $i = 1$ in a sixteen-point compass in which the angle θ is measured counterclockwise starting from the NNW direction (see Eq. 15.4.1).

TABLE 15.4.1. Frequencies of Wind Speeds at 10 m above Grounds in Open Terrain, $f_i^{V_0}$ (in Percentages of Total Time)

i	1	2	3	4	5	6	7	8	9	10	11	12	13	14	15	16	
Direction	NNW	NW	WNW	W	WSW	SW	SSW	S	SSE	SE	ESE	E	ENE	NE	NNE	N	All Directions
$V_0 > 3$ m/sec	3.8	4.2	5.1	6.3	7.4	6.0	3.1	1.8	2.0	2.3	2.3	2.6	3.1	3.1	3.0	3.7	60
$V_0 > 6$ m/sec	1.8	1.7	2.3	2.5	2.5	2.7	1.5	0.8	1.2	1.3	1.0	1.0	2.0	1.8	1.7	1.2	26
$V_0 > 8$ m/sec	0.2	0.2	0.4	0.4	0.3	0.5	0.4	0.3	0.2	0.4	0.4	0.3	0.2	0.2	0.2	0.4	5
$V_0 > 10$ m/sec	0.1	—	0.1	0.1	0.1	0.1	—	—	—	—	—	—	—	—	—	—	0.5

TABLE 15.4.2. Calculation of f^V for $V = 5$ m/sec

i	1	2	3	4	5	6	7	8	9	10	11	12	13	14	15	16
Direction	NNW	NW	WNW	W	WSW	SW	SSW	S	SSE	SE	ESE	E	ENE	NE	NNE	N
$\dfrac{V}{V_H(\theta_i)}$[a]	0.45	0.60	0.63	0.55	0.45	0.23	0.31	0.32	0.58	0.71	0.79	0.93	0.76	0.45	0.23	0.23
$V_0(5, \theta_i)$[b]	16.5	12.5	12.0	13.6	16.5	33.0	24.0	23.0	13.0	10.5	9.5	8.0	9.9	16.5	33.0	33.0
$f_i^{V_0 c}$	<0.1	<0.1	<0.1	<0.1	<0.1	<0.1	<0.1	<0.1	<0.1	<0.1	0.1	0.3	<0.1	<0.1	<0.1	<0.1
$f^V = \sum f_i^{V_0} < 2\%$																

[a]Figure 15.3.21 (location 4).
[b]Equation 15.4.5.
[c]Table 15.4.1.

in Figs. 15.3.5, 15.3.7, and 15.3.11. The ratios V_H/V_0 of mean wind at elevation H in the built environment to mean wind at 10 m above ground in open terrain may be taken from Table 15.3.1. As far as the climatological information is concerned, the data needed are the frequencies of occurrence of all winds with speeds in excess of various values V_0, regardless of direction (in the example of Table 15.4.1, these data are given in the last column). It is noted in [15-11] that this simplified procedure, even though not "exact," provides generally reliable indications on the serviceability of pedestrian areas in a built environment of the type represented in Fig. 15.3.3. It is emphasized, however, that the procedure can only be regarded as useful if applied in conjunction with the comfort criteria proposed in [15-11] (see Sect. 15.2.2).

To illustrate the procedure proposed in [15-11], consider the case of a building complex for which $H = 70$ m, $W = 50$ m, $L = 35$ m, and $h = 10$ m.* From Figs. 15.3.5 and 15.3.7, $V_A/V_H \simeq 0.6$ and $V_B/V_H \simeq 0.95$, where V_A and V_B are the highest mean speeds in the vortex and in the corner flow, respectively. For $H = 70$ m, $V_H/V_0(10) \simeq 1.04$ (Table 15.3.1), so that

$$\frac{V_A}{V_0} \simeq 0.63 \tag{15.4.6a}$$

$$\frac{V_B}{V_0} \simeq 1.00 \tag{15.4.6b}$$

The frequencies of winds $V_A > 5$ m/sec and $V_B > 5$ m/sec are now sought, assuming that the wind climate is described by Table 15.4.1. It follows from Eq. 15.4.6a that, in order that $V_A > 5$ m/sec, $V_0 > 5/0.63 \simeq 8$ m/sec. From Table 15.4.1, the frequency of such winds is 5%. However, to speeds $V_B > 5$ m/sec there correspond speeds $V_0 > 5/1 = 5$ m/sec, which are seen in Table 15.4.1 to occur about 30% of the time.

The comfort criterion proposed in [15-11] and presented in Sect. 15.2.2 states that areas in which wind speeds in excess of 5 m/sec occur more than 20% of the time are generally unsatisfactory from a pedestrian comfort point of view. Therefore, according to this criterion, in the case of the foregoing example wind conditions are unacceptable.

REFERENCES

15-1   F. K. Chang, "Human Response to Motions in Tall Buildings," *J. Struct. Div.*, ASCE, **98**, No. ST6 (June 1973), 1259–1272.

15-2   P. W. Chen and L. E. Robertson, "Human Perception Thresholds of Horizontal Motion," *J. Struct. Div.*, ASCE, **97**, No. ST8 (Aug. 1972), 1681–1695.

15-3   M. Yamada and T. Goto, *Criteria for Motions in Tall Buildings*, College of Engineering, Hosei University, Koganei, Tokyo, Japan, 1975.

15-4   T. Goto, "Human Perception and Tolerance of Motion," *Monograph of Council on Tall Buildings and Urban Habitat*, Vol. PC (1981), 817–849.

*For these notations, see Fig. 15.3.3.

15-5   F. R. Khan and R. A. Parmelee, "Service Criteria for Tall Buildings for Wind Loading," in *Proceedings of the Third International Conference on Wind Effects on Buildings and Structures*, Tokyo, 1971, Saikon, Tokyo, 1972, pp. 401–407.

15-6   R. J. Hansen, J. W. Reed, and E. H. Vanmarcke, "Human Response to Wind-Induced Motion," *J. Struct. Div.*, ASCE, **98**, No. ST7 (July 1973), 1589–1605.

15-7   J. W. Reed, *Wind-Induced Motion and Human Discomfort in Tall Buildings*, Research Report No. R71-42, Department of Civil Engineering, MIT, Cambridge, Mass., 1971.

15-8   T. Goto, "Studies of Wind-Induced Motion of Tall Buildings Based on Occupants' Reaction," *J. Wind Eng. Ind. Aerodyn.*, **13** (1983), 241–252.

15-9   L. Feld, "Superstructure for 1350 ft. World Trade Center," *Civ. Eng.*, ASCE, **41**, 6 (June 1971), 66–70.

15-10   A. D. Penwarden, "Acceptable Wind Speeds in Towns," *Build. Sci.*, **8**, 3 (Sept. 1973), 259–267.

15-11   A. D. Penwarden and A. F. E. Wise, *Wind Environment Around Buildings*, Building Research Establishment Report, Department of the Environment, Building Research Establishment, Her Majesty's Stationary Office, London, 1975.

15-12   T. V. Lawson and A. D. Penwarden," The Effects of Wind on People in the Vicinity of Buildings," in *Proceedings of the Fourth International Conference on Wind Effects on Buildings and Structures*, London, 1975, Cambridge Univ. Press, Cambridge, U.K., 1976, pp. 605–622.

15-13   E. C. Poulton, J. C. R. Hunt, J. C. Mumford, and J. Poulton, "The Mechanical Disturbance Produced by Steady and Gusty Winds of Moderate Strength: Skilled Performance and Semantic Assessments," *Ergonomics*, **18**, 6 (1975), 651–673.

15-14   J. C. R. Hunt, E. C. Poulton, and J. C. Mumford, "The Effects of Wind on People: New Criteria Based on Wind Tunnel Experiments," *Build. Environ.*, **11** (1976), 15–28.

15-15   N. Isyumov and A. G. Davenport, "The Ground Level Wind Environment in Built-up Areas," in *Proceedings of the Fourth International Conference on Wind Effects on Buildings and Structures*, London, 1975, Cambridge Univ. Press, Cambridge, U.K., 1976, pp. 403–422.

15-16   J. Gandemer, "Wind Environment Around Buildings: Aerodynamic Concepts," in *Proceedings of the Fourth International Conference on Wind Effects on Buildings and Structures*, London, 1975, Cambridge Univ. Press, Cambridge, U.K., pp. 423–432.

15-17   S. Murakami and K. Deguchi, "New Criteria for Wind Effects on Pedestrians," *J. Wind Eng. Ind. Aerodyn.*, **7** (1981), 289–309.

15-18   L. W. Apperley and B. J. Vickery, "The Prediction and Evaluation of the Ground Level Wind Environment," in *Proceedings of the Fifth Australasian Conference on Hydraulics and Fluid Mechanics*, University of Canterbury, Christchurch, New Zealand, 1974.

15-19   W. H. Melbourne and P. N. Joubert, "Problems of Wind Flow at the Base of Tall Buildings," in *Proceedings of the Third International Conference on Wind Effects on Building and Structures*, Tokyo, 1971, Saikon, Tokyo, 1972, pp. 105–114.

15-20   E. Arens and D. Ballanti, "Outdoor Comfort of Pedestrians in Cities," in *Proceedings of the Conference on the Urban Physical Environment*, 1975, U.S. Forest Service, American Meteorological Society, and Syracuse University, Syracuse, N.Y., 1975.

15-21   M. O'Hare, "Designing with Wind Tunnels," *Arch. Forum* (April 1968), 60–64.

15-22   R. Poestkoke, *Windtunnelmetingen aan een model van het Transitorium II van de Rijksuniversiteit, Utrecht*, Report No. TR72110L, National Aerospace Laboratory NLR, The Netherlands, 1972.

15-23   N. M. Standen, *A Wind Tunnel Study of Wind Conditions on Scale Models of Place Desjardins, Montreal*, Laboratory Technical Report No. LTR-LA-101, National Research Council of Canada, National Aeronautical Establishment, Ottawa, 1972.

15-24   N. Isyumov and A. G. Davenport, "Comparison of Full-Scale and Wind Tunnel Wind Speed Measurements in the Commerce Court Plaza," *J. Ind. Aerodyn.*, **1**, 2 (Oct. 1975), 201–212.

15-25   A. G. Davenport, C. F. P. Bowen, and N. Isyumov, *A Study of Wind Effects on the Commerce Court Project, Part II, Wind Environment at Pedestrian Level*, Engineering Science

Research Report No. BLWT-3-70, University of Western Ontario, Faculty of Engineering Science, London, Canada, 1970.

15-26   J. Gandemer, *Étude de la tour D.M.A.*, *Partie 2, Détermination du champ de vitesse au voisinage du complexe bâti de la tour D.M.A.*, EN-ADYM-75-4C, Centre Scientifique et Technique du Bâtiment, Nantes, France, 1975.

15-27   M. O'Hare and R. E. Kronauer, "Fence Designs to Keep Wind from Being a Nuisance," *Archit. Rec.* (July 1969), 151–156.

15-28   V. K. Shárán, "Wind Comfort and Wind Shelter," in *Proceedings of the Symposium on External Flows*, University of Bristol, 1972.

15-29   A. D. Penwarden, P. F. Grigg, and R. Rayment, "Measurements of Wind Drag on People Standing in a Wind Tunnel," *Build. Environ.*, 13 (1978), 75–84.

15-30   W. J. Beranek, "Wind Environment around Single Buildings of Rectangular Shape, and Wind Environment around Building Configurations," *Heron*, 29 (1984), 1–70.

15-31   F. H. Durgin and A. W. Chock, "Pedestrian Level Winds: A Brief Review," *J. Struct. Div.*, ASCE, 108 (1982), 1751–1767.

15-32   A. Tallin and B. Ellingwood, "Serviceability Limit States: Wind Induced Vibrations," *J. Struct. Eng.*, 110 (Oct. 1984), 2424–2437.

16

Tornado Effects

Tornadoes are storms containing the most powerful of all winds (see Sect. 1.3); however, their probabilities of occurrence at any one location are low compared to those of other extreme winds (see Sect. 3.5). It has, therefore, been generally considered that the cost of designing structures to withstand tornado effects is significantly higher than the expected loss associated with the risk of a tornado strike (the expected loss being defined as the product of the magnitude of the loss by its probability of occurrence). For this reason, tornado-resistant design requirements are not included in current building codes or standards, for example, the Uniform Building Code [16-1], the Southern Building Code [16-2], or the American National Standard A58.1–1982 [16-3].

However, in designing facilities for which the consequences of failure would be exceptionally grave, the effects of a tornado strike must be explicitly taken into account. Such facilities include nuclear power plants, for which it is required that "structures, systems and components important to safety...be designed to withstand the effects of natural phenomena such as...tornadoes...without loss of capability to perform their safety functions" [16-4]. In the United States, construction permits or operating licenses for nuclear power plants are issued or continued only if this requirement is satisfied in a manner consistent with Regulatory Guides* issued by the U.S. Nuclear Regulatory Commission (e.g., [16-4 and [16-5]) or otherwise acceptable to the Regulatory staff of that agency. It is the purpose of this chapter to describe studies undertaken, as well as design criteria and procedures developed, with a view to ensuring an adequate resistance of nuclear power plants to tornado effects.

Tornado effects may be divided into three groups:

1. Wind pressures, caused by the direct action upon the structure of the air flow.

---

*The Regulatory Guides are reviewed periodically, as needed, to accommodate comments and to reflect new information or experience [16-4].

2. Pressures associated with the variation of the atmospheric pressure field as the tornado moves over the structure (atmospheric pressure change effects).

3. Impactive forces caused by tornado-borne missiles.

To estimate these effects, it is necessary to assume a model of the tornado wind flow. A model currently accepted for use in engineering calculations consists of a vortex characterized by the following parameters: (1) maximum rotational wind speed V_{rot}*; (2) translational speed of the tornado vortex V_{tr}; (3) radius of maximum rotational wind speed R_m; (4) pressure drop p_a; and (5) rate of pressure drop dp_a/dt. (Values of these parameters proposed for the design of nuclear power plants in the United States are listed in Sect. 3.5). The tornado vortex flow model must then be complemented by assumptions on the detailed features of the wind flow. Such features are discussed as needed in the subsequent sections herein.

16.1 WIND PRESSURES

A procedure for calculating wind pressures is now described, which is taken from [16-6], and in which it is assumed that:

1. The wind velocities and, therefore, the wind pressures, do not vary with height above ground.

2. The tangential wind velocity component is given by the expressions

$$V_t = \frac{r}{R_m} V_m \qquad (0 \leqslant r \leqslant R_m) \tag{16.1.1}$$

$$V_t = \frac{R_m}{r} V_m \qquad (R_m \leqslant r < \infty) \tag{16.1.2}$$

where V_m is the maximum tangential wind velocity and R_m is the radius of maximum rotational wind speed.

3. The total horizontal wind speed is

$$V = K V_t \tag{16.1.3}$$

where K is a constant of proportionality. This expression, which is not rigorously correct, is convenient in calculations.

The wind pressure p_w used in designing structures or parts and portions thereof may be written as

$$p_w = q_F C_p + q_M C_{pi} \tag{16.1.4}$$

where C_p is the external pressure coefficient, C_{pi} is the internal pressure co-

---

*The rotational wind speed is defined as the resultant of the tangential and radial wind velocity components [16-13].

efficient, q_F is the basic external pressure,* q_M is the basic internal pressure. Values for the pressure coefficients C_p and C_{pi} are given, for example, in [16-3]. The quantities q_F and q_M may be calculated as follows:

$$q_F = C_s^F p_{max} \qquad (16.1.5)$$

$$q_M = C_s^M p_{max} \qquad (16.1.6)$$

where

$$p_{max} = \tfrac{1}{2}\rho V_{max}^2 \qquad (16.1.7)$$

In Eq. 11.16.7, ρ is the air density and V_{max} is the maximum horizontal wind speed (see Sect. 3.5.1). If V_{max} is expressed in mph and p_{max} in lb/ft^2, $\tfrac{1}{2}\rho = 0.00256$ lb/ft^2/(mph)^2. The quantities C_s^F and C_s^M are reduction (or size) coefficients that account for the nonuniformity in space of the tornado wind field. The size coefficient C_s^F may be determined from Fig. 16.1.1 as a function of the ratio L/R_m, where L is the horizontal dimension, perpendicular to the wind direction, of the tributary area of the structural element concerned (if the wind load is distributed among several structural elements, e.g., by a horizontal diaphragm, L is the horizontal dimension, perpendicular to the wind direction, of the total area tributary to those elements). The size coefficient C_s^M may be determined as follows. If the size and distribution of the openings are relatively uniform around the periphery of the structure, C_s^M is determined in the same way as C_s^F using a value of L equal to the horizontal dimension of the structure perpendicular to the wind direction. If the sizes and distribution of the openings

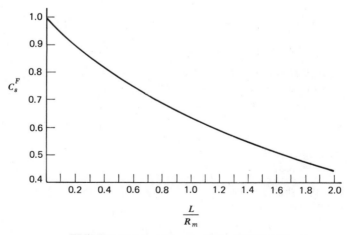

FIGURE 16.1.1.   Size coefficient C_s^F [16-6].

---

*Because no distinction is made in this procedure between basic pressures used in the design of structures, on the one hand, and of parts and portions, on the other hand, the notation q_p used in [16-3] for pressures on parts and portions is not employed herein.

are not uniform, the following weighted averaging procedure is used:

1. Determine quantity r_1/R_m such that

$$\frac{r_1}{R_m} = \frac{R_m}{r_1 + L}$$

(16.1.8)

2. Locate plan of structure drawn at appropriate scale within the non-dimensionalized pressure profile of Fig. 16.1.2, with the left end of the structure at the coordinate r_1/R_m.

3. Determine factor C_q from Fig. 16.1.2 for each exposed opening.

4. Determine C_s^M from Eq. 16.1.9

$$C_s^M = \frac{\sum_1^N A_{0i} C_{q_i}}{\sum_1^N A_{0i}}$$

(16.1.9)

where A_{0_i} is the area of opening at location i, C_{q_i} is the factor C_q at location i, and N is the number of openings. (The coefficient C_q in Fig. 16.1.2 represents nondimensionalized wind pressures and was calculated using Eqs. 16.1.1, 16.1.2, 16.1.3, and 16.1.7. To obtain Fig. 16.1.1 the nondimensionalized pressures of Fig. 16.1.2 were integrated between the limits r_1 and $r_1 + L$, where r_1 is given by Eq. 16.1.8, and the results of the integration were then normalized; the coefficient C_s^F is thus an approximate measure of the average pressure over the interval L [16-6]).

Numerical Example

The building of Fig. 16.1.3 is assumed to be in region I. The sizes and distribution of the openings (not represented in Fig. 16.1.3) are assumed to be uniform around the periphery of the structure. The ratio between area of openings and

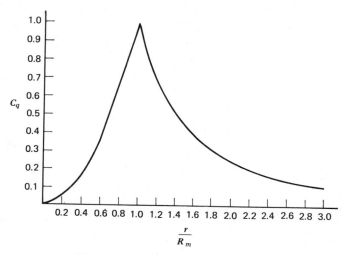

FIGURE 16.1.2. Coefficient C_q [16-6].

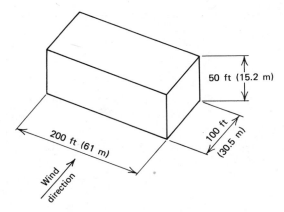

FIGURE 16.1.3.   Schematic view of building.

total wall area is $A_0/A_w = 0.25$. It is assumed $V_{max} = 360$ mph (161 m/sec), $R_m = 150$ ft (46 m) (see Table 3.5.1). The pressures on the 100 ft (30.5 m) side walls induced by the wind blowing in the direction shown in Fig. 16.1.3 are calculated as follows:

$$p_{max} = 0.00256 \times 360^2 = 330 \text{ lb/ft}^2 \quad (15{,}800 \text{ N/m}^2) \qquad \text{(Eq. 16.1.7)}$$

Basic external pressures:

$$L = 200 \text{ ft} \quad (61 \text{ m})$$

$$\frac{L}{R_m} = \frac{200}{150} = 1.33$$

$$C_s^F = 0.56 \quad \text{(Fig. 16.1.1)}$$

$$q_F = 0.56 \times 330 = 185 \text{ lb/ft}^2 \quad (8860 \text{ N/m}^2) \qquad \text{(Eq. 16.1.5)}$$

Basic internal pressures:

$$L = 200 \text{ ft} \quad (61 \text{ m})$$

$$\frac{L}{R_m} = 1.33$$

$$C_s^F = 0.56$$

$$q_M = 0.56 \times 330 = 185 \text{ lb/ft}^2 \quad (8860 \text{ N/m}^2) \qquad \text{(Eq. 16.1.6)}$$

Pressure coefficients:

$$C_p = -0.7 \quad [16\text{-}3]$$

$$C_{pi} = \pm 0.3 \quad \text{(for } A_0/A_w < 0.3 \text{, see } [16\text{-}3])$$

Wind pressure:

$$p_w = -0.7 \times 185 - 0.3 \times 185 = 185 \text{ lb/ft}^2 \quad (8860 \text{ N/m}^2) \quad \text{(Eq. 16.1.4)}$$

16.2   ATMOSPHERIC PRESSURE CHANGE LOADING

Consider the cyclostrophic wind equation (Sect. 1.3) written as

$$\frac{dp_a}{dr} = \rho\,\frac{V_t^2}{r} \tag{16.2.1}$$

where dp_a/dr is the atmospheric pressure gradient at radius r from the center of the tornado vortex. To obtain the pressure drop p_a, Eq. 16.2.1 is integrated from infinity to r. If the expression for V_t given by Eqs. 16.1.1 and 16.1.2 is used [16-6]:

$$p_a(r) = \rho\,\frac{V_m^2}{2}\left(2 - \frac{r^2}{R_m^2}\right) \quad (0 \leqslant r \leqslant R_m) \tag{16.2.2}$$

$$p_a(r) = \rho\,\frac{V_m^2}{2}\frac{R_m^2}{r^2} \quad (R_m \leqslant r < \infty) \tag{16.2.3}$$

In the case of structures with no openings (*unvented structures*), the internal pressure remains equal to the atmospheric pressure before the passage of the tornado. Therefore, during the passage the difference between the internal pressure and the atmospheric pressure is equal to p_a. It follows from Eqs. 16.2.2 and 16.2.3 that the maximum value of p_a, which occurs at $r=0$, is

$$p_a^{\max} = \rho V_m^2 \tag{16.2.3}$$

If the structures are completely open, the internal and external pressures are equalized, for practical purposes, instantaneously, so that the loading due to atmospheric pressure changes approaches zero. In structures with openings (*vented structures*), the internal pressures change during the tornado passage by an amount $p_i(t)$. Denoting the external atmospheric pressure change by $p_a(t)$, the atmospheric differential pressure that acts on the external walls is $p_a(t) - p_i(t)$.

A useful model for $p_a(t)$ can be obtained by assuming, in Eqs. 16.2.2 and 16.2.3, $r = V_{tr}t$, where V_{tr} is the translation speed and t is the time. A simpler model in which the variation of $p_a(t)$ with time is given by the graph of Fig. 16.2.1 may also be used [16-6]. The time-varying internal pressures $p_i(t)$ may be estimated by iteration as follows. Assume that the building consists of a number n of compartments. The air mass in compartment N (where $N \leqslant n$) at time t_{j+1} is denoted by $W_N(t_{j+1})$ and may be written as

$$W_N(t_{j+1}) = W_N(t_j) + [G_{N(\text{in})}(t_j) - G_{N(\text{out})}(t_j)]\,\Delta t \tag{16.2.4}$$

where $G_{N(\text{in})}$ and $G_{N(\text{out})}$ denote the mass of air flowing into and out of compartment N per unit of time, respectively, and Δt is the time increment. The air mass flow rates G_N can be calculated as functions of the pressures outside and within the compartment N and of relevant geometrical parameters, including opening sizes, as shown subsequently. The internal pressure in com-

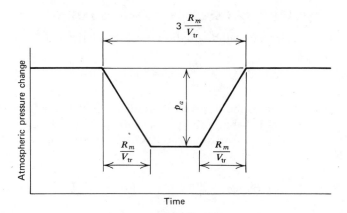

FIGURE 16.2.1. Idealized atmospheric pressure change versus time function [16-6].

partment N at time t_{j+1}, $p_{iN}(t_{j+1})$, is then written as

$$p_{iN}(t_{j+1}) = \left[\frac{W_N(t_{j+1})}{W_N(t_j)} \right]^k p_{jN}(t_j) \qquad (16.2.5)$$

where $k = 1.4$ is the ratio of specific heat of air at constant pressure to specific heat of air at constant volume.

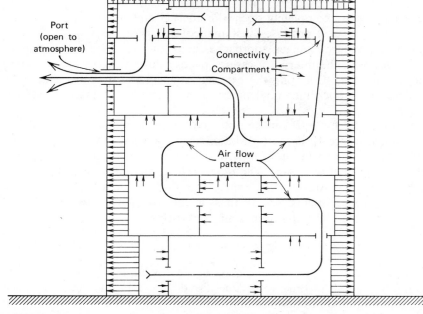

FIGURE 16.2.2. Illustration of pressure distribution and flow pattern during building depressurization [16-6].

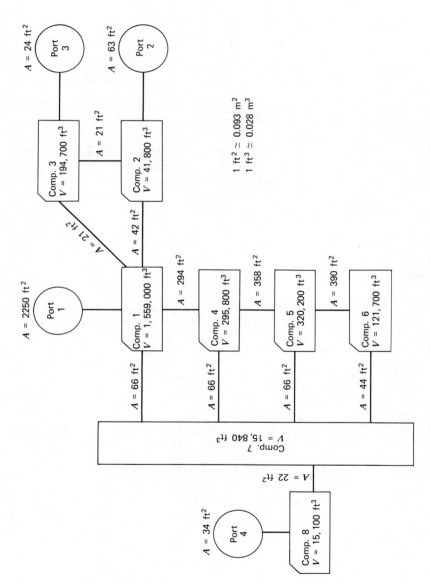

FIGURE 16.2.3. Illustration of a structure depressurization model [16-6].

493

A computer program for calculating loading on vented structures due to atmospheric pressure changes is briefly described in [16-6]. The program incorporates the following model for the air mass flow rate, taken from [16-7]:

$$G = 0.6 C_c A_2 [2g\gamma_1 (p_1 - p_2)]^{1/2} \tag{16.2.6}$$

where

$$C_c = \left\{ \left(\frac{p_2}{p_1} \right)^{2/k} \frac{k}{k-1} \left[\frac{1 - (p_2/p_1)^{(k-1)/k}}{1 - p_2/p_1} \right] \left[\frac{1 - (A_2/A_1)^2}{1 - (A_2/A_1)^2 (p_2/p_1)^{2/k}} \right] \right\}^{1/2} \tag{16.2.7}$$

and A_1 is the area (on the side of compartment 1) of the wall between compartments 1 and 2, A_2 is the area connecting compartments 1 and 2, C_c is the compressibility coefficient, g is the acceleration of gravity, $k = 1.4$, p_1 is the pressure in compartment 1, p_2 is the pressure in compartment 2 ($p_2 < p_1$), and γ_1 is the weight per unit volume of air in compartment 1. If, in compartments provided with a blowout panel, the differential pressure exceeds the design pressure for a panel, a statement in the program transforms the blowout panel area into a wall opening. In view of the presence of three-dimensional effects

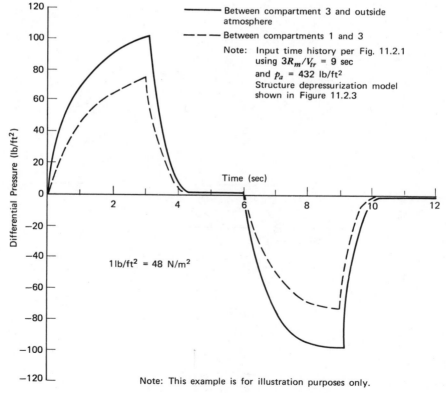

FIGURE 16.2.4. Differential pressure-time history for compartments 1 and 3 [16-6].

not accounted for by Eq. 16.2.6, the atmospheric differential pressures on external walls obtained by the procedure just described are multiplied by a factor of 1.2 [16-6].

An illustration of the pressure distribution and of the flow pattern in a building during depressurization is given in Fig. 16.2.2. An illustration of a structure depressurization model with values of geometric parameters required as input in the computer program, and an example of a corresponding differential pressure-time history calculated using the program, are shown in Figs. 16.2.3 and 16.2.4, respectively.

16.3 TORNADO-BORNE MISSILE SPEEDS

To estimate speeds attained by an object moving under the action of aerodynamic forces induced by tornado winds, a set of assumptions is required on

- the aerodynamic characteristics of the object
- the detailed features of the wind flow field
- the initial position of the object with respect to the ground and to the tornado center, and its initial velocity.

Objects commonly considered as potential missiles in the design of nuclear power plants are bluff bodies such as wooden planks, steel rods, steel pipes, utility poles, and automobiles (see Fig. 11.3.2).

The purpose of this section is to review approaches to the tornado-borne missile problem based on: (1) deterministic modeling, (2) probabilistic modeling involving numerical simulations, and (3) modeling of missile transport as a Markov diffusion process.

16.3.1 Deterministic Modeling of Missile Motions

Equations of Motion and Aerodynamic Modeling. The motion of an object may be described in general by solving a system of three equations of balance of momenta and three equations of balance of moments of momenta. In the case of a bluff body, one major difficulty in writing these six equations is that the aerodynamic forcing functions are not known.

It is possible to measure in the wind tunnel aerodynamic forces and moments acting on a bluff body under static conditions for a sufficient number of positions of the body with respect to the mean direction of the flow. On the basis of such measurements, the dependence of the forces and moments on position and corresponding aerodynamic coefficients can be obtained. Aerodynamic forces and moments can then be calculated following the well-known pattern used in airfoil theory; for example, if an airfoil has a time-dependent vertical motion $h(t)$ in a uniform flow with velocity V, and if the angle of attack is $\alpha = \text{const}$, the lift coefficient is [16-8]

$$C_L = \frac{dC_L}{d\alpha}\left(\alpha + \frac{1}{V}\frac{dh}{dt}\right) \qquad (16.3.1)$$

This procedure for calculating aerodynamic forces and moments may be assumed to be valid if the motions of the body concerned are small. However, in the case of unconstrained bluff bodies moving in a wind flow, the validity of such a procedure remains to be demonstrated.

In the absence of a satisfactory model for the aerodynamic description of the missile as a rigid (six degrees of freedom) body, it is customary to resort to the alternative of describing the missile as a material point acted upon by a drag force

$$\mathbf{D} = \tfrac{1}{2}\rho C_D A |\mathbf{V}_w - \mathbf{V}_M|(\mathbf{V}_w - \mathbf{V}_M) \qquad (16.3.2)$$

where ρ is the air density, $\mathbf{V}_w$ is the wind velocity, $\mathbf{V}_M$ is the missile velocity, A is a suitably chosen area, and C_D is the corresponding drag coefficient. This model is reasonable if, during its motion, the missile either (a) maintains a constant or almost constant attitude with respect to the relative velocity vector $\mathbf{V}_w - \mathbf{V}_M$, or (b) has a tumbling motion such that, with no significant errors, some mean value of the quantity $C_D A$ can be used in the expression for the drag D. The assumption of a constant body attitude with respect to the flow would be credible if the aerodynamic force were applied at all times exactly at the center of mass of the body—which is highly unlikely in the case of a bluff body in a tornado flow—or if the body rotation induced by a nonzero aerodynamic moment with respect to the center of mass were inhibited by aerodynamic forces intrinsic in the body-fluid system. The question thus arises as to whether such forces are present. This question has not been studied exhaustively in the literature. However, simple experiments suggest that in the case of bluff bodies the aerodynamic damping forces have a destabilizing effect. Wind tunnel tests reported in [16-9] tend to confirm this view. The assumption that potential tornado-borne missiles will tumble during their motion appears therefore to be a reasonable one.

Assuming then that Eq. 16.3.2 is valid and that the average lift force vanishes under tumbling conditions, the motion of the missile viewed as a three-degree-of-freedom system is governed by the relation

$$\frac{d\mathbf{V}_M}{dt} = \tfrac{1}{2}\rho \frac{C_D A}{m} |\mathbf{V}_w - \mathbf{V}_M|(\mathbf{V}_w - \mathbf{V}_M) - g\mathbf{k} \qquad (16.3.3)$$

where g is the acceleration of gravity, $\mathbf{k}$ is the unit vector along the vertical axis, and m is the mass of missile. It follows from Eq. 16.3.3 that for a given flow field and given initial conditions the motion depends only upon the value of the parameter $C_D A/m$. For a tumbling body this value can, in principle, be determined experimentally. Unfortunately, little information on this topic appears to be presently available. Reference 16-10 contains information on tumbling motions under flow conditions corresponding to Mach numbers 0.5 to 3.5. The data of [16-10] were extrapolated in [16-11] to lower subsonic speeds; according to this extrapolation, for a randomly tumbling cube the quantity $C_D A$ equals, approximately, the average of the products of the projected areas corresponding to "all positions statistically possible" by the respective static drag coefficients [16-11, pp. 13-17 and 14-16]. In the absence of more experimental information, it appears reasonable to assume that the effective product $C_D A$ is given by the expression

$$C_D A = c(C_{D_1} A_1 + C_{D_2} A_2 + C_{D_3} A_3) \tag{16.3.4}$$

where $C_{D_i} A_i$ $(i = 1, 2, 3)$ are products of the projected areas corresponding to the cases in which the principal axes of the body are parallel to the vector $\mathbf{V}_w - \mathbf{V}_M$ by the respective static drag coefficients, and c is a coefficient assumed to be 0.50 for planks, rods, pipes, and poles and 0.33 for automobiles. In the case of circular cylindrical bodies (rods, pipes, poles), the assumption $c = 0.50$ is clearly conservative.

Computations and Numerical Results. A computer program for calculating and plotting trajectories and velocities of tornado-borne missiles is described in [16-12]. The program includes specialized subroutines incorporating the assumed model for the tornado wind field and the assumed drag coefficients (which may vary as functions of Reynolds number). Input statements include values of relevant parameters and the initial conditions of the missile motion.

In Eq. 16.3.3, both $\mathbf{V}_M$ and $\mathbf{V}_w$ are referred to an absolute frame. The velocity $\mathbf{V}_w$ is usually specified as a sum of two parts. The first part represents the wind velocity of a stationary tornado vortex and is referred to a cylindrical system of coordinates. The second part represents the translation velocity of the tornado vortex with respect to an absolute frame of reference. Transformations required to represent $\mathbf{V}_w$ in an absolute frame are derived in [16-12] and incorporated in the computer program.

For tornadoes with parameters given in Tables 3.5.1 and 3.5.2 for regions I, II, and III, and referred to as type I, type II and type III tornadoes, respectively, calculated values of the maximum horizontal missile speeds $V_H^{\max}$ are given in Fig. 16.3.1 as functions of the parameter $C_D A/m$. These values were obtained on the basis of the following assumptions:

- the tangential velocity of the tornado vortex, V_t, is described by Eqs. 16.1.1 and 16.1.2
- the radial velocity component V_r and the vertical velocity component V_z are given by the expressions* [16-13]:

$$V_r = 0.50 V_t \tag{16.3.5}$$

$$V_z = 0.67 V_t \tag{16.3.6}$$

The radial component is directed toward the center of the vortex (Fig. 16.3.2); the vertical component is directed upwards.

- the translation velocity of the tornado vortex V_{tr} is directed along the x axis (Fig. 16.3.2).

*For alternative models, see [3-46].

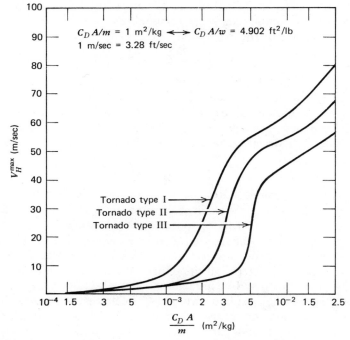

FIGURE 16.3.1. Variation of maximum horizontal missile speed as a function of $C_D A/m$ for various types of tornadoes.

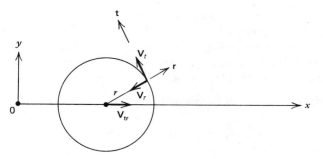

FIGURE 16.3.2. Horizontal components of tornado wind velocity.

- the initial conditions (at time $t=0$) are $x(0) = R_m$, $y(0) = 0$, $z(0) = 40$ m, $V_{M_x}(0) = V_{M_y}(0) = V_{M_z}(0) = 0$, where x, y, z are the coordinates of the center of mass of the missile and V_{M_x}, V_{M_y}, V_{M_z} are the missile velocity components along the x, y, z axes. Also, at $t=0$ the center of the tornado vortex coincides with the origin 0 of the coordinate axes.

Table 16.3.1 lists assumed characteristics of selected missiles and the corresponding horizontal speeds V_H^{max} as obtained from Fig. 16.3.1. A computer

TABLE 16.3.1. Characteristics and Maximum Horizontal Speeds of Selected Missiles.

	Dimensions	Weight (lb/ft)	Mass (kg/m)	C_{D_1}	C_{D_2}	C_{D_3}	$C_D A/w$ (ft²/lb)	$C_D A/m$ (m²/kg)	V_H^{max} Tornado Type I	Tornado Type II	Tornado Type III
1 Wooden Plank	$3\frac{5}{8}'' \times 11\frac{3}{8}'' \times 12'$ (0.092 m × 0.289 m × 3.66 m)	8.2 to 11 (say, 9.6)	12.2 to 16.3 (say, 14.3)	2.0	2.0	2.0	0.132	0.0270	272 ft/sec (83 m/sec)	230 ft/sec (70 m/sec)	190 ft/sec (58 m/sec)
2 6″ Sch. 40 Pipe	6.625″ (diam) × 15′ (length) (0.168 m × 4.58 m)	18.97	28.18	0.7	2.0	0.7	0.0212	0.0043	171 ft/sec (52 m/sec)	138 ft/sec (42 m/sec)	33 ft/sec (10 m/sec)
3 Automobile	16.4′ × 6.6′ × 4.3′ (5 m × 2 m × 1.3 m)	4000 lb (total wt)	1810 kg (total mass)	2.0	2.0	2.0	0.0343	0.0070	193 ft/sec (59 m/sec)	170 ft/sec (52 m/sec)	134 ft/sec (41 m/sec)
4 1″ Solid Steel Rod	1″ (diam) × 3′ (length) (0.0254 m × 0.915 m)	2.67	4.0	1.2	2.0	1.2	0.0190	0.0040	167 ft/sec (51 m/sec)	131 ft/sec (40 m/sec)	26 ft/sec (8 m/sec)
5 13.5″ Utility Pole	13.5″ (diam) × 35′ (length) (0.343 cm × 10.68 m)	27.5–36.5 (say, 32)	40.8–54.2 (say, 47.5)	0.7	2.0	0.7	0.0254	0.0052	180 ft/sec (55 m/sec)	157 ft/sec (48 m/sec)	85 ft/sec (26 m/sec)
6 12″ Sch. 40 Pipe	12.75″ (diam) × 15′ (length) (0.32 m × 4.58 m)	49.56	73.6	0.7	2.0	0.7	0.016	0.0033	154 ft/sec (47 m/sec)	92 ft/sec (28 m/sec)	23 ft/sec (7 m/sec)

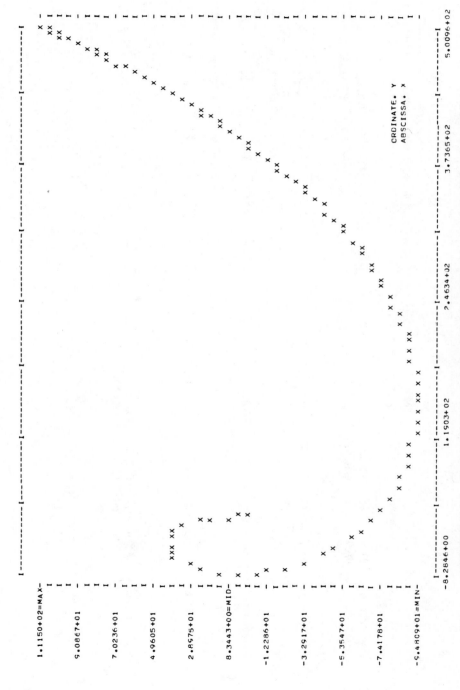

FIGURE 16.3.3. Horizontal projection of calculated missile trajectory (coordinates in meters).

plot of the horizontal projection of the trajectory of a missile with $C_D A/m = 0.1$ in a type I tornado is shown in Fig. 16.3.3.

Sensitivity Study for the Maximum Horizontal Missile Speeds.   In view of the uncertainties involved in the model of the motion, it is of interest to study the sensitivity of the maximum horizontal missile speed V_H^{max} to various changes in the assumptions just described. In each of the cases examined below, all the assumptions—other than those under study as noted—are the same as used to obtain the tornado type I curve of Fig. 16.3.1.

a.   Initial Conditions $x(0)$ and $y(0)$.   Results obtained for $C_D A/m = 0.001$ and $C_D A/m = 0.01$ are shown in Table 16.3.2 for three sets of initial conditions $x(0)$, $y(0)$. In Table 16.3.2 the arrows represent the directions of the tangential and translation wind velocities. It is noted that the initial position corresponding to the largest calculated value of V_H^{max} depends upon $C_D A/m$ [position (c) for $C_D A/m = 0.001$; position (b) for $C_D A/m = 0.01$].

b.   Initial Elevation $z(0)$.   Calculations show that if the parameter $C_D A/m$ corresponds to the middle branch of the S-shaped curves in Fig. 16.3.1, then the values of V_H^{max} decrease as the initial elevation decreases. However, if the missiles are relatively light so that the parameter $C_D A/m$ corresponds to the upper branch of the S-shaped curves, then V_H^{max} is independent of $z(0)$.

c.   Initial Missile Velocity.   If the missile is injected in the flow, for example, by an explosion, the assumption that its initial velocity is zero no longer holds. All other conditions being equal, a nonzero initial velocity does not necessarily result in values of V_H^{max} higher than those corresponding to zero initial velocity. This is illustrated in Table 16.3.3 in which the conditions $V_{M_x}(0) = 0$, $V_{M_x}(0) = 10$ m/sec, and $V_{M_x}(0) = 20$ m/sec; $V_{M_y}(0) = 0$, and $V_{M_z}(0) = 0$ were assumed.

d.   Translation Velocity V_{tr}.   Depending upon the initial conditions $x(0)$, $y(0)$, the speed V_H^{max} may increase or decrease as the translation velocity V_{tr} of the

TABLE 16.3.2. Maximum Missile Speeds V_H^{max} (m/sec) for Various Initial Conditions $x(0)$, $y(0)$

		$x(0)$ (m)	$y(0)$ (m)	$C_D A/m = 0.001$	$C_D A/m = 0.01$
(a)		46	0	7	62
(b)		23	0	15	89
(c)		0	−46	51	67

TABLE 16.3.3. Maximum Horizontal Missile Speeds V_H^{max} (m/sec) Corresponding to Various Initial Velocities

		$x(0)$ (m)	$y(0)$ (m)	$C_D A/m = 0.001$ $V_{M_x}(0)$			$C_D A/m = 0.01$ $V_{M_x}(0)$		
				0	10	20	0	10	20
(a)		46	0	8	9	20	62	58	53
(b)		0	−23	35	45	35	63	59	59

tornado vortex decreases. For example, if $V_{tr} = 0$, for $x(0) = 46$ m, $y(0) = 0$, and $C_D A/m = 0.001$, $V_H^{max} = 25$ m/sec, rather than 7 m/sec, as in Table 16.3.2 (in which it was assumed $V_{tr} = 31$ m/sec). However, for $x(0) = 0$, $y(0) = -46$ m and $C_D A/m = 0.001$, $V_H^{max} = 25$ m/sec rather than 51 m/sec, as in Table 16.3.2.

e. Model of the Vortex Flow. A vortex flow model proposed in [16-14] and [16-15] differs from the model previously described—in which the radius R_m is constant—essentially in that it assumes a significant linear increase of R_m with height above ground. It is shown in [16-12] that if this model is used, the corresponding calculated missile speeds are in most cases higher than those of Fig. 16.3.1.

Some meteorologists have expressed the view that the actual radial velocities V_r are considerably lower over most of the tornado wind field than indicated by Eq. 16.3.5 [16-16]. The radial drag forces available to maintain the trajectory within the region of high winds—where the missile gather momentum at a high rate—would then be comparatively small, and the missile speeds would be considerably lower than those of Fig. 16.3.1. It is also believed that the actual vertical wind speeds are lower than indicated by Eq. 16.3.6 so that missiles would tend to hit the ground sooner than calculated on the basis of this equation, with a consequent reduction of missile speeds V_H^{max} [16-12].

For additional deterministic studies of tornado-borne missile speeds, the reader is referred to [16-17] through [16-23].

Missile Velocities Specified in ANSI/ANS-2.3-192 Standard. Table 16.3.4 lists maximum horizontal missile velocities specified in [3-48] for each tornado wind speed corresponding to probabilities of 10^{-5}, 10^{-6}, and 10^{-7} per year (see Table 3.5.3). The velocities were obtained by methods of the type proposed in [16-12] for two standard missiles "thought to cover the entire range of characteristics associated with all potential objects that can be propelled by a tornado" [3-48]. Also included in Table 16.3.4 are estimates listed in [3-48] of maximum horizontal ranges of the missile paths.

TABLE 16.3.4. Standard Design Missile Velocities and Ranges* (Extracted from American National Standard ANSI/ANS-2.3-1983 with permission of the publisher, the American Nuclear Society)

Missile	Weight (pounds)	Wind Velocity							
		320 mph	260 mph	250 mph	200 mph	180 mph	150 mph	140 mph	100 mph
Automobile	4,000	115 mph 1,100 ft	100 mph 850 ft	95 mph 800 ft	75 mph 600 ft	65 mph 450 ft	50 mph 200 ft	45 mph 180 ft	20 mph 100 ft
Wide-Flange Beam (WF 14×30 or Equivalent)	750	100 mph 700 ft	75 mph 440 ft	70 mph 400 ft	50 mph 200 ft	45 mph 180 ft	30 mph 100 ft	25 mph 80 ft	0 mph 0 ft

*End-on impact is assumed.

Reference 3-48 also provides estimates of the maximum altitude reached by the missiles (not reproduced in Table 16.3.4), and a simple procedure for estimating vertical missile velocities.

16.3.2   Probabilistic Modeling Involving Numerical Simulations

A probabilistic approach to the tornado-borne missile problem was presented in [16-24], which is applicable in situations where the number and geometry of the potential targets is specified, and their location is either specified or determined by Monte Carlo simulation. The maximum speeds of the missiles hitting these targets depend upon the following factors:

1.   maximum tornado wind speed
2.   structure of tornado wind field
3.   tornado path
4.   type, number, and initial positions of missiles
5.   aerodynamic characteristics of missiles
6.   restraining force, that is, force that must be overcome by the aerodynamic force in order for the missile to be injected in the flow field.

We denote the discrete events associated with these six factors by $A_{1j_1}, A_{2j_2}, \ldots,$ A_{6j_6}. The probabilities $p(A_{ij_i})$ are such that, for each $i\,(i=1, 2, \ldots, 6)$,

$$\sum_{j_i=1}^{n_i} p(A_{ij_i})=1 \qquad (j_i=1, 2, \ldots, n_i) \qquad (16.3.7)$$

These probabilities are determined from available information and/or on the basis of judgment.

A computer program [16-25] developed in conjunction with this approach yields the speeds of the missiles that hit each target, given any set of events $A_{ij_i}\,(i=1, 2, \ldots, 6)$. The program automatically performs calculations covering all the combinations of the postulated events A_{ij_i}. Assume, for example, that hits by missiles with speeds larger than some specified value occur for two combinations of events, denoted by $A_{1k}, A_{2l}, \ldots, A_{6m}$ and $A_{1p}, A_{2q}, \ldots, A_{6r}$. The probability of occurrence of those hits is $p_{1k}p_{2l}\cdots p_{6m}+p_{1p}p_{2q}\cdots p_{6r}$.

A more elaborate approach is employed in [16-26], which counts among its features the use of aerodynamic force coefficients that depend upon missile orientation. The latter is assumed to vary randomly with time and is obtained at each time increment by Monte Carlo simulation. For the purpose of modeling missile injection, objects are divided into "minimally restrained missiles" and "sequentially restrained missiles." The first category includes objects restrained only by their weight and by friction, objects lying on the top layer of a stack (or underneath the top layer if the latter is easily removed by wind), and components of buildings that would fail under the action of tornado winds. The second category, which includes all other missiles, is included in the computations for the sake of "completeness in the probabilistic formalism" [16-26]. The restraining forces are sampled by random simulation from probability distributions selected on a subjective basis. Tornado characteristics are also sampled by random simulation. However, the number of potential missiles on the site is specified deterministically. This is done on the basis of surveys of nuclear power plant sites reported in [16-26].

In assessing structural impact, [16-26] accounts for obliqueness of missile with respect to the target surface at the time of impact, missile size relative to target dimensions, susceptibility of concrete to scabbing damage, susceptibility of concrete and steel to perforation, and ricochet motion of missiles that do not perforate the target. All the Monte Carlo simulations are performed by assuming that only one missile is available on the site. The probability of occurrence of an event A_1 affecting a specified target during the j-th strike by a tornado with intensity I_l is denoted by $P(A_1|I_l)_j$. The corresponding probability of event A_1 if N missiles are available on site is then assumed to be:

$$P^{(N)}(A_1|I_l)_j = 1 - [1 - P(A_1|I_l)_j]^N \tag{16.3.8}$$

If the missile population is subdivided into subpopulations M_i (each characterized by the fact that the missiles originate in region R_i and have a specified set of properties π_i),

$$P(A_1|I_l)_j = \sum_i P(A_1|I_l, M_i)_j P(M_i) \tag{16.3.9}$$

where $P(A_1|I_l, M_i)_j$ is the probability of event A_1 due to the action of a single missile belonging to the subpopulation M_i, $P(M_i) = n_i/N$, and n_i is the number of missiles in subpopulation M_i.

16.3.3 Probabilistic Modeling of Missile Motion as a Markov Diffusion Process [16-27]

For various components of nuclear power plant installations the question arises in practice whether protection against tornado-borne missiles is required.

If the probability that a target will be hit by a missile is smaller than some value acceptable to the Nuclear Regulatory Commission (e.g., 10^{-7}/year), then such protection is not needed.

The probability that a target will be damaged by tornado missiles in any one year may be written as

$$P_T = P_O P_H P_D \tag{16.3.10}$$

where P_O is the probability of occurrence of a tornado at the nuclear power plant site, P_H is the conditional probability of hitting a target given a tornado occurrence, and P_D is the probability of damage given a hit. For unprotected targets it may be assumed conservatively $P_D = 1$.

The probability P_H may be written as

$$P_H = n_p A \sum_{F=0}^{6} \phi(F)\eta(F)\psi(z, F) \tag{16.3.11}$$

where n_p is the number of potential missiles per unit area at the site being considered, A is the area of the target, $\phi(F)$ is the relative frequency of tornadoes with intensity F on the Fujita scale in the region being considered, $\eta(F)$ is the probability that a missile will be injected in the wind flow given the occurrence of a tornado with intensity F, and $\psi(z, F)$ is the probability that a unit area whose center is located at elevation z will be hit by an airborne missile during a tornado with intensity F, the site swept by the tornado being assumed to have a potential missile density of one missile per unit of horizontal area.

For the sake of simplicity it is assumed in [16-27] that the density of potential missiles on the site is uniform. It is further assumed that: (a) the wind speed is constant and equal to the maximum tornado wind speed throughout the tornado path, and (b) the angle θ between the vertical and the drag force induced by the maximum tornado wind speed is uniformly distributed between the values $\theta = -\pi/2$ and $\theta = \pi/2$. Both these assumptions are conservative. (For example, when $\theta = 0$ the direction of the maximum tornado wind is assumed to be vertical. This overestimates the probability of injection of the missiles.) Restraint forces are specified in a manner similar to [16-26].

To estimate the function $\psi(z, F)$ it is postulated that the motion of the tornado missile can be represented as a Markov chain.* This postulate is justified by the assumption also used in [16-26] that the missile undergoes purely random tumbling, so that the aerodynamic force it will experience at any one point depends on the random position that the missile has at that point, rather than on its previous geometric attitudes.

Once the Markov chain model for the missile motion is postulated, a probability density function, $G(\mathbf{r}_0, \mathbf{v}_0, t_0, \mathbf{r} - \mathbf{r}_0, \mathbf{v} - \mathbf{v}_0, t - t_0, F, \gamma)$ is defined, such that

$$dP = G\,dx\,dy\,dz\,dv_x\,dv_y\,dv_z \qquad (16.3.12)$$

where x, y, z are the coordinates in space, v_x, v_y, v_z are the missile velocity components, and dP is the probability that, given the occurrence of a tornado with intensity F and characteristics γ, a missile that becomes airborne at moment t_0 hits the volume $dx\,dy\,dz$ around point $\mathbf{r}$ during a unit time interval at the instant t with a velocity between $\mathbf{v}$ and $\mathbf{v} + d\mathbf{v}$ ($\mathbf{r} = x\mathbf{i} + y\mathbf{j} + z\mathbf{k}$; $\mathbf{v} = v_x\mathbf{i} + v_y\mathbf{j} + v_z\mathbf{k}$, where $\mathbf{i}$, $\mathbf{j}$, $\mathbf{k}$ are unit orthogonal vectors). The function G is referred to as the original (fundamental) Green's function of the problem. Modified Green's functions can be derived by integrations and/or averaging of the original Green's function. The modified functions correspond to: (a) the probability that, given the occurrence of a tornado with intensity F and a set of characteristics γ, the missile will hit the volume $dx\,dy\,dz$ around point $\mathbf{r}$ during a unit time interval at the instant t with any velocity; (b) a similar probability for the case where the hit occurs over a unit area with orientation Ω; and (c) a similar probability averaged

---

*A Markov chain is a process in which the probability of transition from one point to another depends only on the coordinates of these points and on the state of the system at the initial point, that is, the probability is independent of the previous history of the system.

over all possible tornadoes having intensity F. By applying the Kolmogorov-Chapman equation to the function G and integrating and averaging the results to obtain the probabilities (b) and (c) just defined, [16-27] derives closed form relations for the function $\psi(z, F)$.

16.4   COMBINED TORNADO EFFECTS

Let the total tornado effects considered in design be denoted by W_t. The following expressions for W_t are specified in [16-6]:

$$W_t = W_{t_q} \tag{16.4.1}$$

$$W_t = W_{t_p} \tag{16.4.2}$$

$$W_t = W_{t_m} \tag{16.4.3}$$

$$W_t = W_{t_q} + 0.5 W_{t_p} \tag{16.4.4}$$

$$W_t = W_{t_q} + W_{t_m} \tag{16.4.5}$$

$$W_t = W_{t_q} + 0.5 W_{t_p} + W_{t_m} \tag{16.4.6}$$

where W_{t_q} is the maximum wind pressure effect, W_{t_p} is the maximum atmospheric pressure change effect, and W_{t_m} is the maximum missile impact effect. Equations 16.4.1 through 16.4.6 are justified in [16-6] as follows.

16.4.1   Wind Pressures Plus Atmospheric Pressure Change Effects

If the structure is unvented, the total pressure p_{wa} due to the direct action of wind and to the atmospheric pressure change can be written as

$$p_{wa} = p_w + p_a \tag{16.4.7}$$

where p_w can be represented in the form

$$p_w = \rho \, \frac{V_m^2}{2} \, \frac{r^2}{R_m^2} \, K^2 C \tag{16.4.8}$$

and C is a constant (see Eqs. 16.1.1 through 16.1.7). Using Eqs. 16.2.2 and 16.2.3, the expression for the combined pressure p_{wa} becomes

$$p_{wa} = \rho \, \frac{V_m^2}{2} \left[2 + \frac{r^2}{R_m^2} (K^2 C - 1) \right] \quad (0 \leqslant r \leqslant R_m) \tag{16.4.9}$$

$$p_{wa} = \rho \, \frac{V_m^2}{2} \, \frac{R_m^2}{r^2} (1 + K^2 C) \quad (R_m \leqslant r < \infty) \tag{16.4.10}$$

From Eq. 16.4.9 it follows that, for $K^2 C < 1$, p_{wa} is a maximum at $r = 0$, where it is equal to the maximum atmospheric pressure change effect (Eq. 16.2.3); for $K^2 C > 1$, p_{wa} is a maximum at $r = R_m$, where it is equal to the maximum wind pressure effect plus one-half the maximum atmospheric pressure effect. Equation 16.4.10 shows that, regardless of the value of $K^2 C$, p_{wa} is maximum at $r = R_m$, where its value is again equal to the maximum wind pressure effect plus one-half the maximum atmospheric pressure effect. These considerations justify Eqs. 16.4.2 and 16.4.4. For completely open structures, the atmospheric pressure change effects approach zero so that the maximum loading is given by Eq. 16.4.1.

16.4.2   Load Combinations Including Missile Effects

It is assumed in [16-6] that the maximum speed is attained by the missile at a distance r from the center of the tornado vortex nearly equal to R_m. It follows then from Eqs. 16.4.9 and 16.2.3 that Eq. 16.4.6 will hold. The case of open structures corresponds to Eq. 16.4.5. Finally, if after attaining its maximum speed near $r = R_m$ the missile is ejected without—or with little—loss of momentum, then the wind pressures and the atmospheric pressure change effects could be negligible at the time of the missile strike. This situation corresponds to Eq. 16.4.3.

REFERENCES

16-1   *Uniform Building Code*, International Conference of Building Officials, Los Angeles, Calif., 1975.

16-2   *Southern Building Code*, Birmingham, Ala., 1965.

16-3   *American National Standard Building Code Requirements for Minimum Design Loads in Buildings and Other Structures, A58.1*, American National Standards Institute, New York, 1982.

16-4   *Code of Federal Regulations*, Title 10, Part 50, Appendix A, Criterion 2 (Design Bases for Protection Against Natural Phenomena), Office of the Federal Register, General Services Administration, Washington, D.C., 1976.

16-5   *Design Basis Tornado for Nuclear Power Plants*, Regulatory Guide 1.76, Directorate of Regulatory Standards, U.S. Atomic Energy Commission, 1974.

16-6   J. V. Rotz, G. C. K. Yeh, and W. Bertwell, *Tornado and Extreme Wind Design Criteria for Nuclear Power Plants*, Topical Report No. BC-TOP-3A Revision 3, Bechtel Power Corporation, San Francisco, Calif., 1974.

16-7   R. C. Binder, *Fluid Mechanics*, 2nd ed., Prentice-Hall, New York, 1949.

16-8   Y. C. Fung, *An Introduction to the Theory of Aeroelasticity*, Dover, New York, 1969.

16-9   R. H. Scanlan, "An Examination of Aerodynamic Response Theories and Model Testing Relative to Suspension Bridges," in *Proceedings of the Third International Conference on Wind Effects on Buildings and Structures*, Tokyo, 1971, Saikon, Tokyo, 1972, pp. 941–951.

16-10   G. E. Hansche and J. S. Rinehart, "Air Drag on Cubes at Mach Numbers 0.5 to 3.5," *J. Aeronaut. Sci.*, **19** (Feb. 1952), 83–84.

16-11   S. F. Hoerner, *Fluid-Dynamic Drag* (published by the author, 1958).

16-12   E. Simiu and M. Cordes, *Tornado-Borne Missile Speeds*, NBSIR 76-1050, National Bureau of Standards, Washington, D.C., 1976.

16-13   J. R. McDonald, K. C. Mehta, and J. E. Minor, "Tornado-Resistant Design of Nuclear Power-Plant Structures," *Nucl. Saf.*, **15**, 4 (July-Aug. 1974), 432–439.

16-14   W. H. Hoecker, "Wind Speed and Air Flow in the Dallas Tornado of April 2, 1975," *Mon. Weather Rev.*, **88**, 5 (1960), 167–180.

16-15   F. C. Bates and A. E. Swanson, "Tornado Considerations for Nuclear Power Plants," *Trans. Am. Nucl. Soc.*, **10** (Nov. 1967), 712–713.

16-16 J. R. Eagleman, V. U. Muirhead, and N. Williams, *Thunderstorms, Tornadoes and Building Damage*, Lexington Books, Lexington, Mass., 1975.

16-17 D. F. Paddleford, *Characteristics of Tornado Generated Missiles*, Report No. WCAP-7897, Westinghouse Electric Corp., Pittsburgh, Pa., 1969.

16-18 A. J. H. Lee, *Design Parameters for Tornado Generated Missiles*, Topical Report No. GAI-TR-102, Gilbert Associates, Inc., Reading, Pa., 1975.

16-19 *The Generation of Missiles by Tornadoes*, Report No. TVA-TR74-1, Tennessee Valley Authority, Knoxville, Tenn., 1974.

16-20 R. C. Lotti, *Velocities of Tornado-Generated Missiles*, Report No. ETR-1003, Ebasco Services, Inc., New York, 1975.

16-21 D. R. Beeth and S. H. Hobbs, Jr., *Analysis of Tornado Generated Missiles*, Report No. B8 R-001, Brown and Root, Inc., Houston, Tex., 1975.

16-22 B. L. Meyers and W. M. Morrow, *Tornado Missile Risk Model*, Report No. BC-TOP-10, Bechtel Power Corp., San Francisco, Calif., 1975.

16-23 A. K. Battacharya, R. C. Boritz, and P. K. Niyogi, *Characteristics of Tornado Generated Missiles*, Report No. VEC-TR-002-0, United Engineers and Constructors, Inc., Philadelphia, Pa., 1975.

16-24 E. Simiu and M. Cordes, *Probabilistic Assessment of Tornado-Borne Missile Speeds*, NBSIR 80-2117, National Bureau of Standards, Washington, D.C., Sept. 1980.

16-25 *Computer Program for Probabilistic Assessment of Tornado-Borne Missile Speeds*, National Bureau of Standards, Computer Tape No. PB81-128423, National Technical Information Service, Springfield, VA 22161, 1981.

16-26 L. A. Twisdale and W. L. Dunn, *Tornado Missile Simulation and Design Methodology*, EPRI NP-2005, Electrical Power Research Institute, Palo Alto, CA, Aug. 1981.

16-27 J. Goodman and J. E. Koch, "The Probability of a Tornado Missile Hitting a Target," *Nuclear Eng. Des.*, **75** (1982), 125–155.

APPENDIX A1

Elements of Probability Theory and Applications

A1.1 INTRODUCTION

Definition and Purpose of Probability Theory

Following Cramér [A1-1], probability theory will be defined as a mathematical model for the description and interpretation of phenomena showing *statistical regularity*. In the field of wind engineering, such phenomena include, for example, the wind intensity at a given location, the turbulent wind speed fluctuations at a point, the pressure fluctuations on the surface of a building, or the fluctuating response of a flexible structure subjected to wind loads.

Relative Frequency and Probability of an Event. Randomness

Consider an experiment that may be repeated an indefinite number of times and the outcome of which can be the occurrence or nonoccurrence of an event A. Let event A occur m times in a sequence (S) of n repetitions of the experiment. If, for large values of n, the ratio m/n, called the *relative frequency* of event A, differs little from some unique limiting value $P(A)$, the number $P(A)$ is defined as the *probability* of occurrence of event A. For example, if a fair coin is tossed, the ratio of the number of heads observed in a very large recorded sequence of H's (heads) and T's (tails) would be close to $\frac{1}{2}$ so that, in any one toss, the probability of occurrence of a head would be $\frac{1}{2}$. Consider, however, the recorded sequence

$$H\,T\,H\,T\,H\,T\,H\,T\,H\,T\,H\,T\,H\,T\ldots$$

consisting of alternating H's and T's. If, in this sequence, the observed outcome of a toss is a head, the probability of a head in the next toss will obviously not be $\frac{1}{2}$ [A1-2].

Indeed, for the definition of probability just advanced to be meaningful, it is required that the sequence (S) previously referred to satisfy the condition of *randomness*. This condition states that the relative frequency of event A must have the same limiting value in the sequence (S) as in any partial sequences that might be selected from it in any arbitrary way, the number of terms in each partial sequence being sufficiently large, and the selection being made in the absence of any information on the outcomes of the experiment [A1-3]. The hypothesis that limiting values of the relative frequencies exist is confirmed for a wide variety of random phenomena by a large body of empirical evidence.

A1.2 FUNDAMENTAL RELATIONS

Addition of Probabilities

Consider two events A_1 and A_2 associated with an experiment. Assume that these events are mutually exclusive (i.e., cannot occur at the same time; an example of mutually exclusive events is, in the case of one die, the throwing of a "five" and of a "six"). The event that either A_1 or A_2 will occur is denoted by $A_1 \cup A_2$. The probability of this event is

$$P(A_1 \cup A_2) = P(A_1) + P(A_2) \tag{A1.1}$$

The empirical basis of the addition rule, Eq. A1.1, lies in the fact that, if the relative frequency of event A_1 is m_1/n and that of event A_2 is m_2/n, the frequency of either A_1 or A_2 is $(m_1 + m_2)/n$. Equation A1.1 then follows from the relation between frequencies and probabilities, and can obviously be extended to any number of mutually exclusive events $A_1, A_2, \ldots, A_n$.

Example

For a fair die, the probability of throwing a "five" is $\frac{1}{6}$ and the probability of throwing a "six" is $\frac{1}{6}$. The probability of throwing either a "five" or a "six" is then $\frac{1}{6} + \frac{1}{6} = \frac{1}{3}$.

Let the nonoccurrence of event A be denoted by $\bar{A}$. Events A and $\bar{A}$ are mutually exclusive. Also, the event that A either occurs or does not occur is certain, that is, its probability is unity:

$$P(A \cup \bar{A}) = 1 \tag{A1.2a}$$

Equation A1.2a follows immediately from the addition rule (Eq. A1.1) applied to the events A and $\bar{A}$, the probabilities of which are the limiting values of the relative frequencies m/n and $(n-m)/n$, respectively. The probability that A does not occur can be written as

$$P(\bar{A}) = 1 - P(A) \tag{A1.2b}$$

Two events A and $\bar{A}$ for which Eq. A1.2b holds are said to be *complementary*.

Compound and Conditional Probabilities. The Multiplication Rule

Consider two events A and B that may occur at the same time. The probability of the event that A and B will occur simultaneously is called the *compound probability* of events A and B and is denoted by $P(A \cap B)$. The probability of event A given that event B has already occurred is denoted by $P(A|B)$ and is known as the *conditional probability* of event A under the condition that event B has occurred. Formally, $P(A|B)$ is defined as follows [A1-1]:

$$P(A|B) = \frac{P(A \cap B)}{P(B)} \qquad (A1.3a)$$

In Eq. A1.3a it is assumed that $P(B)$ is different from zero. Similarly, when $P(A)$ is different from zero,

$$P(B|A) = \frac{P(A \cap B)}{P(A)} \qquad (A1.3b)$$

Example

In a certain region, records kept for a long time show that in an average year 60 days are windy, 200 days are cold, and 50 days are both windy and cold. Let the probability that a day will be windy and the probability that a day will be cold be denoted by $P(A)$ and $P(B)$, respectively. If it is known that condition B (i.e., cold weather) prevails, the probability that a day is windy, or $P(A|B)$, is $50/200 = (50/365)/(200/365)$.

From Eqs. A1.3a and A1.3b, it follows that

$$P(A \cap B) = P(B)P(A|B)$$
$$= P(A)P(B|A) \qquad (A1.4)$$

Equation A1.4 is referred to as the *multiplication rule* of probability theory.

Total Probabilities

If the events $B_1, B_2, \ldots, B_n$ are mutually exclusive and $P(B_1) + P(B_2) + \cdots + P(B_n) = 1$, the probability of event A is

$$P(A) = P(A|B_1)P(B_1) + P(A|B_2)P(B_2) + \cdots + P(A|B_n)P(B_n) \qquad (A1.5)$$

Equation A1.5 is referred to as the theorem of *total probability*.

Example

With reference to the previous example, we now denote the probability of occurrence of winds as $P(A)$, the probability of occurrence of winds given that a day is cold and given that it is not cold as $P(A|B_1)$ and $P(A|B_2)$, respectively, and the probability that a day is cold and that it is not cold as $P(B_1)$ and $P(B_2)$, respectively. From Eq. A1.5 it follows that $P(A) = (50/200)(200/365) + (10/165) \times (165/365) = 60/365$.

Bayes' Rule

If $B_1, B_2, \ldots, B_n$ are n mutually exclusive events, the conditional probability of occurrence of B_i given that the event A has occurred is

$$P(B_i|A) = \frac{P(A|B_i)P(B_i)}{P(A|B_1)P(B_1) + \cdots + P(A|B_n)P(B_n)} \qquad (A1.6)$$

Equation A1.6 follows immediately from Eqs. A1.3b and A1.4 (in which B is replaced by B_i) and Eq. A1.5. Equation A1.6 allows the calculation of the *posterior probabilities* $P(B_i|A)$ in terms of the *prior probabilities* $P(B_1)$, $P(B_2)$, ..., $P(B_n)$ and the conditional probabilities $P(A|B_1)$, $P(A|B_2)$, ..., $P(A|B_n)$.

Example

On the basis of experience with destructive effects of previous tornadoes, it was estimated subjectively that the maximum wind speeds in a tornado were 50 to 70 m/sec. It was further estimated, also subjectively, that the likelihood of the speeds being about 50 m/sec, 60 m/sec, and 70 m/sec is $P(50)=0.3$, $P(60)=0.5$, and $P(70)=0.2$.* These values are prior probabilities. Subsequently a detailed failure investigation was conducted, according to which the speed was 50 m/sec. However, associated with the investigation were uncertainties that were estimated subjectively in terms of conditional probabilities $P(\widehat{50}|V_{\text{true}})$, that is, of probabilities that the speed estimated on the basis of the investigation is 50 m/sec given that the actual speed of the tornado was V_{true}. The estimated values of $P(\widehat{50}|V_{\text{true}})$ were:

$$P(\widehat{50}|50)=0.6$$

$$P(\widehat{50}|60)=0.3$$

$$P(\widehat{50}|70)=0.1$$

It follows from Eq. A1.6 that the posterior probabilities, that is, the probabilities calculated by taking into account the information due to the failure investigation, are

$$P(50|\widehat{50})=\frac{P(\widehat{50}|50)P(50)}{P(\widehat{50}|50)P(50)+P(\widehat{50}|60)P(60)+P(\widehat{50}|70)P(70)}$$

$$=0.51$$

$$P(60|\widehat{50})=0.43$$

$$P(70|\widehat{50})=0.06$$

It is seen that whereas the prior probabilities favored the assumption that the speed was 60 m/sec, according to the calculated posterior probabilities it is more likely that the speed was only 50 m/sec. It should be noted, of course, that this result is useful only to the extent that the various subjective estimates assumed in the calculations are correct.

Independence

In the example following Eq. A1.3b the occurrence of winds and the occurrence of low temperatures are not independent phenomena. Indeed, in the region in question, if the weather is cold, the probability of windiness increases.

---

*Equating such a subjectively estimated likelihood to a probability has philosophical implications that are beyond the scope of this text.

Assume now that event A consists of the occurrence of a rainy day in Pensacola, Florida and that event B consists of an increase in the world market price of gold. It is reasonable to state that the probability of rain in Pensacola is in no way dependent upon whether such a price increase has occurred or not. Consequently, in this case it is natural to assume

$$P(A|B) = P(A) \tag{A1.7}$$

Two events A and B for which Eq. A1.7 holds are called stochastically* *independent*. By virtue of Eqs. A1.3a and A1.7, an alternative definition of independence is

$$P(A \cap B) = P(A)P(B) \tag{A1.8}$$

Example

The probability that one part of a mechanism will be defective is 0.01; for another part, independent of the first, this probability is 0.02. In virtue of Eq. A1.8, the probability that both parts will be defective is 0.0002.

Three events A, B, and C are said to be (stochastically) independent if and only if, in addition to Eq. A1.8, the following relations hold:

$$P(A \cap C) = P(A)P(C)$$
$$P(B \cap C) = P(B)P(C)$$
$$P(A \cap B \cap C) = P(A)P(B)P(C) \tag{A1.9}$$

In general, n events are said to be independent if relations similar to Eqs. A1.9 hold for all combinations of two or more events.

A1.3 RANDOM VARIABLES AND PROBABILITY DISTRIBUTIONS

Random Variables: Definition

Let a numerical value be assigned to each of the events that may occur as a result of an experiment. The resulting set of possible numbers is defined as a *random variable*.

Examples

1. A coin is tossed. The numbers zero and one are assigned to the outcome heads and to the outcome tails, respectively. The set of numbers zero and one constitutes a random variable.

2. To each measurement of a quantity, a number is assigned equal to the result of that measurement. The set of all possible results of the measurements constitutes a random variable.

---

*The word *stochastic* means "connected with random experiments and probability" [A1-1] and is derived from the Greek στοχάζομαι, meaning to aim at, seek after, guess, surmise.

Random variables are called discrete or continuous according as they may take on values restricted to a finite set of numbers (as in Ex. 1) or any value on a segment of the real axis (as in Ex. 2). It is customary to denote random variables by capital Roman letters, for example, X, Y, Z. Specific values that may be taken on by these random variables are then denoted by the corresponding lower case letters x, y, z.

Histograms, Probability Density Functions, Cumulative Distribution Functions

Let the range of a continuous random variable X associated with an experiment be divided into equal intervals ΔX. Assume that if the experiment is carried out n times, the number of times that X has assumed values in the given intervals $X_1 - X_0, \; X_2 - X_1, \ldots, X_i - X_{i-1}, \ldots$ is $n_1, \; n_2, \ldots, n_i, \ldots$, respectively. A graph in which the numbers (or frequencies) n_i are plotted as in Fig. A1.1 is called a *histogram*. (Similar graphs may be plotted in the case of discrete random variables.)

Let the ordinates of the histogram in Fig. A1.1 be divided by $n \Delta X$. The resulting diagram is known as a *frequency density distribution* [A1-4]. The relative frequency of the event $X_{i-1} < X \leqslant X_i$ is then equal to the product of the ordinate of the frequency density distribution, $n_i/(n \Delta X)$ by the interval ΔX. Since the area under the histogram is $(n_1 + n_2 + \ldots + n_i \ldots) \Delta X = n \Delta X$, the total area under the frequency density distribution diagram is unity.

As ΔX becomes very small so that it can be written $\Delta X = dx$, and as n becomes very large, the ordinates of the frequency density distribution approach in the limit values denoted by $f(x)$, where x denotes a value that may be taken on by the random variable X. The function $f(x)$ is known as the *probability density function* of the random variable X (Fig. A1.2a). It follows from this definition of $f(x)$ that the probability of the event $x < X \leqslant x + dx$ is equal to $f(x)\,dx$ and that

$$\int_{-\infty}^{\infty} f(x)\,dx = 1$$

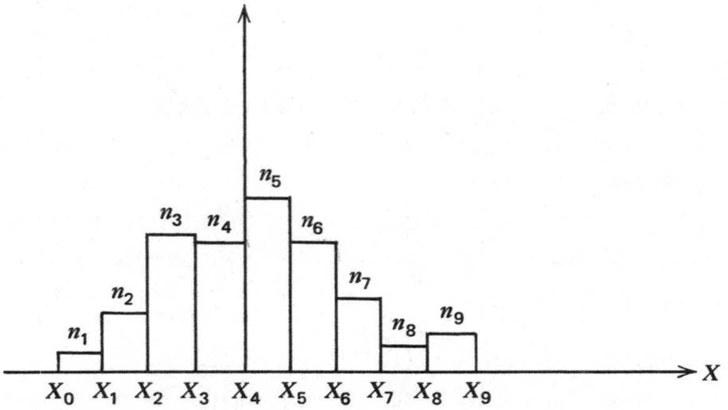

FIGURE A1.1.   Histogram.

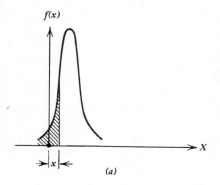

(a)

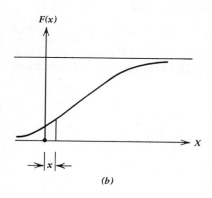

(b)

FIGURE A1.2. (*a*) Probability density function. (*b*) Cumulative distribution function.

In the experiment reflected in Fig. A1.1, the number of times that X has assumed values smaller than X_i is equal to the sum $n_1 + n_2 + \ldots + n_i$. Similarly, the probability that $X \leqslant x$, known as the *cumulative distribution function* of the random variable X and denoted by $F(x)$, can be written as

$$F(x) = \int_{-\infty}^{x} f(x)\, dx \qquad (A1.10)$$

that is, the ordinate at $X = x$ in Fig. A1.2b is equal to the shaded area of Fig. A1.2a.

It follows from Eq. A1.10 that

$$f(x) = \frac{dF(x)}{dx} \qquad (A1.11)$$

Changes of Variable

We consider here only the change of variable $y = (x - a)/b$, where a and b are constants. We assume that the cumulative distribution function $F_X(x)$ is known, and we seek the cumulative distribution function $F_Y(y)$ and the probability density function $f_Y(y)$. We can write

$$F_X(x) = P(X \leqslant x) \tag{A1.12a}$$

$$= P[(X-a)/b \leqslant (x-a)/b] \tag{A1.12b}$$

$$= F_Y(y) \tag{A1.12c}$$

Since Eq. A1.12c implies $dF_X(x) = dF_Y(y)$, or $f_X(x)\,dx = f_Y(y)\,dy$, it follows that

$$f_X(x) = \frac{1}{b} f_Y(y) \tag{A1.13}$$

Joint Probability Distributions

Let X and Y be two continuous random variables and let $f(x, y)\,dx\,dy$ be the probability that $x < X \leqslant x+dx$ and $y < Y \leqslant y+dy$. The quantity $f(x, y)$ is called the *joint probability density function* of the random variables X and Y (Fig. A1.3). The probability that $X \leqslant x$ and $Y \leqslant y$ is called the *joint cumulative probability distribution* of X and Y and is denoted by $F(x, y)$. From the definition of $f(x, y)\,dx\,dy$ it follows

$$F(x, y) = \int_{-\infty}^{x} \int_{-\infty}^{y} f(x, y)\,dx\,dy \tag{A1.14}$$

and

$$\int_{-\infty}^{\infty} \int_{-\infty}^{\infty} f(x, y)\,dx\,dy = 1$$

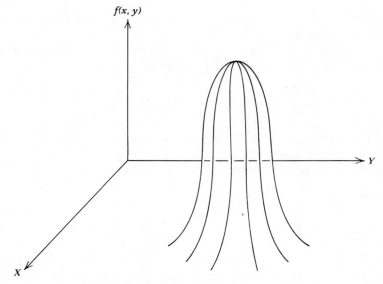

FIGURE A1.3.   Probability density function $f(x, y)$.

It follows from Eq. A1.14

$$f(x, y) = \frac{\partial^2 F(x, y)}{\partial x \, \partial y} \tag{A1.15}$$

If $f(x, y)$ is known, the probability that $x < X \leqslant x + dx$, denoted by $f_X(x) \, dx$, is obtained by applying the addition rule to the probabilities $f(x, y) \, dx \, dy$ over the entire Y domain, that is,

$$f_X(x) = \int_{-\infty}^{\infty} f(x, y) \, dy \tag{A1.16}$$

The function $f_X(x)$ is referred to as the *marginal probability density function* of X.

Finally, the probability that $y < Y \leqslant y + dy$ under the condition that $x < X \leqslant x + dx$ is denoted by $f(y|x) \, dy$. The function $f(y|x)$ is known as the *conditional probability density function* of Y given that $X = x$. If Eq. A1.3a is used it follows that

$$f(y|x) = \frac{f(x, y)}{f_X(x)} \tag{A1.17}$$

If X and Y are independent, $f(y|x) = f_Y(y)$ and

$$f(x, y) = f_X(x) f_Y(y) \tag{A1.18}$$

Similar definitions hold for any number of discrete or continuous random variables.

Application: The Basic Problem of Structural Safety

Consider a simple structure subjected to an external load L. Let S denote the strength of the structure, that is, the value of the external load at which failure, defined as the attainment of some specified limit state, occurs. The basic problem of structural safety consists in determining the probability of failure of the structure if the joint probability distribution $F(l, s) = \text{Prob}(L \leqslant l, S \leqslant s)$ of the load L and strength S is known.

It follows from the definition of the strength S that failure occurs if the inequality $L \geqslant S$ holds. The probability that $l < L \leqslant l + dl$ and $s < S \leqslant s + ds$ will be denoted by $f(l, s) \, dl \, ds$. The probability of failure will then be obtained by summing up the elemental probabilities $f(l, s) \, dl \, ds$ over the entire domain L, S over which $L \geqslant S$ (shaded area in Fig. A1.4):

$$P_f = \int_0^\infty dl \int_0^l f(l, s) \, ds \tag{A1.19}$$

Figure A1.4 and Eq. A1.19 reflect the assumption that $L \geqslant 0$ and $S \geqslant 0$.

It is reasonable to assume that L and S are independent. Then, by virtue of Eq. A1.18,

$$f(l, s) = f_L(l) f_S(s) \tag{A1.20}$$

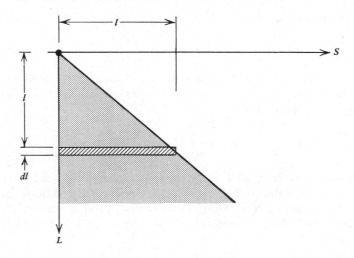

FIGURE A1.4. Domain of integration for calculation of probability of failure.

Substituting Eq. A1.20 into Eq. A1.19

$$P_f = \int_0^\infty f_L(l) \int_0^l f_S(s)\, ds\, dl$$

$$= \int_0^\infty f_L(l) F_S(l)\, dl \tag{A1.21}$$

A1.4   DESCRIPTORS OF RANDOM VARIABLE BEHAVIOR

Mean Value, Median, Mode, Variance, Standard Deviation, and Correlation Coefficient

The complete description of the behavior of a random variable is provided by its probability distribution (in the case of several variables, by the joint probability distribution). Useful if less detailed information is provided by such well-known descriptors as the mean value, the median, the mode, the standard deviation, and, in the case of two variables, the correlation coefficient.

The *mean value*, also known as the *expected value*, or the *expectation* of the discrete random variable X, is defined as

$$E(X) = \sum_{i=1}^m x_i f_i \tag{A1.22}$$

where m is the number of values taken on by x. The counterpart in terms of relative frequencies of the quantity $E(X)$ is

$$\frac{\sum_{i=1}^m x_i n_i}{n} = \sum_{i=1}^m x_i \frac{n_i}{n} \tag{A1.23}$$

If the random variable X is continuous, the expected value of X is written in complete analogy with Eq. A1.22 as

$$E(X) = \int_{-\infty}^{\infty} x f(x)\, dx \qquad (A1.24)$$

The *median* of a continuous random variable X is that value of the variable which corresponds to the value $\frac{1}{2}$ of the cumulative distribution function. The *mode* of a continuous random variable X is that value of the variable that corresponds to the maximum point of the probability density function. It is recalled that $\mathrm{Prob}(x < X < x + dx) = f(x)\, dx$; the mode may thus be interpreted as the value of the variable that has the largest probability of occurrence in any given trial. The mean value, the median, and the mode are referred to as measures of *location*.

The expected value of the quantity $[x - E(X)]^2$ is defined as the *variance* of the random variable X. By virtue of the definition of the expected value (Eq. A1.24), the variance can be written as

$$\mathrm{Var}(X) = E\{[X - E(X)]^2\}$$

$$= \int_{-\infty}^{\infty} [x - E(X)]^2 f(x)\, dx \qquad (A1.25)$$

The quantity $SD(X) = [\mathrm{Var}(X)]^{1/2}$ is known as the *standard deviation* of the random variable X. The ratio $SD(X)/E(X)$ is referred to as the *coefficient of variation* of X. The variance, the standard deviation, and the coefficient of variation are useful measures of the scatter (or dispersion) of the random variable about its mean.

The *correlation coefficient* of two continuous random variables X and Y is defined by the relation

$$\mathrm{Corr}(X, Y) = \frac{\int_{-\infty}^{\infty} \int_{-\infty}^{\infty} [x - E(X)][y - E(Y)] f(x, y)\, dx\, dy}{SD(X) SD(Y)} \qquad (A1.26)$$

The correlation coefficient is similarly defined if the variables are discrete. It can be shown that

$$-1 \leqslant \mathrm{Corr}(X, Y) \leqslant 1 \qquad (A1.27)$$

It can be easily shown that if the two random variables are linearly related, that is,

$$Y = a + bX \qquad (A1.28)$$

then

$$\mathrm{Corr}(X, Y) = \pm 1 \qquad (A1.29)$$

The sign in the right member of Eq. A1.29 is the same as that of the coefficient b in Eq. A1.28. It can be proved that, conversely, Eq. A1.29 implies Eq. A1.28. The correlation coefficient may thus be viewed as an index of the extent to which two variables are linearly related.

It is noted that if X and Y are independent, then $\mathrm{Corr}(X, Y)=0$. This follows immediately from Eqs. A1.26, A1.18, and A1.24. However, the relation $\mathrm{Corr}(X, Y)=0$ does not necessarily imply the independence of X and Y [A1-4].

A1.5   PROBABILITY DISTRIBUTIONS COMMONLY USED IN WIND ENGINEERING

The Geometric Distribution

Consider an experiment of the type known as *Bernoulli trials* in which (a) the only possible outcomes are the occurrence and the nonoccurrence of an event A, (b) the probability p of event A in any one trial is constant, and (c) the outcomes of the trials are independent of each other.

Let the random variable N be equal to the number of the trial in which the event A occurs for the first time. The probability $p(n)$ that event A will first occur on the nth trial is equal to the probability that event A will *not* occur on each of the first $n-1$ trials and *will* occur on the nth trial. Since the probability of nonoccurrence of event A in one trial is $1-p$ (Eq. A1.2) and since the n trials are independent, it follows from the multiplication rule (Eq. A1.8)

$$p(n)=(1-p)^{n-1}p \qquad (n=1, 2, 3, \ldots) \tag{A1.30}$$

This probability distribution is known as the *geometric distribution* with parameter p.

The probability $P(n)$ that event A will occur at least once in n trials can be found in the following manner. The probability that event A will *not* occur in n trials is $(1-p)^n$. The probability that it will occur at least once is therefore

$$P(n)=1-(1-p)^n \tag{A1.31}$$

The expected value of N is, by virtue of Eq. A1.22, in which Eq. A1.30 is used,

$$\bar{N}=\sum_{n=1}^{\infty} n(1-p)^{n-1}p \tag{A1.32}$$

The sum of this series can be shown to be

$$\bar{N}=1/p \tag{A1.33}$$

The quantity $\bar{N}$ is referred to as the *return period*, or the *mean recurrence interval*, of event A.

Examples

1.   For a die, the probability that a "four" occurs is $p=\frac{1}{6}$. If the total number of trials is large, it may be expected that, in the long run, a "four" will appear on the average once in $\bar{N}=1/\frac{1}{6}=6$ trials.

2.   A structure is designed so that the stresses in its members will attain the allowable stress under the action of extreme winds with a 50-year mean re-

currence interval. The probability of occurrence in any one year of winds for which $\bar{N} = 50$ is $p = 1/\bar{N} = 0.02$ (Eq. A1.33). The probability that the allowable stress will be attained at least once in n years is given by Eq. A1.31. For $n = 25$ years, $P(25) = 1 - (1 - 0.02)^{25} \simeq 0.396$; for $n = 50$ years, $P(50) \simeq 0.63$.

The Poisson Distribution

Consider a class of events, each of which may occur independently of the others and with equal likelihood at any time of an interval $0 \leqslant t \leqslant T$. A random variable is defined, which consists of the number N of events that will occur during an arbitrary time interval $\tau = t_2 - t_1$ $(t_1 \geqslant 0, t_1 < t_2 \leqslant T)$. Let $p(n, \tau)$ denote the probability that n events will occur during the interval τ. If it is assumed that $p(n, \tau)$ is not influenced by the occurrence of any number of events at times outside this interval, it can be proved [A1-4] that

$$p(n, \tau) = \frac{(\lambda \tau)^n}{n!} e^{-\lambda \tau} \qquad (n = 0, 1, 2, 3, \ldots) \tag{A1.34}$$

If Eqs. A1.24 and A1.25 are used, it is found that the expected value and the variance of n are both equal to $\lambda \tau$. Since $\lambda \tau$ is the expected number of events occurring during time τ, the parameter λ is called the *average rate of arrival* of the process and represents the expected number of events per unit of time.

The applicability of Poisson's distribution may be illustrated in connection with the question of the incidence of telephone calls in a telephone exchange [A1-5]. Consider an interval of, say, a quarter of an hour, during which the average rate of arrival of calls is constant. During any subinterval, the incidence of a number n of calls is as likely as during any other equal subinterval. In addition, it may be assumed that individual calls are independent of each other. Therefore, Eq. A1.34 applies to any time interval lying within the quarter of an hour.

Normal and Lognormal Distributions

Consider a random variable X which consists of a sum of small, independent contributions $X_1, X_2, \ldots, X_n$. It can be proved [A1-1] that, under very general conditions, if n is large the probability density function of X is

$$f(x) = \frac{1}{\sqrt{2\pi}\sigma_x} \exp\left(\frac{-(x - \mu_x)^2}{2\sigma_x^2}\right) \tag{A1.35}$$

where $\mu_x = E(X)$ and $\sigma_x^2 = \text{Var}(X)$ are the mean value and the variance of X, respectively. This statement is known as the *central limit theorem*. The distribution represented by Eq. A1.35 is called *normal* or *Gaussian*. It can be shown that the probability distribution of a linear function of a normally distributed variable is normal. Also, the sum of two or more independent normally distributed variables is normally distributed.

Normal distributions are used in a wide variety of physical and engineering applications, for example, the description of errors in measurements. At the same time, it should be carefully noted that many phenomena may not be normally distributed, for example, the extreme wind speeds occurring at any given geographical location.

If the distribution of the variable $Z = \log X$ is normal, the distribution of the variable X is said to be *lognormal*.

Type I and Type II Distributions of the Largest Values. Mean Recurrence Intervals

Let the variable X be the maximum of n independent random variables $Y_1, Y_2, \ldots, Y_n$ [A1-6]. Since the inequality $X \leqslant x$ implies $Y_i \leqslant x$ for *all* i ($i = 1, 2, \ldots, n$), it follows that

$$F(X \leqslant x) = \text{Prob}(Y_1 \leqslant x, Y_2 \leqslant x, \ldots, Y_n \leqslant x) \qquad (A1.36a)$$

$$= F_{Y_1}(x) F_{Y_2}(x) \ldots F_{Y_n}(x) \qquad (A1.36b)$$

where, to obtain Eq. A1.36b from Eq. A1.36a, the generalized form of Eq. A1.8 was used. The probabilities $F_{Y_i}(y)$ are referred to as the underlying (or the *initial*) distributions of the variables Y_i. The latter are said to constitute the *parent population* from which the largest values X have been extracted. In the particular case in which all the variables Y_i have the same probability distribution $F_Y(y)$, Eq. A1.36 becomes

$$F_X(x) = [F_Y(x)]^n \qquad (A1.37)$$

In the case in which they are unlimited to the right, the initial variables Y are said to have distributions of the *exponential type* if their cumulative distribution functions converge (with increasing y) toward unity at least as fast as an exponential function; the initial variables Y are said to be of the *Cauchy type* if

$$\lim_{y \to \infty} [1 - F(y)] y^k = A \qquad (A > 0; k > 0) \qquad (A1.38)$$

As the number n becomes very large, the distributions $F_X(x)$ of the largest values approach limits known as the *Type I* and the *Type II* distributions according as the initial distributions are of the exponential and of the Cauchy type, respectively [A1-4, A1-7].

The cumulative distribution function for the *Type I distribution of the largest values* (also referred to as the *Type I Extreme Value* distribution, or the *Gumbel distribution*) is

$$F_I(x) = \exp\{-\exp[-(x-\mu)/\sigma]\} \begin{cases} -\infty < x < \infty \\ -\infty < \mu < \infty \\ 0 < \sigma < \infty \end{cases} \qquad (A1.39)$$

In Eq. A1.39, μ and σ are referred to as the location and the scale parameter, respectively.* It can be shown, using Eqs. A1.24 and A1.25, that the mean

---

*As shown in Eqs. A1.40 and A1.41, these parameters are *not* the expectation and the standard deviation of X.

value and the standard deviation of X are

$$E(X) = \mu + 0.5772\sigma \tag{A1.40}$$

$$SD(X) = \frac{\pi}{\sqrt{6}}\sigma \tag{A1.41}$$

The cumulative distribution function for the *Type II distribution of the largest values* (also referred to as the *Type II Extreme Value* distribution, or the generalized *Fréchet* distribution) is

$$F_{II}(x) = \exp\{-[(x-\mu)/\sigma]^{-\gamma}\}\begin{cases} \mu < x < \infty \\ -\infty < \mu < \infty \\ 0 < \sigma < \infty \\ \gamma > 0 \end{cases} \tag{A1.42}$$

where μ, σ, and γ are the location, the scale, and the shape (or tail length) parameter, respectively [A1-8]. In the particular case $\mu = 0$, Eq. A1.42 is referred to as the Fréchet (as opposed to generalized Fréchet) distribution.

Equations A1.39 and A1.42 may be inverted to yield the so-called *percent point function*, that is, the value x of the random variable that corresponds to any given value of the cumulative distribution function. In the case of the Type I distribution

$$x(F_I) = \mu - \sigma \ln(-\ln F_I) \tag{A1.43}$$

whereas for the Type II distribution

$$x(F_{II}) = \mu + \sigma(-\ln F_{II})^{-1/\gamma} \tag{A1.44}$$

It is convenient to denote the cumulative distribution function value F_I or F_{II} by p and $x(F_I)$ or $x(F_{II})$ by $G_X(p)$. Then, for the Type I distribution

$$G_X(p) = \mu - \sigma \ln(-\ln p) \tag{A1.43a}$$

and for the Type II distribution

$$G_X(p) = \mu + \sigma(-\ln p)^{-1/\gamma} \tag{A1.44a}$$

From the definition of p and Eq. A1.2 it follows that $\text{Prob}(X > x) = 1 - p$. Let the random variable X represent the extreme annual wind speed at some given location. Each year may then be viewed as a trial in which the event that the wind speed X will exceed some value x has the probability of occurrence $1 - p$. By virtue of Eq. A1.33, the mean recurrence interval of this event is

$$\bar{N} = \frac{1}{1-p} \tag{A1.45a}$$

Thus, the wind speed x corresponding to a mean recurrence interval $\bar{N}$ is equal to the value of the percent point function of X corresponding to

$$p = 1 - \frac{1}{\bar{N}} \tag{A1.45b}$$

Relations Between Type I and Type II Extreme Value Distributions

Let the Type II distribution be written as

$$F_{II}(y) = \exp\{-[(y - \mu_{II})/\sigma_{II}]^{-\gamma}\} \qquad (A1.46)$$

(In the present context it is convenient to denote the location and scale parameter of the Type II distribution by μ_{II} and σ_{II}, respectively). If the transformation

$$y - \mu_{II} = \exp x \qquad (A1.47)$$

is applied to Eq. A1.46, the expression obtained is a Type I distribution with parameters

$$\mu = \ln \sigma_{II} \qquad (A1.48)$$

$$\sigma = \frac{1}{\gamma} \qquad (A1.49)$$

It is now shown [A1-12] that as γ approaches infinity, a Type II distribution approaches a Type I distribution.

Consider the distribution of the standardized variate

$$Z = \frac{X - \text{loc}(X)}{\text{scale}(X)} \qquad (A1.50)$$

where $\text{loc}(X)$ and scale (X) are measures of location and scale, respectively, of the distribution of X. Examples of measures of location of a random variable X are its expected value $E(X)$ and its median $G_X(0.5)$. Examples of measures of scale of a random variable X are its standard deviation $SD(X)$, its interquartile difference $\delta_{50} = G_X(0.75) - G_X(0.25)$, and its 95% difference $\delta_{95} = G_X(0.975) - G_X(0.025)$.

The percent point function $G_Z(p)$ is given by

$$G_Z(p) = \frac{G_X(p) - \text{loc}(X)}{\text{scale}(X)} \qquad (0 < p < 1) \qquad (A1.51)$$

With no loss of generality, a reduced variate with $\mu = 0$ and $\sigma = 1$ may be used in the demonstration. Substituting Eq. A1.44a with $\mu = 0$ and $\sigma = 1$ into Eq. A1.51 and choosing, for simplicity, $\text{loc}(X) = G_X(0.5)$ and scale $(X) = \delta_{95}$,

$$G_Z(p) = \frac{[-\ln(p)]^{-1/\gamma} - [-\ln(0.5)]^{-1/\gamma}}{[-\ln(0.975)]^{-1/\gamma} - [-\ln(0.025)]^{-1/\gamma}} \qquad (0 < p < 1) \qquad (A1.52)$$

As $\gamma \to \infty$, this expression becomes indeterminate. However, application of L'Hospital's rule yields, after simplification,

$$G_Z(p) = \frac{-\ln[-\ln(p)] - \{-\ln[-\ln(0.5)]\}}{-\ln[-\ln(0.975)] - \{-\ln[-\ln(0.025)]\}} \qquad (0 < p < 1) \qquad (A1.53)$$

As can be seen from Eqs. A1.43a, the terms in the numerator and denominator of Eq. A1.53 are, respectively, the percent point function, the median, and the 95% difference of the reduced variate for the Type I distribution. It has thus been demonstrated that, as γ approaches infinity, a standardized Type II variate approaches a standardized Type I variate and, hence, a Type II distribution asymptotically approaches a Type I distribution.

Type I Distributions: Mode of the Largest Value from a Sample of Size n as an Approximation of the Percent Point Function $G_X[1/(1-n)]$

Let Z be the largest of a set of n values of a random variable X, each of which has a Type I Extreme Value distribution (Eq. A1.39). The cumulative distribution function of this largest value is

$$F_n(z)=[F_1(z)]^n=\exp[-n\exp(-w)] \tag{A1.54}$$

where

$$w=\frac{z-\mu}{\sigma} \quad \begin{array}{l} -\infty<z<\infty \\ -\infty<\mu<\infty \\ 0<\sigma<\infty \end{array} \tag{A1.55}$$

The corresponding probability density function is

$$f_n(z)=\frac{1}{\sigma}\,n\exp[-ne^{-w}-w] \tag{A1.56}$$

The root of the equation

$$\frac{df_n(z)}{dz}=\frac{1}{\sigma^2}\,n\exp[-ne^{-w}-w][ne^{-w}-1]=0 \tag{A1.57}$$

is, by definition, the mode* of the largest of the set of n values considered. From Eq. A1.57 it follows immediately

$$e^{-w}=\frac{1}{n} \tag{A1.58}$$

or, if Eq. A1.55 is used,

$$\text{mode}(Z)=\mu-\sigma\ln\frac{1}{n} \tag{A1.59}$$

Consider now the initial random variable X. Since X has a Type I distribution, its percent point function is

$$G_X(p)=\mu-\sigma\ln(-\ln p) \tag{A1.43a}$$

or, making use of Eq. A1.45 in which $\bar{N}$ is the mean recurrence interval

$$G_X\left(1-\frac{1}{\bar{N}}\right)=\mu-\sigma\ln\left[-\ln\left(1-\frac{1}{\bar{N}}\right)\right] \tag{A1.60}$$

---

*It is recalled that the mode of a variable X is the value of that variable most likely to occur in any given trial (Sect. A1.4).

In the particular case in which $\bar{N} = n$

$$G_X\left(1 - \frac{1}{n}\right) = \mu - \sigma \ln\left[-\ln\left(1 - \frac{1}{n}\right)\right] \qquad \text{(A1.61)}$$

In Eq. A1.61, $G_X(1 - 1/n)$ is the value of X corresponding to the mean recurrence interval n.

It can be verified that for n sufficiently large, say, $n > 10$,

$$\ln\left[-\ln\left(1 - \frac{1}{n}\right)\right] \simeq \ln\left(\frac{1}{n}\right) \qquad \text{(A1.62)}$$

(For example, for $n = 20$, the right and left members of Eq. A1.63 are equal to -2.970 and -2.996, respectively. For $n = 40$, they are equal to -3.676 and -3.689, respectively.) It follows therefore that

$$G_X\left(1 - \frac{1}{n}\right) \simeq \mu - \sigma \ln\frac{1}{n} = \text{mode}(Z) \qquad \text{(A1.63)}$$

Equation A1.63 shows that if X is a random variable with a Type I distribution, the mode of the largest value in a sample of n values of X is very nearly equal to the value of the random variable corresponding to the mean recurrence interval n [A1-9].

An interesting experimental verification of this statement is provided by the data of [A1-11], which cover a period of 37 years. For example, for the first five sets of [A1-11], the values of the largest of the maximum yearly wind speeds recorded in 37 years, v_{max}, and the values of the estimated 37-year wind, v_{37}, are (in mph)

	Cairo (Ill.)	Alpena (Mich.)	Tatoosh Isl. (Wash.)	Williston (N.D.)	Richmond (Virginia)
v_{max}	51	50	84	50	48
v_{37}	52	51	81	52	50

The probability that the largest of a set of n values of the random variable X with a Type I distribution is contained in a given interval can be easily calculated using Eq. A1.54. For example, from a 37-year record of the largest annual wind speeds at Richmond, Virginia the values of μ and σ were estimated to be 36.8 mph and 3.78 mph, respectively [A1-11]. Using these values, the probability that the largest wind speed $Z = V_{max}$ in a set of $n = 37$ largest annual speeds is contained, say, in the interval $V_{37}(1 \pm 0.24) = 50 \pm 12$ can be estimated as follows:

$$P(38 \leqslant Z \leqslant 62) = \int_{38}^{62} f_{37}(z)\,dz = F_{37}(62) - F_{37}(38) = 0.95 \qquad \text{(A1.63a)}$$

Joint Extreme Value Distributions

The joint Type I Extreme Value probability distribution of two correlated variables X, Y has the expression

$$F_{XY}(x, y) = \exp\left\{-\left[\exp\left(-m\frac{x-\mu_x}{\sigma_x}\right) + \exp\left(-m\frac{y-\mu_y}{\sigma_y}\right)\right]^{1/m}\right\} \quad \text{(A1.64a)}$$

where

$$m = (1 - \rho_{XY})^{-1/2} \quad \text{(A1.64b)}$$

and the correlation coefficient $\rho_{XY} \geqslant 0$ [A1-23]. It can be verified that for probabilities of interest in structural reliability calculations (e.g., $F_{XY}(x, y) > 0.99$) and for values $\rho_{XY} \leqslant 0.7$, say,

$$F_{XY}(x, y) \cong F_X(x)F_Y(y) \quad \text{(A1.64c)}$$

where $F_X(x)$ and $F_Y(y)$ are the Type I Extreme Value distributions of X and Y, respectively, that is, it may be assumed that X and Y are statistically independent.

The Weibull Distribution

The Weibull cumulative distribution function is

$$F(x) = 1 - \exp\left[-\left(\frac{x-\mu}{\sigma}\right)^{\gamma}\right] \quad \text{(A1.65)}$$

The expected value and the standard deviation of the variate $(x - \mu)/\sigma$ are, respectively, $\Gamma(1/\gamma + 1)$ and $\{\Gamma(2/\gamma + 1) - [\Gamma(1/\gamma + 1)]^2\}^{1/2}$, where Γ is the gamma function, and are listed here for various values of γ [A1-8].

γ	1.2	1.6	2.0	2.2	2.6	3.0	3.2	3.6	4.0	6.0
Expected Value	0.9407	0.8966	0.8862	0.8856	0.8882	0.8930	0.8957	0.9011	0.9064	0.9264
Standard Deviation	0.7872	0.5737	0.4632	0.4249	0.3670	0.3245	0.3072	0.2780	0.2543	0.1850

For $\gamma \cong 3.6$, the shape of the Weibull distribution is similar to that of the normal distribution. The Weibull distribution with parameter $\gamma = 2$ is commonly referred to as the Rayleigh distribution.

A1.6   PROBABILITY THEORY AND STATISTICAL DATA

Goodness of Fit

Data obtained—or that may be obtained—from actual observations may be viewed as observed values of random variables. The behavior of the data is then assumed to be described by models governing the behavior of random variables, that is, by such mathematical models as are used in probability theory.

In practical applications two important problems must be dealt with. First, from the nature of the phenomenon being investigated (or on the basis of observations), an inference must be made on the probability distribution that will adequately describe the behavior of the data. Second, the data must be used for drawing inferences on the parameters of the distribution or on some of its characteristics, for example, the mean or the standard deviation.

In practice, given a set of observed data, or a *data sample*, it is hypothesized that its behavior can be modeled by means of some probability distribution believed to be appropriate. This hypothesis must then be tested. Tests incorporate quantitative measures of the degree of agreement, or *goodness of fit*, between the data and the hypothetical distribution or, conversely, of the degree to which the data deviate from that distribution. If the measure of this deviation is appropriately small, then the hypothesis will be accepted, and vice-versa. Associated with the testing of a hypothesis is a *level of significance*, that represents the probability of rejecting the hypothesis when it is in fact true. Tests commonly used in applications, including the well-known χ^2 test, are discussed, for example, in [A1-1] and [A1-4]. Brief mention is made of the probability plot correlation coefficient test [A1-10] that has been used in the study of the behavior of extreme winds [A1-11, A1-12]. The probability plot correlation coefficient is defined as

$$r_D = \frac{\sum (X_i - \bar{X})[M_i(D) - \overline{M(D)}]}{[\sum (X_i - \bar{X})^2 \sum (M_i(D) - \overline{M(D)})^2]^{1/2}} \tag{A1.66}$$

in which $\bar{X} = \sum X_i/n$, $\overline{M(D)} = \sum M_i(D)/n$, n is the sample size, and D is the probability distribution being tested. The quantities X_i are obtained by a rearrangement of the data set: X_1 is the smallest, X_2 the second smallest, ..., X_i the i-th smallest of the observations in the set. The quantities $M_i(D)$ are obtained as follows. Given a random variable X with probability distribution D and given a sample size n, it is possible from probabilistic considerations to derive mathematically the distributions of the smallest, second smallest, and, in general, the i-th smallest values of X in that sample. The quantities $M_i(D)$ are the medians of each of these distributions.

If the data were generated by the distribution D, then, aside from a location and scale factor, X_i will be approximately equal to the theoretical values $M_i(D)$ for all i so that the plot of X_i versus $M_i(D)$ (referred to as probability plot) will be approximately linear. This linearity will, in turn, result in a near-unity value of r_D. Thus, the better fit of the distribution D to the data the closer r_D will be to unity.

To test whether the behavior of a given set of extreme data is better described by a Type I distribution or by a Type II distribution with some unknown value of the tail length parameter γ, the probability plot correlation coefficient r_D is computed for a large number of extreme value distributions, defined by various values of γ suitably spaced from $\gamma = 1$ to $\gamma = \infty$ (it is recalled that $\gamma = \infty$ corresponds to a Type I distribution). The variable in these distributions is written in standardized form so that for any given set of data the coefficients r_D depend solely upon γ, that is, are independent of the location and scale parameters μ and σ on which, therefore, no prior assumptions need to be made [A1-11]. The distribution that best fits the data is that which corresponds to the largest of the calculated values of r_D.

Estimation of Distribution Parameters

From the data of a sample it is, in principle, possible to make inferences on the parameters of the distribution that describes the behavior of the population from which the data are extracted (or on characteristics of the distribution, e.g., the mean). An *estimator* may be defined as a function $\hat{\alpha}(X_1, X_2, \ldots, X_n)$ of the sample values such that $\hat{\alpha}$ is a reasonable approximation to the unknown value α of the distribution parameter (or characteristic). The particular numerical value assumed by an estimator in a given case is referred to as an *estimate*. As a function of random variables, $\hat{\alpha}(X_1, X_2, \ldots, X_n)$ is itself a random variable. This is illustrated by the following example.

Consider the observed sequence of 14 outcomes of an experiment consisting of the tossing of a coin:

$$H\ T\ T\ T\ H\ T\ H\ H\ T\ H\ H\ H\ T\ H \qquad (A1.67a)$$

The random numbers associated with this experiment are the numbers zero and one, which are assigned to the outcome heads and to the outcome tails, respectively. The data sample corresponding to the observed outcomes is then:

$$0, 1, 1, 1, 0, 1, 0, 0, 1, 0, 0, 0, 1, 0 \qquad (A1.67b)$$

This sample is assumed to be extracted from an infinite population that, in the case of an ideally fair coin, will have a mean value, denoted in this case by α, equal to $\frac{1}{2}$. A reasonable estimator for the mean α is the sample mean $\hat{\alpha}$*

$$\hat{\alpha} = \frac{1}{n} \sum_{i=1}^{n} X_i \qquad (A1.68)$$

where n is the sample size (number of observations) and X_i are the observed data. In the case of the sample consisting of all 14 observations in A1.67b, $\hat{\alpha} = \frac{3}{7}$. If the samples consisting of the first seven and of the last seven observations in A1.67b are used, $\hat{\alpha} = \frac{4}{7}$ and $\hat{\alpha} = \frac{2}{7}$, respectively.

As a random variable, an estimator $\hat{\alpha}$ will have a certain probability distribution with a nonzero dispersion about the true value α. Thus, given a sample of statistical data, it is not possible to calculate the true value α of the parameter sought. Rather, *confidence intervals* can be estimated of which it can be stated, with a specified confidence level q (level of significance $1 - q$), that they will contain the unknown value α.

In order that the confidence interval corresponding to a given confidence level q be as narrow as possible, it is desirable that the estimator used be *efficient*.

---

*The symbol $\hat{\ }$ is used to denote estimated value.

Of two different possible estimators $\hat{\alpha}_1$ and $\hat{\alpha}_2$ of the same parameter α, the estimator $\hat{\alpha}_1$ is said to be more efficient if $E[(\hat{\alpha}_1 - \alpha)^2] < E[(\hat{\alpha}_2 - \alpha)^2]$.

Details on procedures for estimating distribution parameters can be found, for example, in [A1-1] and [A1-4] (see also [A1-17] and [A1-22]). The question of parameter estimation for the Type I Extreme Value distribution—which is widely used in the study of extreme wind speeds—will be examined subsequently in this appendix. Before proceeding to this topic it is useful to discuss first the simulation of the behavior of a Type I Extreme Value distribution by means of numerical techniques commonly referred to as Monte Carlo methods.

Monte Carlo Methods. Simulation of a Type I Extreme Value Process

As defined in [A1-13], Monte Carlo methods comprise that branch of experimental mathematics that is concerned with experiments on random numbers. The simulation of the phenomenon of interest is achieved by subjecting available sequences of random numbers to appropriate transformations. The new sequences thus obtained may be viewed as data, the sample statistics of which are representative of the statistical properties of the phenomenon concerned. Examples of engineering applications of Monte Carlo methods can be found in [A1-4] and [A1-14].

The simulation of the behavior of a random variable with a given distribution is a simple application of Monte Carlo techniques that is now discussed. It is assumed that the distribution is Extreme Value Type I with given parameters μ and σ (Eq. A1.39).

Consider a sequence of n uniformly distributed random numbers $0 < Y_i < 1$ ($i = 1, 2, \ldots, n$) such as are listed in [A1-14] or as may be generated by procedures discussed in [A1-2], [A1-13], or [A1-14]. These numbers are viewed as probabilities of occurrence of the data $X(Y_i)$ obtained by the following transformation (Eq. A1.43):

$$X(Y_i) = \mu - \sigma \ln(-\ln Y_i) \qquad (A1.69)$$

From the sample of size n, $X(Y_i)$ ($i = 1, 2, \ldots, n$), it is possible to obtain estimates of μ and σ (i.e., the distribution parameters) and of $G_X(p)$ (the percent point function corresponding to any given value of p, see Eq. A1.43a). Since, as was previously indicated, the estimates are random variables, the estimates will differ, in general, from the known parameters and percent point function of the underlying distribution. The procedure just described can be repeated a large number M of times. Then M sets of values $\hat{\mu}$, $\hat{\sigma}$, and $\hat{G}_X(p)$ and corresponding histograms can be obtained. From those sets it is possible to calculate summary statistics (such as the mean, the variance, the standard deviation) for $\hat{\mu}$, $\hat{\sigma}$, and $\hat{G}_X(p)$.

A Monte Carlo study of the behavior of a random variable with a Type I distribution conducted for the purpose of predicting extreme wind speeds was first reported in [A1-15]. A similar study, subsequently conducted by the writers, is now summarized. The parameter values of Eq. A1.69 used in this study were $\mu = 36.8$ and $\sigma = 3.78$ (these values in mph represent estimates of

Type I distributions found in [A1-11] to best fit the annual extreme wind speeds recorded in Richmond, Virginia between 1912 and 1948). Two sets of 100 samples each were generated, the size of the samples being $n = 25$ for the first set and $n = 50$ for the second. The main results of the study are listed in Table A1.1. For example: $G_X(0.98) = 51.57$ (calculated from the underlying distribution with parameters $\mu = 36.8$, $\sigma = 3.78$); the mean of the 100 estimates $\hat{G}_X(0.98)$ based on the samples of size $n = 25$ is Mean$[\hat{G}_X(0.98)] = 52.58$; the standard deviation of these estimates is $s[\hat{G}_X(0.98)] = 3.46$; the largest of the estimated $\hat{G}_X(0.98)$ is max$[\hat{G}_X(0.98)] = (1 + 16.6/100) \times \{\text{Mean}[\hat{G}_X(0.98)]\} = \text{Mean}[\hat{G}_X(0.98)] + 2.5s[\hat{G}_X(0.98)]$. A histogram of the estimates $\hat{G}_X(0.999)$ for the 100 samples of size $n = 50$ is shown in Fig. A1.5.

The results of Table A1.1 were obtained by fitting a Type I Extreme Value distribution to the data samples generated from sequences of random numbers by Eq. A1.69. However, it is conceivable that, because of the random character of the sampling, the behavior of some of the samples would be better described by Type II Extreme Value distributions rather than by a Type I distribution. To verify whether this is indeed the case, the probability plot correlation coefficient test was applied to each of the samples. The results obtained, which are independent of the parameters μ and σ of the underlying distribution, are shown in Table A1.2.

As shown in Sect.3.2, percentages such as those of Table A1.2 can be compared to similar percentages obtained from the analysis of measured extreme wind speed data in an attempt to draw inferences on the applicability of the Type I distribution to the modeling of extreme wind behavior in certain types of climate. For details on such inferences, see [A1-21].

TABLE A1.1.   Monte Carlo Simulation of a Type I Extreme Value Process

		μ	σ	$G_X(0.98)$	$G_X(0.99)$	$G_X(0.999)$
Original (Underlying) Distribution		36.80	3.78	51.67	54.24	62.97
Mean[a]	$n = 25$	36.90	4.01	52.58	55.38	64.64
	$n = 50$	36.80	3.89	51.92	54.63	63.60
Standard Deviation[a]	$n = 25$	0.86	0.81	3.46	4.00	5.85
	$n = 50$	0.65	0.50	2.14	2.49	3.61
Maximum Deviation Below	$n = 25$	5.90	52.00	19.60	21.30	25.70
Mean[a] (Percent of Mean)	$n = 50$	3.80	32.00	10.20	11.00	13.60
Maximum Deviation Above	$n = 25$	6.60	64.00	16.60	19.00	25.50
Mean[a] (Percent of Mean)	$n = 50$	4.00	32.00	10.70	12.00	14.70
Maximum Deviation Below	$n = 25$	2.50	2.60	3.00	3.00	2.80
Mean[a] (Standard Deviations)	$n = 50$	2.20	2.50	2.50	2.50	2.40
Maximum Deviation Above	$n = 25$	2.80	3.20	2.50	2.60	2.80
Mean[a] (Standard Deviations)	$n = 50$	2.30	2.50	2.60	2.60	2.60

[a]Estimated from 100 samples of size n.

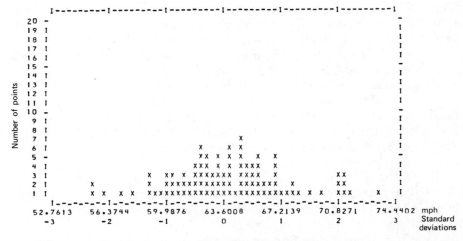

FIGURE A1.5.   Histogram of estimated values $\hat{G}_X$ (0.999) for 100 samples of size $n = 50$.

TABLE A1.2.   Percentage of Samples from a Population with a Type I Distribution that are Best Fit by Type I and Type II Distributions

Extreme Value Distribution	Sample Size	
	$n = 25$	$n = 50$
Type I or Type II ($\gamma \geqslant 13$)	57	77
Type II $7 \leqslant \gamma < 13$	13	12
$2 \leqslant \gamma < 7$	30	11

Estimators for the Type I Extreme Value Distribution

A classical method of approaching the problem of estimation is the *method of moments*. In this method it is assumed that the distribution parameters can be obtained by replacing the expectation and the mean square value of the random variable X by the corresponding statistics of the sample. In the case of the Type I distribution, using Eqs. A1.40 and A1.41,

$$\hat{\sigma} = \frac{\sqrt{6}}{\pi} s \tag{A1.70}$$

$$\hat{\mu} = \bar{X} - 0.5772\hat{\sigma} \tag{A1.71}$$

where $\bar{X}$ and s are the sample mean and the sample standard deviation, respectively, that is,

$$\bar{X} = \frac{1}{n} \sum X_i \tag{A1.72}$$

$$s = \left[\frac{1}{n} \sum (X_i - \bar{X})^2 \right]^{1/2} \tag{A1.73}$$

From the estimators (A1.70) and (A1.71) the following estimator of $G_X(p)$ can be constructed:

$$\hat{G}_X(p) = \bar{X} + s(y - 0.5772)\sqrt{6}/\pi \tag{A1.74}$$

where

$$y = -\ln(-\ln p) \tag{A1.75}$$

Under the assumption that the random variables $\bar{X}$ and s defined by Eqs. A1.72 and A1.73 are, asymptotically, normally distributed, it can be shown [A1-7, pp. 10, 174, and 228] that for large samples of size n

$$SD[\hat{G}_X(p)] = \left[\frac{\pi^2}{6} + 1.1396(y - 0.5772)\frac{\pi}{\sqrt{6}} + 1.1(y - 0.5772)^2 \right]^{1/2} \frac{\sigma}{\sqrt{n}} \tag{A1.76}$$

A more efficient estimator of $G_X(p)$ has been developed by Lieblein on the basis of the *method of order statistics* [A1-7, A1-16, A1-17]. A method frequently used in applications is based on *least squares fitting* of a straight line to the data on probability paper. This method is used in the computer program of [A1-11]. A simplified approximate version of this method is presented in [A1-7, pp. 34, 227, and 228]. For a discussion of other estimation methods used for the Type I distribution, for example, the maximum likelihood method, the reader is referred to [A1-7] and [A1-18].

It can be shown that the standard deviation of any estimator of a parameter is larger than, or at least equal to, a theoretically specified standard deviation known as the *Cramér–Rao lower bound*. In the case of the percent point function of a Type I distribution, the Cramér–Rao lower bound is

$$SD_{CR}[\hat{G}_X(p)] = (0.60793y^2 + 0.51404y + 1.10866)^{1/2} \frac{\sigma}{\sqrt{n}} \tag{A1.77}$$

where y is given by Eq. A1.75 [A1-19]. For $n = 25$ and $n = 50$, the ratio $(1/\sigma)SD_{CR}[\hat{G}_X(p)]$ is now compared with the ratios $(1/\sigma)SD[\hat{G}_X(p)]$, where $SD[\hat{G}_X(p)]$ denotes the standard deviation of the percent point function estimated by the method of moments, by Lieblein's method of order statistics, and by the method of least squares fitting.

In Table A1.3 the quantities of line (1) were calculated by Eq. A1.76. The quantities of line (2) were obtained from [A1-16, p. 131] (through multiplication of corresponding quantities given for $n = 10$ by $\sqrt{10/25}$ and $\sqrt{10/50}$ or of quantities given for $n = 20$ by $\sqrt{20/25}$ and $\sqrt{20/50}$). The quantities of line (3) were

TABLE A1.3. Ratios $(1/\sigma)SD[\hat{G}_X(p)]$ and $(1/\sigma)SD_{CR}[\hat{G}_X(p)]$

	Estimation Method	$\bar{N}$		$n=25$				$n=50$		
			20	50	100	1000	20	50	100	1000
(1)	Moments		0.65	1.02	1.13	1.45	0.46	0.72	0.80	1.03
(2)	Order Statistics (Lieblein)		0.62		0.90	1.32	0.43		0.64	0.93
(3)	Least Squares			0.92	1.06	1.55		0.57	0.66	0.96
(4)	(Cramér–Rao Lower Bound)		0.57	0.70	0.81	1.16	0.40	0.49	0.57	0.82

Row label (1)–(3): $\dfrac{1}{\sigma}SD[\hat{G}_X(p)]$

Row label (4): $(1/\sigma)SD_{CR}[\hat{G}_X(p)]$

obtained from Table A1.1* (as shown in [A1-11], these quantities are independent of the parameters μ and σ used in the calculations). Finally, the quantities of line (4) were calculated by Eq. A1.77.

Assume that $\hat{G}_X(p)$ is normally distributed. The approximate statement can then be made that the interval $\hat{G}_X(p) \pm SD[\hat{G}_X(p)]$ will contain the true unknown parameter $G_X(p)$ in about 68% of the cases. This interval (referred to as the 68% confidence interval) is said to correspond to the 68% *confidence level*. For the interval $\hat{G}_X(p) \pm 2SD[\hat{G}_X(p)]$ the percentage rises to 95%, while for the interval $\hat{G}_X(p) \pm 3SD[\hat{G}_X(p)]$ it rises to over 99% (99.7%). As noted above, these percentages should be viewed as only approximate; however, the approximation is satisfactory for reasonable sample sizes such as are used in the analysis of wind speed data.

Estimation Methods and Reliability of Extreme Wind Speed Predictions

It is of interest to examine the effect of the estimation methods upon the reliability of predictions of extreme wind speeds corresponding to mean recurrence intervals used in structural engineering calculations.† Consider, for example, the case $n = 25$. The 68% confidence interval for the 100-year wind, $x_{100} = G_X(0.99)$, is $\hat{G}_X(0.99) \pm SD[\hat{G}_X(p)]$. If the most reliable method of estimation of Table A1.3—the order statistics method—is used, then the interval is $\hat{G}_X(0.99) \pm 0.90\sigma \simeq \hat{G}_X(0.99) \pm 0.7s$ (Eq. A1.70). If, on the other hand, the least reliable method of Table A1.3—the method of moments—is used, then the estimated interval is $\hat{G}_X(0.99) \pm 0.88s$.

Numerous analyses of wind records show that the ratios $s/\bar{X}$ are of the order of 0.07 to 0.15 [A1-11, A1-15]. Then the 68% confidence intervals obtained by the method of order statistics and by the method of moments are (using the ratio $s/\bar{X} = 0.12$) $\hat{G}_X(0.99)[1 + 0.061]$ and $\hat{G}_X(0.99)[1 + 0.077]$, respectively. The difference between the respective reliabilities of the estimates of the values of X corresponding to $p = 0.99$ (or, in virtue of Eq. A1.45, to a mean recurrence interval $\bar{N} = 100$ years) is seen to be quite small, that is, of the order of 2%. Results of similar calculations carried out for $p = 0.95$, $p = 0.99$, $p = 0.999$; $n = 25$, $n = 50$; and $s/\bar{X} = 0.12$, are shown in Table A1.4. The differences between the reliabilities of the various procedures can be verified to be negligible also for $s/\bar{X} = 0.07$ and $s/\bar{X} = 0.15$.

It is seen from Table A1.4 that any of the methods listed will provide an acceptable estimate of the order of magnitude of the 68% confidence limits. The width of the 95% confidence limits is approximately twice the width of the 68% limits; for example, for $\bar{N} = 20$ and $n = 25$, the nondimensionalized 95% confidence limit estimated by the method of moments is 1 ± 0.098. The dif-

---

*The standard deviation of $\hat{G}_X(p)$ in line (3) is an estimate based on a finite sample. In accordance with the convention adopted herein, the notation s rather than SD should therefore be used for the quantities of line (3). This was not done in Table A1.3 for the sake of clarity.

†Of two different possible estimators $\hat{\alpha}_1$ and $\hat{\alpha}_2$ of the same quantity α, the estimator $\hat{\alpha}_1$ is said to be more reliable than $\hat{\alpha}_2$ if (assuming the estimators to be unbiased) $SD(\hat{\alpha}_1) < SD(\hat{\alpha}_2)$ [A1-16].

TABLE A1.4. 68% Confidence Intervals[a] Based on Various Estimation Methods and on the Cramér–Rao Lower Bounds

$\dfrac{p}{N}$	$n=25$				$n=50$			
	0.95 20	0.98 50	0.99 100	0.999 1000	0.95 20	0.98 50	0.99 100	0.999 1000
Method of Moments	1 ± 0.049	1 ± 0.073	1 ± 0.077	1 ± 0.085	$1+0.035$	1 ± 0.052	1 ± 0.055	1 ± 0.060
Method of Order Statistics (Lieblein)	1 ± 0.047		1 ± 0.061	1 ± 0.078	1 ± 0.033	1 ± 0.044	1 ± 0.044	1 ± 0.056
Least Squares Method		1 ± 0.066	1 ± 0.072	1 ± 0.091		1 ± 0.047	1 ± 0.051	1 ± 0.065
Cramér–Rao Lower Bound	1 ± 0.043	1 ± 0.055	1 ± 0.068	1 ± 0.068	1 ± 0.031	1 ± 0.036	1 ± 0.030	1 ± 0.049

[a]Nondimensionalized with respect to $\hat{G}_x[1-1/\bar{N}]$.

ferences between estimates based on various procedures are seen to remain acceptably small for the 95% confidence limits as well.

It has previously been shown (Eq. A1.64) that if X is a random variable with a Type I distribution, it is possible to view the largest value in a sample of n values of X as an estimator of the value of X corresponding to a mean recurrence interval n. While this estimator has the obvious advantage of extreme simplicity, its reliability is relatively poor. This can be shown by the following example. If Eq. A1.76 is used to estimate the 95% confidence interval for the 37-year wind speed at Richmond, Virginia ($\hat{\mu} = 36.8$ mph, $\hat{\sigma} = 3.78$ mph; see [A1-11]), the interval obtained is (50 ± 5) mph. Using the largest value in a set of 37 values as an estimator of the 37-year wind, the estimated 95% confidence limit interval obtained is (50 ± 12) mph (see Eq. A1.63a), that is, an interval more than twice as wide as the interval estimated by the method of moments.

REFERENCES

A1-1   H. Cramér, *The Elements of Probability Theory*, Wiley, New York, 1955.

A1-2   A. G. Mihram, *Simulation*, Academic, New York, 1972.

A1-3   R. von Mises, *Probability, Statistics and Truth*, Allen and Unwin, London, and Macmillan, New York, 1957.

A1-4   J. R. Benjamin and A. C. Cornell, *Probability, Statistics and Decision for Civil Engineers*, McGraw-Hill, New York, 1970.

A1-5   T. C. Fry, *Probability and Its Engineering Uses*. Van Nostrand, Princeton, 1965.

A1-6   B. Epstein, "Elements of the Theory of Extreme Values," *Technometrics*, **2**, 1 (Feb. 1960), 27–41.

A1-7   E. J. Gumbel, *Statistics of Extremes*, Columbia Univ. Press, New York, 1958.

A1-8   N. L. Johnson and S. Kotz, *Continuous Univariate Distributions*, Vol. 1, Wiley, New York, 1970.

A1-9   E. Simiu and B. R. Ellingwood, "Code Calibration of Extreme Wind Return Periods," Technical Note, *J. Struct. Div.*, ASCE, **103**, No. ST3 (March 1977), 725–729.

A1-10   J. J. Filliben, "The Probability Plot Correlation Coefficient Test For Normality," *Technometrics*, **17**; 1 (Feb. 1975), 111–117.

A1-11   E. Simiu and J. J. Filliben, *Statistical Analysis of Extreme Winds*, Technical Note 868, National Bureau of Standards, Washington, D.C., 1975.

A1-12   E. Simiu and J. J. Filliben, "Probability Distributions of Extreme Wind Speeds," *J. Struct. Div.*, ASCE, **102**, No. ST9, Proc. Paper 12381 (Sept. 1976), 1861–1877.

A1-13   J. M. Hammersley and D. C. Handscomb, *Monte Carlo Methods*, Methuen, London, and Wiley, New York, 1965.

A1-14   J. H. Mize and J. G. Cox, *Essentials of Simulation*, Prentice-Hall, Englewood Cliffs, N.J., 1968.

A1-15   P. Duchêne–Marullaz, "Etude des vitesses maximales annuelles du vent," *Cahiers du Centre Scientifique et Technique du Bâtiment*, No. 131, Cahier 1118, Paris, 1972.

A1-16   J. Lieblein, *A New Method of Analyzing Extreme-Value Data*, National Bureau of Standards Report No. 2190, Washington, D.C., 1953.

A1-17   J. Lieblein, *Efficient Methods of Extreme-Value Methodology*, Report No. NBSIR 74-602, National Bureau of Standards, Washington, D.C., 1974.

A1-18   J. Tiago de Oliveira, "Statistics for Gumbel and Fréchet Distributions" in *Structural Safety and Reliability*, A. Freudenthal (Ed.), Pergamon, Oxford and New York, 1972, pp. 91–105.

A1-19   F. Downton, "Linear Estimates of Parameters in the Extreme Value Distribution," *Technometrics*, **8**, 1 (Feb. 1966), 3–17.

A1-20   G. I. Schuëller and H. Panggabean, "Probabilistic determination of design wind velocity in Germany," in *Proc. Inst. Civ. Eng.*, **61**, Part 2 (1976), 673–683.

A1-21   E. Simiu, J. Biétry and J. J. Filliben, "Sampling Errors in the Estimation of Extreme Winds," *J. Struct. Div.*, ASCE, **104** (1978), 491–501.

A1-22   I. I. Gringorten, "Envelopes for Ordered Observations Applied to Meteorological Extremes," *J. Geophys. Res.*, **68** (1976), 815–826.

A1-23   N. L. Johnson and S. Kotz, *Distributions in Statistics: Continuous Multivariate Distributions*, Wiley, New York, 1972.

A1-24   A. W. Marshall and I. Olkin, "Domains of Attraction of Multivariate Extreme Value Distributions," *The Annals of Probability*, **11** (1983), 168–177.

APPENDIX A2

Random Processes

Consider a process the possible outcomes of which form a collection (or an *ensemble*) of functions of time $\{y(t)\}$. A member of the ensemble is referred to as a sample function. The process is called a *random process* if the values of the member functions of the ensemble at any particular time constitute a random variable. A sample function in a random process is referred to as a *random signal*.

A random process that is a function of time is called *stationary* if its statistical properties are not dependent upon the choice of the time origin [A2-1], that is, if "whatever started to happen at some time could equally likely have started at any other time" [A2-2]. A function belonging to an ensemble that might be generated by a stationary random process, or a *stationary random signal*, is thus assumed to extend over the entire time domain. Its mean and its mean square value do not vary with time (see Fig. A2.1).

The *ensemble average*, or the *expectation*, of a random process is the average of the values of the member functions at any particular time. A stationary random process is said to be *ergodic* if, for that process, time averages equal ensemble averages. Ergodicity requires in effect that every sample function be typical of the entire ensemble (Fig. A2.2).

It is convenient in applications to regard a stationary random signal as a superposition of harmonic oscillations over a continuous range of frequencies. It is the purpose of this appendix to present a description of stationary random signals from this point of view. As a prerequisite to the development of such a description, certain basic results of harmonic analysis are reviewed first. For a more rigorous treatment of the topics discussed herein, the reader is referred, for example, to [A2-1], [A2-2], and [A2-12].

Fourier Series and Fourier Integrals

Consider the case of a *periodic* function $x(t)$ with zero mean and with period T. It can be easily proved that $x(t)$ may be written in the form

$$x(t) = C_0 + \sum_{k=1}^{\infty} C_k \cos(2\pi k n_1 t - \phi_k) \tag{A2.1}$$

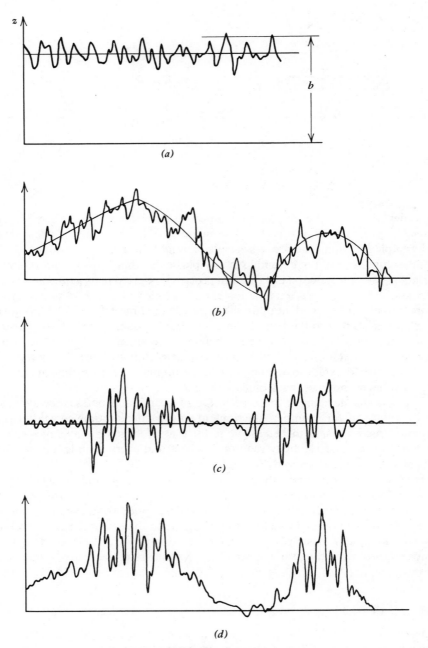

FIGURE A2.1. (*a*) Stationary signal. Nonstationary signals with: (*b*) time varying mean value; (*c*) time varying mean square value; (*d*) time varying mean and mean square value. After J. S. Bendat and A. G. Piersol, *Random Data: Analysis and Measurement Procedures*, Wiley-Interscience, New York, 1971, p. 345.

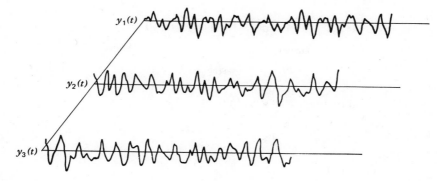

FIGURE A2.2. Sample functions of an ergodic stationary random process.

where $n_1 = 1/T$ is the *fundamental frequency* and

$$C_0 = \frac{1}{T} \int_{-T/2}^{T/2} x(t)\, dt \qquad (A2.1a)$$

$$C_k = (A_k^2 + B_k^2)^{1/2} \qquad (A2.1b)$$

$$\phi_k = \tan^{-1} \frac{B_k}{A_k} \qquad (A2.1c)$$

$$A_k = \frac{2}{T} \int_{-T/2}^{T/2} x(t) \cos 2\pi k n_1 t\, dt \qquad (A2.1d)$$

$$B_k = \frac{2}{T} \int_{-T/2}^{T/2} x(t) \sin 2\pi k n_1 t\, dt \qquad (A2.1e)$$

Equation A2.1 is known as the *Fourier series* expansion of $x(t)$ [A2-3].

If a function $y(t)$ is actually *nonperiodic*, it is still possible to regard it as periodic with infinite period. It can be shown [A2-3, A2-4] that if $y(t)$ is piecewise differentiable in every finite interval and if the integral

$$\int_{-\infty}^{\infty} |y(t)|\, dt \qquad (A2.2)$$

exists, the following relation, analogous to Eq. A2.1, holds:

$$y(t) = \int_{-\infty}^{\infty} C(n) \cos[2\pi n t - \phi(n)]\, dn \qquad (A2.3)$$

In Eq. A2.3, known as the *Fourier integral* of $y(t)$ (in real form), n is a continuously varying frequency and

$$C(n) = [A^2(n) + B^2(n)]^{1/2} \qquad (A2.3a)$$

$$\phi(n) = \tan^{-1} \frac{B(n)}{A(n)} \qquad (A2.3b)$$

$$A(n) = \int_{-\infty}^{\infty} y(t) \cos 2\pi n t \, dt \tag{A2.3c}$$

$$B(n) = \int_{-\infty}^{\infty} y(t) \sin 2\pi n t \, dt \tag{A2.3d}$$

From Eqs. A2.3a through A2.3d and the identities

$$\sin \phi = \frac{\tan \phi}{(1 + \tan^2 \phi)^{1/2}} \tag{A2.4a}$$

$$\cos \phi = \frac{1}{(1 + \tan^2 \phi)^{1/2}} \tag{A2.4b}$$

there follows immediately

$$\int_{-\infty}^{\infty} y(t) \cos[2\pi n t - \phi(n)] \, dt = C(n) \tag{A2.5}$$

The functions $y(t)$ and $C(n)$, which satisfy the symmetrical relations A2.3 and A2.5, are referred to as *Fourier transform pair*.

It is noted that successive differentiation of Eq. A2.3 yields

$$\dot{y}(t) = -\int_{-\infty}^{\infty} 2\pi n C(n) \sin[2\pi n t - \phi(n)] \, dn \tag{A2.6a}$$

$$\ddot{y}(t) = -\int_{-\infty}^{\infty} 4\pi^2 n^2 C(n) \cos[2\pi n t - \phi(n)] \, dn \tag{A2.6b}$$

Parseval's Equality

The mean square value σ_x^2 of the periodic function $x(t)$ with period T (Eq. A2.1) may be written as

$$\sigma_x^2 = \frac{1}{T} \int_{-T/2}^{T/2} x^2(t) \, dt \tag{A2.7}$$

The substitution of Eq. A2.1 into Eq. A2.7 yields

$$\sigma_x^2 = \sum_{k=0}^{\infty} S_k \tag{A2.8}$$

where $S_0 = C_0^2$ and $S_k = \frac{1}{2}C_k^2 \, (k = 1, 2, \ldots)$. *The quantity S_k is the contribution to the mean square value of $x(t)$ of the harmonic component with frequency $k n_1$.* Equation A2.8 is a form of Parseval's equality [A2-3, A2-4, A2-5].

Consider now the case of a nonperiodic function $y(t)$ for which an integral Fourier expression exists. In virtue of Eqs. A2.3 and A2.5,

$$\int_{-\infty}^{\infty} y^2(t)\, dt = \int_{-\infty}^{\infty} y(t) \int_{-\infty}^{\infty} C(n) \cos[2\pi n t - \phi(n)]\, dn\, dt$$

$$= \int_{-\infty}^{\infty} C(n) \int_{-\infty}^{\infty} y(t) \cos[2\pi n t - \phi(n)]\, dt\, dn$$

$$= \int_{-\infty}^{\infty} C^2(n)\, dn$$

$$= 2 \int_{0}^{\infty} C^2(n)\, dn \tag{A2.9}$$

Equation A2.9 is the form taken by Parseval's equality in the case of a non-periodic function [A2-3, A2-4, A2-5].

Spectral Density Function of a Stationary Random Signal

A relation similar to Eq. A2.8 will not be sought for functions generated by stationary processes. The spectral density will be defined as the counterpart, for these functions, of the quantities S_k.

Consider a stationary random signal $z(t)$ with zero mean. Because it does not satisfy the condition A2.2, the function $z(t)$ does not possess a Fourier integral. An auxiliary function $y(t)$ will therefore be defined as follows (Fig. A2.3):

$$y(t) = z(t) \qquad (-T/2 < t < T/2)$$

$$y(t) = 0 \qquad \text{elsewhere} \tag{A2.10}$$

The function $y(t)$ thus defined is nonperiodic, satisfies Eq. A2.2, and thus has a Fourier integral. From the definition of $y(t)$ it follows that

$$\lim_{T \to \infty} y(t) = z(t) \tag{A2.11}$$

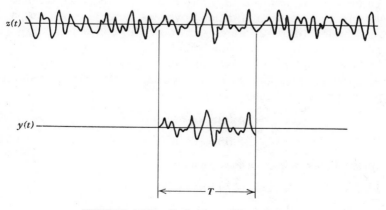

FIGURE A2.3. Definition of function $y(t)$.

By virtue of Eqs. A2.10 and A2.9, the mean square value of $y(t)$ is:

$$\sigma_y^2 = \frac{1}{T} \int_{-T/2}^{T/2} y^2(t)\, dt$$

$$= \frac{1}{T} \int_{-\infty}^{\infty} y^2(t)\, dt$$

$$= \frac{2}{T} \int_0^{\infty} C^2(n)\, dn \tag{A2.12}$$

The mean square value of the function $z(t)$ is then

$$\sigma_z^2 = \lim_{T \to \infty} \sigma_y^2$$

$$= \lim_{T \to \infty} \frac{2}{T} \int_0^{\infty} C^2(n)\, dn \tag{A2.13}$$

With the notation

$$S_z(n) = \lim_{T \to \infty} \frac{2}{T} C^2(n) \tag{A2.14}$$

Equation A2.13 becomes

$$\sigma_z^2 = \int_0^{\infty} S_z(n)\, dn \tag{A2.15}$$

The function $S_z(n)$ is defined as the *spectral density function* of $z(t)$. To each frequency $n\,(0 < n < \infty)$ there corresponds an elemental contribution $S(n)\,dn$ to the mean square value σ_z^2; σ_z^2 is, of course, equal to the area under the spectral density curve $S_z(n)$. Because in Eq. A2.15 the spectrum is defined over the range $0 < n < \infty$, the quantity $S_z(n)$ is referred to as the *one-sided* spectral density function of $z(t)$. It is this definition of the spectrum implicit in Eq. A2.15 that has been used throughout this text. However, a different convention may be used whereby the spectrum is defined over the range $-\infty < n < \infty$ and the integration limits in Eq. A2.15 are $-\infty$ to ∞. This convention yields the so-called *two-sided* spectral density function of $z(t)$ [A2-6, p. 75].

From Eqs. A2.6, following the same steps that led from Eq. A2.3 to Eq. A2.14, there result the expression for the spectral density of the first and second derivative of a random process:

$$S_{\dot{z}}(n) = 4\pi^2 n^2 S_z(n) \tag{A2.16a}$$

$$S_{\ddot{z}}(n) = 16\pi^4 n^4 S_z(n) \tag{A2.16b}$$

Spectral Density and Autocovariance Function

From Eqs. A2.3a, A2.3c, and A2.3d there follows

$$\frac{2}{T} C^2(n) = \frac{2}{T} [A^2(n) + B^2(n)]$$

$$= \frac{2}{T} [A(n) \cdot A(n) + B(n) \cdot B(n)]$$

$$= \frac{2}{T} \left[\int_{-\infty}^{\infty} y(t_1) \cos 2\pi n t_1 \, dt_1 \int_{-\infty}^{\infty} y(t_2) \cos 2\pi n t_2 \, dt_2 \right.$$

$$\left. + \int_{-\infty}^{\infty} y(t_1) \sin 2\pi n t_1 \, dt_1 \int_{-\infty}^{\infty} y(t_2) \sin 2\pi n t_2 \, dt_2 \right]$$

which may be written as

$$\frac{2}{T} \int_{-\infty}^{\infty} \int_{-\infty}^{\infty} y(t_1) y(t_2) \cos 2\pi n (t_2 - t_1) \, dt_1 \, dt_2 \qquad (A2.17)$$

Denoting

$$\tilde{R}(\tau) = \frac{1}{T} \int_{-\infty}^{\infty} y(t_1) y(t_1 + \tau) \, dt_1, \qquad (A2.18)$$

where $\tau = t_2 - t_1$, Eq. A2.17 may be written as

$$\frac{2}{T} C^2(n) = 2 \int_{-\infty}^{\infty} \tilde{R}(\tau) \cos 2\pi n \tau \, d\tau \qquad (A2.19)$$

Equations A2.19, A2.11, and A2.14 thus yield

$$S_z(n) = \int_{-\infty}^{\infty} 2R_z(\tau) \cos 2\pi n \tau \, d\tau \qquad (A2.20)$$

where

$$R_z(\tau) = \lim_{T \to \infty} \frac{1}{T} \int_{-T/2}^{T/2} z(t) z(t + \tau) \, dt \qquad (A2.21)$$

The function $R_z(\tau)$ is defined as the *autocovariance function* of $z(t)$ and provides a measure of the interdependence of the variable z at times t and $t + \tau$.

From the definition of the autocovariance function (Eq. A2.21) and the stationarity of $z(t)$ there follows

$$R_z(\tau) = R_z(-\tau) \qquad (A2.22)$$

Since $R_z(\tau)$ is an even function of τ,

$$\int_{-\infty}^{\infty} 2R_z(\tau) \sin 2\pi n \tau \, d\tau = 0 \qquad (A2.23)$$

It can be seen from a comparison of Eqs. A2.5 and A2.20 that $S_z(n)$ and $2R_z(\tau)$ form a Fourier transform pair. Therefore,

$$R_z(\tau) = \tfrac{1}{2} \int_{-\infty}^{\infty} S_z(n) \cos 2\pi n \tau \, dn \qquad (A2.24)$$

Since, as follows from Eq. A2.20, $S_z(n)$ is an even function of n, Eq. A2.24 may be written as

$$R_z(\tau) = \int_0^\infty S_z(n) \cos 2\pi n\tau \, dn \qquad \text{(A2.24b)}$$

Similarly, by virtue of Eqs. A2.20 and A2.22,

$$S_z(n) = 4 \int_0^\infty R_z(\tau) \cos 2\pi n\tau \, d\tau \qquad \text{(A2.25)}$$

Equation A2.25 permits, in principle, the calculation of the spectral density function corresponding to a given signal $z(t)$. Details concerning the practical computation of the spectra can be found in [A2-1], [A2-6], and [A2-7].

The definition of the autocovariance function (Eq. A2.21) yields

$$R_z(0) = \sigma_z^2 \qquad \text{(A2.26)}$$

For $\tau > 0$, the products $z(t)z(t+\tau)$ in Eq. A2.21 will no longer be always positive as in the case in which $\tau = 0$, and

$$R_z(\tau) < \sigma_z^2 \qquad \text{(A2.27)}$$

Since the variation of $z(t)$ with time is random, it is to be generally expected that for large values of τ the products $z(t)z(t+\tau)$ will be sometimes positive, sometimes negative, and that their mean value will vanish. Thus,

$$\lim_{\tau \to \infty} R_z(\tau) = 0 \qquad \text{(A2.28)}$$

The nondimensional quantity $R_z(\tau)/\sigma_z^2$, known as the *autocorrelation function*, is equal to unity for $\tau = 0$ and vanishes for $\tau = \infty$.

Cross-Covariance Function, Co-Spectrum, Quadrature Spectrum, Coherence

It is useful in applications to employ tools similar to those developed for the analysis of a random signal to describe certain properties of two random stationary signals $z_1(t)$ and $z_2(t)$. Consider two such signals, each with zero mean. The function

$$R_{z_1 z_2}(\tau) = \lim_{T \to \infty} \frac{1}{T} \int_{-T/2}^{T/2} z_1(t)z_2(t+\tau) \, dt \qquad \text{(A2.29)}$$

is defined as the *cross-covariance function* of the signals $z_1(t)$ and $z_2(t)$. From this definition and the stationarity of the two signals, there follows

$$R_{z_1 z_2}(\tau) = R_{z_2 z_1}(-\tau) \qquad \text{(A2.30)}$$

However, it is noted that, in general, $R_{z_1 z_2}(\tau) \neq R_{z_1 z_2}(-\tau)$. For example, if $z_2(t) \equiv z_1(t - \tau_0)$, it can immediately be seen from Fig. A2.4 that

$$R_{z_1 z_2}(\tau_0) = R_{z_1}(0) \qquad \text{(A2.31)}$$

$$R_{z_1 z_2}(-\tau_0) = R_{z_1}(2\tau_0) \qquad \text{(A2.32)}$$

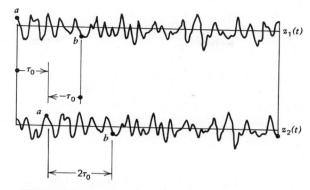

FIGURE A2.4.   Functions $z_1(t)$ and $z_2(t) = z_1(t - \tau_0)$.

The *co-spectrum* and *quadrature spectrum* of the signals $z_1(t)$ and $z_2(t)$ are defined, respectively, as

$$S^C_{z_1z_2}(n) = 2 \int_{-\infty}^{\infty} R_{z_1z_2}(\tau) \cos 2\pi n\tau \, d\tau \qquad (A2.33)$$

and

$$S^Q_{z_1z_2}(n) = 2 \int_{-\infty}^{\infty} R_{z_1z_2}(\tau) \sin 2\pi n\tau \, d\tau \qquad (A2.34)$$

It follows then from Eq. A2.30

$$S^C_{z_1z_2}(n) = S^C_{z_2z_1}(n) \qquad (A2.35)$$

$$S^Q_{z_1z_2}(n) = -S^Q_{z_2z_1}(n) \qquad (A2.35a)$$

The *coherence* function is a convenient measure of the extent to which two signals $z_1(t)$ and $z_2(t)$ are correlated. Its square root is denoted by $\text{Coh}_{z_1z_2}(n)$ and is defined as

$$\text{Coh}_{z_1z_2}(n) = \left\{ \frac{[S^C_{z_1z_2}(n)]^2 + [S^Q_{z_1z_2}(n)]^2}{S_{z_1}(n)S_{z_2}(n)} \right\}^{1/2} \qquad (A2.36)$$

Probability Distribution of the Largest Values of a Normally Distributed Stationary Random Signal

Consider a normally distributed stationary random signal $z(t)$ with zero mean. Let $E(k)$ denote the expected number of peaks per unit of time that are greater than k times the rms value of $z(t)$. It can be shown that the following expression is adequate for use in practical calculations [A2-5, A2-8]:

$$E(k) = \nu \exp\left(-\frac{k^2}{2}\right) \qquad (A2.37)$$

where k is a sufficiently large number, say $k > 3$,

$$v = \left\{ \frac{\int_0^\infty n^2 S_z(n)\, dn}{\int_0^\infty S_z(n)\, dn} \right\}^{1/2} \tag{A2.38}$$

and $S_z(n)$ is the spectral density function of $z(t)$.

Peaks greater than $k\sigma_z$ may be regarded as rare events. Their probability distribution may therefore be assumed to be of the Poisson type. Thus the probability that in the time interval T there will be no peaks equal to or larger than $k\sigma_z$ may be written as

$$p(0, T) = e^{-E(k)T} \tag{A2.39}$$

(see Sect. A1.5). $p(0, T)$ may also be viewed as the probability that, given the interval T, the ratio K of largest peak to the rms value of $z(t)$ is less than k. Equation A2.39 may then be rewritten as

$$P(K < k|T) = e^{-E(k)T} \tag{A2.40}$$

where $P(K < k|T)$ is the cumulative distribution of the random variable K given the interval T. The probability density function of K, that is, the probability that $k < K < k + dk$, is obtained from Eq. A2.40 by differentiation:

$$P_K(k|T) = kTE(k)e^{-E(k)T} \tag{A2.41}$$

The expected value of the largest peak occurring in the interval T may then be calculated as shown in Sect. A1.4:

$$\bar{K} = \int_{-\infty}^{\alpha} kP_K(k|T)\, dk \tag{A2.42}$$

The integral of Eq. A2.42 was evaluated in [A2-9] and is, approximately,

$$\bar{K} = (2 \ln vT)^{1/2} + \frac{0.577}{(2 \ln vT)^{1/2}} \tag{A2.43}$$

where v is given by Eq. A2.38.

Mean Upcrossing and Outcrossing Rates

Consider the stationary random process $z(t)$ represented in Fig. A2.1a. We seek the mean rate of occurrence v_b of the event, denoted by A, that the process crosses the line $z = b$ in an upward direction (that is, the mean upcrossing rate of $z(t)$). The probability of occurrence of this event during a time interval dt in the neighborhood of the time t is equal to the probability that $\dot{z} > 0$ and $b < z < b + dz$, which can be written as

$$\text{Prob}(b < z < b + dz, \dot{z} > 0) = \int_0^\infty f_{z,\dot{z}}(b, \dot{z})\, dz\, d\dot{z}$$

$$= dt \int_0^\infty \dot{z} f_{z,\dot{z}}(b, \dot{z})\, d\dot{z} \tag{A2.44}$$

where $f_{Z,\dot{Z}}(z, \dot{z})$=joint probability density of z and $\dot{z}$. The mean upcrossing rate v_b is by definition the probability of occurrence of the event A per unit time, that is [A2-8]

$$v_b = \int_0^\infty \dot{z} f_{Z,\dot{Z}}(b, \dot{z}) \, d\dot{z} \tag{A2.45}$$

This result was extended to the case where the random process is a vector, $\mathbf{x}$. Let v_D denote the mean rate at which the random process (that is, the tip of a vector with specified origin, O) crosses in an outward direction the boundary F_D of a region containing the point O. It can be shown that the mean outcrossing rate v_D has the expression

$$v_D = \int_{F_D} d\mathbf{x} \int_0^\infty \dot{x}_n f_{\mathbf{X},\dot{X}_n}(\mathbf{x}, \dot{x}_n) \, d\dot{x}_n \tag{A2.46}$$

where $\dot{x}_n$=projection of vector $\dot{\mathbf{x}}$ on the normal to F_D, and $f_{\mathbf{X},\dot{X}_n}(\mathbf{x}, \dot{x}_n)$=joint probability density function of $\mathbf{x}$ and $\dot{x}_n$ [A2-10, A2-11]. Equation A2.46 can be written as

$$v_D = \int_{F_D} \left\{ \int_0^\infty \dot{x}_n f_{\dot{X}_n}[\dot{x}_n | \mathbf{X} = \mathbf{x}] \, d\dot{x}_n \right\} f_{\mathbf{X}}(\mathbf{x}) \, d\mathbf{x} \tag{A2.47a}$$

$$= \int_{F_D} \mathrm{E}_0^\infty [\dot{X}_n | \mathbf{X} = \mathbf{x}] f_{\mathbf{X}}(\mathbf{x}) \, d\mathbf{x} \tag{A2.47b}$$

where $f_{\dot{X}_n}$=conditional probability density of $\dot{X}_n$ given that $\mathbf{X} = \mathbf{x}$, $f_{\mathbf{X}}$=probability density function of the vector $\mathbf{X}$, and $\mathrm{E}_0^\infty [\dot{X}_n | \mathbf{X} = \mathbf{x}]$=average of positive values of $\dot{X}_n$ given that $\mathbf{X} = \mathbf{x}$. If $\dot{X}_n$ and $\mathbf{X}$ are independent,

$$v_D = \mathrm{E}_0^\infty [\dot{X}_n] \int_{F_D} f_{\mathbf{X}}(\mathbf{x}) \, d\mathbf{x} \tag{A2.48}$$

REFERENCES

A2-1 A. Papoulis, *Probability, Random Variables, and Stochastic Processes*, McGraw-Hill, New York, 1975.

A2-2 R. B. Blackman and J. W. Tukey, *The Measurement of Power Spectra*, Dover, New York, 1959.

A2-3 F. B. Hildebrand, *Advanced Calculus for Applications*, Prentice-Hall, Englewood Cliffs, N.J., 1962.

A2-4 D. C. Champeney, *Fourier Transforms and Their Applications*, Academic, London, 1973.

A2-5 J. D. Robson, *An Introduction to Random Vibration*, Elsevier, Amsterdam, The Netherlands, 1964.

A2-6 J. S. Bendat and A. G. Piersol, *Random Data: Analysis and Measurement Procedures*, Wiley-Interscience, New York, 1971.

A2-7 L. D. Enochson and R. K. Otnes, *Programming and Analysis for Digital Time Series Data*, The Shock and Vibration Information Center, U.S. Department of Defense, 1968.

A2-8 S. O. Rice, "Mathematical Analysis of Random Noise" in *Selected Papers on Noise and Stochastic Processes*, N. Wax (Ed.), Dover, New York, 1954.

A2-9   A. G. Davenport, "Note on the Distribution of the Largest Value of a Random Function with Application to Gust Loading," *J. Inst. Civ. Eng.*, **24** (1964), 187–196.

A2-10 Y. K. Belyaev, "On the Number of Exits Across the Boundary of a Region by a Vector Stochastic Process," *Teoriia Veroiatn. Primen.*, **13** (1968), 333–337 (In Russian).

A2-11 D. Veneziano, M. Grigoriu, and C. A. Cornell, "Vector Process Models for System Reliability," *J. Eng. Mech. Div.*, ASCE, **103** (1977), 441–460.

A2-12 E. Vanmarcke, *Random Fields: Analysis and Synthesis*, MIT Press, Cambridge, 1983.

APPENDIX A3

Elements of Structural Reliability

The objective of structural reliability is to develop design criteria and verification procedures aimed at ensuring that structures built according to specifications will perform acceptably from a safety and serviceability viewpoint. This objective could in principle be achieved by meeting the following requirement: *failure probabilities* (i.e., probabilities that structures or members will fail to satisfy certain performance criteria) must be equal to or less than some benchmark values referred to as *target failure probabilities*.* Such an approach would require [A3-1, A3-9]:

1. The probabilistic description of the loads expected to act on the structure.

2. The probabilistic description of the physical properties of the structure which affect its behavior under loads.

3. The physical description of the limit states, that is, the states beyond which the structure is unserviceable (serviceability limit states) or unsafe (ultimate limit states). Examples of limit states include: specified deformations (determined from functional considerations); specified accelerations (determined from studies of equipment performance, or from ergonomic studies on user discomfort in structures experiencing dynamic loads); specified levels of nonstructural damage; structural collapse.

4. Load-structural response relationships covering the range of responses from zero up to the limit state being considered.

5. The estimation of the probabilities of exceeding the various limit states (i.e., of the failure probabilities), based on the elements listed in items 1 through 4 above and on the use of appropriate probabilistic and statistical tools.

6. The specification of maximum acceptable probabilities of exceeding the various limit states (i.e., of target failure probabilities).

*An alternative statement of this requirement is that the reliabilities corresponding to the various limit states must be equal to or exceed the respective target reliabilities (reliability being defined as the difference between unity and the failure probability).

The performance of a structure would be judged acceptable from a safety or serviceability viewpoint if the differences between the target and the failure probabilities were either positive (in which case the structure would be over-designed) or equal to zero.

The approach just described is seldom applicable in practice. Owing to physical and probabilistic modeling difficulties and to the absence of sufficient statistical data it is in general not possible to provide confident probabilistic descriptions of the loads, particularly within the loading range corresponding to ultimate limit states. Comprehensive probabilistic descriptions of the relevant physical properties of the structure are also seldom available. In some cases limit states are difficult to define quantitatively. Difficulties may also arise in attempting to describe relationships between loading and structural response that involve material and/or geometric nonlinearities or contributions by non-structural elements to the total structural capacity. For certain types of structures—e.g., redundant structures subjected to dynamic effects—the estimation of failure probabilities can be analytically unfeasible or computationally prohibitive, at least in the present state of the art. Finally, there are few agreed-upon values of target failure probabilities, particularly for limit states involving loss of life, as opposed to mere economic loss.*

The reliability analyst is therefore forced to accept various compromises. In practice, in the absence of sufficient data and of proved probabilistic and physical models, it may be necessary to use conservative models, or models based at least in part on subjective belief. In addition, definitions of limit states may have to be adopted on the basis of computational convenience, rather than on physical grounds. (For example, ultimate limit states are defined in most cases as the collapse of individual members, rather than as the collapse of the structure as a whole;† member collapse is in certain cases conventionally defined as the attainment of the yield stress at the most highly stressed section of the member, even though this does not usually entail physical collapse.) Finally, to simplify the computations, various approximations may have to be used.

Estimates of *nominal* (or "notional") failure probabilities are thus obtained that can differ—in certain instances significantly—from the "true" probabilities. However, if there are grounds to believe that the ratios of nominal to "true" probabilities for two given designs do not differ significantly, the two designs may be compared from a reliability viewpoint on the basis of the respective nominal, rather than "true," probabilities.

It would, of course, be desirable to establish target (i.e., maximum acceptable) nominal failure probabilities. In principle this could be done if the "true" target failure probability and the ratio between "true" and nominal failure probability were known. This is not the case, and attempts are therefore made to infer target nominal failure probabilities from the reliability analysis of

---

*For questions pertaining to safety goals for the operation of nuclear power plants, see [A3-19].
†In recent years a number of studies concerned with structural systems have been reported. For useful reviews of these developments, see [A3-1] and [A3-25].

exemplary designs, that is, designs that are regarded by professional consensus as acceptably but not overly safe. Such inferences are part of the process referred to as safety calibration against accepted practice.

While there are instances where such a process can be carried out successfully, difficulties arise in many practical situations. For example, structural reliability calculations suggest that current design practice as embodied in the American National Standard A58.1 [2-49] and other building standards and codes is not risk-consistent. In particular, estimated reliabilities of members designed in accordance with current practice are considerably lower for members subjected to dead, live, and wind loads then for members subjected only to dead and live loads [A3-3, A3-4], especially when the effect of wind is large compared to the effect of dead and live loads.* Whether these differences are real or only apparent, that is, due to shortcomings of current reliability analyses, remains to be established. Thus, it is not possible in the present state of the art to determine whether it is the lower or the higher estimated reliabilities that should be adopted as target values.

In spite of both theoretical and practical difficulties, structural reliability tools can in a number of cases be used to advantage in design and for code development purposes. An important example is the design of cladding for wind loading (see Sect. 9.5). The objective of this appendix is to present a review of fundamental topics in structural reliability, as applied to individual members, which have found application or are potentially applicable to wind engineering problems. These topics include: the estimation of failure probabilities; safety indices; and safety (or load and resistance) factors [A3-24].

Nominal Failure Probabilities, Safety Indices, and Load and Resistance Factors

The estimation of nominal failure probabilities† is the core of structural reliability. Safety indices are, at least in theory, measures of failure probabilities. The use of load and resistance factors in design criteria is intended to ensure that the members to which the criteria are applied have acceptable failure probabilities within any specified period of interest (e.g., one year, or the lifetime of a structure).

Modeling of Loads as Random Processes and Random Variables.
Quantities that vary continuously and randomly with time (e.g., the wind speed, the wind velocity vector, or the wave height at a given location) can be modeled as random processes. Quantities that are constant in time (e.g., the dead weight), or whose variation in time follows a deterministic law, can be modeled more simply as random variables.

---

*The earliest justification of current design practice with respect to wind loading was traced by the authors to Fleming's 1915 monograph *Wind Stresses* [A3-2], which states: "Maximum wind loading comes seldom and lasts but a short time. The working stresses used for this loading may therefore be increased 50% above those used for ordinary live- and dead-loads."
†For brevity, nominal failure probabilities are henceforth referred to simply as failure probabilities.

In problems involving combinations of two or more randomly time-dependent loads, it is in general necessary to estimate failure probabilities by resorting to models and techniques drawn from the theory of random processes. If the system being considered depends, in addition to the randomly time-dependent loads, upon the random variables, $X_1, X_2, \ldots, X_m$, estimates of failure probabilities $P(\text{failure} \,|\, X_1, X_2, \ldots, X_m)$ are obtained which are conditional upon the values $X_1, X_2, \ldots, X_m$ taken on by these variables. The probability of failure of the structure, P_f, is then estimated by applying the theorem of total probability as follows:

$$P_f = \int P(\text{failure}|x_1, x_2, \ldots, x_k) f_{X_1, X_2, \ldots, X_k}(x_1, x_2, \ldots, x_k) \, dx_1 \, dx_2 \ldots dx_k$$

(A3.1)

where $f_{X_1, X_2, \ldots, X_k}(x_1, x_2, \ldots, x_k) =$ joint probability density function of X_1, $X_2, \ldots, X_k$. For treatments of load combination problems based on random process representations see [A3-5] to [A3-8].

In problems involving only one randomly time-dependent parameter, $\zeta(t)$, the question of combining time-dependent random processes no longer arises. It is therefore convenient in applications to use, in lieu of the random process $\zeta(t)$, its largest value during the lifetime of the structure, denoted by $X^{(n)}$. By substituting the random variable $X^{(n)}$ for the random process $\zeta(t)$ the treatment of the reliability problem is considerably simplified. The largest lifetime value $X^{(n)}$ can be characterized probabilistically as follows:

$$F_{X^{(n)}}(x) = [F_{X^{(1)}}(x)]^n$$

(A3.2)

where $X^{(1)} =$ extreme value of $\zeta(t)$ during a time interval $t_1 = T/n$, $T =$ lifetime of structure, $n =$ integer and $F_{X^{(n)}} =$ cumulative distribution function of $X^{(n)}$. The cumulative distribution function of $X^{(1)}$, $F_{X^{(1)}}$, is referred to as the *underlying distribution* of $X^{(n)}$. Equation A3.2 holds if successive values of $X^{(1)}$ are identically distributed and statistically independent. An application of Eq. A3.2 is presented in the following example.

Example

Let $X^{(1)} \equiv U_a$ denote the largest yearly wind speed at a given location. Then $X^{(n)} \equiv U$ denotes the largest wind speed occurring at that location during an n-year period (equal to the lifetime of the structure). It is assumed that the largest yearly wind speed U_a has an Extreme Value Type I distribution, that is,

$$F_{U_a}(u_a) = \exp\left[-\exp\left(-\frac{u_a - \mu}{\sigma}\right)\right]$$

(A3.3)

It can be shown that

$$\mu \simeq \bar{U}_a - 0.45\sigma_{U_a}$$

(A3.4)

$$\sigma \simeq 0.78\sigma_{U_a}$$

(A3.5)

where $\bar{U}_a$ and σ_{U_a} = sample mean and sample standard deviation of the largest annual wind speed data, U_a, recorded at the given location over a sufficient number of consecutive years (say, 20 years or more) (Eqs. A1.40 and A1.41). From Eqs. A3.2 through A3.5 it follows that the probability distribution of the largest lifetime wind speed U is:

$$F_U(u) = \exp\left[-\exp\left(-\frac{u-\mu_n}{\sigma_n} \right) \right] \qquad (A3.6)$$

where

$$\mu_n \simeq \bar{U} - 0.45\sigma_U \qquad (A3.7)$$

$$\sigma_n \simeq 0.78\sigma_U \qquad (A3.7a)$$

$$\bar{U} = \bar{U}_a + 0.78\sigma_{U_a}\ln n \qquad (A3.8)$$

$$\sigma_U = \sigma_{U_a} \qquad (A3.9)$$

and n = lifetime of structure in years.

Finally, we consider the case where the structure is acted upon by two randomly time-dependent loads with the following properties: (1) their extreme values have negligible probability of simultaneous occurrence, (2) their most unfavorable combination occurs when one of the loads reaches its largest lifetime value, while the other has an "ordinary" (also termed "arbitrary-point-in time"), rather than an extreme, value.* Since the "arbitrary-point-in-time" loading can be modeled by an appropriately chosen time-independent probability distribution [A3-9], the reliability problem can in this case also be reduced to one involving only random variables.

Failure Region, Safe Region, and Failure Boundary.   Consider a structure or member subjected to a load Q, and let the value of the loading that induces a certain limit state in the structure (e.g., the yield stress) be denoted by R. It is assumed that both Q and R are random variables. The space defined by these variables is referred to as the *load space*. By definition, failure occurs for any pair of values, Q, R, satisfying the relation

$$R - Q < 0 \qquad (A3.10)$$

Equation A3.10 defines the *failure region* in the load space. The survival region, or *safe region*, is defined by the relation

$$R - Q > 0 \qquad (A3.11)$$

The *failure boundary*, which separates the failure and safe regions, is defined by the equation

$$R - Q = 0 \qquad (A3.12)$$

*For example, it may be assumed that these properties characterize the wind load and the live load acting on members of high rise buildings frames.

Relations similar to Eqs. A3.10, A3.11, and A3.12 can be written in the *load effect space*, defined by the variables Q_e, R_e, where Q_e is the effect, or state, induced in the structure by the load Q (e.g., a state of stress or deformation), and R_e is the corresponding limit state (e.g., the yield stress or a specified deformation). The equation of the failure boundary in the load effect space is

$$R_e - Q_e = 0 \qquad (A3.13)$$

In general, Q and R are functions of random variables $X_1, X_2, \ldots, X_n$ (e.g., aerodynamic coefficients, terrain roughness, cross-sectional area, modulus of elasticity, breaking strength),* that is,

$$Q = Q(X_1, X_2, \ldots, X_n) \qquad (A3.14)$$

$$R = R(X_1, X_2, \ldots, X_n) \qquad (A3.15)$$

Substitution of Eqs. A3.14 and A3.15 into Eqs. A3.10, A3.11, and A3.12 yields the mapping of the failure region, safe region, and failure boundary onto the space of the variables $X_1, X_2, \ldots, X_n$. The equation of the failure boundary can thus be written as

$$g(X_1, X_2, \ldots, X_n) = 0 \qquad (A3.16)$$

The well-behaved nature of structural mechanics relations generally ensures that Eq. A3.13 is the mapping of Eq. A3.12 onto the load effect space. Equation A3.16 is thus the mapping onto the space $X_1, X_2, \ldots, X_n$ not only of Eq. A3.12, but of Eq. A3.13 as well. Therefore, once it is made clear at the outset that the problem is formulated in the load, or in the load effect, space, it is common practice to refer generically to Q and Q_e as "loads" and to R and R_e as "resistances," and to omit the index "e" in Eq. A3.13.

It is useful in various applications to map the failure region, the safe region, and the failure boundary onto the space of the variables $Y_1, Y_2, \ldots, Y_r$, defined by transformations

$$Y_i = Y_i(X_1, X_2, \ldots, X_n)(i = 1, 2, \ldots, r) \qquad (A3.17)$$

For example, if in Eq. A3.16 $X_1 = \rho$, and $X_2 = U$, where $\rho = $ air density and $U = $ wind speed, a variable representing the dynamic pressure may be defined by the transformation $Y_1 = \frac{1}{2}\rho U^2$, and Eq. A3.16 may be mapped onto the space of the variables $Y_1, X_3, X_4, \ldots, X_n$. Another example is the frequently used set of transformations

$$Y_1 = \ln R \qquad (A3.18)$$

$$Y_2 = \ln Q \qquad (A3.19)$$

---

*These are sometimes referred to as basic variables. We will use here simply the term "variables," since what constitutes a basic variable is in many instances a matter of convention. For example, the hourly wind speed at 10 m above ground in open terrain, which is regarded in most applications as a basic variable, depends in turn upon various random storm characteristics, such as the difference between atmospheric pressures at the center and the periphery of the storm, the radius of maximum storm winds, and so forth.

It follows immediately from Eqs. A3.12, A3.18, and A3.19 that the mapping of the failure boundary onto the space Y_1, Y_2, is

$$Y_1 - Y_2 = 0 \qquad \text{(A3.20)}$$

or

$$\ln R - \ln Q = 0 \qquad \text{(A3.21)}$$

The failure boundary is a point, a curve, a surface, or a hypersurface according to whether the problem at hand is formulated in a space of one, two, three, or more than three random variables.

General Expression for Estimation of Failure Probability. Let the failure region region in the space of reduced variables* $x_{1_r}, x_{2_r}, \ldots x_{n_r}$, be denoted by Ω. The probability of failure P_f can be written as:

$$P_f = \int_\Omega f_{x_{1_r}, x_{2_r}, \ldots, x_{n_r}}(x_{1_r}, x_{2_r}, \ldots, x_{n_r}) \, dx_{1_r} \, dx_{2_r} \ldots dx_{n_r} \qquad \text{(A3.22)}$$

where the integrand is the joint probability density function of the reduced variables.

In most cases the estimation of failure probabilities by Eq. A3.22 is computationally unwieldy, if not prohibitive, and the use of alternative methods is attempted instead. Various such methods, whose applicability depends upon the characteristics of the problem at hand, are described subsequently in this appendix. However, we first discuss in the next section the useful notion of the safety index.

Safety Indices. The safety index is a statistic which, under certain conditions that will be illustrated subsequently, can provide a simple and convenient means of assessing structural reliability.

We consider a failure boundary in the space of a given set of variables, and denote by S its mapping in the space of the corresponding reduced variables. The safety index, β, is defined as the shortest distance in this space between the origin and the boundary S [A3-10].† The point on the boundary S that is closest to the origin, as well as its mapping in the space of the original variables, is referred to as the *checking point*. For any given structural problem, *the*

---

*The reduced (or standardized) variable x_r corresponding to a variable X is defined as

$$x_r = \frac{X - \bar{X}}{\sigma_X}$$

where $\bar{X}$ and σ_X are the mean and standard deviation of X, respectively.

†This definition is applicable to statistically independent variables. If the variables of the problem are correlated, they can be transformed by a linear operator into a set of uncorrelated variables [A3-10]. Note that an alternative, generalized safety index was proposed in [A3-22], whose performance is superior in situations where the failure boundaries are nonlinear (see also Chapter 9 of [A3-23]).

numerical value of the safety index depends upon the set of variables in which the problem is formulated. The examples that follow illustrate the meaning of the safety index and the dependence of its numerical value upon the set of variables being used.

Example 1

It is assumed that the only random variable of the problem is the load (effect) Q. The resistance—a deterministic quantity—is denoted by $\bar{R}$. The mapping of the failure boundary

$$Q - \bar{R} = 0 \qquad\qquad (A3.23)$$

onto the space of the reduced variable $q_r = (Q - \bar{Q})/\sigma_Q$ (i.e., onto the axis Oq_r— see Fig. A3.1) is a point, q_r^*, whose distance from the origin O is $\beta = (\bar{R} - \bar{Q})/\sigma_Q$. The safety index represents in this case the difference between the values $\bar{R}$ and $\bar{Q}$ measured in terms of standard deviations σ_Q. It is clear that the larger the safety index β (i.e., the larger the difference $\bar{R} - \bar{Q}$ for any given σ_Q, or the smaller σ_Q for any given difference $\bar{R} - \bar{Q}$), the smaller the probability that $Q \geqslant \bar{R}$.

Example 2

Consider the failure boundary in the load space (Eq. A3.10), and assume that both R and Q are random variables. The mapping of Eq. A3.12 onto the space of the reduced variables $q_r = (Q - \bar{Q})/\sigma_Q$ and $r_r = (R - \bar{R})/\sigma_R$ is the line [A3-9]:

$$\sigma_Q q_r + \bar{Q} - \sigma_R r_r - \bar{R} = 0 \qquad\qquad (A3.24)$$

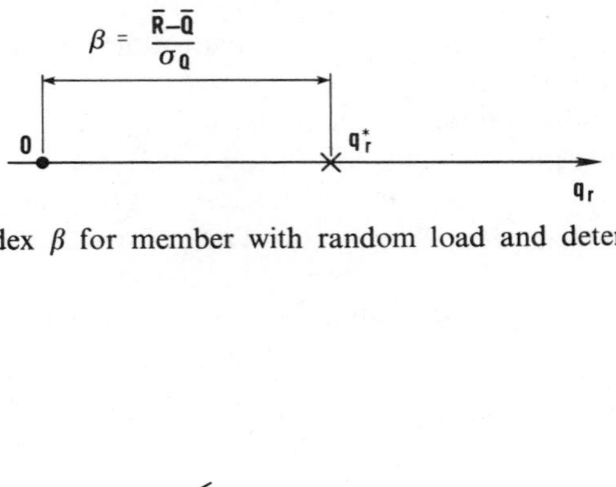

FIGURE A3.1. Index β for member with random load and deterministic resistance [A3-24].

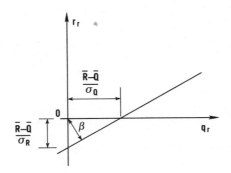

FIGURE A3.2. Index β for member with random load and random resistance [A3-24].

(Fig. A3.2). The distance between the origin and this line is

$$\beta = \frac{\bar{R} - \bar{Q}}{(\sigma_R^2 + \sigma_Q^2)^{1/2}} \tag{A3.25}$$

Example 3

Instead of operating in the load space R, Q, we consider Eq. A3.20, that is, the failure boundary in the space Y_1, Y_2, defined by Eqs. A3.18 and A3.19. Following exactly the same steps as in the preceding example, but applying them to the variables Y_1 and Y_2, the safety index is in this case

$$\beta = \frac{\bar{Y}_1 - \bar{Y}_2}{(\sigma_{Y_1}^2 + \sigma_{Y_2}^2)^{1/2}} \tag{A3.26}$$

Expansion in a Taylor series yields

$$Y_1 = \ln \bar{R} + (R - \bar{R}) \frac{1}{\bar{R}} - \frac{1}{2} (R - \bar{R})^2 \frac{1}{\bar{R}^2} + \cdots \tag{A3.27}$$

and a similar expression for Y_2. It then follows that if R and Q are uncorrelated,

$$\beta \simeq \frac{\ln \bar{R} - \frac{1}{2} V_R^2 - (\ln \bar{Q} - \frac{1}{2} V_Q^2)}{(V_R^2 + V_Q^2)^{1/2}} \tag{A3.28}$$

where V_R and V_Q denote the coefficient of variation of R and Q, respectively.* If higher order terms in the numerator of Eq. A3.28 are neglected,

$$\beta \simeq \frac{\ln(\bar{R}/\bar{Q})}{(V_R^2 + V_Q^2)^{1/2}} \tag{A3.29}$$

Example 4

Consider a linearly elastic member whose stresses Q can be written as

$$Q = aU^2 \tag{A3.30}$$

where a = deterministic influence coefficient and U = wind speed. We assume that: $a = 0.00267$ ksi/(mph)^2; the mean and standard deviation of the largest annual wind speed are $\bar{U}_a = 43.73$ mph and $\sigma_{U_a} = 8.61$ mph; the mean and standard deviation of the resistance are $\bar{R} = 35.3$ ksi and $\sigma_R = 3.39$ ksi; and the lifetime of the member is $n = 50$ years. From Eqs. A3.8 and A3.9, it follows that the mean and standard deviation of the largest lifetime wind speed are $\bar{U} = 70$ mph, $\sigma_U = 8.61$ mph. An expansion of Eq. A3.30 in a Taylor series yields

$$\bar{Q} = a\bar{U}^2(1 + V_U^2) \tag{A3.31}$$

and

$$V_Q \simeq 2V_U \tag{A3.32}$$

*For the definition of the coefficient of variation, see Sect. A1.4.

that is, $\bar{Q} \simeq 13.3$ ksi and $\sigma_Q \simeq 3.27$ ksi. The equation of the failure surface in the space of the variables U, R is

$$aU^2 - R = 0 \qquad (A3.33)$$

and its mapping in the space of the reduced variables u_r, r_r is

$$\left(u_r + \frac{\bar{U}}{\sigma_U}\right)^2 = \frac{\sigma_R}{a\sigma_U^2}\left(r_r + \frac{\bar{R}}{\sigma_R}\right) \qquad (A3.34)$$

The value of the safety index being sought is $\beta = 4.31$ (Fig. A3.3). The coordinates of the checking point are $r_r^* = -2.51$, $u_r^* = 3.50$, to which there correspond in the U, R space the coordinates $U^* = 100.14$ mph, $R^* = 26.76$ ksi. It can be verified that the values of the safety index corresponding to the variables Q, R (Eq. A3.25) and $\ln Q$, $\ln R$ (Eq. A3.29) are $\beta \simeq 4.66$ and $\beta \simeq 3.69$, respectively.

Note that the mean and standard deviation of the largest lifetime wind speed or of the largest lifetime load, which are needed for the calculation of the safety index, cannot be estimated directly from measured data, but must be obtained from the probability distribution of the lifetime extreme. This distribution is estimated from the underlying distribution that best fits the measured data. *Knowledge of, or an assumption concerning, the underlying probability distribution is required for the estimation of the safety index in all cases involving a random variable that represents a lifetime extreme.*

Safety Indices and Failure Probabilities: The Case of Normal Variables.
Consider the space of the variables Q and R, and assume that both variables are

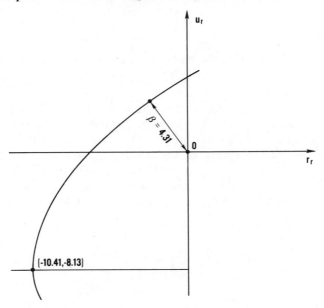

FIGURE A3.3. Index β in space of variables r_r, u_r [A3-24].

normally distributed. Note that the failure boundary (Eq. A3.12) is linear. Since the variate $R-Q$ is normally distributed, the probability of failure can be written as

$$P_f = F(R-Q<0)$$

$$= \frac{1}{\sqrt{2\pi}\sigma_{R-Q}} \int_{-\infty}^{0} \exp\left[-\frac{1}{2}\left(\frac{x-(\overline{R-Q})}{\sigma_{R-Q}}\right)^2\right] dx$$

$$= 1 - \Phi\left(\frac{\bar{R}-\bar{Q}}{\sqrt{\sigma_R^2+\sigma_Q^2}}\right) \tag{A3.35}$$

where Φ = standardized normal cumulative distribution function, and the quantity between parentheses is the safety index, β, corresponding to the space of the variables R, Q (Eq. A3.25).

More generally, it can similarly be shown that the relationship

$$P_f = 1 - \Phi(\beta) \tag{A3.36}$$

is valid if all the independent variables, $X_1, X_2, \ldots, X_n$ are normally distributed, the failure boundary (Eq. A3.16) is a linear function, and β is the safety index in the space of the reduced variables $x_1, x_2, \ldots, x_{n_r}$. Equation A3.36, of which Eq. A3.35 is a particular case, can be used even if the variables $X_1, X_2, \ldots, X_n$ are not normally distributed and the failure boundary $g(X_1, X_2, \ldots, X_n) = 0$ is nonlinear, provided that a transformation of variables $Y_i = f_i(X_i)(i = 1, 2, \ldots, n)$ can be found such that Y_i are normally distributed and the failure boundary in the space $Y_1, Y_2, \ldots, Y_n$ is linear. For example, assume that R and Q have lognormal distributions. Then $Y_1 = \ln Q$ and $Y_2 = \ln R$ are normally distributed, and the equation of the failure boundary is $Y_1 - Y_2 = 0$ (Eq. A3.20). Applying Eq. A3.35 to the variables Y_1 and Y_2,

$$P_f = 1 - \Phi\left(\frac{\overline{\ln R} - \overline{\ln Q}}{\sqrt{\sigma_{\ln R}^2 + \sigma_{\ln Q}^2}}\right)$$

$$\simeq 1 - \Phi\left(\frac{\ln(\bar{R}/\bar{Q})}{\sqrt{V_R^2 + V_Q^2}}\right) \tag{A3.37}$$

The quantities between parentheses in Eqs. A3.37 are recognized as the exact and approximate expression for the safety index β corresponding to the space of the variables $\ln R$, $\ln Q$ (Eqs. A3.26 and A3.29).

Safety Indices and Failure Probabilities: The Case of Nonnormal Variables. If the variables $X_1, X_2, \ldots, X_n$, or functions thereof, are nonnormal, Eq. A3.36 is not applicable. The fact that Eq. A3.36 does not hold means that members having the same safety index will, in general, have different failure probabilities. To illustrate the relationship between safety index and failure probability we consider the four members for which the means and standard deviations of the load and resistance are listed in Table A3.1. Members I and II have the same

TABLE A3.1. Probabilities of Failure of Four Members Corresponding to Various Distributions of the Variables

Member (1)	$\bar{Q}$ (ksi) (2)	σ_Q (ksi) (3)	$\bar{R}$ (ksi) (4)	σ_R (ksi) (5)	β (Eq. A3.25) (6)	β (Eq. A3.29) (7)	P_f (Eq. A3.35)[a] (8)	P_f (Eq. A3.37)[b] (9)	P_f (Eq. A3.22)[c] (10)
I	13.3	3.27	35.27	3.39	4.66	3.69	1.6×10^{-6}	1.1×10^{-4}	1.2×10^{-3}
II	18.0	2.79	35.27	3.39	3.93	3.69	4.2×10^{-5}	1.1×10^{-4}	1.4×10^{-3}
III	19.6	2.10	35.27	3.39	3.93	4.08	4.2×10^{-5}	2.5×10^{-5}	3.0×10^{-4}
IV	15.5	3.70	35.27	3.39	3.93	3.20	4.2×10^{-5}	6.9×10^{-4}	3.0×10^{-3}

[a]Based on the assumption that R and Q are normally distributed.
[b]Based on the assumption that R and Q are lognormally distributed.
[c]Based on the assumption that R has the distribution of Fig. A3.5 and that $Q^{1/2}$ has an Extreme Value Type I distribution.

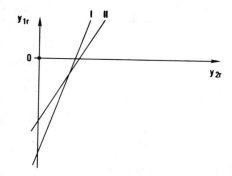

FIGURE A3.4. Failure boundaries for members I and II [A3-24].

value of β in the space of the reduced variables y_{1_r}, y_{2_r} (corresponding to Eqs. A3.18 and A3.19); their failure boundaries in that space are shown in Fig. A3.4. Members II, III, and IV have the same value of β in the space of the reduced variables r_r, q_r, representing the resistance and the load, respectively. The probabilities of failure of the four members based on the assumption that all the variables are normal are shown in column 8, Table A3.1. Those corresponding to the assumption that all the variables are lognormally distributed are listed in column 9. Column 10 shows the failure probabilities calculated by Eq. A3.22 and based on the assumptions that (a) the load is given by Eq. A3.30, (b) the distribution of the wind speed is Extreme Value Type I, and (c) the reduced variable representing the resistance has the probability density function $f_\rho(r_r)$ shown in Fig. A3.5.*

It is seen from Table A3.1 that to equal values of the safety index β, calculated by Eq. A3.25 (i.e., based on the assumption that the probability distributions of Q and R are normal), there correspond failure probabilities obtained by quadrature (column 10) that can differ from each other by as much as one order of magnitude (members II, III, and IV). On the other hand, numerical studies reported in [A3-12] show that the probability of failure P_f of column 10 is uniquely determined, to within an approximation of about 15%, by the safety index β calculated by Eq. A3.29, regardless of the relative values of $\bar{Q}$, V_Q, $\bar{R}$, and V_R. This interesting conclusion—which, as seen previously, does not hold for the safety index β calculated by Eq. A3.25—is explained by the approximate similarity between the shapes of the lognormal distribution on the one hand, and the distributions used in Eq. A3.22 on the other hand.

Note, however, that while this conclusion is valid in the particular case just examined, it does not necessarily hold in other situations. For example, it is possible that two members, one subjected to gravity loads and the other to wind loads, will have widely different failure probabilities even if their safety indices calculated by Eq. A3.29 are nearly equal. In this case, or in similar cases, a

---

*Figure A3.5 corresponds approximately to published data on the yield stress of A33 steel, for which the nominal value, the mean value, and the coefficient of variation are $F_y = 33$ ksi, $\bar{R} \simeq 1.07 F_y$, and $V_R \simeq 0.096$ [A3-12; A3-13, p. 237].

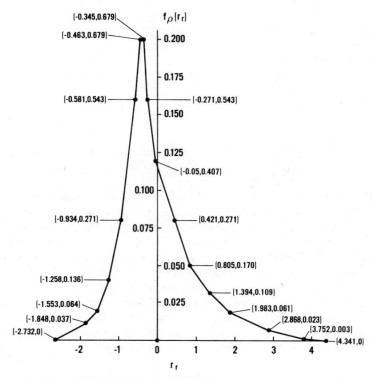

FIGURE A3.5.   Probability density function $f_\rho(r_r)$ [A3-24].

comparative reliability analysis would require the estimation of the failure probability by Eq. A3.22, or by alternative, approximate methods. A few such methods are briefly described in the following section.

Approximate Methods for Estimating Failure Probabilities.   We first describe the method referred to as *normalization at the checking point*. The principle of the method is to transform the variables, X_i, into a set of approximately equivalent normal variables, X_i^n, having the following property:

$$p_{X_i}(X_i^*) = f(X_i^{n*})$$

$$= \frac{1}{\sigma_{X_i^n}} \phi(x_{i_r}^{n*})$$

(A3.38)

$$P_{X_i}(X_i^*) = F(X_i^{n*})$$

$$= \Phi(x_{i_r}^{n*})$$

(A3.39)

where $i = 1, 2, \ldots, m$, the asterisk denotes the checking point, p_{X_i} and $P_{X_i} =$ probability density function and cumulative distribution function of X_i, respectively, f and $\phi =$ normal and standardized normal probability density function, respectively, F and Φ = normal and standardized normal cumulative distribution function, respectively, $x_{i_r}^n$ = reduced variable corresponding to X_i^n, and $\sigma_{X_i^n}$ = standard deviation of X_i^n (see Eqs. A1.12 and A1.13). From Eqs. A3.38 and A3.39, it follows

$$\sigma_{X_i^n} = \frac{\phi(\Phi^{-1}[P(X_i^*)])}{p_{X_i}(X_i^*)} \tag{A3.40}$$

$$\bar{X}_i^n = X_i^* - \Phi^{-1}[P_{X_i}(X_i^*)]\sigma_{X_i^n} \tag{A3.41}$$

where the bar denotes mean value. Once $\bar{X}_i^n$ and $\sigma_{X_i^n}$ are obtained from Eqs A3.40 and A3.41, the problem can be restated in the space of the reduced variables $x_{i_r}^n$. The safety index β is the distance in this space between the origin and the failure boundary. A computer program for calculating this distance, based on an algorithm proposed in [A3-14], is listed in [A3-9]. An alternative method for calculating β was proposed in [A3-15]. More recently it has been noted that β can be obtained by nonlinear programming methods [A3-11]. Following the calculation of β, the probability of failure is estimated by Eq. A3.36. (For an alternative normalization procedure, see [A3-26].)

The procedure just described is approximate because Eq. A3.36 does not hold if the failure boundary is nonlinear. A method for reducing the errors due to the nonlinearity of the failure boundary was proposed in [A3-16]. Following normalization at the checking point, relations similar in principle to Eq. A3.36 are used that correspond to the case where the failure boundary is a quadric, rather than a (hyper) plane. The failure probability depends upon the safety index corresponding to the space of the coordinates X_i^n and upon the characteristics of the quadric. The latter are determined from the condition that the quadric approximates the nonlinear failure boundary as closely as possible at the checking point.

Alternative approaches to the estimation of failure probabilities have been proposed in [A3-17] and [A3-18].

Safety Factors. Consider a structure characterized by a set of variables with means and standard deviations $\bar{X}_i$ and σ_{X_i}, and checking points with coordinates X_i^* and $x_{i_r}^*(i=1, 2, \ldots, n)$ in the space of the original and the reduced variables, respectively. By definition:

$$X_i^* = \bar{X}_i + \sigma_{\bar{X}_i} x_{i_r}^* \tag{A3.42}$$

Equation A3.42 can be written as

$$X_i^* = \gamma_{\bar{X}_i} \bar{X}_i \tag{A3.43}$$

where

$$\gamma_{\bar{X}_i} = 1 + V_{X_i} x_{i_r}^* \tag{A3.44}$$

The quantity $\gamma_{\bar{X}_i}$ is termed the partial safety factor applicable to the mean of the variable X_i.

In design applications the means, $\bar{X}_i$, are seldom used, and nominal design values, such as the 50-year wind, the allowable steel stress F_a, or the nominal yield stress F_y, are employed instead. Let these nominal values be denoted by $\tilde{X}_i$. Equation A3.43 can be rewritten as [A3-9]:

$$X_i^* = \gamma_{\tilde{X}_i} \tilde{X}_i \qquad (i = 1, 2, \ldots, n) \tag{A3.45}$$

where

$$\gamma_{\tilde{X}_i} = \frac{\bar{X}_i}{\tilde{X}_i} \gamma_{\bar{X}_i} \tag{A3.46}$$

The factor $\gamma_{\tilde{X}_i}$ is the partial safety factor applicable to the nominal design value of the variable X_i.

In the particular case in which the variables of concern are the load Q and the resistance R, the partial safety factors are referred to as the load and resistance factor. For the resistance factor the notation $\phi_{\bar{R}}$ or $\phi_{\tilde{R}}$ is used in lieu of $\gamma_{\bar{R}}$ or $\gamma_{\tilde{R}}$.

From the definition of the partial safety factor (Eq. A3.44) and the definition of the checking point in the space of the reduced variables corresponding to $Y_1 = \ln R$ and $Y_2 = \ln Q$, it follows that if higher order terms (see Eq. A3.27) are neglected

$$\phi_{\bar{R}} \simeq \exp(-\alpha_R \beta V_R) \tag{A3.47}$$

$$\gamma_{\bar{Q}} \simeq \exp(\alpha_Q \beta V_Q) \tag{A3.48}$$

$$\alpha_R = \cos[\tan^{-1}(V_Q/V_R)] \tag{A3.49}$$

$$\alpha_Q = \sin[\tan^{-1}(V_Q/V_R)] \tag{A3.50}$$

where β is the safety index given by Eq. A3.29.

Care must be exercised in using simplified approximate expressions for partial safety factors. Consider, for example, the following expression for the load factor $\gamma_{\bar{Q}}$, proposed in [A3-20] as an approximation to Eq. A3.48 for members subjected to wind loads:

$$\gamma_{\bar{Q}} = 1 + 0.55 \beta V_Q \tag{A3.51}$$

where β is given by Eq. A3.29. For members I and II of Table A3.1, it would follow from Eq. A3.51 that $\gamma_{\bar{Q}_I} = 1.50$ and $\gamma_{\bar{Q}_{II}} = 1.31$, respectively. However, if Eq. A3.48 is used, $\gamma_{\bar{Q}_I} = 2.32$ and $\gamma_{\bar{Q}_{II}} = 1.63$. It is concluded that Eq. A3.51 may result in misleading estimates of the load factor $\gamma_{\bar{Q}}$.

Equations A3.47 and A3.48 show that load and resistance factors depend not only on the safety index β, but on the coefficients of variation V_Q, V_R as well. For this reason, to members having the same safety index β there can correspond widely different sets of load and resistance factors as calculated by Eqs. A3.47 and A3.48. This is illustrated in the following example.

Example

Members I and II of Table A3.1 have the same safety index calculated by Eq. A3.29, $\beta = 3.69$, as well as approximately the same failure probabilities—

see column 10 of Table A3.1. The values of V_R for the two members are the same; however, the respective values of V_Q differ (Table A3.1). It can be verified that, owing to this difference, the load and resistance factors for the two members (given by Eqs. A3.47 through A3.50) are $\phi_{\bar{R}I} \simeq 0.88$ versus $\phi_{\bar{R}II} \simeq 0.83$ and, as indicated earlier, $\gamma_{\bar{Q}I} = 2.32$ versus $\gamma_{\bar{Q}II} = 1.63$.

Conversely, the use in the design of various members of the same set of load and resistance factors does not necessarily ensure that those members will have the same probabilities of failure. This creates difficulties in the development of risk-consistent load and resistance factors for codified design. These difficulties, and proposed approaches for dealing with them, are discussed in [A3-9] and [A3-21]. Problems related to codes of practice are also discussed in [A3-22].

REFERENCES

A3-1 P. Thoft-Christensen and M. J. Baker, *Structural Reliability Theory and Applications*, Springer-Verlag, Berlin, 1982.

A3-2 R. Fleming, *Wind Stresses*, Engineering News, New York, 1915 (reprinted from *Engineering News*, 1915).

A3-3 E. Simiu, J. R. Shaver and J. J. Filliben, "Wind Speed Distributions and Reliability Estimates," *J. Struct. Div.*, ASCE, **107** (May 1981), 1003–1007, and (errata) Oct. 1981, 2052.

A3-4 T. V. Galambos et al., "Probability-Based Load Criteria: Assessment of Current Design Practice," *J. Struct. Div.*, ASCE, **108** (May 1982), 959–977.

A3-5 Y. K. Wen, "Statistical Combination of Extreme Loads," *J. Struct. Div.*, ASCE, **103** (May 1977), 1079–1095.

A3-6 Y. K. Wen, "Clustering Model for Correlated Load Processes," *J. Struct. Div.*, ASCE, **107** (May 1981), 965–983.

A3-7 R. D. Larrabee and C. A. Cornell, "Combination of Various Load Processes," *J. Struct. Div.*, ASCE, **107** (Jan. 1981), 223–239.

A3-8 R. D. Larrabee, *Approximate Stochastic Analysis of Combined Loading*, Dept. of Civil Engineering, MIT, MIT CE R78-28, Order No. 629, Cambridge, Mass., September 1978.

A3-9 B. Ellingwood et al., *Development of a Probability-Based Load Criterion for American National Standard A58*, NBS Special Publication 577, National Bureau of Standards, Washington, D.C., June 1980.

A3-10 A. M. Hasofer and N. C. Lind, "Exact and Invariant Second-Moment Code Format," *J. Eng. Mech. Div.*, ASCE, **100** (Feb. 1974), 829–844.

A3-11 M. Shinozuka, "Basic Analysis of Structural Safety," *J. Struct. Eng.*, **109** (March 1983), 721–740.

A3-12 E. Simiu and J. R. Shaver, "Wind Loading and Reliability-Based Design," in *Wind Engineering*, Proceedings of the Fifth International Conference, Fort Collins, CO, July 1979, Vol. 2, Pergamon Press, New York, 1980.

A3-13 W. McGuire, *Steel Structures*, Prentice Hall, Englewood Cliffs, N.J., 1968.

A3-14 R. Rackwitz and B. Fiessler, *Nonnormal Distributions in Structural Reliability*, SFB 96, Technical University of Munich, Ber. zur Sicherheitstheorie der Bauwerke, No. 29, 1978, pp. 1–22.

A3-15 O. Ditlevsen, "Principle of Normal Tail Approximation," *J. Eng. Mech. Div.*, ASCE, **107** (Dec. 1981), 1191–1208.

A3-16 B. Fiessler, H.-J. Neumann and R. Rackwitz, "Quadratic Limit States in Structural Reliability," *J. Eng. Mech. Div.*, ASCE, **105** (Aug. 1979), 661–676.

A3-17 M. Grigoriu, "Methods for Approximate Reliability Analysis," *Struct. Safety*, **1** (1982/1983), 155–165.

A3-18 O. Ditlevsen, "Gaussian Safety Margins," in *Proceedings Fourth International Conference*,

Applications of Statistics and Probability in Soil and Structural Engineering, University of Florence, June 13–17 1983, Pitagora Editrice, Bologna, 1983.

A3-19 *Safety Goals for Nuclear Power Plant Operation*, NUREG-0880, Revision 1, U.S. Nucl. Reg. Comm., Office of Policy Evaluation, Washington, D.C., May 1983.

A3-20 M. K. Ravindra, C. A. Cornell, and T. V. Galambos, "Wind and Snow Load Factors for Use in LRFD," *J. Struct. Div.*, ASCE, **104** (1978), 1443–1457.

A3-21 F. Casciati and L. Faravelli, "Load Combination by Partial Safety Factors," *Nuclear Eng. Des.*, **75** (1982), 432–452.

A3-22 O. Ditlevsen, "Generalized Second Moment Reliability Index," *J. Struct. Mech.*, **7** (1979), 435–451.

A3-23 O. Ditlevsen, *Uncertainty Modeling*, McGraw-Hill, New York, 1981.

A3-24 E. Simiu and C. E. Smith, *Structural Reliability Fundamentals and Their Application to Offshore Structures*, NBSIR 84-2921, Nat. Bureau of Standards, Gaithersburg, MD, 1984.

A3-25 G. I. Schuëller, "Current Trends in Systems Reliability," *Proceedings, International Conference on Structural Safety and Reliability*, May 27–29, 1985, I. Konishi and M. Shinozuka (Eds.), Kobe, Japan.

A3-26 R. Rackwitz and B. Fiessler, "Structural Reliability Under Combined Load Sequences," *Computers and Structures*, **9** (1978), 489–494.

APPENDIX A4

Pressure Coefficients for Buildings and Structures

This appendix presents pressure coefficients for various buildings and structures [4-10].* These were obtained from tests conducted in *uniform, smooth flow*. (For information and comments on the limitations of results obtained in such tests, see Sections 4.5 and 4.6.) In the tables that follow local pressure coefficients are denoted by C_{pe}^*.

*Excerpted from *Wind Forces on Structures, Trans. ASCE*, **126**(1961), 1124–1198.

WIND PRESSURE COEFFICIENTS FOR HOUSES AND CLOSED HALLS

Structure	α	External Pressure Coefficient, C_{pe}								Internal Pressure Coefficient, C_{pi}			
		A	B	C	D	E	F	G	H	Openings uniformly distributed	Openings Mainly on Side		
											A	B	C
House Roofs 0° to 10°, $h{:}b{:}L = 1{:}1{:}1$	0°	0.9	-0.5	-0.6	-0.6	-0.7	-0.7	-0.5	-0.5	+0.2	0.8	-0.4	-0.5
	15°	0.8	-0.5	-0.7	-0.5	-0.7	-0.6	-0.5	-0.6	+0.2	0.7	-0.4	-0.6
	45°	0.5	-0.5	0.5	-0.5	-0.8	-0.5	-0.5	-0.4	+−0.2	0.4	-0.4	0.4
	45°	For m, $C_{pe}^{*} = -1.2$; for n, $C_{pe}^{*} = -0.8$											
House $h{:}b{:}L = 2.5{:}2.5{:}5$	0°	0.9	-0.5	-0.7	-0.7	-0.6	-0.6	-0.5	-0.5	+0.2	0.8	-0.4	-0.6
	45°	0.6	-0.5	0.4	-0.5	-0.9	-0.7	-0.6	-0.7	+0.2	0.5	-0.4	0.3
	90°	-0.5	-0.5	0.9	-0.4	-0.8	-0.2	-0.8	-0.2	+−0.2	-0.4	-0.4	0.8
	45°	For m, $C_{pe}^{*} = -1.5$											

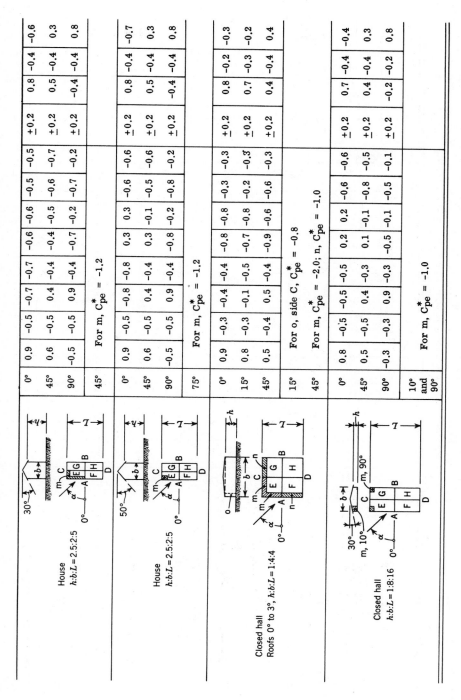

Description	Angle											
House h:b:L = 2.5:2:5 (30°)	0°	0.9	-0.5	-0.7	-0.6	-0.6	-0.5	-0.5	±0.2	0.8	-0.4	-0.6
	45°	0.6	-0.5	0.4	-0.4	-0.5	-0.6	-0.7	±0.2	0.5	-0.4	0.3
	90°	-0.5	-0.5	0.9	-0.7	-0.2	-0.7	-0.2	±0.2	-0.4	-0.4	0.8
	45°	For m, $C_{pe}^* = -1.2$										
House h:b:L = 2.5:2:5 (50°)	0°	0.9	-0.5	-0.8	0.3	0.3	-0.6	-0.6	±0.2	0.8	-0.4	-0.7
	45°	0.6	-0.5	0.4	0.3	-0.1	-0.5	-0.6	±0.2	0.5	-0.4	0.3
	90°	-0.5	-0.5	0.9	-0.8	-0.2	-0.8	-0.2	±0.2	-0.4	-0.4	0.8
	75°	For m, $C_{pe}^* = -1.2$										
Closed hall Roofs 0° to 3°, h:b:L = 1:4:4	0°	0.9	-0.3	-0.4	-0.8	-0.8	-0.3	-0.3	±0.2	0.8	-0.2	-0.3
	15°	0.8	-0.3	-0.1	-0.7	-0.8	-0.2	-0.3	±0.2	0.7	-0.3	-0.2
	45°	0.5	-0.4	0.5	-0.9	-0.6	-0.6	-0.3	±0.2	0.4	-0.4	0.4
	15°	For o, side C, $C_{pe}^* = -0.8$										
	45°	For m, $C_{pe}^* = -2.0$; n, $C_{pe}^* = -1.0$										
Closed hall h:b:L = 1:8:16	0°	0.8	-0.5	-0.5	0.2	0.2	-0.6	-0.6	±0.2	0.7	-0.4	-0.4
	45°	0.5	-0.5	0.4	0.1	-0.1	-0.8	-0.5	±0.2	0.4	-0.4	0.3
	90°	-0.3	-0.3	0.9	-0.5	-0.1	-0.5	-0.1	±0.2	-0.2	-0.2	0.8
	10° and 90°	For m, $C_{pe}^* = -1.0$										

WIND PRESSURE COEFFICIENTS FOR OPEN AND CLOSED BUILDINGS

Structure	α	External Pressure Coefficient, C_{pe}									
		A	B	C	D	E	F	G	H	J	K
		(I) One long wall open									
Building open on one side Roof 30°, $h{:}b{:}L = 1{:}2{:}4$	0°	0.8	-0.5	-0.7	0.8	0.8	-0.7	-0.3	0.8	-0.4	0.8
	45°	0.7	-0.6	0.4	0.6	0.8	-0.4	-0.2	0.6	-0.7	0.7
	60°	0.3	-0.7	0.7	0.3	0.4	-0.4	-0.3	0.2	-0.6	0.2
	180°	-0.5	0.9	-0.8	-0.5	-0.5	-0.8	-0.4	-0.5	-0.2	-0.5
		(II) One end wall open									
	0°	0.9	-0.7	-0.7	-0.4	-0.7	-0.8	-0.2	-0.7	-0.4	-0.7
	45°	0.5	0.7	0.8	-0.5	0.7	-0.4	-0.3	0.7	-0.6	0.8
	60°	0.1	0.9	0.9	-0.6	0.9	-0.4	-0.3	0.9	-0.7	0.9
	90°	-0.5	0.8	0.8	-0.5	0.8	-0.3	-0.4	0.8	-0.4	0.8

Buildings open on two sides
Roofs 30°, $h:b:L = 1:2:4$

(I) Two long walls open

0°	−0.2	−0.7	−0.7	−0.2	0.4	−0.9	−0.5	−0.8
45°	0.5	−0.4	0.5	−0.4	0	−0.3	−0.6	0
60°	0.7	−0.6	0.5	−0.4	−0.3	−0.1	−0.7	−0.3

(II) Two end walls open

0°	0.9	−0.7	−0.7	−0.4	−0.2	−0.7	−0.4	−0.7
45°	0.5	−0.4	−0.1	−0.8	−0.3	−0.4	−0.8	−0.3
60°	0.3	−0.2	0.1	−0.5	−0.3	−0.1	−0.8	0.1

Closed buildings with roof vents
Roofs 20°, $h:b:L = 1:4:8$

0°	0.8	−0.5	−0.7	−0.2	−1.0	−0.6	−0.5	−0.6
45°	0.4	−0.5	0.4	−0.3	−1.3	−1.4	−1.0	−0.7
90°	−0.4	−0.4	0.8	−0.4	−0.3	−0.3	−0.2	−0.4
0° and 45°	For m, $C_{pe}^{*} = -1.2$; n, $C_{pe}^{*} = -2.4$							

Internal Pressure Coefficient, C_{pi}, with Vents

	F & J open	F & J closed	F only open	J only open
0°	−0.2	±0.2	0.5	−0.4
45°	−0.5	±0.2	0.1	−0.9
90°	−0.3	±0.2	−0.2	−0.2

WIND PRESSURE COEFFICIENTS FOR TALL BUILDINGS, ROOFS, AND PASSAGEWAYS

Structure	α	External Pressure Coefficient, C_{pe}								Internal Pressure Coefficient, C_{pi}				
		A	B	C	D	E	F	G	H	Openings uniformly distributed	Openings Mainly in			Roof EF
											A	B	C	
Tall buildings closed Roofs 0° to 15° $h{:}b{:}L = 2.5{:}1{:}1$	0°	0.9	-0.6	-0.7	-0.7	-0.8	-0.8	-0.8	-0.8	±0.2	0.8	-0.5	-0.6	
	15°	0.8	-0.5	-0.9	-0.6	-0.8	-0.8	-0.7	-0.7	±0.2	0.7	-0.5	-0.8	
	45°	0.5	-0.5	0.5	-0.5	-0.8	-0.7	-0.7	-0.5	±0.2	0.4	-0.4	0.4	
	45°	For m, $C_{pe}^* = -1.0$; n, $C_{pe}^* = -0.8$												
Tall buildings closed $h{:}b{:}L = 2{:}1{:}2$	0°	0.9	-0.5	-0.8	-0.8	-1.0	-1.0	-0.5	-0.5	±0.2	0.8	-0.4	-0.7	
	45°	0.6	-0.5	0.4	-0.4	-0.3	-0.4	-0.5	-0.6	±0.2	0.5	-0.4	0.3	
	90°	-0.6	-0.6	0.9	-0.4	-0.7	-0.5	-0.7	-0.5	±0.2	-0.5	-0.5	0.8	
	0°	For m, $C_{pe}^* = -1.2$												

Shed roof
$h:b:L = 1:2.4:12$

α													
0°	0.9	−0.5	−0.6	−0.6	−0.5	−0.5	−0.5	−0.5	±0.2	0.8	−0.4	−0.5	−0.4
45°	0.5	−0.6	0.4	−0.4	−1.2	−0.7	−1.1	−0.7	±0.2	0.4	−0.5	0.3	−0.8
90°	−0.4	−0.3	0.9	−0.2	−0.3	0	−0.3	0	±0.2	−0.2	−0.1	0.8	0
180°	−0.4	0.8	−0.7	−0.7	0.1	0.1	0.2	0.2	±0.2	−0.3	0.7	−0.6	0
45°	For m, $C_{pe}^{*} = -1.4$												

Peaked roof
$h:b:L = 1:1:5$

α													
0°	0.9	−0.5	−0.6	−0.6	−0.6	0.6	−0.5	−0.5	±0.2	0.8	−0.4	−0.5	0.5
45°	0.5	−0.8	0.4	−0.5	0.2	−0.1	−1.0	−0.8	±0.2	0.4	−0.7	0.3	0
90°	−0.4	−0.4	0.9	−0.3	−0.4	0	−0.4	0	±0.2	−0.1	−0.1	0.8	−0.1
180°	−0.5	0.9	−0.6	−0.6	−0.5	−0.5	−0.1	−0.1	±0.2	−0.4	0.8	−0.5	−0.4
45°	For m, $C_{pe}^{*} = -1.3$												

Closed connecting passage between large walls
$h:b:L = 1:1:10$

α													
0°	0.8	−1.2	−1.4	−1.5					−0.5	0.7	−1.1	−1.3	

Access doors closed

WIND PRESSURE COEFFICIENTS FOR SHELTER ROOFS

Structure	α	External Pressure Coefficient, C_{pe}				End Surfaces			
		A	B	C	D	J	K	L	M
Roof +30° α = 0°-45°, A-D full length α = 90°, A-D part length L' 	0°	0.6	-1.0	-0.5	-0.9				
	45°	0.1	-0.3	-0.6	-0.3				
	90°	-0.3	-0.4	-0.3	-0.4	0.8	-0.4	0.3	-0.3
	45°	For m, $C^*_{pe,top}$ = -1.0; $C_{pe,bottom}$ = -0.2							
	90°	Tangential acting friction, R = 0.05 q; L:b							
Effect of trains or stored material Roof +30° α = 0°-45°-180°, A-D full length α = 90°, A-D part length L' 	0°	0.1	0.8	-0.7	0.9				
	45°	-0.1	0.5	-0.8	0.5				
	90°	-0.4	-0.5	-0.4	-0.5				
	180°	-0.3	-0.6	0.4	-0.6				
	45°	For m, $C^*_{pe,top}$ = -1.5; $C_{pe,bottom}$ = 0.5							
	90°	End surface friction load, see above							

Roof +10°
α=0°-45°, A-D full length
α=90°, A-D part length L'

Diagram labels: $h=0.5b$, $R_{0°}$, A, C, D, B, J&K, $L'=b$, $R_{90°}$, L, M, m, K, A, C, $L=5b$, α, 0°

0°	-1.0	0.3	-0.5	0.2				
45°	-0.3	0.1	-0.3	0.1				
90°	-0.3	0	-0.3		0.8	-0.6	0.3	-0.4
0°	For m, $C_{pe,top}^* = -1.0$; $C_{pe,bottom}^* = 0.4$							
0° and 90°	Tangential acting friction, R = 0.1 q: L: b							

Effect of trains or stored material
Roof +10°
α=0°-45°-180°, A-D full length
α=90°, A-D part length L'

Diagram labels: $h=0.5b$, $R_{0°}$, A, C, D, B, J&K, $h'=0.8h$, $L'=b$, $R_{90°}$, L, J, A, C, $L=5b$, m, M, b, α, 0°

0°	-1.3	0.8	-0.6	0.7
45°	-0.5	0.4	-0.3	0.3
90°	-0.3	0	-0.3	0
180°	-0.4	-0.3	-0.6	-0.3
0°	For m, $C_{pe,top}^* = -1.6$; $C_{pe,bottom}^* = 0.9$			
0° and 180°	End surface friction load, see above			

Roof −10°
α=0°-45°, A-D full length
α=90°, A-D part length L'

Diagram labels: $h=0.5b$, $R_{0°}$, A, C, D, B, $L'=b$, $R_{90°}$, A, C, $L=5b$, b, m, α, 0°

0°	0.3	-0.7	0.2	-0.9
45°	0	-0.2	0.1	-0.3
90°	-0.1	0.1	-0.1	0.1
0°	For m, $C_{pe,top}^* = 0.4$; $C_{pe,bottom}^* = -1.5$			
0° and 90°	Tangential acting friction, R = 0.1 q: L: b			

Effect of trains or stored material
Roof −10°
α=0°-45°-180°, A-D full length
α=90°, A-D part length L'

Diagram labels: $h=0.5b$, $R_{0°}$, A, C, D, B, $h'=0.8h$, $L'=b$, $R_{90°}$, A, C, $L=5b$, b, m, α, 0°

0°	-0.7	0.8	-0.6	0.6
45°	-0.4	0.3	-0.2	0.2
90°	-0.1	0.1	-0.1	0.1
180°	-0.4	-0.2	-0.6	-0.3
0°	For m, $C_{pe,top}^* = -1.1$; $C_{pe,bottom}^* = 0.9$			
0° and 180°	Tangential acting friction, R = 0.1 q: L: b			

WIND PRESSURE COEFFICIENTS ON ROOFS AND GRANDSTANDS

Structure	α	External Pressure Coefficient, Cpe																Internal Pressure Coefficient, Cpi			
																		Openings uniformly distributed	Openings mainly in		
		A	B	C	D	E	F	G	H	J	K	L	M	N	O	P	Q		A	B	C
Sawtooth roof. Friction in wind direction $R_\alpha = 0.1\,q.b.L$ $h.b.L = 1:4.5$	0°	0.9	-0.3	-0.4	-0.4	0.6	-0.6	-0.6	-0.5	-0.5	-0.4	-0.3	-0.3					±0.2	0.8	-0.2	-0.3
	45°	0.5	-0.4	0.5	-0.3	0.2	-0.8	-0.5	-0.4	-0.2	-0.4	-0.2	-0.5					±0.2	0.4	-0.3	0.4
	90°	-0.4	-0.4	0.9	-0.3	-0.3	-0.4	-0.4	-0.4	-0.4	-0.4	-0.4	-0.3					±0.2	-0.3	-0.3	0.8
	180°	-0.3	0.9	-0.3	-0.3	-0.2	-0.3	-0.3	-0.4	-0.4	-0.6	-0.6	-0.1					±0.2	-0.2	0.8	-0.2
	0° and 180°	For m, $C_{pe}^* = -1.3$; n, $C_{p3}^* = -2.0$																			
Flat roof $h.b.L = 1:3.4$	0°	0.9	-0.5	-0.6	-0.6	-0.8	-0.4	-0.4	-0.4	-1.0	-0.4	-0.5	-0.5					±0.2	0.8	-0.4	-0.5
	45°	0.5	-0.5	0.5	-0.4	-0.6	-0.5	-0.5	-0.5	-0.5	-0.5	-0.5	-0.5					±0.2	0.4	-0.4	0.4
	90°	-0.5	-0.5	0.9	-0.4	-0.8	-0.4	-0.8	-0.4	-0.4	-0.4	-1.0	-0.4					±0.2	-0.4	-0.4	0.8
	0° and 90°	For m, $C_{pe}^* = -1.1$; n, $C_{pe}^* = -1.5$																			

Hanger arch roof

Smooth surface
$h:b:L = 1:12:12$
$r = 5/6\,b$
$Y = 0.1b$
$X = 0.1b$

	A	B	C	D	E	F	G	H	J	K	L	M	N	O	P	Q	Openings uniformly distributed with windows and doors closed	Openings mainly in — A with window Y open	Openings mainly in — C with all gates open	Openings mainly in — C with only gate X open
0°	0.7	-0.2	-0.3	-0.1	-0.5	-0.8	-0.8	-0.4	-0.1								± 0.2	0.4	-0.1	-1.5
30°	0.6	-0.3	0.2	-0.1	-0.4	-0.7	-0.9	-0.7	-0.4								± 0.2	0.7	0.6	0.7
90°	-0.3	-0.3	0.9	-0.3							-0.8	-0.7	-0.5	-0.3	-0.1	-0.1	± 0.2	-1.0	0.8	0.4

30°	For m, $C^*_{pe} = -1.8$ with $C^*_{pe,min} = -2.5$

Grandstands open three sides

Roof 5°
$h' = 0.4\,h$
$h = 0.8\,b$
$L = 2.2\,b$

	C_{pe} at top and bottom of roof					C_{pe} at front and back of wall	
0°	-1.0	0.9	-0.7	0.9		0.9	-0.5
45°	-1.0	0.7	-0.5	0.9		0.4	-0.4
135°	-0.4	-1.1	-0.9	-1.1		0.6	0.4
180°	-0.6	-0.3	-0.6	-0.3		0.9	0.9

45°	For m_D, $C^*_{pe,top} = -2.0$; $C^*_{pe,bottom} = 1.0$
60°	For m_w, $C^*_{pe,k} = -1.0$; $C^*_{pe,j} = 1.0$
90°	Tangential acting friction, $R_D = 0.1\ q:b:L$; $R_w = 0.1\ q:h:L$

Smooth closed reservoir

$$F_D = (P_i - P_a)\,A$$

in which P_i = working pressure, $A = \dfrac{\pi}{4}d^2$, $P_a = C_{pe}\,q$, and

$$C_{pe} = -1.0$$

About the Authors

EMIL SIMIU earned his Ph.D. degree from Princeton University, and joined the National Bureau of Standards in 1971. He has served in a research and consulting capacity on a wide variety of wind, earthquake, and ocean engineering projects for various government agencies, the World Bank, and industry. He is Chairman of the ASCE Committee on Dynamic Effects (Structural Division), past Chairman of the ASCE Committee on Wind Effects, and Chairman of the Subcommittee on Environmental Forces of the ASCE Committee on the Reliability of Offshore Structures. Dr. Simiu was the 1984 recipient of the Federal Engineer of the Year award from the National Society of Professional Engineers.

ROBERT H. SCANLAN holds doctorates from the Massachusetts Institute of Technology and the University of Paris. He is emeritus Professor of Civil Engineering at Princeton University and is presently Professor of Civil Engineering at Johns Hopkins University. He has held a variety of positions in industry and government and has been a consultant to the United States government and various foreign governments and private firms. He is past Chairman of the ASCE Committee on Wind Effects, and of the ASCE Dynamics Committee (Engineering Mechanics Division). He is co-author of two other texts, both in aeroelasticity. He has received national awards from ASCE and AISC for his work on the aeroelastic problems of suspension bridges.

Index

VSL Corporation
11925 - 12th Ave. S.
P. O. Box 1228
Burnsville, MN 55337